PHOTONIC SENSORS FOR DETECTION OF VOC BIOMARKERS

AF752343

PHOTONIC SENSORS FOR DETECTION OF VOC BIOMARKERS

Technological Principles and Medical Applications

Akhilesh Kumar Pathak
Mechanical Engineering, Northwestern University, IL, United States; Department of Engineering, School of Science & Technology, City, St George's, University of London, London, United Kingdom

Charusluk Viphavakit
International school of engineering, Chulalongkorn University, Bangkok, Thailand

Sridhar Krishnaswamy
Mechanical Engineering, Northwestern University, IL, United States

ELSEVIER

Elsevier
Radarweg 29, PO Box 211, 1000 AE Amsterdam, Netherlands
50 Hampshire Street, 5th Floor, Cambridge, MA 02139, United States

Copyright © 2026 Elsevier Inc. All rights are reserved, including those for text and data mining, AI training, and similar technologies.

Publisher's note: Elsevier takes a neutral position with respect to territorial disputes or jurisdictional claims in its published content, including in maps and institutional affiliations.

Books and Journals published by Elsevier comply with applicable product safety requirements. For any product safety concerns or queries, please contact our authorised representative, Elsevier B.V., at productsafety@elsevier.com.

For accessibility purposes, images in electronic versions of this book are accompanied by alt text descriptions provided by Elsevier. For more information, see https://www.elsevier.com/about/accessibility.

No part of this publication may be reproduced or transmitted in any form or by any means, electronic or mechanical, including photocopying, recording, or any information storage and retrieval system, without permission in writing from the publisher. Details on how to seek permission, further information about the Publisher's permissions policies and our arrangements with organizations such as the Copyright Clearance Center and the Copyright Licensing Agency, can be found at our website: www.elsevier.com/permissions.

This book and the individual contributions contained in it are protected under copyright by the Publisher (other than as may be noted herein).

Notices
Knowledge and best practice in this field are constantly changing. As new research and experience broaden our understanding, changes in research methods, professional practices, or medical treatment may become necessary.

Practitioners and researchers must always rely on their own experience and knowledge in evaluating and using any information, methods, compounds, or experiments described herein. In using such information or methods they should be mindful of their own safety and the safety of others, including parties for whom they have a professional responsibility.

To the fullest extent of the law, neither the Publisher nor the authors, contributors, or editors, assume any liability for any injury and/or damage to persons or property as a matter of products liability, negligence or otherwise, or from any use or operation of any methods, products, instructions, or ideas contained in the material herein.

ISBN: 978-0-443-29166-1

For Information on all Elsevier publications visit our website at
https://www.elsevier.com/books-and-journals

Publisher: Peter Llewellyn
Acquisitions Editor: Yashila Girdhar
Editorial Project Manager: Ashi Jain
Production Project Manager: Vijayaraj Purushothaman
Cover Designer: Gopalakrishnan Venkatraman

Typeset by Aptara, New Delhi, India

Working together to grow libraries in developing countries
www.elsevier.com • www.bookaid.org

Contents

About the Authors

Akhilesh Kumar Pathak (FMSCA, SMIEEE, SMOPTICA) is a Marie Skłodowska-Curie Actions Postdoctoral Fellow (Grant ID: 101210686) in the Department of Engineering, School of Science & Technology, City, St George's University of London, United Kingdom. He received his master's degree in physics from the University of Allahabad, India, in 2012, and a PhD in Fiber Optics and Photonics from the Indian Institute of Technology (Indian School of Mines), Dhanbad, India, in 2019, followed by postdoctoral positions at Chulalongkorn University, Thailand (2020–2022), and Northwestern University, United States (2023–2025). His research expertise includes optical fiber modeling using FV-FEM, photonic sensing, fs-laser densification, two-photon polymerization-based 3D printing, and the synthesis and controlled growth of nanoparticles and smart materials. He has been awarded several prestigious international fellowships, including the Erasmus Mundus Leaders Fellowship, the MSCA Postdoctoral Fellowship, and the C2F Postdoctoral Fellowship.

Charusluk Viphavakit was born in Bangkok, Thailand. She received the Bachelor's degree in Nanoelectronic Engineering from Chulalongkorn University, Thailand, in 2010, the Master's degree in nanotechnology from the Asian Institute of Technology, Thailand, in 2012, and the PhD degree from Frederick University, Cyprus, with an Erasmus Mundus scholarship in 2015 and from City, University of London, United Kingdom, in 2018. She was a Postdoctoral Researcher in the Optical Fibre Research Centre, University of Limerick, Ireland, supported by an Irish Research Council Fellowship for 2 years. During her postdoctoral career in 2016, she received Harvard Medical School Scholarships and University of Limerick Graduate Entry Medical School funding to enroll in the Global Clinical Scholars Research Training Program (GCSRT) 2016–2017 from Harvard Medical School (HMS), Boston, MA, United States, for advanced training and conducting of clinical research. She is a Senior Member of IEEE and the IEEE Photonics Society. She is also a member of OPTICA and SPIE. She is currently a full-time Assistant Professor and Head of the Nanoengineering program at the International School of Engineering, Chulalongkorn University. She is on the editorial board of the *Engineering Journal*. Her research interests include optical waveguides and optical fiber sensors for biomedical applications.

Sridhar Krishnaswamy has been a member of the Faculty of Mechanical Engineering at Northwestern University since 1990. He received his MS and PhD degrees in aeronautics from the California Institute of Technology in 1984 and 1989, respectively. Currently, he is the director of the Center for Smart Structures and Materials at Northwestern University. He has been actively involved in the areas of nondestructive materials characterization, optical metrology, structural health management, and fiber-optic and 3D-printed photonic sensors. He is a fellow of SPIE and ASME.

CHAPTER

1

Introduction to volatile organic compounds

1.1 Introduction

Volatile organic compounds (VOCs) are a class of organic compounds that quickly evaporate at room temperature due to high vapor pressure, allowing them to easily transition into the gas phase under these conditions. The United States Environmental Protection Agency (EPA) describes VOCs as "organic compounds whose chemical characteristics allow them to evaporate even at indoor atmospheric conditions" (Khan et al., 2020; *Technical Overview of Volatile Organic Compounds*, no date). The definition of VOCs varies depending on their applications or the country in which they are defined. Conventionally, VOCs have been associated with indoor air pollutants, which can greatly affect human health, as they contribute to the formation of ground-level ozone and particulate matter. These indoor VOCs are considered primary constituents of smog, leading to atmospheric and environmental concerns. Consequently, VOCs have gained significant attention, and various regulations exist, e.g., EU CE, 2010; US EPA, 2014.

A team of experts from the World Health Organization (WHO) provided an internationally acceptable definition of VOCs. According to the WHO, VOCs are characterized by high vapor pressure at room temperature and can be categorized based on their boiling points. VOCs with lower boiling points possess higher volatility, and hence have a higher tendency to be released into the atmosphere. Therefore, VOCs with boiling points ranging from 50–100°C to 240–260°C can be collected by adsorption and tested using liquid chromatography (LC) or gas chromatography (GC). VOCs with boiling points outside this range are classified differently: those with lower boiling points are termed very volatile organic compounds (VVOC), which range from <0°C to 50–100°C, and are typically found almost entirely as gases due to their high volatility. On the other hand, compounds

Photonic Sensors for Detection of VOC Biomarkers: Technological Principles and Medical Applications
DOI: https://doi.org/10.1016/B978-0-443-29166-1.00010-X

Copyright © 2026 Elsevier Inc. All rights are reserved, including those for text and data mining, AI training, and similar technologies.

TABLE 1.1 Summary of various volatility types along with their boiling points and examples.

Volatility	Boiling point (°C)	Example VOCs
Volatile organic compounds (VOCs)	50–100 to 240–260	Toluene, acetone, formaldehyde, ethanol, 2-propanol, hexanal
Semi-volatile organic compounds (SVOCs)	240–260 to 380–400	Dichloro-diphenyl-trichloroethane (DDT), Polychlorinated biphenyls (PCBs), polybrominated biphenyls (PBBs), plasticizers, chlordane
Very volatile organic compounds (VVOCs)	<0 to 50–100	Butane, propane, methyl chloride

From Killeit, U. (2023). On-site detection of volatile organic compounds Nachweis flüchtiger organischer Verbindungen vor Ort. Deutsche Lebensmittel-Rundschau, *119(4), 1598. https://doi.org/10.3390/molecules28041598.*

with higher boiling points are known as semi-volatile organic compounds (SVOC), ranging from 240–260°C to 380°C. Lastly, the compounds with boiling points greater than 380°C are known as particulate organic matter (POM). Table 1.1 shows different types of volatility along with their boiling points and provides selected examples of such VOCs.

The rise in economic growth and rapid urbanization has led to severe air pollution, particularly indoor air pollution. The main sources of indoor VOCs generally include personal care products, tobacco smoking, heating, and various building materials such as paint, wooden furniture, and carpet (Liu et al., 2022). Excessive exposure to indoor VOCs can cause several health issues, including allergies and serious chronic diseases such as cancer. For instance, formaldehyde has been associated not only with asthma (McGwin et al., 2010) but also with leukemia and nasopharyngeal cancer (Bosetti et al., 2008). The composition of VOCs in exhaled breath is highly complex and is influenced by numerous internal and external factors, such as metabolic processes, inflammatory responses, environmental conditions, and oxidative stress (Ruzsanyi et al., 2022). These VOCs are produced from various sources such as the microbiome, respiratory tract, organs, and tissues. Based on the sources, these VOCs can be classified into two categories: (1) exogenous VOCs, which are related to external environmental exposure, and (2) endogenous VOCs, associated with internal physiological processes. The exogenous VOCs are generally released from a vast range of sources, which are often overlooked as environmental contaminants in biological samples (Bosetti et al., 2008). However, exogenous VOCs are now gaining significant attention due to

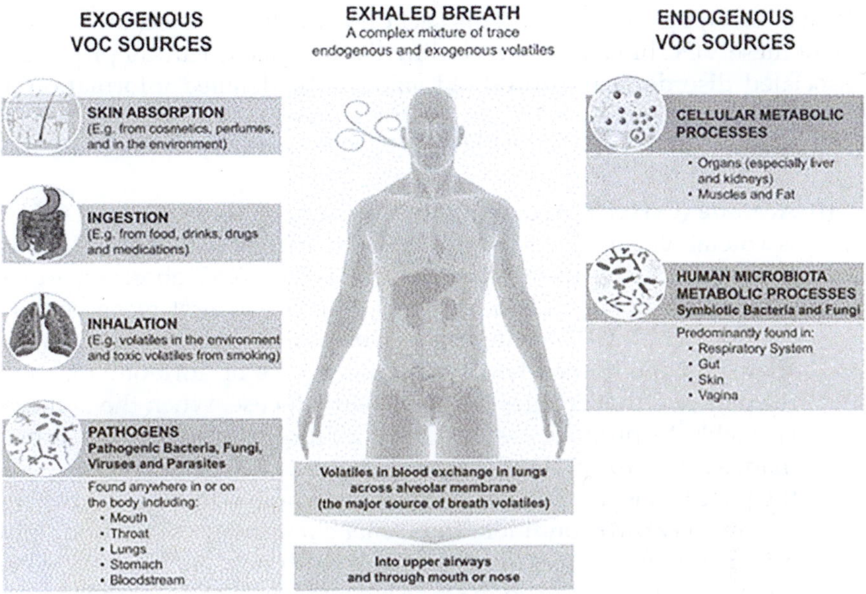

FIGURE 1.1 Origins of endogenous and exogenous VOCs found in exhaled breath. The schematic diagram shows the origin of endogenous and exogenous VOCs observed in the exhaled breath. *From Ruzsanyi, V., Lochmann, F., Jürschik, S., Mochalski, P., Unterkofler, K., & Mayhew, C. A. (2022). Origin and emission of volatile biomarkers in breath: End-tidal perspective. In* Breathborne biomarkers and the human volatilome *(pp. 1–21). Royal Society of Chemistry (RSC). https://doi.org/10.1039/9781839166990-00001.*

their interaction with biological systems and their potential to provide useful health-related information. Meanwhile, endogenous VOCs are generated within the human body and distributed in the bloodstream, and are eventually exchanged through the air into the lungs. From there, they are released into the environment in the form of respiratory droplets and gases. VOCs present in the exhaled breath are commonly known as a major source of multiple useful biomarkers with strong correlation with the body's metabolism. Breath biopsy offers a non-invasive way to collect and analyze these exhaled VOC biomarkers. Typically, it takes about 1 minute for blood to circulate through the entire circulatory system. Fig. 1.1 illustrates a schematic of the various sources of VOCs that contribute to their presence in exhaled human breath.

1.2 Types of volatile organic compounds

The primary VOC biomarkers for disease detection in the human body include acetone (C_3H_6O), isoprene (C_5H_8), ammonia (NH_3), aldehyde (R–CH=O), methane (CH_4), and hydrogen sulfide (H_2S)

(Pathak & Viphavakit, 2022). This section provides detailed information about these VOC biomarkers, including their physicochemical properties, associated disorders, and threshold values. The detailed information on these VOCs is adopted from the author's previously published articles (Pathak et al., 2023; Pathak & Viphavakit, 2022):

(i) **Acetone (C_3H_3O):** Acetone, with the chemical formula CH_3COCH_3, is a metabolite of glucose and human fat (Pathak et al., 2023). Acetone is known as the most frequently detected VOC observed in the exhaled breath of diabetic patients, which was first recognized by Crofford et al. (1977). Glucose is well-known as a primary energy source for the human body. Insulin is a type of hormone made by the pancreas that enables cells to absorb glucose. When the pancreas is unable to produce insulin, the condition is classified as Type 1 diabetes mellitus (T1DM). In contrast, when the insulin produced by the pancreas is not utilized fully, it may result in the Type 2 diabetes mellitus (T2DM) condition. In normal circumstances, the body produces energy by breaking down carbohydrates into glucose. When glucose is limited, the body starts metabolizing fat through ketogenesis to produce ketone as an alternative energy source (Akram, 2013). This process is called ketogenesis. In the human body, all ketones, including acetone, are produced from the ketogenesis process. The level of acetone concentration in exhaled breath is directly linked to blood sugar levels, exhibiting a strong linear relationship between the concentration of acetone in the exhaled breath and blood, which is utilized to assess and monitor diabetes conditions (Fan, 2021). Increased concentration levels of acetone indicate the severity and stages of diabetes and can serve as an early sign of a critical situation. On the other hand, in non-diabetic individuals, the concentration of acetone is typically present at or below ≤ 0.9 parts per million (ppm). For individuals affected with moderate diabetes, this acetone concentration ranges from 0.9 to 1.8 ppm, whereas in the case of severe-stage diabetes, this value can rise to several tens of ppm.

(ii) **Isoprene (C_5H_8):** Isoprene generally appears in the breath of individuals suffering from specific chronic conditions (Pathak et al., 2023). In 2001, McGrath et al. diagnosed heart failure by monitoring the concentration of isoprene in the exhaled breath (McGrath et al., 2001). They found that the concentration of isoprene in heart patients or patients suffering from chronic heart failure was notably reduced, measuring 83 pmol/min_hkg as compared with healthy individuals, which measured 168 pmol/min_hkg. In 2009, Amel et al. investigated the concentration of isoprene in the exhaled breath to assess the lung cancer stages (Bajtarevic et al., 2009). According to the published data, lung cancer patients typically exhale lower concentrations of acetone, isoprene, and methanol compared with healthy individuals. In 2019, Phillip et al. conducted a historical investigation of 53 pediatric Type 1 diabetes mellitus (T1DM) patients, comparing them to children suffering from chronic kidney disease and healthy individuals (Trefz, 2019). The obtained results suggested that the T1DM patients exhaled extremely high concentrations of isoprene, ethanol, and pentanal in comparison with the control group. Furthermore, the alveolar concentrations of pentanal and isopropanol differed significantly between patients with good and poor long-term metabolic control. The study also indicated that isoprene is mostly stored in the fat tissue and is directly linked to cholesterol biosynthesis.

(iii) Ammonia (NH_3): Ammonia, with the chemical formula NH_3, is composed of hydrogen and nitrogen. It is associated with a sharp, penetrating odor and has a colorless appearance. Ammonia plays a key role in various nutritional processes, which include the formation of amino sugars, pyrimidines, and purines, the production of amino acids, and the maintenance of acid-base balance within the human body (Pathak et al., 2023). However, the same ammonia becomes hazardous to the human body when its concentration exceeds the normal range. This excess amount of ammonia must be removed from the body through the ornithine or urea cycle, where it is converted into urea and excreted in urine via the kidneys. This conversion process occurs in the kidneys and liver. Therefore, any impairment in the function of these organs can lead to elevated ammonia levels in exhaled breath. Consequently, high concentrations of ammonia levels in exhaled breath may indicate various diseases, including hepatic encephalopathy, kidney failure, liver dysfunction, type II Alzheimer's disease, and brain swelling (Adlimoghaddam, 2016; Gallego, 2025; Jo, 2021).

(iv) Methane (CH_4): In human breath, the methane gas is produced within the intestine via methane-producing bacteria in the intestine under anaerobic condition (Chaudhary, 2018). Normally, methane is not found in the breath, but it can appear when the body starts producing it excessively. Variations in the concentration of methane can also be detected under specific conditions, including obesity, inflammatory bowel disease, and irritable bowel syndrome (Polag & Keppler, 2019). Methane can be considered a potential biomarker for various diseases and cancers, such as hepatic disease, breast cancer, and heart disease. Moreover, other hydrocarbons such as methane, ethane, and pentane can offer useful information for several diseases linked with oxidative stress, including head and neck cancer (Dharmawardana, 2020).

(v) Aldehydes (–RCHO): Aldehydes are organic molecules characterized by a functional group containing hydrogen, carbon, and oxygen. The chemical formula of an aldehyde is written as –RCHO or –CHO, where "R" denotes the remainder of the organic molecule. Aldehydes, including dialdehydes, alkenals, and hydroxy alkenals, are products of alcohol oxidation and lipid peroxidation. Variations in the concentration of aldehydes can be utilized in cancer diagnosis (Ayala et al., 2014). Aldehydes can be produced when polyunsaturated fatty acids interact with free radicals. Increased levels of aldehydes have been reported in individuals with Wilson's disease, liver-related diseases, cancer, and oxidative stress. These elevated levels have been detected in both blood samples and exhaled breath (Pizzimenti et al., 2013). Different concentrations of aldehydes, specifically formaldehyde and glyoxal, have been observed in urine, blood, and exhaled breath in several metabolic and genetic disorders, including aging, Parkinson's and Alzheimer's diseases, and diabetes (Skoumalová & Hort, 2012).

(vi) Hydrogen sulfide (H_2S): Hydrogen sulfide, with the chemical formula H_2S and a boiling point of −60.2°C, is a widely known colorless VOC with a strong and unpleasant odor. H_2S can be produced endogenously and exogenously in animals and the humans. As a notable gasotransmitter in living organisms, it plays a crucial role in signaling various physiological processes, including inflammation, neuromodulation, apoptosis, cytoprotection, and the regulation of vascular tone (Wang, 2002; Zaichko, 2014). H_2S present in exhaled breath is considered a potential marker to identify asthma, airway inflammation, and oral health issues (Rosenberg & McCulloch, 1992; Suarez et al., 2000; Toombs et al., 2010).

(vii) Cyclohexene (C_6H_{10}): Cyclohexene is a cyclic alkane, primarily utilized in the synthesis of adipic acid, maleic acid, and butadiene, as well as in inorganic synthesis and oil extraction, where it plays the role of a stabilizing agent for high-octane gasoline. Similar to acetone, the excessive exposure to cyclohexane, even at low concentrations, can cause serious health issues, including irritation to the eyes, throat, and respiratory tract, and sometimes it can also cause drowsiness (Campos-Ordonez & Gonzalez-Perez, 2016).

(viii) Acetylene (C_2H_2): Acetylene is recognized as a highly hazardous VOC due to its intrinsic instability, particularly under high temperature, liquefied, pressurized, or mixed with air. The combustion of acetylene generates substantial heat, which can cause explosions if the pressure exceeds the gas's absolute pressure limit of 103 kPa (Iftekhar Uddin et al., 2015; Oliveira et al., 2019). Direct exposure to acetylene can lead to dizziness, nausea, headache, tachycardia, fatigue, and vomiting. In case of excessive exposure, acetylene can also cause loss of consciousness and death (Newhouse, 1988).

(ix) **Benzene (C_6H_6):** Benzene, with the chemical formula C_6H_6, is a natural component of both refined and crude petroleum. Approximately 98% of benzene is generated from the petrochemical and associated refining industries (Duarte-Davidson et al., 2001). Additionally, benzene is one of the most extensively used VOCs in the production of various resins, synthetic fibers, and polymers. Inhalation of benzene, even at low concentrations of 200 ppm, can induce narcotic effects. Exposure to higher concentrations may provoke convulsions and lead to fatal respiratory failure. Prolonged exposure to benzene can also contribute to the development of leukemia and lymphomas (Ke et al., 2009). In urban environments, benzene primarily released from vehicle exhaust, posing a potential threat to residents of cities. Notably, in 1994, benzene was ranked among the top 20 most toxic chemicals in the United States (Kirschner, 1995). To prevent health issues, the maximum permissible exposure for individuals has been set to 16.25 $\mu g/m^3$ (Mabrook & Hawkins, 2002).

(x) **Ethanol (C_2H_5OH):** Ethanol is known as one of the less hazardous VOCs. However, its vapor can still lead to serious health issues such as headaches, drowsiness, eye irritation, and breathing difficulty (Ahmed, 1995). Recently, the use of ethanol-gasoline blends in vehicles has become a popular trend for cost-saving. However, this practice can lead to an increased release of VOCs into the environment (Tibaquirá et al., 2018).

(xi) **Methanol (CH_3OH):** Methanol is a well-recognized highly toxic VOC that can cause a range of serious health issues, such as airway narrowing, headaches, vertigo, bronchial constriction, visual impairment, and nausea (Rong et al., 2018). Despite its high toxicity, methanol is widely used in the synthesis of automotive fuel, dyes, drugs, formaldehyde, and perfumes (Zhao et al., 2020). It is also extensively employed in the manufacturing of several alcohols distributed worldwide. However, due to its structural similarity to other aliphatic alcohols, the detection of methanol can be challenging.

(xii) **Toluene (C_7H_8):** Toluene, with molecular formula C_7H_8, is an important industrial VOC widely used in the synthesis of various chemical intermediates, including cosmetics, adhesives, solvents, dyes, pharmaceuticals, medicines, explosives, and agricultural chemicals (Hermawan et al., 2020). The toxicity of toluene is similar to that of benzene. Exposure to high concentrations of toluene can damage human organs such as the skin, liver, kidneys, and central nervous system (Patnaik, 2007). To control and monitor air quality, the WHO has established a guideline for the indoor air concentration of toluene at 0.07 ppm at 25°C.

(xiii) **n-Butanol (C_4H_9OH):** n-butanol or Butan-1-ol is a primary alcohol with the C_4H_9OH commonly used as a solvent, extraction agent, and in the preparation of plasticizer, ester, and other industrial products (Wang et al., 2020b). Exposure to high concentrations of n-butanol may result in drowsiness, dizziness, loss of vision, skin damage, and respiratory irritation due to its narcotic features and stimulating nature (Liu et al., 2019). Prolonged exposure to higher concentrations can also cause photophobia, lacrimation, and blurred vision.

(xiv) **2-propanol (C_3H_8O):** 2-propanol is extensively utilized in the production of isopropyl halides and acetone. Additionally, it is applied in various industrial products, including paints, polishes, insecticides, and rubbing alcohol (Lomonaco et al., 2020). At high concentrations, ranging from 400 to 5000 ppm, 2-propanol can cause deterioration of ciliary activity in guinea pigs. Inhalation of 2-propanol may lead to nausea, dizziness, irritation of the eyes and nose, and drowsiness, and excessive exposure may also increase the risk of coma (Smeets et al., 2002).

In Table 1.2, various types of VOCs are summarized in detail along with their molecular formula and associated disorders (Pathak et al., 2023).

1.3 Sources of volatile organic compound emissions

VOCs are primarily derived from natural sources, such as the transformation of biogenic precursors and forest fires; however, human-induced activities have emerged as significant contributors to harmful VOC emissions (Montero-Montoya et al., 2018). These activities contribute to 25% of the total VOCs present in the surrounding atmosphere. The primary sources of VOCs include the extraction of natural gas and petroleum, petrochemical operations, and the combustion of fossil fuels in various sectors, including industries, households, and transportation. Additionally, significant contributions are made by chemical and industrial processes involved in the production of paints, lubricants, adhesives, and oil derivatives, as well as activities related to mining, commercial operations, gas leaks from cooking appliances, boilers, and the use of agricultural pesticides. VOCs can be released from various products and activities, both indoors and outdoors. Some common sources of VOCs are discussed in the following subsections.

1.3.1 Indoor sources

More than 85% of VOC exposure to humans primarily occurs in indoor environments (Buonanno et al., 2014). Gaseous organic compounds commonly found indoors can be categorized as VOCs, intermediate VOCs, semi-VOCs, low VOCs, and extremely low VOCs based on their vapor pressures. Among these, VOCs with an effective saturation concentration

TABLE 1.2 Summary of all VOCs along with their molecular formula, associated disorder, and their concentration in health and diseased person (Pathak & Viphavakit, 2022).

Exhaled VOC	Molecular formula	Associated disorder	Concentration (healthy person)	Concentration (diseased)
Benzene	C_6H_6	Cancer	NA	300 and 3000 ppm
Isoprene	C_5H_8	Diabetes, heart failure, lung cancer	5.8–274.9 ppb	NA
Acetone	C_3H_6O	Diabetes,	0.39–0.85 ppm	T1D: >2.2 ppm, typically >10 ppm
Ammonia	NH_3	Liver dysfunction, kidney failure, hepatic encephalopathy, type-II Alzheimer, swelling of the brain	0.25–2.9 ppm	NA
Methane	CH_4	Oxidative stress, heart disease, breast cancer	4–10 ppm	8–50 ppm
Aldehyde	R–CH=O	Liver cancer, Alzheimer, Parkinson's Disease	13 ± 5 ppb	NA
Hydrogen sulfide	H_2S	Asthma, oral, and dental health	8–16 ppb	NA

(continued on next page)

TABLE 1.2 Summary of all VOCs along with their molecular formula, associated disorder, and their concentration in health and diseased person (Pathak & Viphavakit, 2022)—cont'd

Exhaled VOC	Molecular formula	Associated disorder	Concentration (healthy person)	Concentration (diseased)
Nitric Oxide	NO	Asthma, acute lung injury, lung infection, lung cancer	<35 ppb	NA
Ethane	C_2H_6	Scleroderma, cystic fibrosis	0–12 ppb	NA
Pentane	C_5H_{12}	Myocardial infarction	0.3–0.8 nmol/L	NA
Carbon monoxide	CO	Smoking	0.4–0.8 ppm (non-smokers)	2–20 ppm
8-isoprostane	–	Obstructive sleep apnea	4–5 pg/mL	6.7–7.1 pg/mL
Interleukin-6	IL-6	Non-small cell lung cancer	3.3–3.7 pg/mL	9.3–9.9 pg/mL

From Pathak, A. K., & Viphavakit, C. (2022). A review on all-optical fiber-based VOC sensors: Heading towards the development of promising technology. Sensors and Actuators A: Physical, 338, *113455. https://doi.org/10.1016/j.sna.2022.113455.*

(C^*, $\mu g\ m^{-3}$) of greater than $3 \times 10^6\ \mu g\ m^{-3}$ and semi-VOCs/intermediate-VOCs ($C^* = 0.3–3 \times 10^6\ \mu g\ m^{-3}$) are particularly noted for their adverse health effects (Li et al., 2022). Ventilation and temperature are considered crucial environmental factors that regulate VOC concentrations indoors. Common indoor sources of VOCs include cooking (Ditto et al., 2022), personal care products (McDonald et al., 2018), cleaning products (Finewax et al., 2021), and volatile gases emitted from building materials (Stinson et al., 2022). The primary sources of indoor VOCs, grouped by their origin, are provided in Table 1.3. The emission of VOCs from these sources is influenced by several factors (Paciência et al., 2016).

Despite the high exposure to VOCs in indoor environments, especially in areas like educational institutions, research on VOCs within such settings remains extremely limited (Jin et al., 2023). In 2013, a health risk assessment of indoor VOCs was performed by Kumar et al. (2014) at Jawaharlal Nehru University, India, during both summer and winter (2011–12). The study focused on the presence of benzene, toluene, ethylbenzene, m/p-xylene, and o-xylene (BTEX) (Kumar et al., 2014). It was observed that the indoor concentration of BTEX (excluding benzene) was significantly higher than the outdoor environment during both seasons. The BTEX indoor-to-outdoor (I/O) ratios were typically higher than one, with winter values ranging from 0.2 to 8.7 and summer values from 0.2 to 4.3. The study reported an excessive benzene value (8.93×10^{-6}) compared to the recommended limit (1×10^{-6}) set by the United States Environmental Protection Agency (USEPA) in the indoor library of Jawaharlal Nehru University, India. In 2023, Jin et al. monitored indoor VOCs across three different universities in China (Jin et al., 2023). The study aimed to determine the VOC concentrations on different college campuses by collecting VOC samples from various campuses across various seasons, along with assessing the exposure duration of students at each site. To evaluate the potential health risks, the researcher conducted a questionnaire survey. The highest VOC concentration of $254 \pm 101\ \mu g/m^3$ was observed in the student dormitory. Health risk assessments associated with VOCs were conducted by examining non-carcinogenic [expressed as hazard quotient (HQ)] and carcinogenic [expressed as lifetime cancer risk (LCR)] risk values. Residential houses exhibited the greatest carcinogenic risk, while the carcinogenic risk on the college campus was found to be low LCR ($< 1.0 \times 10^{-6}$). Additionally, 1,2-dichloroethane was recognized as a potential carcinogenic risk substance in the dormitory because of its high LCR ($\sim 1.95 \times 10^{-6}$). In 2017, Kang et al. investigated air quality in a university's dormitory, library, teaching classes, and building, and observed that the total VOC concentration in the dormitory area exceeded the allowed standard Chinese indoor environmental guidelines ($0.6\ mg/m^3$) (Kang et al., 2017). Moreover, it was noted that universities possess more

TABLE 1.3 Summary of different sources of VOCs and associated examples.

Sources of Indoor VOC Emissions	
Endogenous	**External**
Originating from the building: • furnishing materials, paints, oil, etc. • building construction and renovations • indoor equipment and related elements • fungal and bacterial activities • office appliances (printers, copiers)	Chemical and petrochemical industry Transport: • means of transportation • fuel loading/unloading stations and terminals
Originating from humans and human activities inside the building: • household and consumption products such as hygiene goods and aerosols • processes associated with indoor cleaning, e.g., preserving and cleaning chemicals • making indoor foods (frying, baking, brewing) • pets • recreational activities • indoor plants • human and pet waste • household cooking equipment, including stoves, ovens, heaters, etc. • tobacco smoke	Small businesses of dry cleaners, hardware stores, printing shops, etc. Short-term emissions: • break-down of pipelines, valves, and pumps • cleaning of tanks Emissions of biological origin: • agriculture • green plants • fossil fuels • bioaerosol • global ocean surface Utilization and disposal of wastes: • waste storage sites • wastewater treatment plants
Additional factors to consider: • industrial plants or production facilities situated in residential areas • occupational exposure (release of pollutants from clothing and the body) • associated with motorization (residential use of a garage within a building) • combustion sources (heating systems, water heating systems) • water sourced from a water supply system.	

indoor pollution sources, highlighting the need for comprehensive indoor air quality assessments in such environments (Goodman et al., 2018).

In 2014, Chin et al. conducted a study examining VOC levels in 126 households of children diagnosed with asthma in Detroit, Michigan, United States (Chin et al., 2014). The study identified a range of VOC concentrations and associated them with potential health effects. The sources of VOC emissions include renovations, cigarette smoking, household products, and pesticides. While the concentrations were generally lower than those observed in other North American studies, several homes still had increased levels of VOCs. A total of 56 different VOCs were identified across 126 households, predominantly in the densely populated regions of Detroit. The mean total target concentration of VOCs (ΣVOC) in households was 150 $\mu g/m^3$, with a median of 91 $\mu g/m^3$. Concentrations varied significantly across households, ranging from 14 to 2274 $\mu g/m^3$, and exhibited a pronounced right-skewed distribution. Among the highest average concentrations were alkanes (n-C7-13 and methylcyclohexane), aromatics (benzene, toluene, and xylenes), tetrachloroethane, and terpenes (*d*-limonene and α-pinene). Several other VOCs were also detected in most homes, e.g., α-pinene and *d*-limonene, which are produced from air fresheners, fragrances, and cleaning products (Nazaroff & Weschler, 2004), and toluene and other aromatic compounds, which are emitted from adhesives, paints, synthetic fragrances, vehicle emissions, and evaporated fuel. Ethyl acetate was detected in 20 households, approximately 71% of visits, with a concentration of 10 $\mu g/m^3$, mostly from glue, nail polish remover, and other products (*Hazardous substances data bank (HSDB)*, no date).

Given these findings, special attention should be given to monitoring VOCs and semi-VOCs in universities, schools, and their dormitories. Children, who are particularly vulnerable to the adverse health effects associated with these compounds, according to the USEPA, are especially at risk (Ninyà et al., 2022). Ensuring proper ventilation and controlling indoor VOC sources are crucial steps to mitigate health risks in these educational environments. Regular assessments and adherence to safety guidelines can help protect students from the potential harmful effects of VOC exposure.

1.3.2 Outdoor sources

While indoor VOCs significantly impact human health, their presence in outdoor environments also demands attention. The movement of air between indoor and outdoor spaces means that outdoor pollutants can contribute to indoor VOC levels. As previously reported, over 70 VOCs have been detected in indoor environments, while 42 VOCs and semi-VOCs have been monitored in outdoor environments (Ninyà et al., 2022). Over the past decades, several studies have monitored VOCs and semi-VOCs in outdoor environments, particularly in urban and industrial atmospheres (Maceira et al., 2017; Wania & Shunthirasingham, 1925; Xuan et al., 2021). VOCs are considered a critical class of atmospheric pollutants in both urban and rural areas due to their adverse health effects (Lin et al., 2016). Urbanization and increasing traffic mobility have led to significant challenges for city development, particularly in the form of intense urban traffic congestion. The emission of vehicle exhaust plays a crucial role in escalating the level of air pollution, specifically in urban areas, releasing various pollutants, including ultrafine particles, hydrocarbons, nitrogen oxide, and VOCs. Liao (Liao et al., 2014) and Marty (Marty et al., 2010) reported the crucial role of VOCs in atmospheric chemistry. VOCs supply OH radicals in the air, which act as strong ozone (O_3) precursors, contributing to the formation of ground-level O_3. These compounds are widely recognized for their role in the formation of stratospheric O_3, photochemical smog, and organic acids, which leads to acid rain by reducing the pH level of rainwater.

In 2014, Do et al. conducted a study on three transportation methods (auto vehicle, tram, and bicycle) in Ghent, Belgium, to analyze the emission of VOCs (Do et al., 2014). A total of 84 VOCs were observed, including oxygen-containing hydrocarbons, hydrocarbons, aromatic hydrocarbons, terpenes, and halogenated compounds. The concentration levels of these VOCs were measured using active sampling with Tenax TA sorbent tubes, followed by analysis utilizing thermal desorption gas chromatography-mass spectrometry (TD-GC-MS) with deuterated toluene working as internal standard. The reported median concentrations of total VOCs for tram commuters were found to be 1.7 times higher than for cyclists (20 $\mu g/m^3$) and 1.5 times higher than for car users (22 $\mu g/m^3$). Aromatic hydrocarbons account for most of the total VOC concentration, ranging from 41% to 57% for the tram, 58% to 72% for the bicycle, and 59% to 72% for the car. A high correlation was observed among BTEX compounds, n-propylbenzene, 1,3,5-trimethylbenzene, 1,2,4-trimethylbenzene, and iso-propylbenzene across all transportation modes. Compared with other modes of transport on fixed routes in Ghent, commuters using the tram were found to be exposed to the highest total VOC levels.

Another significant contributor to outdoor VOCs originates directly from the earth, impacting global warming. This source is a notable originator of trace gases from the soil into the troposphere (Potard et al., 2017). The release generally includes carbon dioxide (CO_2) along with other low-concentration gases such as methane (CH_4), nitrous oxide (N_2O), and various VOCs (Lashof & Ahuja, 1990). One of the main sources of VOC emissions into the ambient environment is natural biogenic processes (Atkinson & Arey, 2003b; Lamb et al., 1993). It has been reported that approximately 1200 Tg of VOCs are released into the atmosphere annually, which is three times the emissions of CH_4, with around 1100 Tg coming from biogenic VOCs released by vegetation (Atkinson & Arey, 2003a). The emission of VOCs from living plants and related senescent litter has been documented (Gray & Fierer, 2012; Rinnan et al., 2013). Plant vegetation in situ releases a large number of VOC molecules, specifically monoterpenes and isoprene (Kesselmeier & Staudt, 1999), with isoprene fluxes ranging from 0 to 738 $\mu g\ m^{-2}\ h^{-1}$, in subarctic wet heathland (Faubert et al., 2010;

Tiiva et al., 2008) and from 2.0 to 9.7 mg m^{-2} h^{-1} in oak forests of the French Mediterranean (Kalogridis et al., 2014). Konig et al. also reported that VOC generation varies according to plant species and their phenology across the season (Konig, 1995). The petrochemical industry is another crucial source of excessive VOC emission (Zheng et al., 2020). The risks associated with VOCs released from petrochemical industries have gained significant attention from environmentalists and scientists (Hsu et al., 2018). For example, the hazard index around industrial areas was observed to be 1.07, which is significantly higher than the urban areas with a hazard index of 0.96 (Hu et al., 2018). In 2018 Shuai et al. reported a cross-sectional study of VOC exposure near the Daegu Dyeing industrial area in South Korea (Shuai et al., 2018). It was observed that health risks, including anaphylactic, respiratory, and cardiovascular diseases associated with VOC species such as chloroform and toluene, were higher in industrial areas compared with control areas. Epidemiological investigations also suggest that people residing within a 5 km radius of any petrochemical industry are more prone to the risk of respiratory symptoms than those living farther away (Zheng et al., 2020). Table 1.4 summarizes the sources of outdoor VOCs.

1.3.3 Common consumer products

A large number of VOCs are emitted from various consumer products, which can significantly impact the atmosphere, air quality, and human health. However, the associated risks from these VOCs are not fully understood due to limited information on emissions from specific products (Steinemann, 2015). In various developed countries, such as the United States, consumer products have been recognized as one of the major contributors to human VOC exposure (EDWARDS et al., 2006). For instance, fragranced products release terpenes such as α-pinene and limonene, which dominate indoor VOC concentrations (Maisey et al., 2013; Steinemann et al., 2011). These terpenes may undergo reactions with ozone to produce a broad range of secondary pollutants, such as acetaldehyde, formaldehyde, ultrafine particles, and organic aerosols (Rossignol et al., 2013; Singer et al., 2006). Some of these indoor VOCs also migrate to outdoor environments, further affecting the ambient atmosphere and human health, even in the absence of direct product use (Chen & Luo, 2012). In 2013, Steinemann et al. found that the dryer sheets and fragranced laundry detergents can release certain chemicals into the environment via wastewater, air vents, and other routes (Steinemann et al., 2013). Fragrances alone in consumer products can contain thousands of harmful chemicals. With over 2600 listed as fragrance ingredients (Bickers et al., 2003), some of which can be classified as highly toxic under the federal laws (Steinemann et al., 2011). Some studies have analyzed VOCs emitted directly from

TABLE 1.4 Summary of sources of outdoor VOCs emissions.

Chemical and petrochemical industry: In the chemical and petrochemical industry VOC emissions occur due to the chemical synthesis, added solvent, and refining procedure. Storage tanks, loading operations, and leakage of equipment contribute to significant VOC emissions.

Transport:

- **Means of transportation:** Combustion of fossil fuels in internal combustion engines releases VOCs through exhaust emissions.
- **Fuel stations and terminals:** Loading/unloading of fuels between storage tanks, tanker trucks, and pipelines can release a significant amount of VOCs into the air due to the displacement of vapors.

Small businesses of dry cleaners, hardware stores, printing shops, etc.: Perchloroethylene is the most common solvent used in dry cleaning, which releases a significant amount of VOCs into the environment. In the hardware stores the paint mixing and selling also contribute to the VOC emission. Printing processes often involve inks containing VOCs, and the use of certain solvents in the printing industry can contribute to emissions.

Short-term emissions:

- break-down of pipelines, valves, and pumps
- cleaning of tanks
- smoke released from automobiles and vehicle

Other

- **Agriculture:** Pesticides and fertilizers used in agriculture can contribute to VOC emissions, particularly when sprayed or applied to crops.
- **Fossil fuels:** The combustion of fossil fuels releases VOCs, contributing to air pollution and environmental concerns.
- **Power plants:** Combustion of fossil fuels in power plants, specifically using oil or coal, releases VOCs along with other hazardous pollutants.
- **Waste treatment and disposal:** Decomposition of organic waste in landfills and waste treatment facilities can produce VOC emissions.
- **Wood burning:** Whether in any industrial region or residential, the combustion of wood releases VOCs due to the combustion process.

products, whereas others have investigated the formation of secondary pollutants through reactions between VOCs released from the consumer product and ozone (Destaillats et al., 2006; Jo et al., 2008). Other research has also explored the presence of synthetic musks in various environments, including water, air, sediments, aquatic organisms, wastewater treatment systems, and household items (Peck & Hornbuckle, 2006; Reiner & Kannan, 2006). In 2015, Steinemann reported 156 different VOCs emitted from 37 common consumer products, with a minimum of 15 VOCs per product

(Steinemann, 2015). Headspace GC analysis was utilized to monitor VOCs emitted from 37 types of consumer products, categorized into four groups: air fresheners, cleaning products, personal and beauty care products, and laundry products. Among these, 9 products are classified as air fresheners (e.g., gels, sprays, disks, solids, and oils), 11 as laundry products (e.g., dryer sheets, detergents, and fabric softeners), 7 as cleaning products (e.g., dish detergent and disinfectants), and 10 as personal and beauty care products (e.g., soaps, lotions, deodorants, hand sanitizers, and shampoo).

1.4 Environmental impacts of volatile organic compounds and climate change

The global expansion of agricultural activities involves increased fuel use, agricultural waste burning, and the application of VOCs in pesticides, leading to the widespread distribution of VOCs in the environment (David & Niculescu, 2021). Additionally, several studies have reported the presence of numerous VOCs from sources such as tobacco smoke, chlorinated water, paint removers, perfumes, adhesives, plastics, kerosene heaters, and new clothing (Spengler & Chen, 2000). Some researchers have also detected VOCs in drinking water (Diduch et al., 2011), and microbial VOCs in the air, likely produced from spores of fungus or airborne microbial metabolites (Fischer & Dott, 2003). Certain VOCs are recognized as greenhouse gases (Lelieveld et al., 2009), which absorb radiated energy from the Earth, contributing to global warming. Methane, the second most abundant greenhouse gas in the United States (National Academies of Sciences, 2018), is often released simultaneously with VOCs. Greenhouse gases trap heat in the atmosphere, leading to rising global temperatures. Research by the Environmental Protection Agency (EPA) indicates that the industrial sector is responsible for 22% of all greenhouse gases emitted. Moreover, VOCs in wastewater reduce the feasibility of water reuse for activities such as irrigation, placing additional strain on already scarce primary water resources (Abdullahi et al., 2014). VOCs also play a crucial role in the formation of fine particulates and the ozone layer in the atmosphere. In the presence of sunlight, atmospheric VOCs react with nitrogen oxides, primarily from power plants, vehicles, and industrial activities, resulting in ozone formation and contributing to the generation of fine particulates. The accumulation of ozone, fine particulates, and other gaseous pollutants leads to smog, which reduces visibility and can cause eye, nose, and throat irritation (*Volatile Organic Compounds and Smog*, no date). Smog becomes particularly severe during periods of intense sunlight and stagnant weather conditions, which generally occur when there is little wind or atmospheric movement, most commonly seen in autumn. Severe smog can affect individuals with heart conditions and respiratory issues such as asthma, posing a higher risk to children, the elderly, and those engaged in vigorous outdoor activities. Prolonged exposure to severe smog can lead to degradation of the immune system or permanent damage to lung tissues. Dense smog can also impact tourism by impairing visibility. Another significant environmental impact of VOCs is acid rain, which results from high concentrations of acidic compounds. When atmospheric water reacts with sulfur oxide and nitrogen oxides, the pH of rainwater can drop to 4.2 or 4.4. This acidic rainfall is distributed across the cityscapes, oceans, and natural habitats, resulting in the disruption of fragile ecosystems.

1.5 Origin of volatile organic compounds in the human body

In the above section, we introduced the endogenous and exogenous sources of VOCs and their importance in early diagnosis. Breath biopsy provides a non-invasive way to collect and analyze these VOCs from exhaled breath. Typically, the blood circulates through the entire body in approximately 1 minute. Sampling breath for a minute or longer enables the detection and characterization of VOC biomarkers, even at low concentrations. Extending the sample collection time can enhance sensitivity, allowing for the detection of early-stage disease. This section provides a detailed discussion of the endogenous sources of VOCs within the human body.

1.5.1 Endogenous source of volatile organic compounds

A recent study identifies approximately 1849 VOCs detected in various human biofluids, including interstitial fluids, blood, breath, saliva, sweat, urine, breast milk, serum, and tears (Amann et al., 2014). The pie diagram in Fig. 1.2 illustrates the common sources of biofluids VOCs and their respective production percentages across biofluids. Another study reported that, around 2577 VOCs were identified from these biofluids, among which most of the VOCs percentage was recorded from the exhaled breath (34%) and skin (20%), while feces, saliva, urine, and blood contribute 15%, 14%, 11%, and 6%, respectively, as shown in Fig. 1.2 (Broza & Haick, 2013). These biofluidics serve as sources for either invasive or non-invasive monitoring of VOCs, facilitating the early diagnosis of various diseases.

1.5.1.1 Origins used in invasive detection

Blood and Plasma: Blood and plasma are considered the most trusted sources used for characterizing an individual's physiological state

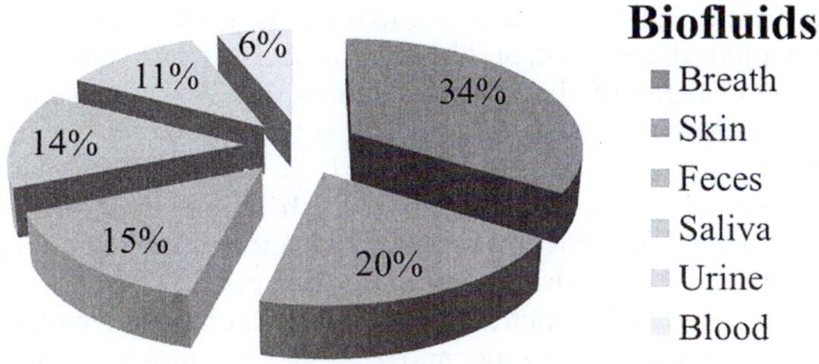

FIGURE 1.2 Pie diagram showing the source of VOCs. The pie diagram illustrates the common sources of biofluids VOCs along with their percentage contribution. *From Broza, Y. Y., Mochalski, P., Ruzsanyi, V., Amann, A., & Haick, H. (2015). Hybrid volatolomics and disease detection.* Angewandte Chemie – International Edition, 54*(38), 11036–11048. https://doi.org/10.1002/anie.201500153.*

(Jalal et al., 2018). To date, 154 different types of VOCs have been reported in blood samples (de Lacy Costello et al., 2014). These VOCs act as potential biomarkers for various diseases and reflect the internal health conditions related to metabolic, nutritional, and immune states (Broza et al., 2015). Human blood contains a variety of VOCs, such as ammonia, acetonitrile, ethers, alcohols, alkanes, phenyl hydrocarbons, and benzyl (de Lacy Costello et al., 2014). The health condition of an individual can be reliably assessed by analyzing changes in the concentration of specific VOCs in blood samples. For instance, an increased concentration of toluene and benzene is observed in smokers, while swimmers exhibit elevated levels of trihalomethane and chloroform in their blood. A rise in VOC concentration beyond the normal values for these exogenous VOCs can play a vital role in human health. Excessive amounts of chloroform can potentially cause liver damage, creating a potential pathway for various other life-threatening diseases, along with DNA damage and anemia. In 2010, Horvat et al. suggested that certain aromas in the blood could be suitable tools for screening various internal disorders, such as ovarian carcinoma, hepatic encephalopathy, and lung cancer (Horvath et al., 2010). Researchers have focused on endogenous VOCs derived from plasma in blood for the early detection of cancers, including gynecological and lung cancers. In 2013, Selyanchyn et al. reported an analytical approach to evaluate VOC levels in blood for the early detection of cancer (Selyanchyn et al., 2013). In the study, VOCs released from 40 diseased blood samples and 10 control samples were compared to examine differences between the groups, revealing similar qualitative results across both groups regardless of disease presence. Although blood analysis provides more consistent and stable results, the sampling process from other biofluids is invasive, inefficient, and unsuitable for rapid point-of-care applications.

Interstitial Fluids: Various interstitial fluids, such as peritoneal, cerebrospinal, pleural, synovial, and biliary fluids, have been studied to identify different stages of disease conditions. The VOCs present in these interstitial fluids have been reported by several researchers (Jalal et al., 2018). In 2014, Liu et al. investigated nearly 76 samples from pleural fluid and identified 3 major VOC groups, which mainly included alcohols, ketones, and benzene, directly linked with cancer malignancy (Liu et al., 2014). Among the 76 VOCs identified, 9 compounds, such as ethyl acetate, n-heptane, dichloromethane, xylene, ethylbenzene, cyclohexane, cyclohexanol, tetramethylbenzene, and 2-ethyl-1-hexanol, were recognized as potential markers for diagnosing lung cancer, distinguishing between malignant and benign pleural effusions. Various health conditions, including tuberculosis, heart failure, cirrhosis, and pneumonia, are major causes of benign effusions, with 30% of malignant effusions directly related to lung cancer. VOC analysis of biliary fluid can also help differentiate malignant cholangiocarcinoma from benign biliary strictures. It serves as a crucial biomarker to diagnose pancreatic cancer, particularly through the concentration spectra of acetone, tetramethyl acetate, carbon disulfide, acetaldehyde, pentane, and benzene. In 2013, Tominaga et al. conducted toxicological analyses on several cases of serial forensic autopsies and compared pericardial fluid and peripheral blood to reexamine the ethanol and other VOCs distribution (Tominaga et al., 2013).

1.5.1.2 Sources for non-invasive detection

Urinary Bio-fluid: Urine testing is a standard invasive approach utilized for the detection of various diseases and health conditions. The urinary matrix is reported to contain a minor proportion of VOCs (Bouatra et al., 2013). Both exogenous and endogenous VOCs in urine are influenced by renal activities. Endogenous VOCs include alcohols, ketones, heterocyclic compounds, amines, organic acids, aldehydes, and various hydrocarbons. Among these, ketones are the most common biomarkers, produced through the oxidation of fatty acids. The concentration of ketones can increase due to factors such as starvation (fasting ketosis), a low carbohydrate-high fat diet, or alcohol consumption. The state of nutritional ketosis provides energy to the brain of epileptic patients. On the other hand, reduced ketone levels are observed in conditions such as diabetic ketoacidosis or hyperglycemia. Insufficient ketogenesis can lead to hypoglycemia, while an excessive presence of ketones can result in ketoacidosis. Diabetic ketoacidosis represents a severe condition, with ketone levels exceeding those in other ketotic states by more than twofold. Ketone concentrations are influenced by multiple physiological factors, including energy balance, physical activity, diet composition, and certain diseases. Besides ketones, several other VOCs present in urine serve as biomarkers for various diseases, providing a specific VOC profile depending on factors such as diet, dehydration, drug intake, and fluid consumption. A reduced concentration of acetonitrile in urine can confirm the smoking habit of an individual (McKee et al., 1962). Urinary metabolomics have been widely employed in cancer research, including breast, colorectal, esophageal, pancreatic ductal adenocarcinoma, liver, and prostate cancers. Khalid et al. identified four VOC biomarkers, i.e., pentanal, 2-octanone, 2,6-dimethyl-7-octen-2-ol, and 3-octanone, linked with prostate cancer (Khalid et al., 2015). In 2010, Matsumura et al. reported an ex vivo study to investigate urinary VOC biomarkers related to lung cancer (Matsumura et al., 2010). While non-invasive urine testing is reliable, challenges such as sample collection, handling, preservation, and analysis complicate the process. Consequently, the technical complexity of diagnosing VOC biomarkers from urine samples makes this biofluid unsuitable for continuous monitoring (Wang et al., 2015).

Saliva: Saliva is reported to contain around 360 types of exogenous and endogenous VOCs (Broza et al., 2015). Unlike breath, various factors such as food debris, xenobiotics, drugs, and environmental pollutants can significantly influence the detection of VOCs in saliva. Saliva VOC detection frequently relies on common methods such as active and passive diffusion from blood, along with ultrafiltration techniques (Soini et al., 2010). Additionally, 300 different types of bacterial species have been identified as sources of VOC production in saliva. Some of the key VOC biomarkers present in saliva include ester, acetic acid, acetonitrile, mercaptan, diene, various alkanes, toluene, benzene, xylene, polycyclic hydrocarbons, propanol, methyl sulfide, ethanol, butanol, and aldehydes (de Lacy Costello et al., 2014). Recent research highlights the potential of salivary fluid for continuous, non-invasive, and real-time monitoring of various diseases. Various research groups have integrated specially designed electrochemical sensors into mouthguards to selectively detect such VOC biomarkers (Kim et al., 2015). Due to the abundant presence of VOCs in saliva, it is considered a valuable biofluid for the real-time and continuous monitoring of various diseases.

Sweat and Skin: The human skin, recognized as the largest organ, facilitates the prolonged emission of VOCs. Costello et al. reported that 532 different types of VOCs are released from the skin secretions through the diffusion of sweat (de Lacy Costello et al., 2014). The primary VOCs emitted from a healthy skin typically include carboxylic acid, ammonia, alcohols, ketones, hydrocarbons, terpenes, esters, heterocyclic compounds, aldehydes, and sulfur compounds (Gallagher et al., 2008). The non-uniform arrangement of sweat glands under the skin significantly contribute to the complexity of VOC profiles. Additionally, the bacteria present on the skin induce variation in the odor profiles among individuals (Sanford & Gallo, 2013). The release of VOCs via the skin is primarily regulated by three major glands: (1) apocrine glands, (2) eccrine glands, and (3) sebaceous glands. The emission of VOCs can also vary with other factors such as xenobiotics, body temperature, diet, wounds, psychological stress, drugs, dehydration, shock, age, ambient parameters, and menstrual cycle (Gajjar et al., 2013). In 2013, Boman and Maibach identified the transmission of VOCs through the skin using diffusion studies on various VOCs, employing the Franz cell diffusion technique (Boman & Maibach, 2000). In addition to naturally occurring VOCs, various foreign substances, such as the skin, can produce xenobiotics once they undergo metabolism within the body. For instance, 1% of the total alcohol intake is expelled through a diffusive process from the exocrine sweat glands in the skin, governed by a specific partition ratio (Henry's law).

Breath: Breath monitoring has been considered a promising tool for diagnosing various diseases at an early stage for several decades (Pathak & Viphavakit, 2021). Historically, physicians have observed specific odors from exhaled breath linked to particular diseases. For example, a "fruity" odor is associated with diabetic patients, while a "fishy" smell may indicate renal failure (Minh et al., 2012). Monitoring VOC biomarkers in exhaled breath is a well-established technique in medical applications and diabetes diagnostics due to its potential to develop non-invasive, rapid, and cost-effective screening and measurement tools. In 1971, Pauling et al. reported nearly 250 different types of complex mixture of VOCs present within the exhaled human breath. To date, 874 types of VOCs have been detected in exhaled breath, linked with both exogenous and endogenous compounds (Pauling et al., 1971). Major VOCs in exhaled human breath include ethanol (13–1000 ppb), acetone (1.2–900 ppb), isoprene (12–580 ppb), methanol (160–2000 ppb), ammonia, as well as minor compounds such as pentane and higher chains of aldehydes, alcohols, and ketones (Fenske & Paulson, 1999). The concentration of VOCs in an individual's exhaled breath, reflecting the individual's "exposome," can be influenced by diet, environmental exposure, lifestyle factors such as alcohol consumption, and other environmental factors. These exposomes vary from person to person. For instance, individuals who smoke tend to have higher levels of furans and acetonitrile concentrations in their exhaled breath. Exertion leads to an increase in alveolar ventilation and a subsequent 3–4-fold rise in isoprene concentration (de Lacy Costello et al., 2014). Similarly, an acetonic odor may suggest diabetes, while a pungent odor might be associated with liver disease (Tangerman et al., 1983; Wang & Wang, 2013).

1.6 Biomedical diagnosis applications

VOCs present in exhaled breath exist in a gaseous state at ambient temperature (Dima et al., 2021). These VOCs are considered valuable non-invasive biomarkers, detectable in various human discharge products such as breath, saliva, urine, and blood (Janssens et al., 2020).

1.6.1 Cancer diagnosis

The origin of certain cancer-associated VOCs remains unclear, as it is uncertain whether they are generated from cancer cells or from the surrounding environment, including elements such as the extracellular matrix and other cell types that support tissues and the immune response against cancer. It is important to understand that cancer is composed of various cells with different proliferative characteristics and resistance to external treatments such as radiotherapy, chemotherapy, and targeted therapy. Additionally, tumors consist of a mix of proliferating cells, quiescent cells, and dying or dead cells, with various combinations of these cell types over time. These tumor characteristics can affect the VOC pattern unique to each tumor. Moreover, the immune system's response to cancer cells may vary between patients and tumor types, influencing the VOC profile, such as those measured in breath. Various researchers have investigated the emission of VOCs from cancerous cells under different environmental conditions, including before and after the administration of drugs (Brunner et al., 2010; Rutter et al., 2014; Sponring et al., 2011). However, the majority of these investigations have not considered intra-tumor cell variability while analyzing entire populations of cancer cells. One study observed that VOC levels vary not only among different cell lines of the same cancer but also within the cells of the same line (Schmidt & Podmore, 2015). Therefore, it is crucial to monitor how VOC profiles vary among various tumor cells within tumor cell populations. This issue can be addressed by cloning tumor cells. Cloning a cancer cell from a single cell makes it possible to identify variations in VOC patterns among different cells within the given cancer cell population (Rutter & Sulé-Suso, 2018). In-vitro analysis can also be utilized to evaluate how the VOCs emitted from the tumor microenvironment are influenced by the extracellular matrix and other non-malignant cells. Cancer cell growth in 3D models, such as collagen gels, is more accurate and nearly replicates the physiological environment of in vivo cells compared with 2D models. The ability to identify cancer in its early stages is critical for diagnosis using VOC measurement. It is well known that the stage of the tumor is significantly associated with prognosis; early identification of the tumor and its effective treatment lead to positive outcomes. Therefore, it is crucial for VOC analysis to identify tumors before patients present symptoms. However, the least quantity of cells required to release detectable VOC concentrations using current techniques remains unclear for specific tumors. In vitro analysis utilizing 3D models could provide valuable insights into this issue. Although in vivo analysis of VOCs is crucial, it is equally important to conduct in vitro studies simultaneously to enhance our understanding of clinical study outcomes. However, it is important to recognize that in vitro studies do not always ensure that the measured VOCs are of endogenous origin.

1.6.2 Infectious diseases

Exhaled breath comprises a vast range of VOCs that can be employed to diagnose various infections and their stages, i.e., from early to intermediate stages. For example, ongoing research, particularly in developed countries where the social impact is significant, focuses on the detection of infectious diseases such as tuberculosis (TB). A broad range of peptides, proteins, and VOCs has been identified in respiratory droplets, water condensates, and aerosols from exhaled breath, which can serve as measurable biomarkers for diagnosing inflammation, carcinogens, and microbial infection (Timm et al., 2018). The respiratory tract microbiomes are usually dominated by the Bacteroidetes and Firmicutes, along with bacterial species from the genera Haemophilus, Streptococcus, Fusobacterium, Prevotella, and Veillonella (Huffnagle et al., 2017). These lung microbiomes play a crucial role in various inflammatory conditions and respiratory diseases, including asthma, chronic obstructive pulmonary disease (COPD), and acute respiratory distress syndrome (ARDS) (Barnes, 2017). Asthma and COPD are recognized as respiratory diseases comprising multiple endotypes and phenotypes, largely influenced by factors such as exposure to respiratory viruses, bacterial co-infection, genetic background, and inhaled environmental pollutants (Mendez et al., 2019). ARDS, on the other hand, is commonly seen in severely ill patients and results from the sudden onset of tissue hypoxia. This leads to lung infiltration, diffuse alveolar damage, intense inflammation, coagulation activation, and eventual lung fibrosis. Following the initial infection, pathogenic bacteria and viruses induce inflammation, leading to mucus production. This airway inflammation increases the generation of reactive nitrogen and oxygen species in the immune cells. This reactive nitrogen is generated by inducible nitric oxide synthase (Belizário et al., 2021). These reactive compounds are known to cause cell membrane damage, tissue destruction, and degeneration. In 2004, Zetola et al. diagnosed pulmonary TB by analyzing the differences in the VOC patterns obtained from exhaled breath (Zetola et al., 2017). The study exploited the sensing capability of metalloporphyrin-based VOC sensors, which demonstrated sensitive, stable, and selective diagnostic capabilities for various pathologies, including lung cancer (D'Amico et al., 2010). The authors hypothesized that patients suffering from TB would exhale distinct VOC signatures compared with healthy individuals. To validate this hypothesis, the authors first analyzed the diagnostic performance of an electronic nose system consisting of metalloporphyrins-coated quartz microbalances. An electronic nose is a gas-sensing system that consists of three main elements, i.e., (1) a sampling system, (2) an array of gas sensors (which generates a series of signals under the exposure of different vapors, gases, and odors), and (3) an appropriate pattern recognition system (Ahluwalia & De Rossi, 2001). In 2022,

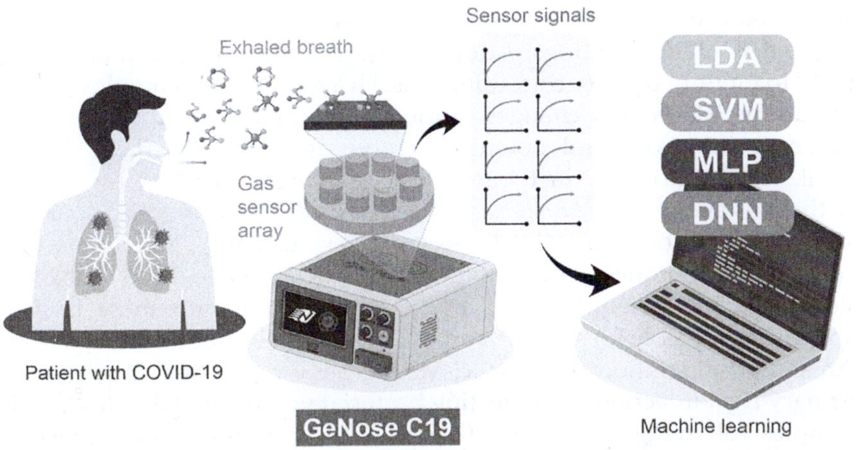

FIGURE 1.3 Fast and non-invasive COVID-19 detection utilizing portable GeNose C19. Schematic diagram illustrates non-invasive monitoring of COVID-19 by utilizing portable electronic nose (GeNose C19) combined with artificial intelligence (AI). *From Nurputra, D. K., Kusumaatmaja, A., Hakim, M. S., Hidayat, S. N., Julian, T., Sumanto, B., Mahendradhata, Y., Saktiawati, A. M. I., Wasisto, H. S., & Triyana, K. (2022). Fast and noninvasive electronic nose for sniffing out COVID-19 based on exhaled breath-print recognition.* NPJ Digital Medicine, 5*(1), 101. https://doi.org/10.1038/s41746-022-00661-2.*

Nurputra et al. reported the development of the first electronic nose capable of detecting COVID-19 from exhaled breath (Nurputra et al., 2022). In this work, the authors fabricated a portable breathalyzer (GeNose C19) by integrating machine learning (ML) analysis and a breath sampling tool and an array of metal oxide semiconductor gas sensors, as shown in Fig. 1.3. The sensor was also integrated with the four different ML algorithms [i.e., linear discriminant analysis (LDA), support vector machine (SVM), stacked multilayer perceptron (MLP), and deep neural network (DNN)] to differentiate and identify the patterns of the exhaled breath in the patients. The system was tested on two different groups: one consisting of RT-qPCR-confirmed COVID-19-positive patients and another comprising COVID-19-negative patients in Indonesia. The sensor baseline values were first monitored in the absence of exhaled breath in the sample chamber, and these values were later used as subtrahends for the output response signals to the VOC-containing breath. In the presence of gas molecules, the chemoresistive sensors exhibited an increase in output voltage due to reduced resistance.

Pseudomonas aeruginosa is considered a significant pulmonary pathogen, specifically in children with cystic fibrosis, a genetic disorder resulting from mutations in the gene encoding the cystic fibrosis transmembrane regulator (Sethi et al., 2013). The lungs of cystic fibrosis patients are commonly infected or colonized by the *P. aeruginosa* bacterium, leading to an unfavorable prognosis of disease progression. Early detection of *P.*

aeruginosa is essential for timely and effective treatment, making it highly relevant in clinical settings. Additionally, VOC sampling can also be applied to identify urinary tract infections. Urine contains a complex mixture of various VOCs, and changes in their concentrations can be utilized to diagnose infections and potentially serve as disease markers (Pauling et al., 1971). Urinary tract infections are typically caused by bacterial pathogens such as *Proteus species, Escherichia coli, enterococci, Staphylococcus saprophyticus*, and *Klebsiella species* (Sethi et al., 2013). Numerous urine sampling techniques and analytical approaches have been utilized to determine the potential VOC markers associated with these bacterial species responsible for such urinary tract infections (Guernion et al., 2001). An analytical method such as GC-MS has shown potential in monitoring urinary tract infections by detecting bacterial volatile metabolites in urine samples following brief incubation periods with appropriate precursor compounds (Davies & Hayward, 1984). Although this technique is effective in identifying characteristic high-abundance volatiles generated by specific pathogens, it lacks the accuracy and practicality needed for routine clinical use. Among the commercially available tools, the electronic nose based on conducting polymer technology (Bloodhound BH-114) has demonstrated promising results in detecting urinary tract infections in 45 suspected cases (Kodogiannis & Wadge, 2005).

1.7 Conclusion

In conclusion, this chapter provides in-depth information related to the basics of VOCs, along with their sources in both indoor and outdoor environments, their origins within the human body, and their applications in biomedical diagnosis. VOCs, a diverse group of organic compounds, are encountered daily through multiple sources such as wall paints, furniture, building materials, cleaning reagents, and cosmetics. Despite the presence of thousands of VOCs in the ambient environment, the precise health effects of such exposure remain largely unknown. This chapter also explores the role of VOCs in human health, including their association with diseases and threshold values. Additionally, it highlights the potential of VOCs in biomedical diagnostics for identifying cancers and infectious diseases.

References

Abdullahi, M. E., Hassan, M. A. A., Noor, Z. Z., & Ibrahim, R. K. R. (2014). Application of a packed column air stripper in the removal of volatile organic compounds from wastewater. *Reviews in Chemical Engineering*, *30*(5), 431–451. http://www.reference-global.com/loi/revce. doi:10.1515/revce-2014-0003.

Adlimoghaddam, A., et al. (2016). Ammonia as a Potential Neurotoxic Factor in Alzheimer's Disease. *Frontiers in Molecular Neuroscience*. doi:10.3389/fnmol.2016.00057.

Ahluwalia, A., & De Rossi, D. (2001). Artificial noses and tongues. *Encyclopedia of materials: Science and technology* (pp. 344–347). Elsevier BV. doi:10.1016/b0-08-043152-6/00071-1.

Ahmed, F. E. (1995). Toxicological effects of ethanol on human health. *Critical Reviews in Toxicology, 25*(4), 347–367. doi:10.3109/10408449509021614.

Akram, M., et al. (2013). A Focused Review of the Role of Ketone Bodies in Health and Disease. *Journal of Medicinal Food, 16*. doi:10.1089/jmf.2012.2592.

Amann, A., Costello, B. L., Miekisch, W., Schubert, J., Buszewski, B. Ł., Pleil, J., Ratcliffe, N., & Risby, T. (2014). The human volatilome: Volatile organic compounds (VOCs) in exhaled breath, skin emanations, urine, feces and saliva. *Journal of Breath Research, 8*(3), 034001. doi:10.1088/1752-7155/8/3/034001.

Atkinson, R., & Arey, J. (2003a). Atmospheric degradation of volatile organic compounds. *Chemical Reviews, 103*(12), 4605–4638. doi:10.1021/cr0206420.

Atkinson, R., & Arey, J. (2003b). Gas-phase tropospheric chemistry of biogenic volatile organic compounds: A review. *Atmospheric Environment, 37*(Supplement 2), 197–219. www.elsevier.com/locate/atmosenv. doi:10.1016/S1352-2310(03)00391-1.

Ayala, A., Muñoz, M. F., & Argüelles, S. (2014). Lipid peroxidation: Production, metabolism, and signaling mechanisms of malondialdehyde and 4-hydroxy-2-nonenal. *Oxidative Medicine and Cellular Longevity, 2014*. http://www.landesbioscience.com/journals/oximed. doi:10.1155/2014/360438.

Bajtarevic, A., Ager, C., Pienz, M., Klieber, M., Schwarz, K., Ligor, M., Ligor, T., et al. (2009). Noninvasive detection of lung cancer by analysis of exhaled breath. *BMC Cancer, 9*(1), 348. doi:10.1186/1471-2407-9-348.

Barnes, P. J. (2017). Cellular and molecular mechanisms of asthma and COPD. *Clinical Science, 131*(13), 1541–1558. http://www.clinsci.org/content/ppclinsci/131/13/1541.full.pdf#zoom=75. doi:10.1042/CS20160487.

Belizário, J. E., Faintuch, J., & Malpartida, M. G. (2021). Breath biopsy and discovery of exclusive volatile organic compounds for diagnosis of infectious diseases. *Frontiers in Cellular and Infection Microbiology, 10*. http://www.frontiersin.org/Cellular_and_Infection_Microbiology/archive. doi:10.3389/fcimb.2020.564194.

Bickers, D. R., Calow, P., Greim, H. A., Hanifin, J. M., Rogers, A. E., Saurat, J. H., Sipes, I. G., Smith, R. L., & Tagami, H. (2003). The safety assessment of fragrance materials. *Regulatory Toxicology and Pharmacology, 37*(2), 218–273. http://www.elsevier.com/inca/publications/store/6/2/2/9/3/9/index.htt. doi:10.1016/S0273-2300(03)00003-5.

Boman, A., & Maibach, H. I. (2000). Percutaneous absorption of organic solvents. *International Journal of Occupational and Environmental Health, 6*(2), 93–95. http://www.tandfonline.com/loi/yjoh20#.Vxm2Kk1f3cs. doi:10.1179/oeh.2000.6.2.93.

Bosetti, M. L., Tarone, P., & Vecchia, L. A. (2008). Formaldehyde and cancer risk: A quantitative review of cohort studies through 2006. *Annals of Oncology, 19*(1), 29–43. doi:10.1093/annonc/mdm202.

Bouatra, S., Aziat, F., Mandal, R., Guo, A. C., Wilson, M. R., Knox, C., Bjorndahl, T. C., Krishnamurthy, R., Saleem, F., Liu, P., Dame, Z. T., Poelzer, J., Huynh, J., Yallou, F. S., Psychogios, N., Dong, E., Bogumil, R., Roehring, C., & Wishart, D. S. (2013). The human urine metabolome. *PLoS ONE, 8*(9). http://www.plosone.org/article/fetchObjectAttachment.action;jsessionid=0F589D2B027F92343C11F367F8560B06?uri=info%3Adoi%2F10.1371%2Fjournal.pone.0073076&representation=PDF. doi:10.1371/journal.pone.0073076Canada.

Broza, Y. Y., & Haick, H. (2013). Nanomaterial-based sensors for detection of disease by volatile organic compounds. *Nanomedicine, 8*, 785–806. doi:10.2217/nnm.13.64.

Broza, Y. Y., Mochalski, P., Ruzsanyi, V., Amann, A., & Haick, H. (2015). Hybrid volatolomics and disease detection. *Angewandte Chemie – International Edition, 54*(38), 11036–11048. http://onlinelibrary.wiley.com/journal/10.1002/(ISSN)1521-3773. doi:10.1002/anie.201500153.

Brunner, C., Szymczak, W., Höllriegl, V., Mörtl, S., Oelmez, H., Bergner, A., Huber, R. M., Hoeschen, C., & Oeh, U. (2010). Discrimination of cancerous and non-cancerous cell lines by headspace-analysis with PTR-MS. *Analytical and Bioanalytical Chemistry, 397*(6), 2315–2324. doi:10.1007/s00216-010-3838-x.

Buonanno, G., Stabile, L., & Morawska, L. (2014). Personal exposure to ultrafine particles: The influence of time-activity patterns. *Science of the Total Environment, 468–469*, 903–907. doi:10.1016/j.scitotenv.2013.09.016.

Campos-Ordonez, T., & Gonzalez-Perez, O. (2016). Cyclohexane, a Potential Drug of Abuse with Pernicious Effects on the Brain. *Frontiers in Pharmacology*. doi:10.3389/fphar.2015.00291.

Chaudhary, P. P., et al. (2018). Methanogens in humans: potentially beneficial or harmful for health. *Applied Microbiology and Biotechnology, 102*, 3095–3104. doi:10.1007/s00253-018-8871-2.

Chen, J., & Luo, D. (2012). Ozone formation potentials of organic compounds from different emission sources in the South Coast Air Basin of California. *Atmospheric Environment, 55*, 448–455. doi:10.1016/j.atmosenv.2012.02.082.

Chin, J.-Y., Godwin, C., Parker, E., Robins, T., Lewis, T., Harbin, P., & Batterman, S. (2014). Levels and sources of volatile organic compounds in homes of children with asthma. *Indoor Air, 24*(4), 403–415. doi:10.1111/ina.12086.

Crofford, O. B., Mallard, R. E., & Winton, R. E. (1977). Acetone in breath and blood. *Transactions of the American Clinical and Climatological Association, 88*, 128–139.

D'Amico, A., Pennazza, G., Santonico, M., Martinelli, E., Roscioni, C., Galluccio, G., Paolesse, R., & Di Natale, C. (2010). An investigation on electronic nose diagnosis of lung cancer. *Lung Cancer, 68*(2), 170–176. doi:10.1016/j.lungcan.2009.11.003.

David, E., & Niculescu, V. C. (2021). Volatile organic compounds (Vocs) as environmental pollutants: Occurrence and mitigation using nanomaterials. *International Journal of Environmental Research and Public Health, 18*(24). https://www.mdpi.com/1660-4601/18/24/13147/pdf. doi:10.3390/ijerph182413147.

Davies, T. J., & Hayward, N. J. (1984). Volatile products from acetylcholine as markers in the rapid urine test using head-space gas—liquid chromatography. *Journal of Chromatography B: Biomedical Sciences and Applications, 307*(C), 11–21. doi:10.1016/s0378-4347(00)84068-2.

Destaillats, H., Lunden, M. M., Singer, B. C., Coleman, B. K., Hodgson, A. T., Weschler, C. J., & Nazaroff, W. W. (2006). Indoor secondary pollutants from household product emissions in the presence of ozone: A bench-scale chamber study. *Environmental Science and Technology, 40*(14), 4421–4428. doi:10.1021/es052198z.

Dharmawardana, N., et al. (2020). Breath methane to hydrogen ratio as a surrogate marker of intestinal dysbiosis in head and neck cancer. *Scientific Reports, 10*, 15010. doi:10.1038/s41598-020-72115-2.

Diduch, M., Polkowska, Z., & Namieśnik, J. (2011). Chemical quality of bottled waters: A review. *Journal of Food Science, 76*(9), R178. doi:10.1111/j.1750-3841.2011.02386.x.

Dima, A. C., Balaban, D. V., & Dima, A. (2021). Diagnostic application of volatile organic compounds as potential biomarkers for detecting digestive neoplasia: A systematic review. *Diagnostics, 11*(12). https://www.mdpi.com/2075-4418/11/12/2317/pdf. doi:10.3390/diagnostics11122317.

Ditto, J. C., Abbatt, J. P. D., & Chan, A. W. H. (2022). Gas- and particle-phase amide emissions from cooking: Mechanisms and air quality impacts. *Environmental Science and Technology, 56*(12), 7741–7750. http://pubs.acs.org/journal/esthag. doi:10.1021/acs.est.2c01409.

Do, D. H., Van Langenhove, H., Chigbo, S. I., Amare, A. N., Demeestere, K., & Walgraeve, C. (2014). Exposure to volatile organic compounds: Comparison among different transportation modes. *Atmospheric Environment, 94*, 53–62. www.elsevier.com/locate/atmosenv. doi:10.1016/j.atmosenv.2014.05.019.

Duarte-Davidson, R., Courage, C., Rushton, L., & Levy, L. (2001). Benzene in the environment: An assessment of the potential risks to the health of the population. *Occupational and Environmental Medicine, 58*(1), 2–13. doi:10.1136/oem.58.1.2.

Edwards, R., Schweizer, C., Llacqua, V., Lai, H., Jantunen, M., Bayeroglesby, L., & Kunzli, N. (2006). Time–activity relationships to VOC personal exposure factors. *Atmospheric Environment, 40*(29), 5685–5700. doi:10.1016/j.atmosenv.2006.04.057.

Fan, X., et al. (2021). In-situ growth of Co3O4 nanoparticles based on electrospray for an acetone gas sensor. *Journal of Alloys and Compounds*, 157234. doi:10.1016/j.jallcom.2020.157234.

Faubert, P., Tiiva, P., Rinnan, A., Michelsen, A., Holopainen, J. K., & Rinnan, R. (2010). Doubled volatile organic compound emissions from subarctic tundra under simulated climate warming. *New Phytologist, 187*(1), 199–208. doi:10.1111/j.1469-8137.2010.03270.x.

Fenske, J. D., & Paulson, S. E. (1999). Human breath emissions of VOCs. *Journal of the Air and Waste Management Association, 49*(5), 594–598. doi:10.1080/10473289.1999.10463831.

Finewax, Z., Pagonis, D., Claflin, M. S., Handschy, A. V., Brown, W. L., Jenks, O., Nault, B. A., Day, D. A., Lerner, B. M., Jimenez, J. L., Ziemann, P. J., & de Gouw, J. A. (2021). Quantification and source characterization of volatile organic compounds from exercising and application of chlorine-based cleaning products in a university athletic center. *Indoor Air, 31*(5), 1323–1339. http://onlinelibrary.wiley.com/journal/10.1111/. doi:10.1111/ina.12781. (ISSN)1600-0668.

Fischer, G., & Dott, W. (2003). Relevance of airborne fungi and their secondary metabolites for environmental, occupational and indoor hygiene. *Archives of Microbiology, 179*(2), 75–82. link.springer.de/link/service/journals/00203/index.htm. doi:10.1007/s00203-002-0495-2.

Gajjar, R. M., Miller, M. A., & Kasting, G. B. (2013). Evaporation of volatile organic compounds from human skin in vitro. *Annals of Occupational Hygiene, 57*(7), 853–865. doi:10.1093/annhyg/met004.

Gallagher, M., Wysocki, C. J., Leyden, J. J., Spielman, A. I., Sun, X., & Preti, G. (2008). Analyses of volatile organic compounds from human skin. *British Journal of Dermatology, 159*(4), 780–791. doi:10.1111/j.1365-2133.2008.08748.x.

Gallego, J.-J., et al. (2025). Ammonia and beyond – biomarkers of hepatic encephalopathy. *Metabolic Brain Disease, 40*, 100. doi:10.1007/s11011-024-01512-7.

Goodman, N. B., Wheeler, A. J., Paevere, P. J., Selleck, P. W., Cheng, M., & Steinemann, A. (2018). Indoor volatile organic compounds at an Australian university. *Building and Environment, 135*, 344–351. http://www.elsevier.com/inca/publications/store/2/9/6/index.htt. doi:10.1016/j.buildenv.2018.02.035.

Gray, C. M., & Fierer, N. (2012). Impacts of nitrogen fertilization on volatile organic compound emissions from decomposing plant litter. *Global Change Biology, 18*(2), 739–748. doi:10.1111/j.1365-2486.2011.02569.x.

Guernion, N., Ratcliffe, N. M., Spencer-Phillips, P. T. N., & Howe, R. A. (2001). Identifying bacteria in human urine: Current practice and the potential for rapid, near-patient diagnosis by sensing volatile organic compounds. *Clinical Chemistry and Laboratory Medicine, 39*(10), 893–906. doi:10.1515/CCLM.2001.146.

Hazardous Substances Data Bank (HSDB). n.d. https://www.nlm.nih.gov/toxnet/index.html. https://pubchem.ncbi.nlm.nih.gov/source/11933.

Hermawan, A., Asakura, Y., Inada, M., & Yin, S. (2020). A facile method for preparation of uniformly decorated-spherical SnO2 by CuO nanoparticles for highly responsive toluene detection at high temperature. *Journal of Materials Science and Technology, 51*, 119–129. http://www.sciencedirect.com/science/journal/10050302. doi:10.1016/j.jmst.2020.02.041.

Horvath, G., Andersson, H., & Paulsson, G. (2010). Characteristic odour in the blood reveals ovarian carcinoma. *BMC Cancer, 10*. http://www.biomedcentral.com/1471-2407/10/643. doi:10.1186/1471-2407-10-643Sweden.

Hsu, C. Y., Chiang, H. C., Shie, R. H., Ku, C. H., Lin, T. Y., Chen, M. J., Chen, N. T., & Chen, Y. C. (2018). Ambient VOCs in residential areas near a large-scale petrochemical complex: Spatiotemporal variation, source apportionment and health risk. *Environmental Pollution, 240*, 95–104. www.elsevier.com/inca/publications/store/4/0/5/8/5/6. doi:10.1016/j.envpol.2018.04.076.

Hu, R., Liu, G., Zhang, H., Xue, H., & Wang, X. (2018). Levels, characteristics and health risk assessment of VOCs in different functional zones of Hefei. *Ecotoxicology and Environmental Safety, 160*, 301–307. http://www.elsevier.com/inca/publications/store/6/2/2/8/1/9/index.htt. doi:10.1016/j.ecoenv.2018.05.056.

Huffnagle, G. B., Dickson, R. P., & Lukacs, N. W. (2017). The respiratory tract microbiome and lung inflammation: A two-way street. *Mucosal Immunology, 10*(2), 299–306. doi:10.1038/mi.2016.108.

Iftekhar Uddin, A. S. M., Phan, D. T., & Chung, G. S. (2015). Low temperature acetylene gas sensor based on Ag nanoparticles-loaded ZnO-reduced graphene oxide hybrid. *Sensors and Actuators, B: Chemical, 207*, 362–369. doi:10.1016/j.snb.2014.10.091.

Jalal, A. H., Alam, F., Roychoudhury, S., Umasankar, Y., Pala, N., & Bhansali, S. (2018). Prospects and challenges of volatile organic compound sensors in human healthcare. *ACS Sensors, 3*(7), 1246–1263. http://pubs.acs.org/journal/ascefj. doi:10.1021/acssensors.8b00400.

Janssens, E., van Meerbeeck, J. P., & Lamote, K. (2020). Volatile organic compounds in human matrices as lung cancer biomarkers: A systematic review. *Critical Reviews in Oncology/Hematology, 153*. www.elsevier.com/locate/critrevonc. doi:10.1016/j.critrevonc.2020.103037.

Jin, S., Zhong, L., Zhang, X., Li, X., Li, B., & Fang, X. (2023). Indoor volatile organic compounds: Concentration characteristics and health risk analysis on a university campus. *International Journal of Environmental Research and Public Health, 20*(10). http://www.mdpi.com/journal/ijerph. doi:10.3390/ijerph20105829.

Jo, D., et al. (2021). The Cerebral Effect of Ammonia in Brain Aging: Blood–Brain Barrier Breakdown, Mitochondrial Dysfunction, and Neuroinflammation. *Journal of Clinical Medicine, 10*, 2773. doi:10.3390/jcm10132773.

Jo, W. K., Lee, J. H., & Kim, M. K. (2008). Head-space, small-chamber and in-vehicle tests for volatile organic compounds (VOCs) emitted from air fresheners for the Korean market. *Chemosphere, 70*(10), 1827–1834. doi:10.1016/j.chemosphere.2007.08.021.

Kalogridis, C., Gros, V., Sarda-Esteve, R., Langford, B., Loubet, B., Bonsang, B., Bonnaire, N., Nemitz, E., Genard, A.-C., Boissard, C., Fernandez, C., Ormeño, E., Baisnée, D., Reiter, I., & Lathière, J. (2014). Concentrations and fluxes of isoprene and oxygenated VOCs at a French Mediterranean oak forest. *Atmospheric Chemistry and Physics, 14*(18), 10085–10102. doi:10.5194/acp-14-10085-2014.

Kang, J., Liu, J., & Pei, J. (2017). The indoor volatile organic compound (VOC) characteristics and source identification in a new university campus in Tianjin, China. *Journal of the Air and Waste Management Association., 67*(6), 725–737. http://www.tandfonline.com/loi/uawm20?. doi:10.1080/10962247.2017.1280561.

Ke, M. T., Lee, M. T., Lee, C. Y., & Fu, L. M. (2009). A MEMS-based benzene gas sensor with a self-heating WO3 sensing layer. *Sensors, 9*(4), 2895–2906. https://www.mdpi.com/1424-8220/9/4/2895/pdf. doi:10.3390/s90402895.

Kesselmeier, J., & Staudt, M. (1999). Biogenic volatile organic compounds (VOC): An overview on emission, physiology and ecology. *Journal of Atmospheric Chemistry, 33*(1), 23–88. doi:10.1023/A:1006127516791.

Khalid, T., Aggio, R., White, P., De Lacy Costello, B., Persad, R., Al-Kateb, H., Jones, P., Probert, C. S., & Ratcliffe, N. (2015). Urinary volatile organic compounds for the detection of prostate cancer. *PLoS ONE, 10*(11). http://www.plosone.org/article/fetchObject.action?uri=info:doi/10.1371/journal.pone.0143283&representation=PDF. doi:10.1371/journal.pone.0143283.

Khan, S., Le Calvé, S. T., & Newport, D. (2020). A review of optical interferometry techniques for VOC detection. *Sensors and Actuators A: Physical, 302*, 111782. doi:10.1016/j.sna.2019.111782.

Kim, J., Imani, S., de Araujo, W. R., Warchall, J., Valdés-Ramírez, G., Paixão, T. R. L. C., Mercier, P. P., & Wang, J. (2015). Wearable salivary uric acid mouthguard biosensor with integrated wireless electronics. *Biosensors and Bioelectronics, 74*, 1061–1068. www.elsevier.com/locate/bios. doi:10.1016/j.bios.2015.07.039.

Kirschner, E. M. (1995). Production of top 50 chemicals increased substantially in 1994. *Chemical & Engineering News Archive, 73*(15), 16–20. doi:10.1021/cen-v073n015.p016.

Konig, G. (1995). Relative contribution of oxygenated hydrocarbons to the total biogenic VOC emissions of selected mid-European agricultural and natural plant species. *Atmospheric Environment, 29*(8), 861–874. doi:10.1016/1352-2310(95)00026-u.

Kumar, A., Singh, B. P., Punia, M., Singh, D., Kumar, K., & Jain, V. K. (2014). Assessment of indoor air concentrations of VOCs and their associated health risks in the library of Jawaharlal Nehru University, New Delhi. *Environmental Science and Pollution Research, 21*(3), 2240–2248. doi:10.1007/s11356-013-2150-7.

de Lacy Costello, B., Amann, A., Al-Kateb, H., Flynn, C., Filipiak, W., Khalid, T., Osborne, D., & Ratcliffe, N. M. (2014). A review of the volatiles from the healthy human body. *Journal of Breath Research, 8*(1), 014001. doi:10.1088/1752-7155/8/1/014001.

Lamb, B., Gay, D., Westberg, H., & Pierce, T. (1993). A biogenic hydrocarbon emission inventory for the U.S.A. using a simple forest canopy model. *Atmospheric Environment. Part A. General Topics, 27*(11), 1673–1690. doi:10.1016/0960-1686(93)90230-v.

Lashof, D. A., & Ahuja, D. R. (1990). Relative contributions of greenhouse gas emissions to global warming. *Nature, 344*(6266), 529–531. doi:10.1038/344529a0.

Lelieveld, J., Hoor, P., Jöckel, P., Pozzer, A., Hadjinicolaou, P., Cammas, J.-P., & Beirle, S. (2009). Severe ozone air pollution in the Persian Gulf region. *Atmospheric Chemistry and Physics, 9*(4), 1393–1406. doi:10.5194/acp-9-1393-2009.

Li, Y., He, L., Xie, D., Zhao, A., Wang, L., Kreisberg, N. M., Jayne, J., & Liu, Y. (2022). Strong temperature influence and indiscernible ventilation effect on dynamics of some semivolatile organic compounds in the indoor air of an office. *Environment International, 165*. www.elsevier.com/locate/envint. doi:10.1016/j.envint.2022.107305.

Liao, K. J., Hou, X., & Baker, D. R. (2014). Impacts of interstate transport of pollutants on high ozone events over the Mid-Atlantic United States. *Atmospheric Environment, 84*, 100–112. www.elsevier.com/locate/atmosenv. doi:10.1016/j.atmosenv.2013.10.062.

Lin, S., Hung, W., & Leng, Z. (2016). Air pollutant emissions and acoustic performance of hot mix asphalts. *Construction and Building Materials, 129*, 1–10. doi:10.1016/j.conbuildmat.2016.11.013.

Liu, H., Wang, H., Li, C., Wang, L., Pan, Z., & Wang, L. (2014). Investigation of volatile organic metabolites in lung cancer pleural effusions by solid-phase microextraction and gas chromatography/mass spectrometry. *Journal of Chromatography B: Analytical Technologies in the Biomedical and Life Sciences, 945-946*, 53–59. www.elsevier.com/inca/publications/store/5/0/2/6/8/9. doi:10.1016/j.jchromb.2013.11.038.

Liu, W., Zhang, X., Wang, Z., Wang, R., Chen, C., & Dong, C. (2019). Nanoparticles assembled CdIn2O4 spheres with high sensing properties towards n-butanol. *Nanomaterials, 9*(12). https://www.mdpi.com/2079-4991/9/12/1714/pdf. doi:10.3390/nano9121714.

Liu, N., Bu, Z., Liu, W., Kan, H., Zhao, Z., Deng, F., Huang, C., Zhao, B., Zeng, X., Sun, Y., Qian, H., Mo, J., Sun, C., Guo, J., Zheng, X., Weschler, L. B., & Zhang, Y. (2022). Health effects of exposure to indoor volatile organic compounds from 1980 to 2017: A systematic review and meta-analysis. *Indoor Air, 32*(5). doi:10.1111/ina.13038.

Lomonaco, T., Manco, E., Corti, A., Nasa, J. L., Ghimenti, S., Biagini, D., Francesco, F. D., Modugno, F., Ceccarini, A., Fuoco, R., & Castelvetro, V. (2020). Release of harmful volatile organic compounds (VOCs) from photo-degraded plastic debris: A neglected source of environmental pollution. *Journal of Hazardous Materials, 394*, 122596. doi:10.1016/j.jhazmat.2020.122596.

Mabrook, M., & Hawkins, P. (2002). Benzene sensing using thin films of titanium dioxide operating at room temperature. *Sensors, 2*(9), 374–382. http://www.mdpi.com/journal/sensors. doi:10.3390/s20900374.

Maceira, A., Vallecillos, L., Borrull, F., & Marcé, R. M. (2017). New approach to resolve the humidity problem in VOC determination in outdoor air samples using solid adsorbent tubes followed by TD-GC–MS. *Science of the Total Environment, 599-600*, 1718–1727. www.elsevier.com/locate/scitotenv. doi:10.1016/j.scitotenv.2017.05.141.

Maisey, S. J., Saunders, S. M., West, N., & Franklin, P. J. (2013). An extended baseline examination of indoor VOCs in a city of low ambient pollution: Perth, Western Australia. *Atmospheric Environment, 81*, 546–553. doi:10.1016/j.atmosenv.2013.09.008.

Marty, M., Spurlock, F., & Barry, T. (2010). Volatile organic compounds from pesticide application and contribution to tropospheric ozone. *Hayes' handbook of pesticide toxicology* (pp. 571–585). Elsevier Inc. http://www.sciencedirect.com/science/book/9780123743671. doi:10.1016/B978-0-12-374367-1.00019-7.

Matsumura, K., Opiekun, M., Oka, H., Vachani, A., Albelda, S. M., Yamazaki, K., & Beauchamp, G. K. (2010). Urinary volatile compounds as biomarkers for lung cancer: A proof of principle study using odor signatures in mouse models of lung cancer. *PLoS ONE, 5*(1). http://www.plosone.org/article/fetchObjectAttachment.action?uri=info%3Adoi%2F10.1371%2Fjournal.pone.0008819&representation=PDF. doi:10.1371/journal.pone.0008819. United States.

McDonald, B. C., De Gouw, J. A., Gilman, J. B., Jathar, S. H., Akherati, A., Cappa, C. D., Jimenez, J. L., Lee-Taylor, J., Hayes, P. L., McKeen, S. A., Cui, Y. Y., Kim, S. W., Gentner, D. R., Isaacman-VanWertz, G., Goldstein, A. H., Harley, R. A., Frost, G. J., Roberts, J. M., Ryerson, T. B., & Trainer, M. (2018). Volatile chemical products emerging as largest petrochemical source of urban organic emissions. *Science, 359*(6377), 760–764. http://science.sciencemag.org/content/359/6377/760/tab-pdf. doi:10.1126/science.aaq0524.

McGrath, Lawrence T., Robin, P., & Bernard, S. (2001). Breath isoprene in patients with heart failure. *European Journal of Heart Failure, 3*, 423–427. doi:10.1016/S1388-9842(01)00128-3.

McGwin, G., Lienert, J., & Kennedy, J. I. (2010). Formaldehyde exposure and asthma in children: A systematic review. *Environmental Health Perspectives, 118*(3), 313–317. doi:10.1289/ehp.0901143.

McKee, H. C., Rhoades, J. W., Campbell, J., & Gross, A. L. (1962). Acetonitrile in body fluids related to smoking. *Public Health Reports (1896–1970), 77*(7), 553. doi:10.2307/4591551.

Mendez, R., Banerjee, S., Bhattacharya, S. K., & Banerjee, S. (2019). Lung inflammation and disease: A perspective on microbial homeostasis and metabolism. *IUBMB Life, 71*(2), 152–165. http://iubmb.onlinelibrary.wiley.com/hub/journal/10.1002/(ISSN)1521-6551/. doi:10.1002/iub.1969.

Minh, T. D. C., Blake, D. R., & Galassetti, P. R. (2012). The clinical potential of exhaled breath analysis for diabetes mellitus. *Diabetes Research and Clinical Practice, 97*(2), 195–205. doi:10.1016/j.diabres.2012.02.006.

Montero-Montoya, R., López-Vargas, R., & Arellano-Aguilar, O. (2018). Volatile organic compounds in air: Sources, distribution, exposure and associated illnesses in children. *Annals of Global Health, 84*(2), 225–238. https://agh.ubiquitypress.com/articles/10.29024/aogh.910/galley/2309/download/. doi:10.29024/aogh.910.

National Academies of Sciences, Engineering, and Medicine. (2018). Improving characterization of anthropogenic methane emissions in the United States. National Academies Press. https://doi.org/10.17226/24987

Nazaroff, W. W., & Weschler, C. J. (2004). Cleaning products and air fresheners: Exposure to primary and secondary air pollutants. *Atmospheric Environment, 38*(18), 2841–2865. www.elsevier.com/locate/atmosenv. doi:10.1016/j.atmosenv.2004.02.040.

Newhouse, M. L., et al. (1988). Mortality of workers at acetylene production plants. *Occupational & Environmental Medicine, 45*(1). doi:10.1136/oem.45.1.63.

Ninyà, N., Vallecillos, L., Marcé, R. M., & Borrull, F. (2022). Evaluation of air quality in indoor and outdoor environments: Impact of anti-COVID-19 measures. *Science of the Total Environment, 836,* 155611. doi:10.1016/j.scitotenv.2022.155611.

Nurputra, D. K., Kusumaatmaja, A., Hakim, M. S., Hidayat, S. N., Julian, T., Sumanto, B., Mahendradhata, Y., Saktiawati, A. M. I., Wasisto, H. S., & Triyana, K. (2022). Fast and noninvasive electronic nose for sniffing out COVID-19 based on exhaled breath-print recognition. *NPJ Digital Medicine, 5*(1). https://www.nature.com/npjdigitalmed/. doi:10.1038/s41746-022-00661-2.

Oliveira, R. M., Vieira, M. S., & Silva, M. M. N. F. (2019). Enhancement of acetylene gas sensing properties for a ZnO-based gas sensor produced by plasma immersion ion implantation and deposition. *Materials Science in Semiconductor Processing, 93,* 339–344. doi:10.1016/j.mssp.2018.12.031.

Paciência, I., Madureira, J., Rufo, J., Moreira, A., & Fernandes, E. D. O. (2016). A systematic review of evidence and implications of spatial and seasonal variations of volatile organic compounds (VOC) in indoor human environments. *Journal of Toxicology and Environmental Health – Part B: Critical Reviews, 19*(2), 47–64. http://www.tandf.co.uk/journals/titles/10937404.asp. doi:10.1080/10937404.2015.1134371.

Pathak, A. K., & Viphavakit, C. (2022). A review on all-optical fiber-based VOC sensors: Heading towards the development of promising technology. *Sensors and Actuators A: Physical, 338,* 113455. doi:10.1016/j.sna.2022.113455.

Pathak, A. K., & Viphavakit, C. (2021). VOC biomarker monitoring for diabetes through exhaled breath using Ag/P-TiO_2 composite plasmonic sensor. *IEEE Sensors Journal, 21*(20), 22631–22637. http://ieeexplore.ieee.org/xpl/RecentIssue.jsp?punumber=7361. doi:10.1109/JSEN.2021.3104766.

Pathak, A., Swargiary, K., Kongsawang, N., Jitpratak, P., Ajchareeyasoontorn, N., Udomkittivorakul, J., & Viphavakit, C. (2023). Recent advances in sensing materials targeting clinical volatile organic compound (VOC) biomarkers: A review. *Biosensors, 13*(1), 114. doi:10.3390/bios13010114.

Patnaik, P. (2007). *A comprehensive guide to the hazardous properties of chemical substances*. John Wiley & Sons, Inc.

Pauling, L., Robinson, A. B., Teranishi, R., & Cary, P. (1971). Quantitative analysis of urine vapor and breath by gas-liquid partition chromatography. *Proceedings of the National Academy of Sciences of the United States of America, 68*(10), 2374–2376. doi:10.1073/pnas.68.10.2374.

Peck, A., & Hornbuckle, K. (2006). Synthetic musk fragrances in urban and rural air of Iowa and the Great Lakes. *Atmospheric Environment, 40*(32), 6101–6111. doi:10.1016/j.atmosenv.2006.05.058.

Pizzimenti, S., Ciamporcero, E., Daga, M., Pettazzoni, P., Arcaro, A., Cetrangolo, G., Minelli, R., Dianzani, C., Lepore, A., Gentile, F., & Barrera, G. (2013). Interaction of aldehydes derived from lipid peroxidation and membrane proteins. *Frontiers in Physiology, 4.* http://www.frontiersin.org/Journal/FullText.aspx?ART_. doi:10.3389/fphys.2013.00242. Italy.

Polag, D., & Keppler, F. (2019). Global methane emissions from the human body: Past, present and future. *Atmospheric Environment,* 116823. doi:10.1016/j.atmosenv.2019.116823.

Potard, K., Monard, C., Le Garrec, J. L., Caudal, J. P., Bris, N. L., & Binet, F. (2017). Organic amendment practices as possible drivers of biogenic Volatile Organic Compounds emitted by soils in agrosystems. *Agriculture, Ecosystems and Environment, 250,* 25–36. www.elsevier.com/inca/publications/store/5/0/3/2/9/8. doi:10.1016/j.agee.2017.09.007.

Reiner, J. L., & Kannan, K. (2006). A survey of polycyclic musks in selected household commodities from the United States. *Chemosphere, 62*(6), 867–873. www.elsevier.com/locate/chemosphere. doi:10.1016/j.chemosphere.2005.10.006.

Rinnan, R., Gierth, D., Bilde, M., Rosenørn, T., & Michelsen, A. (2013). Off-season biogenic volatile organic compound emissions from heath mesocosms: Responses to vegetation cutting. *Frontiers in Microbiology, 4,* 224. http://www.frontiersin.org/Journal/FullText.aspx?ART_DOI=10.3389/fmicb.2013.00224&name=Terrestrial_Microbiology&x=y#h9. doi:10.3389/fmicb.2013.00224.

Rong, Q., Zhang, Y., Lv, T., Shen, K., Zi, B., Zhu, Z., Zhang, J., & Liu, Q. (2018). Highly selective and sensitive methanol gas sensor based on molecular imprinted silver-doped $LaFeO_3$ core–shell and cage structures. *Nanotechnology, 29*(14), 145503. doi:10.1088/1361-6528/aaabd0.

Rosenberg, M., & McCulloch, C. A. (1992). Measurement of oral malodor: Current methods and future prospects. *Journal of Periodontology, 63*(9), 776–782. doi:10.1902/jop.1992.63.9.776.

Rossignol, S., Rio, C., Ustache, A., Fable, S., Nicolle, J., Même, A., D'Anna, B., Nicolas, M., Leoz, E., & Chiappini, L. (2013). The use of a housecleaning product in an indoor environment leading to oxygenated polar compounds and SOA formation: Gas and particulate phase chemical characterization. *Atmospheric Environment, 75,* 196–205. doi:10.1016/j.atmosenv.2013.03.045.

Rutter, A. V., & Sulé-Suso, J. (2018). Analysis of volatile organic compounds for cancer diagnosis. *Volatile organic compounds and exhaled breath biomarkers* (pp. 53–78). CRC Press. doi:10.1201/9780429433580-3.

Rutter, A. V., Siddique, M. R., Filik, J., Sandt, C., Dumas, P., Cinque, G., Sockalingum, G. D., Yang, Y., & Sulé-Suso, J. (2014). Study of gemcitabine-sensitive/resistant cancer cells by cell cloning and synchrotron FTIR microspectroscopy. *Cytometry Part A, 85*(8), 688–697. http://onlinelibrary.wiley.com/journal/10.1002/. doi:10.1002/cyto.a.22488. (ISSN)1552-4930.

Ruzsanyi, V., Lochmann, F., Jürschik, S., Mochalski, P., Unterkofler, K., & Mayhew, C. A. (2022). Origin and emission of volatile biomarkers in breath: End-tidal perspective. *Breathborne biomarkers and the human volatilome* (pp. 1–21). Royal Society of Chemistry (RSC). doi:10.1039/9781839166990-00001.

Sanford, J. A., & Gallo, R. L. (2013). Functions of the skin microbiota in health and disease. *Seminars in Immunology, 25*(5), 370–377. doi:10.1016/j.smim.2013.09.005.

Schmidt, K., & Podmore, I. (2015). Current challenges in volatile organic compounds analysis as potential biomarkers of cancer. *Journal of Biomarkers, 2015,* 1–16. doi:10.1155/2015/981458.

Schulz, S., & Dickschat, J. S. (2007). Bacterial volatiles: The smell of small organisms. *Natural Product Reports, 24*(4), 814–842. doi:10.1039/b507392h.

Selyanchyn, R., Nozoe, T., Matsui, H., Kadosawa, T., & Lee, S. W. (2013). TD-GC-MS investigation of the VOCs released from blood plasma of dogs with cancer. *Diagnostics, 3*(1), 68–83. http://www.mdpi.com/2075-4418/3/1/68/pdf. doi:10.3390/diagnostics3010068.

Sethi, S., Nanda, R., & Chakraborty, T. (2013). Clinical application of volatile organic compound analysis for detecting infectious diseases. *Clinical Microbiology Reviews, 26*(3), 462–475. http://cmr.asm.org/content/26/3/462.full.pdf+html. doi:10.1128/CMR.00020-13Germany.

Shuai, J., Kim, S., Ryu, H., Park, J., Lee, C. K., Kim, G. B., Ultra, V. U., & Yang, W. (2018). Health risk assessment of volatile organic compounds exposure near Daegu dyeing industrial complex in South Korea. *BMC Public Health, 18*(1). http://www.biomedcentral.com/bmcpublichealth. doi:10.1186/s12889-018-5454-1.

Singer, B. C., Coleman, B. K., Destaillats, H., Hodgson, A. T., Lunden, M. M., Weschler, C. J., & Nazaroff, W. W. (2006). Indoor secondary pollutants from cleaning product and air freshener use in the presence of ozone. *Atmospheric Environment, 40*(35), 6696–6710. doi:10.1016/j.atmosenv.2006.06.005.

Skoumalová, A., & Hort, J. (2012). Blood markers of oxidative stress in Alzheimer's disease. *Journal of Cellular and Molecular Medicine, 16*(10), 2291–2300. doi:10.1111/j.1582-4934.2012.01585.x.

Smeets, M. A. M., Mauté, C., & Dalton, P. H. (2002). Acute sensory irritation from exposure to isopropanol (2-propanol) at TLV in workers and controls: objective versus subjective effects. *Annals of Occupational Hygiene, 46*, 359–373. doi:10.1093/annhyg/mef054.

Soini, H. A., Klouckova, I., Wiesler, D., Oberzaucher, E., Grammer, K., Dixon, S. J., Xu, Y., Brereton, R. G., Penn, D. J., & Novotny, M. V. (2010). Analysis of volatile organic compounds in human saliva by a static sorptive extraction method and gas chromatography-mass spectrometry. *Journal of Chemical Ecology, 36*(9), 1035–1042. doi:10.1007/s10886-010-9846-7.

Spengler, J. D., & Chen, Q. (2000). Indoor air quality factors in designing a healthy building. *Annual Review of Energy and the Environment, 25*, 567–600.

Sponring, A., Filipiak, W., Ager, C., Schubert, J., Miekisch, W., Amann, A., & Troppmair, J. (2011). Analysis of volatile organic compounds (VOCs) in the headspace of NCI-H1666 lung cancer cells. *Cancer Biomarkers, 7*(3), 153–161. doi:10.3233/cbm-2010-0182.

Steinemann, A. (2015). Volatile emissions from common consumer products. *Air Quality, Atmosphere and Health, 8*(3), 273–281. http://www.springer.com/environment/environmental+health+-+public+health/journal/11869. doi:10.1007/s11869-015-0327-6.

Steinemann, A. C., Gallagher, L. G., Davis, A. L., & MacGregor, I. C. (2013). Chemical emissions from residential dryer vents during use of fragranced laundry products. *Air Quality, Atmosphere and Health, 6*(1), 151–156. http://www.springer.com/environment/environmental+health+-+public+health/journal/11869. doi:10.1007/s11869-011-0156-1.

Steinemann, A. C., MacGregor, I. C., Gordon, S. M., Gallagher, L. G., Davis, A. L., Ribeiro, D. S., & Wallace, L. A. (2011). Fragranced consumer products: Chemicals emitted, ingredients unlisted. *Environmental Impact Assessment Review, 31*(3), 328–333. doi:10.1016/j.eiar.2010.08.002.

Stinson, B., Laguerre, A., & Gall, E. T. (2022). Per-person and whole-building VOC emission factors in an occupied school with gas-phase air cleaning. *Environmental Science and Technology, 56*(6), 3354–3364. http://pubs.acs.org/journal/esthag. doi:10.1021/acs.est.1c06767.

Suarez, F. L., Furne, J. K., Springfield, J., & Levitt, M. D. (2000). Morning breath odor: Influence of treatments on sulfur gases. *Journal of Dental Research, 79*(10), 1773–1777. doi:10.1177/00220345000790100701.

Tangerman, A., Meuwese-Arends, M. T., & van Tongeren, J. H. M. (1983). A new sensitive assay for measuring volatile sulphur compounds in human breath by Tenax trapping and gas chromatography and its application in liver cirrhosis. *Clinica Chimica Acta, 130*(1), 103–110. doi:10.1016/0009-8981(83)90263-2.

Technical Overview of Volatile Organic Compounds. n.d. https://www.epa.gov/indoor-air-quality-iaq/technical-overview-volatile-organic-compounds.

Tibaquirá, J., Huertas, J. É., Ospina, S. Á., Quirama, L., & Niño, J. É. (2018). The effect of using ethanol-gasoline blends on the mechanical, energy and environmental performance of in-use vehicles. *Energies, 11*(1), 221. doi:10.3390/en11010221.

Tiiva, P., Faubert, P., Michelsen, A., Holopainen, T., Holopainen, J. K., & Rinnan, R. (2008). Climatic warming increases isoprene emission from a subarctic heath. *New Phytologist, 180*(4), 853–863. doi:10.1111/j.1469-8137.2008.02587.x.

Timm, C. M., Lloyd, E. P., Egan, A., Mariner, R., & Karig, D. (2018). Direct growth of bacteria in headspace vials allows for screening of volatiles by gas chromatography mass spectrometry. *Frontiers in Microbiology, 9*, 491. https://www.frontiersin.org/articles/10.3389/fmicb.2018.00491/full. doi:10.3389/fmicb.2018.00491.

Tominaga, M., Ishikawa, T., Michiue, T., Oritani, S., Koide, I., Kuramoto, Y., Ogawa, M., & Maeda, H. (2013). Postmortem analyses of gaseous and volatile substances in pericardial fluid and bone marrow aspirate. *Journal of Analytical Toxicology, 37*(3), 147–151. doi:10.1093/jat/bkt004.

Toombs, C. F., Insko, M. A., Wintner, E. A., Deckwerth, T. L., Usansky, H., Jamil, K., Goldstein, B., Cooreman, M., & Szabo, C. (2010). Detection of exhaled hydrogen sulphide gas in healthy human volunteers during intravenous administration of sodium sulphide. *British Journal of Clinical Pharmacology, 69*(6), 626–636. doi:10.1111/j.1365-2125.2010.03636.x.

Kodogiannis, V., and Wadge, E. "The use of gas-sensor arrays to diagnose urinary tract infections." International journal of neural systems 15.05 (2005): 363-376.

Volatile Organic Compounds and Smog. n.d. https://www.epd.gov.hk/epd/english/environmentinhk/air/prob_solutions/vocs_smog.html

Trefz, P., et al. (2019). Exhaled volatile substances in children suffering from type 1 diabetes mellitus: results from a cross-sectional study. *Scientific Reports, 9,* 15707. doi:10.1038/s41598-019-52165-x.

Wang, C., Feng, Y., Wang, M., Pi, X., Tong, H., Wang, Y., Zhu, L., & Li, E. (2015). Volatile organic metabolites identify patients with mesangial proliferative glomerulonephritis, IgA nephropathy and normal controls. *Scientific Reports., 5,* 14744. www.nature.com/srep/index.html. doi:10.1038/srep14744.

Wang, R. (2002). Two's company, three's a crowd: Can H_2S be the third endogenous gaseous transmitter? *FASEB Journal, 16*(13), 1792–1798. doi:10.1096/fj.02-0211hyp.

Wang, W., He, Y., & Qi, L. (2020a). High-efficiency colorful perovskite solar cells using TiO2 nanobowl arrays as a structured electron transport layer. *Science China Materials, 63*(1), 35–46. http://www.springer.com/journal/40843/about. doi:10.1007/s40843-019-9452-1.

Wang, X., Liu, F., Chen, X., Song, X., Xu, G., Han, Y. E., Tian, J., & Cui, H. (2020b). In_2O_3 nanoparticles decorated ZnO hierarchical structures for n-butanol sensor. *ACS Applied Nano Materials, 3*(4), 3295–3304. doi:10.1021/acsanm.0c00025.

Wang, Z., & Wang, C. (2013). Is breath acetone a biomarker of diabetes? A historical review on breath acetone measurements. *Journal of Breath Research, 7*(3), 037109. doi:10.1088/1752-7155/7/3/037109.

Wania, F., & Shunthirasingham, C. (1925). Passive air sampling for semi-volatile organic chemicals. *Environmental Science: Processes & Impacts, 22,* 1925.

Xuan, L., Ma, Y., Xing, Y., Meng, Q., Song, J., Chen, T., Wang, H., Wang, P., Zhang, Y., & Gao, P. (2021). Source, temporal variation and health risk of volatile organic compounds (VOCs) from urban traffic in harbin, China. *Environmental Pollution., 270,* 116074. doi:10.1016/j.envpol.2020.116074.

Zaichko, N. V. (2014). Hydrogen sulfide: Metabolism, biological and medical role. *The Ukrainian Biochemical Journal, 86*(5), 5–25. doi:10.15407/ubj86.05.005.

Zetola, N. M., Modongo, C., Matsiri, O., Tamuhla, T., Mbongwe, B., Matlhagela, K., Sepako, E., Catini, A., Sirugo, G., Martinelli, E., Paolesse, R., & Di Natale, C. (2017). Diagnosis of pulmonary tuberculosis and assessment of treatment response through analyses of volatile compound patterns in exhaled breath samples. *Journal of Infection, 74*(4), 367–376. http://www.elsevier.com/inca/publications/store/6/2/3/0/5/4/index.htt. doi:10.1016/j.jinf.2016.12.006.

Zhao, R., Zhang, M., Liu, Y., Zhang, X., Duan, Y., & Han, T. (2020). Fabricating D-A type AIE luminogen into film sensor for turn-on detection of methanol vapour. *Sensors and Actuators B: Chemical, 319,* 128323. doi:10.1016/j.snb.2020.128323.

Zheng, H., Kong, S., Yan, Y., Chen, N., Yao, L., Liu, X. I., Wu, F., Cheng, Y. I., Niu, Z., Zheng, S., Zeng, X., Yan, Q., Wu, J., Zheng, M., Liu, D., & Zhao, D. (2020). Shihua Qi, Compositions, sources and health risks of ambient volatile organic compounds (VOCs) at a petrochemical industrial park along the Yangtze River. *Science of the Total Environment, 703,* 135505. doi:10.1016/j.scitotenv.2019.135505.

CHAPTER

2

Classical approaches to breath VOC monitoring for disease diagnosis

2.1 Introduction

In the field of medical science, various diagnostic procedures are utilized to understand the condition of any disease. Among the various methods available, clinicians generally decide whether to proceed with an invasive or non-invasive approach, which generally relies on the sample collection method. The invasive method is based on devices that are used to diagnose the condition through skin penetration or internal insertion. The primary examples of invasive tests include biopsies, blood tests, and colonoscopies. On the other hand, the non-invasive approach is a painless procedure that does not require any surgery or physical penetration of the body but relies on advanced diagnostic techniques for the detection of various diseases, including computed tomography (CT) scans, X-rays, magnetic resonance imaging (MRI), electrocardiograph (EKG or ECG), urine tests, and monitoring of exhaled breath (Castro & Russo, 2019). Non-invasive diagnostic techniques have become increasingly significant in the field of medical diagnostics due to their ability to offer rapid, accurate, and minimally disruptive results. These methods not only enhance patient comfort but also reduce the risks associated with traditional invasive procedures, such as infections, bleeding, and other complications.

The X-ray test was the first non-invasive diagnostic test to be reported, offering several advantages but also presenting some limitations. It provides ease of use, cost-effectiveness, and rapid analysis; however, it is limited in its ability to generate three-dimensional (3D) visualizations, as it only produces two-dimensional (2D) results. Furthermore, the emission

Photonic Sensors for Detection of VOC Biomarkers: Technological Principles and Medical Applications
DOI: https://doi.org/10.1016/B978-0-443-29166-1.00011-1

Copyright © 2026 Elsevier Inc. All rights are reserved, including those for text and data mining, AI training, and similar technologies.

of ionizing radiation in X-ray imaging raises safety concerns, particularly for pregnant women and pediatric patients. Another type of non-invasive testing is a CT scan, which is utilized to diagnose several types of diseases in the body via a medical imaging procedure that uses X-ray radiation and a computer to produce a detailed visualization in the form of 3D images from the inside of the body. It provides high anatomical resolution and has several advantages, such as 3D imaging capability and fast processing time. However, it is highly expensive, and its exposure to ionizing radiation poses potential adverse effects on the human body (Einoch Amor et al., 2019). The third technique is MRI, which produces four-dimensional (4D) imaging of almost every internal organ present in the human body, including muscles, bones, blood vessels, and other soft tissues. Unlike X-ray and CT, MRI does not use ionizing radiation but instead employs strong magnetic fields and radiofrequency pulses, making it safer for long-term use, especially in sensitive populations. Similarly, ECG signals are widely used to identify medical conditions related to cardiovascular diseases. This process involves complex and challenging domain activities concerning appropriate data selection and equipment validation. These aspects present significant challenges not only for the operator responsible for conducting the measurement but also for the physician tasked with interpreting and endorsing the data for clinical analysis (Lyon et al., 2018). When the ECG technology is integrated with smartphone applications for diagnostic purposes, it requires additional central processing unit (CPU) power and memory resources to generate accurate results. Consequently, this increased computational demand can potentially lead to reduced performance and slower response time of the embedded device. The possibility of integrating ECG signals into a wireless sensor within a Body Area Network (BAN) is currently being actively explored. To achieve this integration effectively, the ECG signals are processed through the Discrete Wavelet Transform (DWT) and the Pan-Tompkins algorithm methods, which are recognized for their ability to perform reliable signal decomposition in real-time biomedical monitoring systems.

In order to comply with international guidelines set by the World Health Organization (WHO), it is crucial to conduct diagnostic tests in accordance with standard operating procedures. Failure to adhere to these guidelines accurately can lead to misdiagnosis or false-negative and false-positive results. Skilled professionals are required to effectively carry out these testing procedures to ensure the reliability and reproducibility of the results. In the absence of qualified technicians, diagnostic outcomes may lack accuracy, and improper handling can also lead to compromised hygiene and contamination risks. Therefore, to minimize the complexity of diagnostic devices and reduce dependency on specialized personnel, several researchers have reported the development of various affordable, rapid-responsive breath monitoring tools designed to identify multiple diseases efficiently and non-invasively. Taking into account the aforementioned testing procedures along with their advantages, limitations, and prerequisites, the analysis of exhaled breath has gained significant attention in diagnosing various diseases due to its non-invasive nature and biochemical richness (Zhang et al., 2017). The identification of volatile organic compounds (VOCs) in exhaled breath has a well-established historical basis in medical diagnosis (Kwak et al., 2014). Healthcare professionals have recognized that the odor present in exhaled breath may provide valuable insights into disease identification and its underlying causes. Human exhaled breath comprises with a variety of biochemical constituents that reflect physiological and metabolic processes within the body. These biochemical factors are closely linked to the hygiene status or dysfunction of bodily organs, indicating the overall health of internal body systems.

This chapter discusses various non-invasive analytical tools used for exhaled breath analysis. Among the various tools available today, gas chromatography-mass spectrometry (GC-MS), electric nose (E-nose), ion mobility spectrometry (IMS), proton transfer reaction-mass spectrometry (PTR-MS), secondary electrospray ionization mass spectrometry (SESI-MS), and selected-ion flow-tube mass spectrometry (SIFT-MS), are considered some of the most powerful techniques to identify several VOC biomarkers associated with various diseases. Collecting breath samples from patients is painless, non-invasive, and virtually limitless, offering significant advantages, specifically in pediatric treatments where invasive methods are limited. Through breath analysis, illnesses in children can be effectively diagnosed. Key disease biomarkers found in human breath include acetone, ammonia, isoprene, hydrogen sulfide, nitric oxide, ethane, methane, and pentane. Fig. 2.1 illustrates the offline breath sampling and analysis procedure.

2.2 Breath collection

The exhaled breath can be classified into three categories: gases, condensates, and volatiles. These categories can be obtained non-invasively using various approaches, such as collecting gaseous samples with gas bags, obtaining condensates via cold condensation chambers, and capturing volatile compounds through adsorbent traps (Khoubnasabjafari et al., 2022). The most frequent techniques utilized in order to collect the gaseous exhaled breath involve the use of three primary methods (Preti et al., 1988). First, containers or sampling bags are commonly used. These types of bags comprise a mouthpiece through which the breath is exhaled into the bag. This straightforward approach allows samples to be collected and stored directly in bags of different volumes (typically ranging from 1 to 10 L)

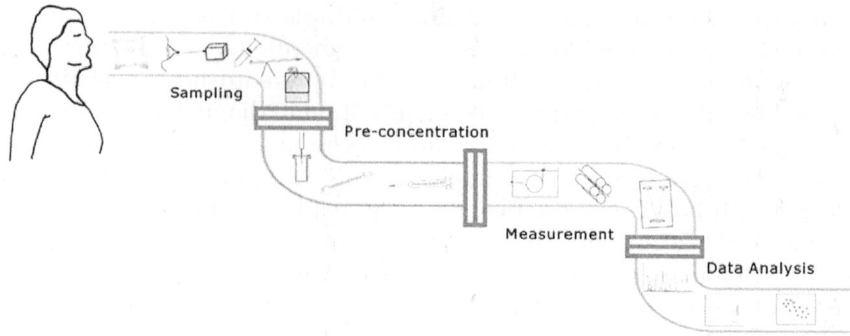

FIGURE 2.1 Off-line analysis of exhaled breath. Schematic diagram illustrating the off-line breath sampling pipeline. *From Lawal, O., Ahmed, W. M., Nijsen, T. M. E., Goodacre, R., & Fowler, S. J. (2017). Exhaled breath analysis: A review of "breath-taking" methods for off-line analysis.* Metabolomics, 13*(10), 110. https://doi.org/10.1007/s11306-017-1241-8.*

via the mouthpiece, thereby preventing rebreathing as well as the entry of any impurities. Second, reservoir or buffered sampling techniques are employed to segregate air from the upper and lower respiratory tracts by capturing a portion of breath during late expiration. Straightforward solutions such as BioVOC, a handheld device designed to collect alveolar breath with minimal contamination, and RTubeVOC, a disposable sampler used to capture exhaled breath condensate and (VOCs) through a cooled collection system, involve venting the initial portion of the breath and preserving a predetermined volume of end-tidal air (127 mL and 65 mL, respectively) (Preti et al., 1988). A breath collection tool offers a comprehensive solution that specifically focuses on capturing breath from the end of expiration (Khoubnasabjafari et al., 2022). Third, the respiratory-cycle-based method involves monitoring respiration and sampling breath in synchrony with signals from pressure or carbon dioxide (CO_2) sensors, in order to capture late-expiratory air for analysis (Cope et al., 2004).

These techniques enable a more personalized diagnostic approach and can accommodate significant variations in lung volume due to factors such as gender, age, race, ethnicity, and underlying disease. The analysis of gaseous breath can be performed using either online or offline methods, with specific detection and analytical strategies tailored for each technique. Online measurement enables real-time analysis and eliminates the requirement for sample storage, thereby reducing potential sample degradation. On the other hand, offline monitoring requires a more complex procedure, including sample collection, storage, and transportation to the analytical tool (Lawal et al., 2017). The next section provides a detailed discussion of the various analytical techniques used to analyze exhaled breath samples based on both online and offline approaches.

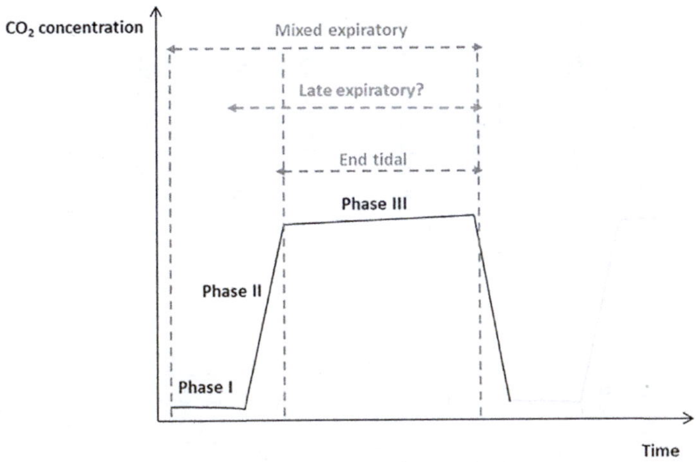

FIGURE 2.2 Single exhaled breath analyzed by capnography. The graph illustrates a single exhaled breath phase recorded by capnography representing Phase I, Phase II and Phase III for dead space, transition, and alveolar, respectively. *From Lawal, O., Ahmed, W. M., Nijsen, T. M. E., Goodacre, R., & Fowler, S. J. (2017). Exhaled breath analysis: A review of "breath-taking" methods for off-line analysis.* Metabolomics, *13(10), 110. https://doi.org/10.1007/s11306-017-1241-8.*

2.2.1 Sampling exhaled breath

During breath sampling, it is necessary to determine the specific phase of the exhaled breath to be collected, broadly categorized as late-expiratory, end-tidal, and mixed-expiratory fractions (Lawal et al., 2017). Mixed-expiratory breath sampling generally comprises the whole exhaled breath, including so-called "dead space air," which refers to air not involved in gas exchange, such as that from the nose, mouth, and upper airways. In contrast, other sampling strategies aim to minimize contamination from dead space air. Further discussion of these approaches is provided in the following subsections.

2.2.1.1 *Late expiratory breath*

Sampling of late expiratory breath requires excluding the initial portion of the exhaled breath and capturing air toward the end of the breath cycle (Lawal et al., 2017). This specific breath sample has been the subject of numerous investigations, as demonstrated in Table 2.1 (Fig. 2.2). Reducing the sampling of dead space air improves the relative presence of endogenous VOCs in the collected breath sample and lowers the concentration of exogenous VOCs. To improve the sampling, some techniques require the removal of the initial portion of exhaled air before capturing the breath for analysis (Castellanos et al., 2016). Another approach involves continuous

TABLE 2.1 Studies that collected late expiratory breath.

Brief description	Breath collection container	Pre-concentration method	References
Dead space air collected in one bag, late expired air in separate bag	Tedlar bag	TD tube—Multi-bed ORBO 420/Tenax TA TD tube—Tenax	Amal et al. (2016), Barash et al. (2015)
First 10 s excluded from bag	Tedlar bag	TD tube—Carboxen 1000/Carbopack X/Carbopack B	Sanchez and Sacks (2006)
Deep breath, held for 10 s, exhaled slowly for 10 s prior to filling bag	Tedlar bag	TD tube—Carbopack Y/B/X/Carboxen 1000	Libardoni et al. (2006)
Exhale air into tube connected to bag	Tedlar bag	TD tube—Tenax GC	Preti et al. (1988)
Forced expiratory breath with first 2–3 s of expiration not collected	Tedlar bag	In house sorbent microtrap—Carboxen 1000, Carbopack X, Carbopack B	Castellanos et al. (2016)
Final portion of exhaled breath collected	ALTEF polypropylene bag	TD tube—Tenax TA	Kwak et al. (2014)
Last portion of breath collected using ha-pause method	Aluminium gas bag	TD tube—Tenax TA/Unicarb	Berna et al. (2015)

(continued on next page)

Second half of exhaled breath collected	FlexFilm bag	TD tube—Tenax TA/Carbotrap B	Bigazzi et al. (2016)
First third of each breath not collected	Tedlar bag	SPME—75 µM CAR/PDMS	Chen et al. (2016)
Dead space air collected in one bag, late expired air in separate bag Mixed expiratory	Tedlar bag	SPME—75 µM CAR/PDMS	Capuano et al. (2015)
Last portion of exhaled breath using breath collecting device	Tedlar bag	SPME—100 µM PDMS	Wang et al. (2012)
Dead space air collected in one bag, late expired air in separate bag	Tedlar bag	SPME—fiber not specified	Santonico et al. (2012)
Dead space air collected in one bag, late expired air in separate bag	Mylar bag	SPME—PDMS/DVB	Hakim et al. (2011), Peng et al. (2010)
Last of exhaled breath (150 mL)	Bio-VOC	TD tube—Carbotrap 300	Das et al. (2014)
Deep breath and slowly exhale as fully as possible	Bio-VOC	TD tube—Carbograph 1TD/Carbopack X	Kwak et al. (2014)
Retained last of exhaled breath (~ 100 mL)	Bio-VOC	TD tube—Tenax TA/Unicarb	Zaric et al. (2014)

(*continued on next page*)

TABLE 2.1 Studies that collected late expiratory breath—cont'd

Brief description	Breath collection container	Pre-concentration method	References
End of forced vital capacity collected	Bio-VOC	TD tube—Tenax TA/Graphitized Carbon black/Carbonized molecular sieve	Jareño-Esteban et al. (2013)
Last portion of exhaled breath (~150 mL)	Bio-VOC	TD tube—Tenax TA/Unicarb	Dadamio et al. (2012)
Last 150 mL of single slow vital capacity Mouth air	Bio-VOC	TD tube—Tenax TA/Unicarb	van den Velde et al. (2007)
Exhale until lungs are empty	Bio-VOC	TD tube—sorbent not specified	Henderson and Matthews (2002)
Blow deeply and slowly through sampler (150 mL)	Bio-VOC	SPME—75 μM CAR/PDMS	Raninen et al. (2016)
Last 150 mL of single slow vital capacity	Bio-VOC	SPME—75 μM CAR/PDMS, 65 μM PDMS/DVB	Corradi et al. (2015)
Last portion of exhaled breath (~100 mL)	Bio-VOC	SPME—75 μM CAR/PDMS	Kramer et al. (2015)
Exhale into tubular structure with dead space air flowing downstream and air collected upstream	BCA	TD tube—Carbotrap/Carbosieve SIII	Zeliger et al. (2012)

(continued on next page)

18 mL of one single end-tidal exhalation portion	Gas-tight syringe, SPME vial	SPME—75 μM CAR/PDMS	King et al. (2010)
Single exhalations with first 750 mL discarded	Gas-tight syringe, glass vial	SPME—65 μM PDMS/DVB	Svensson et al. (2007)
Take 2–3 deep breaths, inhale and hold breath for 10–15 s, exhale into glass tube and collect an aliquot of end-expired air	Glass tube, aluminum tube	SPME—100 μM CAR/PDMS TD tube—Tenax TA	Prado et al. (2003)
Pressure sensors to estimate breath phases	Face mask	TD tube—Tenax TA/Carbograph 1TD	Kang and Paul Thomas (2016)
Pressure sensors to estimate breath phases	Face mask	TD tube—Tenax TA/Carbotrap	Basanta et al. (2010)
First portion of breath removed	Breath device connected to desorption tube	TD tube—Carbopack B/Carbopack C	Khalid et al. (2013)
Two tidal volume ventilations, a deep inspiration and slow exhalation for 10 s, first 3 s discarded	Stainless steel canisters	Glass beads	Minh et al. (2011)
Deep inspiration, 5 s breathhold, slow and complete exhalation over 10 s. First 2 s discarded	Electro-polished stainless steel	–	Barker et al. (2006)

From Lawal, O., Ahmed, W. M., Nijsen, T. M. E., Goodacre, R., & Fowler, S. J. (2017). Exhaled breath analysis: A review of "breath-taking" methods for off-line analysis. Metabolomics, *13(10), 110. https://doi.org/10.1007/s11306-017-1241-8.*

exhalation into a reservoir, which allows the breath to flow downstream, and only the portion of air closest to the donor is selectively collected. This method may introduce additional complexity and is not as straightforward to implement (Phillips, 1997). Reproducibility concerns also emerge due to the various physiological characteristics among individuals, including pulmonary ventilation and cardiac output. These factors can cause unwanted variations, even in the same individual, when the samples are collected repeatedly under varying physiological conditions (Sukul et al., 2016). Other contributing factors to this variability include breath-holding and changes in the rate of exhalation, both of which can significantly affect the VOC concentrations in breath samples (Sukul et al., 2014). Therefore, despite multiple challenges linked to the late expiratory breath sampling method, further refinement is highly necessary before it can be considered for clinical applications. Ideally, the system should be modified to the current physiological status of individuals during the sampling period to reduce the impact of dead space contamination. However, this level of adaptation could compromise the simplicity and broad applicability of several existing systems (Kwak et al., 2014).

2.2.1.2 *End-tidal breath*

The end-tidal breath, also known as "alveolar" breath, refers to the portion of air collected from the start to the end of phase III of the breath cycle, as shown in Fig. 2.2 (Lawal et al., 2017). This type of air is characterized by minimal contamination from dead space and exhibits high concentrations of endogenous VOCs, making it particularly valuable for diagnostic purposes. However, the use of the term "alveolar" can be confusing, as the gas exchange models suggest that the VOC exchange may occur within the airways, meaning that the sampled breath may not fully represent the actual alveolar concentrations (Anderson & Hlastala, 2007). To address this issue, the term "alveolar" has been replaced by more specific terms such as end-exhaled, end-tidal, and end-expired breath (Lourenço & Turner, 2014). End-tidal breath samples generally vary from late expiratory ones in the level of confidence they offer in capturing a representative and personalized sample. This is often achieved by utilizing a visual cue to precisely collect air solely from phase III, where the concentration of VOCs is considered most stable and reflective of systemic physiological conditions (Lawal et al., 2017).

In previously reported studies CO_2 visual control has consistently emerged as the most common method (as shown in Table 2.2) utilized to capture end-tidal breath, relying on real-time monitoring of CO_2 concentration during exhalation. As can be seen from Fig. 2.2, during phase I of exhalation, the CO_2 levels are initially low. However, they begin to rise during phase II and then reach a plateau during phase III, indicating the start of the alveolar phase. The concentration level of CO_2 in the

TABLE 2.2 Studies that collected end-tidal or "alveolar" breath.

Brief description	Breath collection container	Pre-concentration method	References
CO_2 visual control & mixed expiratory using plastic straws	Tedlar bag	TD tube—Tenax TA/Carboxen 569/Carboxen 1000	Filipiak et al. (2014)
CO_2 visual control	Tedlar bag	TD tube—Tenax TA	Grabowska-Polanowska et al. (2013)
CO_2 visual control	Nalophan bag	TD tube—Tenax GR	Salvo et al. (2015)
CO_2 visual control	Tedlar bag	SPME—75 μM CAR/PDMS	Mochalski et al. (2014)
CO_2 visual control	Tedlar bag Gas-tight syringe, glass vial	SPME—PDMS, PDMS/DVB, PA, CAR/PDMS, CW/DVB, DVB/CAR/PDMS SPME—75 μM CAR/PDMS	Bajtarevic et al. (2009)
CO_2 visual control	Tedlar bag Gas-tight syringe, glass vial	SPME—CAR/PDMS	Miekisch et al. (2008)
CO_2 visual control	–	NTD—Tenax TA/Carbopack X/Carboxen 1000	Phillips et al. (2012)

(continued on next page)

TABLE 2.2 Studies that collected end-tidal or "alveolar" breath—cont'd

Brief description	Breath collection container	Pre-concentration method	References
CO_2 visual control	–	NTD—Tenax TA/Carbopack X/Carboxen 1000	Miekisch et al. (2010)
CO_2 visual control	Gas-tight syringe, glass vial	SPME—75 μM CAR/PDMS	Guo et al. (2015)
CO_2 visual control	Gas-tight syringe, glass vial	SPME—CAR/PDMS SPME—75 μM CAR/PDMS SPME—65 μM PDMS/DVB	Fuchs et al. (2010), Fuchs et al. (2010), Goerl et al. (2013), Kischkel et al. (2012), Pabst et al. (2007), Schubert et al. (2005)
Rebreathed air	Tedlar bag	Freeze-trap breath in glass U-tube	Jones et al. (1983)
*CO_2 visual control	glass syringe	TD tube—Carbotrap B/Carbopack X	Filipiak et al. (2015)

From Lawal, O., Ahmed, W. M., Nijsen, T. M. E., Goodacre, R., & Fowler, S. J. (2017). Exhaled breath analysis: A review of "breath-taking" methods for off-line analysis. Metabolomics, *13(10), 110. https://doi.org/10.1007/s11306-017-1241-8.*

exhaled breath can be monitored using a capnometer, which enables clear visualization of the different exhalation phases and provides guidance on the optimal moment to initiate end-tidal breath collection. Various devices have been reported in the past decade that support both manual exclusion of the initial breath phases and automatic sampling once the alveolar phase has been detected (Miekisch et al., 2010; Salvo et al., 2015). Additionally, in 2005, Schubert et al. reported an extensive study on CO_2 visual control breath sampling, involving 46 mechanically ventilated patients (Schubert et al., 2005). Results indicated that patients with sepsis exhibited a higher concentration of n-pentane and a lower concentration of acetone in mixed venous blood, whereas concentration levels were observed to be within normal ranges in patients without sepsis. In mechanically ventilated patients, the VOC concentration profiles in exhaled breath can vary greatly from those in blood, influenced by the relative amounts of these compounds in the inspired air. The expectation of a consistent VOC profile in the endogenous samples increases the likelihood of identifying potentially robust markers, thereby reducing the risk of type I error. Similar to late-expiratory breath sampling, airway VOCs may be lost during end-tidal collection. Furthermore, since not all endogenous biomarkers have a systemic origin, this type of exhaled breath is generally not suitable for studying diseases associated with the airways. Among the many VOCs, nitric oxide is one of the most widely known VOC biomarkers, typically utilized to characterize asthma (Dweik et al., 2011). The modification of breath collection to suit each individual, combined with sample acquisition during phase III of exhalation, minimizes variability in sample collection. This consistency enhances its suitability for clinical applications by providing a physiological benchmark. Although the method involves several procedural steps before it can be adopted for clinical trials, this approach facilitates the comparison of data across different laboratories and studies.

Rebreathing has been proposed as an alternative technique for collecting end-tidal breath samples. This technique aims to attain equilibrium between the sample container, the airways, and the bloodstream as the subject breathes (Lawal et al., 2017). Generally, no external air source is introduced during this process. Under equilibrium conditions, it is generally considered that the VOC concentrations inside the container are indicative of the concentration observed in the blood. In 1983, Jones utilized the approach to compare ethanol concentrations in blood and breath samples (Jones, 1983a). However, this comparison was conducted using a single VOC and may not be applicable to other metabolites. In 1990, Ohlsson et al. reported similar findings using a modified rebreathing approach also referred to as "isothermal rebreathing" (Ohlsson et al., 1990). Despite those promising results, the method may not be suitable for clinical applications due to potential adverse effects on human health, including patient discomfort caused by low oxygen concentrations and elevated levels of

CO_2. This technique is particularly unsuitable for individuals with acute respiratory conditions. A summary of previously published studies on this topic is provided in Table 2.2.

2.2.1.3 Mixed expiratory breath

Mixed expiratory breath is known as one of the simplest types of breath samples to obtain. It comprises all phases of exhaled air, as depicted in Fig. 2.2. Its operational simplicity makes it a promising candidate for breath sampling; however, it may not provide high-quality samples due to the elevated risk of contamination from environmental sources, as well as nasal and oral substances. Even when feature selection models are applied to identify potential biomarkers, there remains a significant risk of false positives, such as mistaking an exogenous VOC for an endogenous candidate marker, unless a very large sample size is used. This concern has been raised by various investigations, which emphasize the requirements of strict regulatory protocols and highly reproducible collection methods when capturing exhaled breath samples for analytical purposes (O'Hara et al., 2008; Thekedar et al., 2011).

2.3 Breath analysis techniques

Breath analysis is considered one of the most crucial elements in non-invasive detection, and effective investigation requires proper breath sampling utilizing specialized equipment, followed by data analysis with appropriate analytical tools. The VOCs present in the exhaled breath can be collected and analyzed using either online or offline methods. To conduct offline analysis, the breath sample must be collected, stored, and transported to the analytical laboratory for processing. In contrast, online breath analysis techniques utilize real-time measurement tools, such as IMS, including SIFT–MS and PTR-MS, which do not require the isolation of air from VOCs present in exhaled breath samples. Offline breath examination techniques typically involve expensive analytical instruments (such as GC-MS), which are costly, time-consuming, and require skilled personnel for operation and interpretation. These instruments employ ion-pair extraction and mass spectrometry for quantifying VOC concentrations, capable of detecting levels as low as parts per trillion (ppt) (Pathak et al., 2023). For rapid detection, several researchers have developed alternative techniques, such as the Tor Vergata electronic nose (E-nose) and Cyranose, which utilize arrays of VOC sensors to enable quick responses to fluctuations in VOC concentrations. A comprehensive discussion is included in this chapter on the commonly used conventional techniques for detecting VOCs. A summary of conventional tools used in breath monitoring is provided in Table 2.3, along with the detection limit, mode of operation, advantages, and limitations.

TABLE 2.3 Various conventional tools used to analyze exhaled breath.

Technique	Principle	Detection limit	Mode of operation	Advantages	Disadvantages
GC-MS	Separate chemical mixtures using GC and identify the components via MS	ppb and ppt	Pre-concentration	Highly sensitive and selective	Bulky size, expensive, required skilled technician
SIFT-MS	Ionizing VOCs a flow tube and measure their mass	ppb and ppt	Online/real time	Real-time quantification, detection of low concentration ppb and ppt., wide range	Expensive, bulky in size, selectivity, not suitable for semi-VOCs and non-VOCs
PTR-MS	Ionizing VOCs in a reaction chamber and measure their mass.	ppb	Online/real time	Real time analysis	Narrow range of detection, bulky and required trained professionals, selectivity
E-nose	Detect and differentiate complex odors	ppb	Reference to a database	Rapid detection, portable, useful for complex VOC environments	Less sensitive for low-concentration, require calibration, deteriorate with age, poisoning is possible in presence of other gases
SESI-MS	Generate Ions for MS	ppbv	Real-Time	Robust, sensitive, ionize non-volatile solutions, and multiple ionization modes	Calibration is needed

From Mule, N. M., Patil, D. D., & Kaur, M. (2021). A comprehensive survey on investigation techniques of exhaled breath (EB) for diagnosis of diseases in human body. Informatics in Medicine Unlocked, *26, 100715. https://doi.org/10.1016/j.imu.2021.100715.*

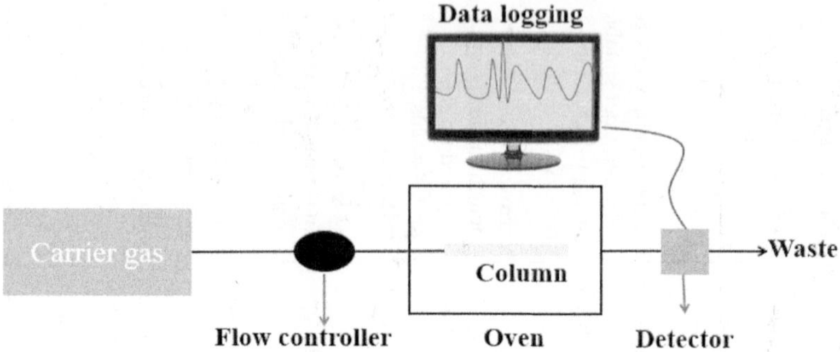

FIGURE 2.3 Gas chromatography-mass spectrometry. Schematic block diagram of a GC-MS. *From Pathak, A., Swargiary, K., Kongsawang, N., Jitpratak, P., Ajchareeyasoontorn, N., Udomkittivorakul, J., & Viphavakit, C. (2023). Recent advances in sensing materials targeting clinical volatile organic compound (VOC) biomarkers: A review.* Biosensors, *13(1), 114. https://doi.org/10.3390/bios13010114.*

2.3.1 Offline analysis

2.3.1.1 Gas chromatography-mass spectrometry

Gas chromatography-mass spectrometry (GC-MS) is an analytical tool that integrates the separating capability of gas chromatography (GC) with the ability of mass spectrometry (MS) to detect and identify individual compounds. It serves a crucial role in recognizing multiple VOCs and their respective concentrations within a mixture. MS is frequently employed to obtain the characteristic "fingerprint" spectrum of molecules. Mass spectrometric analysis reveals in-depth information, including elemental structure, molecular weight, functional groups, and sometimes the spatial isomerism of the molecule. The GC-MS technique was introduced in 1951–52 and first applied by James and Martin for gas detection (James & Martin, 1952). The GC technique has undergone remarkable improvement over recent decades, spanning from 1951 to 2020. In the GC-MS analysis, the complex mixture of VOC gases generally travels through a column that separates the VOC molecules based on their chemical properties, and the concentrations are monitored using a detector. Helium is generally used as the carrier gas in this process. GC-MS techniques have been widely used for several decades to analyze the VOC distribution in exhaled breath, particularly for investigating health conditions and enabling early diagnosis of various diseases. Fig. 2.3 illustrates the schematic block diagram of the GC-MS. In 2003, Sanchez et al. discovered the presence of 25 different types of VOCs in exhaled breath by utilizing a series of GC columns, which identified several crucial VOC biomarkers such as isoprene, acetone, ethanol, methanol, and pentane (Sanchez & Sacks, 2003). The limit of detection (LOD) of GC-MS is reported to range from 1 to 5 ppb in 0.8 L of exhaled human breath. In 2002, Lord et al. reported a GC-MS-based detection system that selectively detects ethanol and acetone in exhaled breath while significantly reducing the effect of moisture, which is commonly present in substantial quantities in exhaled breath (Lord et al., 2002). In 2003, Giardina and Olesik introduced a low-temperature, glassy carbon-based solid-phase microextraction fiber for breath analysis using GC-MS (Giardina & Olesik, 2003). This approach enabled the detection of five different cancer-related VOC biomarkers from collected breath samples. In 2017, Lamote et al. proposed the combined use of an electronic nose (e-Nose) and GC-MS to identify pleural mesothelioma (Lamote et al., 2017). Malignant pleural mesothelioma is primarily caused by prior exposure to asbestos fibers and is associated with a poor prognosis. The approach reported by Lamote successfully differentiated malignant pleural mesothelioma patients from asymptomatic individuals previously exposed to asbestos, with 97% accuracy, using VOCs such as limonene, diethyl ether, nonanal, cyclohexane, and methylcyclopentane as key diagnostic markers. In GC-MS, the first step involves feature extraction, which is then organized into compound profiles. In the second phase, the ion profiles of these compounds are determined. The output from the GC is fed into the MS, where molecules are ionized and the resulting ions are monitored through a suitable ionization method. VOC analysis in exhaled breath using GC-MS has also been carried out using advanced configurations of mass spectrometry, such as GC-time-of-flight mass spectrometry (GC-TOF MS) (Rudnicka, 2011), comprehensive 2D GC coupled to time-of-flight mass spectrometry (GCxGC-TOF MS), and mobility spectroscopy (Pesesse, 2019).

2.3.2 Online analysis

2.3.2.1 Selected-ion flow-tube mass spectrometry

Selected-ion flow-tube mass spectrometry (SIFT-MS) is an analytical technique comparable to gas chromatography for the quantitative analysis of VOC gases (Adams & Smith, 1976). This method involves ionizing VOC samples using reagent ions like H_3O^+, NO^+, and O^{2+}, which are subsequently monitored and analyzed by a quadrupole mass spectrometer (Pathak et al., 2023). SIFT-MS was initially documented for two primary purposes: (1) to identify VOCs in exhaled human breath for disease prognosis and (2) to understand the physiological and pathophysiological states of the human body. SIFT-MS was first conceived and developed by Adams and Smith in 1976, with the original aim of investigating ion-neutral reactions at thermal interaction energies. Initially, its development was driven to fulfill the requirement for kinetic data concerning gas-phase ion-neutral reactions occurring in cold interstellar clouds. The use of SIFT-MS was exploited in clinical applications to identify hazardous VOC gases

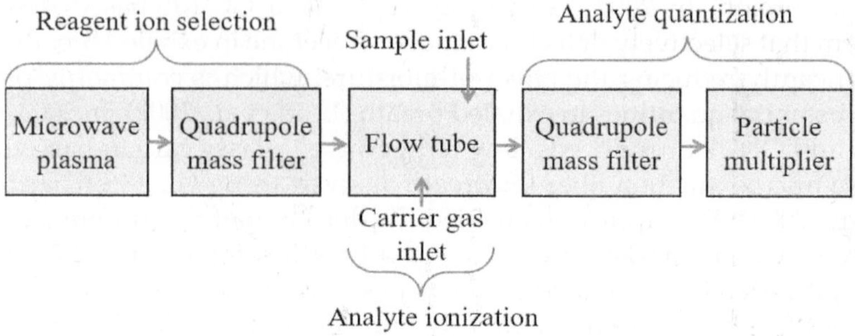

FIGURE 2.4 Selected-ion flow-tube mass spectrometry. Block diagram and working principle of SIFT-MS. *From Pathak, A., Swargiary, K., Kongsawang, N., Jitpratak, P., Ajchareeyasoontorn, N., Udomkittivorakul, J., & Viphavakit, C. (2023). Recent advances in sensing materials targeting clinical volatile organic compound (VOC) biomarkers: A review.* Biosensors, *13(1), 114. https://doi.org/10.3390/bios13010114.*

through online investigation (Španěl & Smith, 2020). The technique enables the analysis of various clinical conditions, physiological responses, and treatment monitoring. SIFT-MS is capable of differentiating and isolating isomers by using three distinct precursor ions in a full-scan mode. A simplified block diagram of SIFT-MS is illustrated in Fig. 2.4. To analyze VOCs in exhaled breath, the samples are collected at a specific flow rate into the flow tube, where they interact with the ionizing agents. These VOC molecules undergo ionization upon collision with the precursor ions within the stream tube, forming product ions specific to each compound of interest (Kim et al., 2012). This approach leads to the production of ionized particles, which are then measured by the mass spectrometer for quantitative and qualitative analysis (Pereira et al., 2015). SIFT-MS represents a novel technique that combines rapid flow-tube chemistry with mass spectrometry. In 2003, Diskin et al. studied the fluctuations in the concentration of VOC biomarkers in exhaled breath using SIFT-MS (Diskin et al., 2003). The investigation focused on determining the concentrations of isoprene, ammonia, ethanol, acetone, and acetaldehyde over a 30-day period in healthy individuals. In the same year, Abbott et al. utilized the same technique to investigate the concentration of acetonitrile in both urinary samples and the exhaled breath of groups of smokers and non-smokers (Abbott et al., 2003). The results illustrated that the concentration of acetonitrile in exhaled breath was monitored within the range of 17–124 ppb, while concentrations in urinary samples were measured in the range of 0–150 µg/L.

2.3.2.2 *Proton-transfer-reaction mass spectrometry*

Proton transfer reaction-mass spectrometry (PTR-MS) is one of the most sensitive analytical tools used for real-time monitoring of various VOCs,

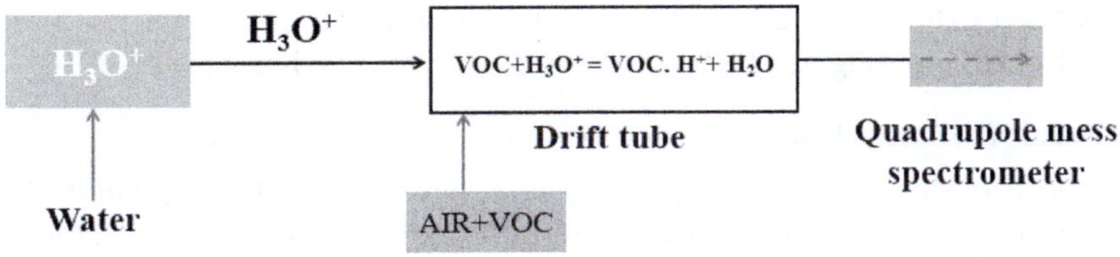

FIGURE 2.5 Proton-transfer-reaction mass spectrometry. Block diagram showing the working principle for PTR-MS. *From Pathak, A., Swargiary, K., Kongsawang, N., Jitpratak, P., Ajchareeyasoontorn, N., Udomkittivorakul, J., & Viphavakit, C. (2023). Recent advances in sensing materials targeting clinical volatile organic compound (VOC) biomarkers: A review.* Biosensors, *13(1), 114. https://doi.org/10.3390/bios13010114.*

such as acetaldehyde, acetone, ethanol, methanol, xylene, and benzene, along with certain inorganic compounds present in the atmosphere. PTR-MS was invented and developed by Lindinger and co-workers in 1990 (Calamai et al., 2012). This method enables continuous detection of complex, low-concentration mixtures of VOCs with higher detection sensitivity than SIFT-MS. PTR-MS offers several advantages over GC-MS and SIFT-MS, including exceptional analytical efficiency, high sensitivity, and a low detection limit of approximately 1 part per trillion by volume (pptv) (Hansel et al., 1995). Additionally, this technique does not require pre-separation or sample preparation, which are typically necessary for GC-MS. Unlike other detection tools such as SIFT-MS, PTR-MS prevents dilution of low-concentration samples by avoiding the use of carrier gas, and it also minimizes the loss of precursor ions by eliminating the need for a mass filter between the ion source and the drift tube. PTR-MS operates using gas-phase hydronium ion (H_3O^+) with a source reagent purity of greater than 99.5%. H_3O^+ ions are chemically inert with respect to the predominant components in ambient air, owing to their relatively low proton affinity. Fig. 2.5 illustrates the schematic block diagram of PTR-MS. In addition to several advantages, PTR-MS has certain limitations. These include constraints in detection range and a limited number of VOCs that can be simultaneously identified and quantified. Breath samples analyzed by PTR-MS can be collected under various atmospheric conditions, at different locations, and in both controlled and uncontrolled laboratory environments. However, these variables may significantly influence the VOCs and should be carefully managed during sampling and interpretation. In 2004, Amann et al. employed PTR-MS to investigate fluctuations in the concentration of various VOCs exhaled by patients experiencing carbohydrate malabsorption during sleep, as well as to analyze the inter- and intra-subject variability of a specific mass signal (Amann et al., 2004). Karl et al. demonstrated PTR-MS for monitoring cholesterologenesis by measuring the concentration of isoprene in exhaled breath (Karl et al., 2001). Overall, this technique has been extensively utilized to identify a wide range of VOCs, thereby facilitating the diagnosis and monitoring of various disease stages.

2.3.2.3 *Ion mobility spectrometry*

Ion mobility spectrometry (IMS) is a powerful and rapid separation technique originally developed to identify controlled substances, including illegal drugs and explosives. In recent years, the application of IMS has expanded to include a range of modern and environmental friendly purposes, including food quality assessments and air quality monitoring. Technically, IMS employs an external electric field within a surrounding medium to separate ionized particles derived from the target analyte based on their mobility in the gas phase (Pereira et al., 2015). IMS can be implemented in various formats, including high-field asymmetric waveform ion mobility spectrometers (FAIMS), differential mobility spectrometers (DMS), custom-built laboratory IMS systems, and commercially available IMS (Mule et al., 2021). IMS technique is capable of detecting VOCs in exhaled breath within the range of ppb to ppt, making it a suitable tool for non-invasive medical diagnostics and trace-level analysis.

2.3.2.4 *Secondary electrospray ionization mass spectrometry*

Secondary electrospray ionization mass spectrometry (SEI-MS) is an ambient ionization method used to detect VOCs without the need for preprocessing or sample preparation. Ionization in SEI-MS occurs through proton transfer reactions between VOC analytes and the electrospray solution. Unlike conventional electrospray ionization, the proton transfer in SESI takes place in the vapor phase rather than in solution, making SEI-MS a promising tool for the detection of VOCs in exhaled breath. Additionally, the electrospray ionization process utilizes electrical energy to facilitate this transition. Researchers in 1986 and 1994 identified two key aspects associated with the electrospray ionization mechanism: first, it ionizes molecules mixed within the spray solvent, and second, it possesses the ability to interact with molecules in the surrounding ionized environment. Later, in 1998, Wu et al. demonstrated the potential of SEI-MS as a high-resolution mass spectrometry technique for VOC detections (Wu et al., 1998). Prior to MS analysis, molecular fractionation is considered essential to improve selectivity and resolution. To achieve this, high-performance liquid chromatography (HPLC) is employed along with the SEI-MS. The HPLC-SEI-MS platform is recognized as a highly capable analytical tool for detecting both small and large molecules with varying polarities within complex biological samples.

2.3.2.5 *Electronic nose*

The electric nose (E-nose) is a diagnostic system developed using a range of sensor technologies, particularly non-selective gas sensors

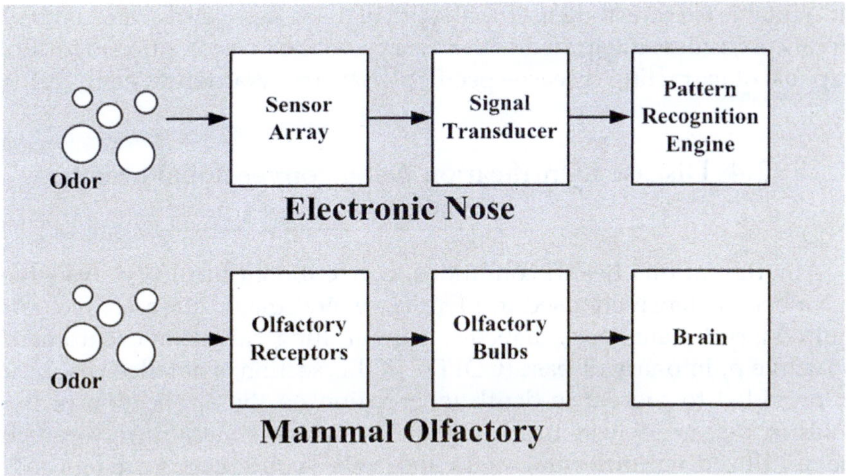

FIGURE 2.6 Electronic nose. The basic gas identification system blocks for an electronic nose. *From Chiu, S.-W., & Tang, K.-T. (2013). Towards a chemiresistive sensor-integrated electronic nose: A review.* Sensors (Switzerland), *13(10), 14214–14247. https://doi.org/10.3390/s131014214.*

combined with pattern recognition algorithms. A basic block diagram of the E-nose system is shown in Fig. 2.6. Additionally, other analytical methods such as IMS, GC, MS, and optical sensing have been integrated into E-nose configurations. The development of E-noses is primarily focused on recognizing a specific type of mixture or compound, such as isoprene, NO, CO, and alkanes. The recognition capability of E-noses is generally more practical and straightforward compared to conventional instruments such as MS-based systems (Pereira et al., 2015). In various applications, multiple sensors and sensing materials have been integrated into a single platform to enable the detection of different classes of VOCs (Lourenço & Turner, 2014). These advanced E-noses are designed for target-specific detection by observing changes in the electrical properties of the sensing material in response to specific VOCs. Commercially available E-noses generally utilize conducting polymers or metal oxide semiconductors as sensing materials (Nidheesh et al., 2021). MS-based VOC detection techniques are associated with several limitations, including bulky instrument size, high cost, time-consuming operation, and the need for trained personnel. These limitations prevent their widespread clinical implementation. In contrast, E-noses offer several advantages, including cost-effectiveness, compact design, portability, and plug-and-play operation. However, developing and optimizing a reliable E-nose system requires a clearly defined dataset for training and validation to enable accurate disease diagnosis. E-noses typically exhibit defined detection limits and selectivity. Some devices are assembled with selective sensors to detect specific target VOCs, but they are not suitable for screening multiple types of diseases, such as various classes of cancers, unless integrated with advanced pattern recognition algorithms capable of extracting disease-specific signatures (Wardencki et al., 2013).

2.4 Disease identification using conventional breath analyzers

Among various health conditions, conventional breath analysis tools have been extensively used to identify several major diseases, including pulmonary tuberculosis, diabetes, asthma, lung cancer, and chronic obstructive pulmonary disease (COPD). In this section, a detailed discussion is provided to present in-depth information on the application of these tools in disease detection. In previously reported literature, researchers have utilized various parameters and VOC biomarkers to diagnose the associated diseases. As discussed in Chapter 1, the VOC biomarker shows a strong association with particular diseases. Every disease produces a unique odor profile comprised of multiple VOCs, which can be used as diagnostic biomarkers. For instance, asthma patients typically exhale elevated levels of nitric oxide, whereas diabetic patients exhibit increased concentrations of acetone in their breath.

2.4.1 Diabetes monitoring

Diabetes is a serious health condition characterized by impaired regulation of blood glucose generated during food digestion. In this condition, the body produces various types of chemical compounds originating from the digestion of food, altered metabolism of carbohydrates, proteins, and fats, as well as oxidative stress. Diabetes Mellitus is directly associated with an increased concentration of acetone in exhaled breath. Moreover, the concentration of acetone also varies depending on dietary factors. Acetone is one of the key VOCs present in diabetic patients, with an increased concentration level compared to healthy individuals, which serves as a significant parameter for diagnosing diabetes (Lourenço & Turner, 2014). Researchers have undertaken several investigations and efforts to develop methods that utilize various VOCs for diabetes monitoring. This is especially critical given the need for real-time monitoring and diagnosis of diabetes mellitus for daily healthcare purposes (Kim et al., 2015). Analyzing acetone levels in exhaled breath using GC-MS requires an experienced technician to handle sample preparation, which typically involves preparation of acetone standards and derivatization on the solid-phase microextraction (SPME) needle. SPME derivatization is predominantly utilized due to the small mass and highly volatile nature of acetone (Deng et al., 2004). In 2009, Ueta et al. reported a novel technique utilizing a particle-packed sample preparation needle for detecting human breath acetone, as shown in (Ueta et al., 2009).

The technique involves collecting exhaled breath in a sampling bag, extracting a 50 mL breath sample with the needle-type device, and analyzing it using GC-MS. Results from controlled type 2 diabetic patients indicated a correlation between breath acetone concentration and HbA1c levels, with levels ranging from 0.19 to 0.66 ppmv, while medically untreated patients exhibited higher concentrations. Additionally, fasting tests on healthy individuals showed increased breath acetone levels, suggesting potential diagnostic applications for diabetes and ketoacidosis. It has been reported that sample preparation is a time-consuming process that takes nearly 40 minutes to complete. The preparation of the acetone standard takes 20 minutes, while its derivatization on the SPME fiber takes an additional 20 minutes (Ueta et al., 2009). Along with the sample preparation, another drawback of the technique is associated with the breath sample collection, which requires Tedlar bags. Tedlar bags are highly expensive and are not capable of storing the sample for more than 6 hours. In 2010, Wang et al. developed another prototype to monitor acetone concentration levels in diabetic patients (Wang et al., 2010). The study investigated the relation between the concentration of acetone in exhaled breath and commonly used diabetes monitoring parameters. Using a cavity ring-down spectroscopy (CRDS)-based breath analyzer, the researchers detected acetone in 34 Type 1 diabetes (T1D) patients, 10 Type 2 diabetes (T2D) patients, and 15 healthy individuals, revealing correlations with blood glucose and glycohemoglobin A1C levels when grouped. The mean acetone concentration in T1D subjects was found to be 2.19 ppmv, which is higher compared to healthy individuals (0.48 ppmv). In 2012, Righettoni et al. reported a Si-doped WO_3 nanoparticle-based portable acetone sensor for diabetes analysis (Righettoni et al., 2012). The sensor design included miniaturization of the chamber volume, identification of reaction-limited and transport-limited gas flow rates, and optimization of sensing temperatures, achieving a low acetone detection limit (~20 ppb) with quick response (10–15 seconds) and recovery times (35–70 seconds). The utilization of multiple exhaled VOC biomarkers to determine blood glucose levels has been widely investigated and reported by Galassetti et al.(Galassetti et al., 2005). In the reported work, the authors not only identified acetone as a key biomarker for diabetes but also explored other VOCs present in the exhaled breath, supporting the identification of diabetic patients. In another study reported by Righettoni et al. (Righettoni et al., 2013), a portable acetone sensor was developed and employed alongside PTR-MS to analyze breath samples from eight healthy volunteers before and after meals. The sensor demonstrated fast response times of less than 12 seconds, along with high accuracy in detecting acetone, correlating strongly with blood glucose levels, especially after overnight fasting (i.e., in the morning). Statistical analysis depicted strong correlations among PTR-TOF-MS measurements of exhaled VOCs (such as isoprene, methanol, acetone, and ethanol), sensor performance, and blood glucose level across both sampling periods.

Table 2.4 summarizes the identification techniques of diabetes based on exhaled breath, including statistical techniques, algorithms, and biomarkers.

2.4.2 Asthma

Asthma is considered a chronic inflammatory disease leading to reversible airway obstruction and is marked by episodes of bronchospasms. Bronchospasms refer to an unexpected and rapid tightening of the muscles present in the walls of the airways, bronchioles, or lungs. Short episodes of wheezing, chest tightness, shortness of breath, and coughing are considered common symptoms of asthma (Kharitonov, 2002). Some patients experience these symptoms and discomfort more frequently at night or after rigorous physical activities, including exercise and climbing stairs (Lärstad et al., 2005). Although the exact causes of asthma are unknown, it is often believed to be triggered by allergens, environmental pollution, irritants, or specific medications (including aspirin and beta-blockers) (Ratiu et al., 2021). Asthma cannot be cured, but it can be controlled by avoiding irritants and allergenic agents by utilizing inhaled corticosteroids. Traditional asthma diagnostic tests are fundamentally limited due to their time-consuming and invasive nature (Stefanuto et al., 2020). In order to identify hundreds of VOCs from various chemical families, Caldeira et al. designed and reported an efficient approach using multidimensional gas chromatography (Caldeira et al., 2012). In the reported work, the authors utilized partial least squares-discriminant analysis (PLS-DA) in tandem with Monte Carlo cross-validation, using data obtained from exhaled breath samples analyzed by GC × GC–(TOF)–MS, to determine the predictive power and aid in the interpretation of the recovered VOCs possibly related to inflammation processes, oxidative stress, or other cellular mechanisms characteristic of asthma. Silvano Dragonieri et al. reported that the E-Nose can differentiate asthma patients from healthy individuals by examining exhaled air, irrespective of disease severity (Dragonieri et al., 2007). Although the reported E-nose can identify the presence of asthma-related VOCs in exhaled breath, its accuracy in determining the severity of asthma remains limited. Vliet et al. choose to combine the characteristics of predictive biomarkers with a range of exhaled VOCs to improve the accuracy of asthma diagnosis (Van Vliet et al., 2017). Montuschi et al. utilized E-Nose-based analysis, achieving a diagnostic accuracy of 87%. In comparison, FeNO and lung function testing achieved accuracies of 70.2% and 70.8%, respectively (Montuschi et al., 2010).

2.4.3 Lung cancer detection

Lung cancers fall into the category of malignant tumors, which can be developed due to the uncontrolled growth of cells within lung tissues.

TABLE 2.4 Summary of analytical tools used in diabetes diagnosis.

Method	Algorithm/statistical approach	Detection range	Database	References
Cavity ring down spectroscopy (CRDS)	Linear correlation	34-Patients, 15-Healthy	2.19 ppmv	Wang et al. (2010)
PTR-MS along with Si-doped epsilon-WO3 nanostructured films	Linear regression	~ 20 ppb	5 samples	Righettoni et al. (2012)
PTR-TOF-MS, Chemo-resistive sensor	Pearson's correlations, Spearman's rank correlations	1600–1400 ppb	8 samples	Righettoni et al. (2013)
GC-MS	Linear regression	Ethanol-9.6 ± 3.1 ppbv Acetone-392 ± 85 ppbv	10 samples	Galassetti et al. (2005)
GC-MS, SPME	Linear regression	0.049 ppbv	15-Patients, 15-Healthy	Deng et al. (2004)
GC	Linear regression, and Pearson's correlations	0.641–0.962	8-Patients, 17-Healthy	Minh et al. (2011)
HT-GC/MS	Cyclic S-matrix	1–10 ppmv, 0.1–1 ppmv	4-Patients, 8-Healthy	Fan et al. (2014)
MWCNTs (Multiwall carbon nanotubes)/SnO_2 Sensor	Brunauer-Emmett-Teller (BET)	0.5–5 ppm	–	Salehi et al. (2014)

(continued on next page)

TABLE 2.4 Summary of analytical tools used in diabetes diagnosis—cont'd

Method	Algorithm/statistical approach	Detection range	Database	References
C-doped WO_3 nanomaterial's	FFT (Fast Fourier Transform), Brunauer-Emmett-Teller (BET) surface area, X-ray Photoelectron Spectroscopy (XPS)	0.2–5 ppm	–	Xiao et al. (2014)
SIFT-MS	Linear correlation	8 Patients	1–21 ppm	Turner et al. (2009)
Chemical Sensor, 12-metal-oxide semiconductor gas sensor	SVM, PCA	90-Patients, 108-Healthy	–	Guo et al. (2010)
Chemical Sensor	PCA, SVM (support vector machine)	279-Patients, 295-Healthy	0.1–20 ppm	Yan et al. (2014)

From Mule, N. M., Patil, D. D., & Kaur, M. (2021). A comprehensive survey on investigation techniques of exhaled breath (EB) for diagnosis of diseases in human body. Informatics in Medicine Unlocked, *26, 100715. https://doi.org/10.1016/j.imu.2021.100715.*

Their primary symptoms include frequent coughing, chest pain or tightness, shortness of breath, and weight loss. Approximately 85% of lung cancer cases arise from direct or indirect (secondhand) exposure to tobacco smoking. Exposure to secondhand smoke, also known as passive smoking or environmental tobacco smoke, occurs when any individual inhales the smoke released from burning tobacco products or from the exhalation of a nearby smoker. The remaining 15% of lung cancers are attributed to other exposures, such as radiation, asbestos, radon, and various air pollutants (Ratiu et al., 2021; Thun et al., 2008). The main types of primary lung cancers are classified as lung carcinomas (LC), which can be divided into two primary groups based on histological characteristics: small-cell lung carcinoma (SCLC) and non-small-cell lung carcinoma (NSCLC). SCLC is composed of dense cells containing neurosecretory granules, which are vesicles filled with neuroendocrine hormones, a characteristic that explains why these tumors are often associated with endocrine or paraneoplastic syndromes (Ratiu et al., 2021). SCLC accounts for approximately 15% of all lung cancers worldwide, while the remaining 85% are associated with NSCLC. The most common subtypes of NSCLC include squamous cell carcinoma, non-squamous cell carcinoma, large cell carcinoma, and several other less frequent forms (Ratiu et al., 2021). Among these, adenocarcinoma stands out as the most frequent subtype, typically located in the peripheral regions of the lung (Kenfield et al., 2008). Molecular testing enables the detection of potential mutations in adenocarcinomas. The most commonly occurring mutations are outlined in Fig. 2.7.

CT or chest radiographs are most often used for lung cancer detection. However, confirmation requires an additional biopsy, which is a time-consuming, expensive, and invasive procedure. Nowadays, several patients suffering from lung cancer prefer a non-invasive approach to the detection of VOC biomarkers from the exhaled breath. The non-invasive detection of VOCs associated with lung cancer has gained significant attention due to its painless detection, cost-effectiveness, and high detection accuracy. In 2009, Peng et al. reported a sensor array based on gold nanoparticles to differentiate lung cancer patients from healthy individuals (Peng et al., 2009). A colorimetric device sensor array for lung cancer has been reported by Mazzone et al. (Mazzone et al., 2007). In this approach, the VOCs related to lung cancers are statistically analyzed using the random forest method. The Random Forest technique is a machine learning algorithm used to classify and analyze data obtained from GC-MS or similar tools. Generally, this approach is based on an ensemble learning technique that combines multiple decision trees to produce an accurate prediction model. In 2017, Sakumura et al. reported an innovative approach to identify lung cancer from the exhaled breath using GC-MS. In the reported work, the authors identified nearly 68 potential VOC biomarkers (Sakumura et al., 2017). Using a support vector

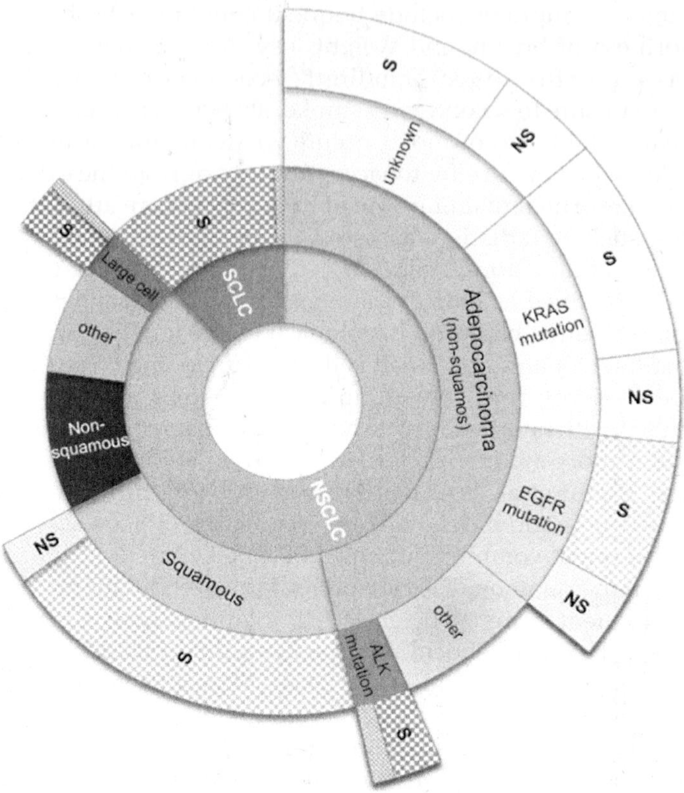

FIGURE 2.7 Classification of lung cancer subtypes. Multistep classification of lung cancer subtypes; common mutations that occur in adenocarcinoma tumors and the occurrence ratio caused by smoking tobacco for each subtype. *From Ratiu, I. A., Ligor, T., Bocos-Bintintan, V., Mayhew, C. A., & Buszewski, B. (2021). Volatile organic compounds in exhaled breath as fingerprints of lung cancer, asthma and COPD.* Journal of Clinical Medicine, *10(1), 1–41. https://doi.org/10.3390/jcm10010032.*

machine (SVM) algorithm, the authors pinpointed five crucial VOCs, i.e., hydrogen cyanide, methanol, isoprene, acetonitrile, and 1-propanol, observing around 89% screening accuracy, which could potentially be used to design and develop a desktop GC-sensor. Fig. 2.8 illustrates the actual GC-MS system used for breath sampling and diagnosis. The study involved 107 patients suffering from lung cancer, along with 29 healthy individuals. At Aichi Cancer Center, human breath and room air were collected using an analytic barrier bag (Omi Odor-Air Service Corp., Omihachiman, Japan). To ensure reliable sampling conditions, volunteers were instructed to refrain from eating or smoking for several hours prior to breath collection and were required to spend at least 10 minutes in the room

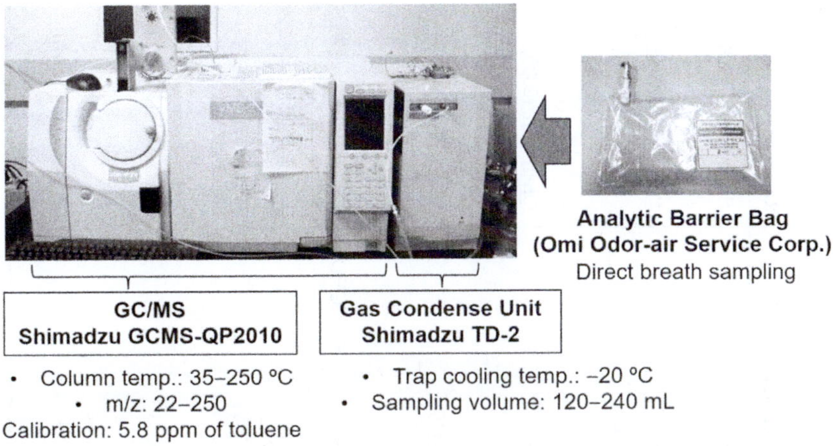

FIGURE 2.8 Breath sampling and gas analysis by GC/MS. Actual device utilized to study and analyze the collected breath sample via GCMS-QP2010, consisting of a TD-2 gas-condensing unit. *From Sakumura, Y., Koyama, Y., Tokutake, H., Hida, T., Sato, K., Itoh, T., Akamatsu, T., & Shin, W. (2017). Diagnosis by volatile organic compounds in exhaled breath from lung cancer patients using support vector machine algorithm.* Sensors, *17(2), 287. https://doi.org/10.3390/s17020287.*

beforehand. Alveolar breath samples were collected from all volunteers in 1 L analytical barrier bags following the exhalation of respiratory tract air in a consultation room. The exhaled breath was then analyzed using a GC-MS-QP2010 instrument (Shimadzu, Kyoto, Japan) equipped with a TD-2 gas-condensing unit (Shimadzu) (Sakumura et al., 2017).

PTR-MS is also recognized as a valuable tool for VOC detection and has been widely utilized in identifying lung cancer from exhaled breath (Wehinger et al., 2007). The use of an E-nose comprising eight quartz microbalance gas sensors functionalized with various metalloporphyrins was proposed by Natale et al. (Di Natale et al., 2003). These reported sensors exhibit extremely high sensitivity towards the VOCs associated with lung cancer patients, with an accuracy of 99.9%. SPME-GC/MS and PTR-MS have been employed as non-invasive tools for the analysis of inhaled and exhaled breath samples (Bajtarevic et al., 2009). A summary of the identification of lung cancer using various analytical tools, algorithms, and biomarkers is given in Table 2.5.

2.4.4 Detection of pulmonary tuberculosis

Mycobacterium tuberculosis is the causative agent of contagious lung infections. Ziehl-Neelsen staining is a bacteriological technique used to

TABLE 2.5 Summary of various detection tools used to identify lung cancer.

Method	Sensor	Database	Algorithms/statistical techniques	Accuracy	References
GC-MS, SPME	Nanomaterial based sensor array	24 LC patients	DFA as statistical pattern recognition	80%	Broza et al. (2013)
GC-MS, SPME	–	36-LC, 25-COPD	PCA, Kruskal –Wallis, ANOVA tests, Multinomial logistic regression analysis	82.5%	Poli et al. (2005)
E-nose, SPME	SAW gas sensor	20-LC, 15-Healthy, 7-chronic bronchitis	ANN used a back-propagation algorithm	–	Chen et al. (2005)
E-nose	Colorimetric Sensor	92-LC, 137-Healthy	Logistics Regression, Pearson's Chi-square	81%	Mazzone et al. (2012)
E-nose	Nano-composite array with 32 Polymer Sensors	10-LC, 10-COPD, 10-Healthy	PCA,CDA	80%-LC, 85%-COPD	Dragonieri et al. (2009)
GC-MS, SPME	Gold Nanoparticles	40-LC, 56-Healthy	PCA	83%	Peng et al. (2009)
E-nose	10-MOSFET, 12-MOS	24-LC	ANN	–	Schmekel et al. (2014)
GC-MS	Colorimetric sensor Array, carbon Polymer Sensor	49-LC, 21-Healthy, 20Sarcoidosis, 18-COPD	Random Forest	82%	Mazzone et al. (2007)
GC-MS, SPME	–	31-LC, 31-Smokers, 31-Healthy	PCA	69%	Kischkel et al. (2010)

(continued on next page)

TABLE 2.5 Summary of various detection tools used to identify lung cancer—cont'd

Method	Sensor	Database	Algorithms/statistical techniques	Accuracy	References
PTR-MS, GC-MS	–	17-LC, 170-Healthy	Wilcoxon Rank Sum Tests	0.96 ± 0.02	Wehinger et al. (2007)
GC-MS, SPME-OFD	–	12-LC, 12-Smokers, 12-Healthy	Linear Regression, Kruskal –Wallis,	–	Fuchs et al. (2010)
E-nose	QMB	35-LC, 25-Healthy	PLS-DA	–	Di Natale et al. (2003)
GC-MS	Gold Nanoparticle	24- LC	SVM	90%	Barash et al. (2012)
FT-ICR-MS/MS	Silicon Microreactors	129-LC, 88-Healthy	Wilcoxon Tests	–	Fu et al. (2014)
E-nose	32 Polymer Composite Sensors	14-LC, 62-Healthy	SVM, Principal components and canonic discriminant analysis	85%	Machado et al. (2005)
GC-MS, SPME	Chemical Nanoarray, PAH/SWCNT	72-LC	Wilcoxon/Kruskal Wallis	88 ± 3%	Peled et al. (2012)
PTR-MS, SPME-GCMS	–	For PTR-MS 220-LC, 441-Healthy For GCMS 65-LC, 31-Healthy	Kruskal Wallis	–	Bajtarevic et al. (2009)
GC-MS	–	107-LC, 29-Healthy	SVM	89%	Sakumura et al. (2017)
GC-MS	–	60-LC, 48-Healthy	Discriminant Analysis	69.4%	Phillips et al. (1999)

From Mule, N. M., Patil, D. D., & Kaur, M. (2021). A comprehensive survey on investigation techniques of exhaled breath (EB) for diagnosis of diseases in human body. Informatics in Medicine Unlocked, 26, 100715. *https://doi.org/10.1016/j.imu.2021.100715.*

detect tuberculosis infections (McNerney et al., 2010). It can only identify advanced pulmonary infections. Typically, Mycobacterium tuberculosis grows slowly and remains undetectable until symptoms emerge (Lourenço & Turner, 2014). Several innovations have been developed and commercialized by researchers and physicians to detect pulmonary TB (Phillips et al., 2010). These attempts consider multiple factors, including biomarkers, a VOC-monitoring device, and sputum culture, which play crucial roles in the identification of TB (Chen et al., 2020; Houben & Dodd, 2016). In 2010, Phillips et al. reported a detailed study on the potential biomarkers that can be utilized to identify active pulmonary tuberculosis among symptomatic high-risk individuals (Phillips et al., 2010). The study was performed on 226 patients. Breath analysis detected specific markers for active pulmonary tuberculosis, including substances indicating oxidative stress and VOCs generated by the tuberculosis bacterium. These biomarkers enabled the identification of active tuberculosis with an overall accuracy of 85%, sensitivity of 84.0%, and specificity of 64.7%, particularly when traditional diagnostic methods yield inconsistent or even false-positive or false-negative results. In comparison, individual diagnostic methods had lower accuracy scores, with smear microscopy at 0.76, sputum culture at 0.68, and chest radiography at 0.66. Another type of study was performed by the same research group to examine VOCs in cases of dynamic pulmonary tuberculosis, using a web-connected and rapid point-of-care breath analyzer (Phillips et al., 2012). Recently, Sahota et al. analyzed the exhaled breath of 21 patients diagnosed with pulmonary or extra-pulmonary tuberculosis, recruited from a UK teaching hospital either prior to or within 1 week of initiating anti-tuberculous treatment (Sahota et al., 2016). The diagnosis of tuberculosis was confirmed via reference tests such as mycobacterial culture, histology, or radiology. To calculate specificity, 19 controls were enrolled, all of whom tested negative on interferon-gamma release assays. Whole-breath samples were collected and analyzed using IMS. The findings demonstrated a high sensitivity of 81% and specificity of 79% for all TB cases, including both pulmonary and extra-pulmonary forms. Another attempt focused on the use of specific VOC biomarkers as diagnostic indicators for active pulmonary tuberculosis (Phillips et al., 2007). In this approach, VOCs were captured using sorbent traps and analyzed by GC-MS to diagnose Mycobacterium TB (Phillips et al., 2007). Breath VOCs were investigated using GC-MS in 42 individuals admitted to the hospital due to suspected pulmonary tuberculosis, alongside 59 healthy individuals serving as reference controls. Among the hospitalized patients, 23 tested positive for Mycobacteria in sputum cultures, while 19 tested negative. Elevated levels of breath markers linked to oxidative stress were detected in all hospitalized individuals. Utilizing pattern recognition and fuzzy logic analyses, the breath VOCs effectively differentiated healthy controls from hospitalized patients with 100% sensitivity and 100% specificity.

2.4.5 Detection of chronic obstructive pulmonary disease

Chronic obstructive pulmonary disease (COPD) is a chronic inflammatory lung disease characterized by persistent airflow obstruction in the lungs. The primary causes of COPD are long-term tobacco use and exposure to cigarette smoke (Kharitonov, 2005). While COPD poses a significant health risk, it is often underestimated in terms of timely and accurate diagnosis. In medical practice, diagnosing COPD typically involves three main elements: primary presenting symptoms, results from lung assessment, and analyzing patients responses to inhaled medications (Gaida et al., 2016). Several research studies have explored various aspects of COPD, including VOCs, diagnostic tools, and the intricate relationships between various compounds and disease stages. While conventional testing techniques for COPD provide valuable information, they are often time-consuming and heavily reliant on the experience of clinical personnel. The aim of the research is to analyze, monitor, and identify key aspects that are deemed relevant in the current context. Allers et al. have reported a closed-loop GC-ions mobility spectrometer (GC-IMS) system for the identification of COPD (Allers et al., 2016). In this study, the authors investigated the performance of this GC-IMS under clinical conditions to determine the relation between exhaled VOCs ranging from 10 ppbv to 1 ppmv, and COPD. The GC-IMS system was utilized to analyze samples from 58 individuals, including 21 smokers with moderate COPD, 12 ex-smokers with COPD, 16 healthy smokers, and 9 non-smokers. Separately, GC-APCI-MS data were obtained from 94 participants, comprising 21 smokers with moderate COPD, 25 ex-smokers with COPD, 25 healthy smokers, and 23 non-smokers. A comparison between GC-IMS and gas chromatography–atmospheric pressure chemical ionization–mass spectrometry (GC-APCI-MS) data could be conducted for 44 subjects. In the reported work, only 2-pentanone was identified as a distinctive VOC based on GC-APCI-MS data. Later, Aguilar et al. reported that the ultrafast GC coupled with an E-nose detector proved to be another potential tool for detecting VOCs as biomarkers in exhaled breath (Rodríguez-Aguilar et al., 2019). A systematic method was reported for the comprehensive investigation of concentrated exhaled breath samples from patients experiencing hindered respiratory capacity. In 2010, Basanta et al. reported a two-stage GC-DMS analysis to differentiate among healthy individuals and COPD patients (Basanta et al., 2010). The published work exhibit the identification of COPD using available conventional tools, algorithms, and databases, along with the sensitivity and accuracy of each approach.

2.5 Challenges

The examination of the exhaled breath is a non-invasive technique used to identify and diagnose various health conditions and diseases. This strategy generally involves several activities, including breath collection and sampling, selection of an appropriate analytical technique, identifying appropriate VOCs, analysis of the collected data, and establishing the correlation between the investigated exhaled breath and reported disease. Various kinds of containers can be used for breath sampling and collection, such as Mylar bags and Tedlar bags. Preserving the collected breath samples without any external contamination and ensuring their proper storage pose significant challenges. Extended storage of collected breath can lead to degradation of its original composition.

This chapter discusses various non-invasive methods, including GC-MS, PTR-MS, SIFT-MS, and E-nose, for quantitative analysis of exhaled breath. These non-invasive techniques offer alternatives for patients unwilling to undergo other invasive diagnostic procedures. Despite their ease of use, the popularity of these tools remains limited due to equipment costs, transportation challenges, and the need for skilled technicians to collect breath samples and analyze data. The modern automated and computerized devices utilized in exhaled breath analysis are equipped with various types of sensor arrays these days. These sensors serve to selectively recognize various diseases, such as diabetes, TB, and lung cancer. Other techniques outlined in this chapter for exhaled breath analysis are based on data samples. Ensuring the quality, storage, and analysis of these samples in a controlled environment is crucial for obtaining accurate results.

Overall, the conventional analytical tools used for disease diagnosis present several challenges, preventing their widespread adoption and applicability, particularly in clinical applications. One major challenge is achieving sufficient selectivity due to the complexity of VOC profiles, which can vary greatly depending on several factors such as individual diet, country, race, method of breath collection, sample storage, sample type, environmental conditions, and individual physiology. Analyzing these profiles accurately requires advanced techniques and methodologies, as discussed above. Additionally, the sensitivity and selectivity of conventional analytical tools may not always meet the requirements for detecting low concentrations of VOC biomarkers in collected exhaled breath samples. This limitation can compromise the reliability of results, particularly in clinical diagnostics where precise and accurate measurements are essential. Two other major challenges in VOC biomarker analysis are interference from external sources, such as contaminants or any foreign elements present in internal compounds, which can significantly distort results. Finally, the cost and accessibility of such analytical tools and expertise represent a real-world challenge for implementing VOC biomarker analysis in routine clinical practice or large-scale research studies.

2.6 Conclusion

Analytical breath detection is a well-established and widely adopted approach for accurately identifying various health conditions through the selective detection of disease-related VOC biomarkers. Employing this technique, especially when integrated with computerized algorithms, makes this instrument more user-friendly and suitable for non-invasive detection. This chapter provides a detailed discussion of various conventional tools used in exhaled breath monitoring, along with their advantages and limitations in clinical and research practices. It covers all aspects related to exhaled breath sampling, collection, and analysis using both offline and online approaches. A clear picture is drawn of the online and offline detection tools and their respective roles in diagnosing various health conditions. This chapter aims to guide readers in selecting the most appropriate diagnostic tools while also helping them understand the inherent limitations and disadvantages of other available tools.

References

Abbott, S. M., Elder, J. B., Španěl, P., & Smith, D. (2003). Quantification of acetonitrile in exhaled breath and urinary headspace using selected ion flow tube mass spectrometry. *International Journal of Mass Spectrometry, 228*(2–3), 655–665. http://www.journals.elsevier.com/international-journal-of-mass-spectrometry/. doi:10.1016/S1387-3806(03)00212-4.

Adams, N. G., & Smith, D. (1976). The selected ion flow tube (SIFT); A technique for studying ion-neutral reactions. *International Journal of Mass Spectrometry and Ion Physics, 21*(3–4), 349–359. doi:10.1016/0020-7381(76)80133-7.

Allers, M., Langejuergen, J., Gaida, A., Holz, O., Schuchardt, S., Hohlfeld, J. M., & Zimmermann, S. (2016). Measurement of exhaled volatile organic compounds from patients with chronic obstructive pulmonary disease (COPD) using closed gas loop GC-IMS and GC-APCI-MS. *Journal of Breath Research, 10*(2), 026004. doi:10.1088/1752-7155/10/2/026004.

Amal, H., Leja, M., Funka, K., Lasina, I., Skapars, R., Sivins, A., Ancans, G., Kikuste, I., Vanags, A., Tolmanis, I., Kirsners, A., Kupcinskas, L., & Haick, H. (2016). Breath testing as potential colorectal cancer screening tool. *International Journal of Cancer, 138*(1), 229–236. http://onlinelibrary.wiley.com/journal/10.1002/. doi:10.1002/ijc.29701. (ISSN)1097-0215.

Amann, A., Poupart, G., Telser, S., Ledochowski, M., Schmid, A., & Mechtcheriakov, S. (2004). Austria applications of breath gas analysis in medicine. *International Journal of Mass Spectrometry, 239*(2–3), 227–233. doi:10.1016/j.ijms.2004.08.010.

Anderson, J. C., & Hlastala, M. P. (2007). Breath tests and airway gas exchange. *Pulmonary Pharmacology and Therapeutics, 20*(2), 112–117. doi:10.1016/j.pupt.2005.12.002.

Bajtarevic, A., Ager, C., Pienz, M., Klieber, M., Schwarz, K., Ligor, M., Ligor, T., Filipiak, W., Denz, H., Fiegl, M., Hilbe, W., Weiss, W., Lukas, P., Jamnig, H., Hackl, M., Haidenberger, A., Buszewski, B., Miekisch, W., Schubert, J., & Amann, A. (2009). Noninvasive detection of lung cancer by analysis of exhaled breath. *BMC Cancer, 9*, 348. http://www.biomedcentral.com/bmccancer/. doi:10.1186/1471-2407-9-348.

Barash, O., Peled, N., Tisch, U., Bunn, P. A., Hirsch, F. R., & Haick, H. (2012). Classification of lung cancer histology by gold nanoparticle sensors. *Nanomedicine: Nanotechnology, Biology, and Medicine, 8*(5), 580–589. doi:10.1016/j.nano.2011.10.001.

Barash, O., Zhang, W., Halpern, J. M., Hua, Q. L., Pan, Y. Y., Kayal, H., Khoury, K., Liu, H., Davies, M. P. A., & Haick, H. (2015). Differentiation between genetic mutations of breast cancer by breath volatolomics. *Oncotarget, 6*(42), 44864–44876. http://www.impactjournals.com/oncotarget/index.php?journal=oncotarget&page=article&op=download&path%5B%5D=6269&path%5B%5D=16270. doi:10.18632/oncotarget.6269.

Barker, M., Hengst, M., Schmid, J., Buers, H.-J., Mittermaier, B., Klemp, D., & Koppmann, R. (2006). Volatile organic compounds in the exhaled breath of young patients with cystic fibrosis. *European Respiratory Journal, 27*(5), 929–936. doi:10.1183/09031936.06.00085105.

Basanta, M., Jarvis, R. M., Xu, Y., Blackburn, G., Tal-Singer, R., Woodcock, A., Singh, D., Goodacre, R., Paul Thomas, C. L., & Fowler, S. J. (2010). Non-invasive metabolomic analysis of breath using differential mobility spectrometry in patients with chronic obstructive pulmonary disease and healthy smokers. *Analyst, 135*(2), 315–320. http://pubs.rsc.org/en/journals/journal/an. doi:10.1039/b916374c.

Berna, A. Z., McCarthy, J. S., Wang, R. X., Saliba, K. J., Bravo, F. G., Cassells, J., Padovan, B., & Trowell, S. C. (2015). Analysis of breath specimens for biomarkers of plasmodium falciparum infection. *Journal of Infectious Diseases, 212*(7), 1120–1128. http://jid.oxfordjournals.org/content/current. doi:10.1093/infdis/jiv176.

Bigazzi, A. Y., Figliozzi, M. A., Luo, W., & Pankow, J. F. (2016). Breath biomarkers to measure uptake of volatile organic compounds by bicyclists. *Environmental Science and Technology, 50*(10), 5357–5363. http://pubs.acs.org/journal/esthag. doi:10.1021/acs.est.6b01159.

Broza, Y. Y., Kremer, R., Tisch, U., Gevorkyan, A., Shiban, A., Best, L. A., & Haick, H. (2013). A nanomaterial-based breath test for short-term follow-up after lung tumor resection. *Nanomedicine: Nanotechnology, Biology, and Medicine, 9*(1), 15–21. doi:10.1016/j.nano.2012.07.009.

Calamai, L., Villanelli, F., Bartolucci, G., Pieraccini, G., & Moneti, G. (2012). Sample preparation for direct ms analysis of food. *Comprehensive Sampling and Sample Preparation, 4*, 535–557. http://www.sciencedirect.com/science/referenceworks/9780123813749. doi:10.1016/B978-0-12-381373-2.00148-4.

Caldeira, M., Perestrelo, R., Barros, A. S., Bilelo, M. J., Morête, A., Câmara, J. S., & Rocha, S. M. (2012). Allergic asthma exhaled breath metabolome: A challenge for comprehensive two-dimensional gas chromatography. *Journal of Chromatography A, 1254*, 87–97. doi:10.1016/j.chroma.2012.07.023.

Capuano, R., Santonico, M., Pennazza, G., Ghezzi, S., Martinelli, E., Roscioni, C., Lucantoni, G., Galluccio, G., Paolesse, R., Di Natale, C., & D'Amico, A. (2015). The lung cancer breath signature: A comparative analysis of exhaled breath and air sampled from inside the lungs. *Scientific Reports, 5*, 16491. www.nature.com/srep/index.html. doi:10.1038/srep16491.

Castellanos, M., Xifra, G., Fernández-Real, J. M., & Sánchez, J. M. (2016). Breath gas concentrations mirror exposure to sevoflurane and isopropyl alcohol in hospital environments in non-occupational conditions. *Journal of Breath Research, 10*(1), 016001. http://iopscience.iop.org/article/10.1088/1752-7155/10/1/016001/pdf. doi:10.1088/1752-7155/10/1/016001.

Castro, J. M. A., & Russo, M. (2019). Use and limitations of noninvasive and invasive methods for studying pulmonary function. *Drug Discovery Today: Disease Models, 29–30*, 3–9. doi:10.1016/j.ddmod.2019.07.001.

Chen, D., Bryden, W. A., & Wood, R. (2020). Detection of tuberculosis by the analysis of exhaled breath particles with high-resolution mass spectrometry. *Scientific Reports, 10*(1), 64637. www.nature.com/srep/index.html. doi:10.1038/s41598-020-64637-6.

Chen, X., Cao, M., Li, Y., Hu, W., Wang, P., Ying, K., & Pan, H. (2005). A study of an electronic nose for detection of lung cancer based on a virtual SAW gas sensors array and imaging recognition method. *Measurement Science and Technology, 16*(8), 1535–1546. http://www.iop.org/EJ/journal/0957-0233. doi:10.1088/0957-0233/16/8/001.

Chen, Y., Zhang, Y., Pan, F., Liu, J., Wang, K., Zhang, C., Cheng, S., Lu, L., Zhang, W., Zhang, Z., Zhi, X., Zhang, Q., Alfranca, G., De La Fuente, J. M., Chen, D., & Cui, D. (2016). Breath analysis based on surface-enhanced Raman scattering sensors distinguishes early and advanced gastric cancer patients from healthy persons. *ACS Nano, 10*(9), 8169–8179. http://pubs.acs.org/journal/ancac3. doi:10.1021/acsnano.6b01441.

Cope, K. A., Watson, M. T., Foster, W. M., Sehnert, S. S., & Risby, T. H. (2004). Effects of ventilation on the collection of exhaled breath in humans. *Journal of Applied Physiology, 96*(4), 1371–1379. doi:10.1152/japplphysiol.01034.2003.

Corradi, M., Poli, D., Banda, I., Bonini, S., Mozzoni, P., Pinelli, S., Alinovi, R., Andreoli, R., Ampollini, L., Casalini, A., Carbognani, P., Goldoni, M., & Mutti, A. (2015). Exhaled breath analysis in suspected cases of non-small-cell lung cancer: a cross-sectional study. *Journal of Breath Research, 9*(2), 027101. doi:10.1088/1752-7155/9/2/027101.

Dadamio, J., Velde, S. V., Laleman, W., Hee, P. V., Coucke, W., Nevens, F., & Quirynen, M. (2012). Breath biomarkers of liver cirrhosis. *Journal of Chromatography B, 905*, 17–22. doi:10.1016/j.jchromb.2012.07.025.

Das, M. K., Bishwal, S. C., Das, A., Dabral, D., Varshney, A., Badireddy, V. K., & Nanda, R. (2014). Investigation of gender-specific exhaled breath volatome in humans by GCxGC-TOF-MS. *Analytical Chemistry, 86*(2), 1229–1237. doi:10.1021/ac403541a.

Deng, C., Zhang, J., Yu, X., Zhang, W., & Zhang, X. (2004). Determination of acetone in human breath by gas chromatography–mass spectrometry and solid-phase microextraction with on-fiber derivatization. *Journal of Chromatography B, 810*(2), 269–275. doi:10.1016/j.jchromb.2004.08.013.

Di Natale, C., Macagnano, A., Martinelli, E., Paolesse, R., D'Arcangelo, G., Roscioni, C., Finazzi-Agrò, A., & D'Amico, A. (2003). Lung cancer identification by the analysis of breath by means of an array of non-selective gas sensors. *Biosensors and Bioelectronics, 18*(10), 1209–1218. www.elsevier.com/locate/bios. doi:10.1016/S0956-5663(03)00086-1.

Diskin, A. M., Pan, L. P., & Smith, D. (2003). Time variation of ammonia, acetone, isoprene and ethanol in breath: A quantitative SIFT-MS study over 30 days. *Physiological Measurement, 24*(1), 107–119. doi:10.1088/0967-3334/24/1/308.

Dragonieri, S., Annema, J. T., Schot, R., van der Schee, M. P. C., Spanevello, A., Carratú, P., Resta, O., Rabe, K. F., & Sterk, P. J. (2009). An electronic nose in the discrimination of patients with non-small cell lung cancer and COPD. *Lung Cancer, 64*(2), 166–170. doi:10.1016/j.lungcan.2008.08.008.

Dragonieri, S., Schot, R., Mertens, B. J. A., Le Cessie, S., Gauw, S. A., Spanevello, A., Resta, O., Willard, N. P., Vink, T. J., Rabe, K. F., Bel, E. H., & Sterk, P. J. (2007). An electronic nose in the discrimination of patients with asthma and controls. *Journal of Allergy and Clinical Immunology, 120*(4), 856–862. doi:10.1016/j.jaci.2007.05.043.

Dweik, R. A., Boggs, P. B., Erzurum, S. C., Irvin, C. G., Leigh, M. W., Lundberg, J. O., Olin, A. C., Plummer, A. L., & Taylor, D. R. (2011). An official ATS clinical practice guideline: Interpretation of exhaled nitric oxide levels (FENO) for clinical applications. *American Journal of Respiratory and Critical Care Medicine, 184*(5), 602–615. http://ajrccm.atsjournals.org/cgi/reprint/184/5/602. doi:10.1164/rccm.9120-11STundefined.

Einoch Amor, R., Nakhleh, M. K., Barash, O., & Haick, H. (2019). Breath analysis of cancer in the present and the future. *European Respiratory Review, 28*(152), 190002. doi:10.1183/16000617.0002-2019.

Fan, G. T., Yang, C. L., Lin, C. H., Chen, C. C., & Shih, C. H. (2014). Applications of Hadamard transform-gas chromatography/mass spectrometry to the detection of acetone in healthy human and diabetes mellitus patient breath. *Talanta, 120,* 386–390. https://www.journals.elsevier.com/talanta. doi:10.1016/j.talanta.2013.12.025.

Filipiak, W., Filipiak, A., Sponring, A., Schmid, T., Zelger, B., Ager, C., Klodzinska, E., Denz, H., Pizzini, A., Lucciarini, P., Jamnig, H., Troppmair, J., & Amann, A. (2014). Comparative analyses of volatile organic compounds (VOCs) from patients, tumors and transformed cell lines for the validation of lung cancer-derived breath markers. *Journal of Breath Research, 8*(2), 027111. doi:10.1088/1752-7155/8/2/027111.

Filipiak, W., Beer, R., Sponring, A., Filipiak, A., Ager, C., Schiefecker, A., Lanthaler, S., Helbok, R., Nagl, M., Troppmair, J., et al. (2015). Breath analysis for in vivo detection of pathogens related to ventilator-associated pneumonia in intensive care patients: a prospective pilot study. *Journal of Breath Research, 9,* 016004.

Fu, X.-A., Li, M., Knipp, R. J., Nantz, M. H., & Bousamra, M. (2014). Noninvasive detection of lung cancer using exhaled breath. *Cancer Medicine, 3*(1), 174–181. doi:10.1002/cam4.162.

Fuchs, P., Loeseken, C., Schubert, J. K., & Miekisch, W. (2010). Breath gas aldehydes as biomarkers of lung cancer. *International Journal of Cancer, 126*(11), 2663–2670. http://www3.interscience.wiley.com/cgi-bin/fulltext/122653294/PDFSTART. doi:10.1002/ijc.24970Germany.

Gaida, A., Holz, O., Nell, C., Schuchardt, S., Lavae-Mokhtari, B., Kruse, L., Boas, U., Langejuergen, J., Allers, M., Zimmermann, S., Vogelmeier, C., Koczulla, A. R., & Hohlfeld, J. M. (2016). A dual center study to compare breath volatile organic compounds from smokers and non-smokers with and without COPD. *Journal of Breath Research, 10*(2), 026006. doi:10.1088/1752-7155/10/2/026006.

Galassetti, P. R., Novak, B., Nemet, D., Rose-Gottron, C., Cooper, D. M., Meinardi, S., Newcomb, R., Zaldivar, F., & Blake, D. R. (2005). Breath ethanol and acetone as indicators of serum glucose levels: An initial report. *Diabetes Technology and Therapeutics, 7*(1), 115–123. doi:10.1089/dia.2005.7.115.

Giardina, M., & Olesik, S. V. (2003). Application of low-temperature glassy carbon-coated macrofibers for solid-phase microextraction analysis of simulated breath volatiles. *Analytical Chemistry, 75*(7), 1604–1614. doi:10.1021/ac025984k.

Goerl, T., Kischkel, S., Sawacki, A., Fuchs, P., Miekisch, W., & Schubert, J. K. (2013). Volatile breath biomarkers for patient monitoring during haemodialysis. *Journal of Breath Research, 7,* 017116.

Grabowska-Polanowska, B., Faber, J., Skowron, M., Miarka, P., Pietrzycka, A., Śliwka, I., & Amann, A. (2013). Detection of potential chronic kidney disease markers in breath using gas chromatography with mass-spectral detection coupled with thermal desorption method. *Journal of Chromatography A, 1301,* 179–189. doi:10.1016/j.chroma.2013.05.012.

Guo, D., Zhang, D., Li, N., Zhang, L., & Yang, J. (2010). Hong Kong Diabetes identification and classification by means of a breath analysis system, *Lecture Notes in Computer Science (including subseries Lecture Notes in Artificial Intelligence and Lecture Notes in Bioinformatics)* (6165, pp. 52–63). Springer. doi:10.1007/978-3-642-13923-9_6.

Guo, L., Wang, C., Chi, C., Wang, X., Liu, S., Zhao, W., Ke, C., Xu, G., & Li, E. (2015). Exhaled breath volatile biomarker analysis for thyroid cancer. *Translational Research, 166*(2), 188–195. doi:10.1016/j.trsl.2015.01.005.

Hakim, M., Billan, S., Tisch, U., Peng, G., Dvrokind, I., Marom, O., Abdah-Bortnyak, R., Kuten, A., & Haick, H. (2011). Diagnosis of head-and-neck cancer from exhaled breath. *British Journal of Cancer, 104*(10), 1649–1655. doi:10.1038/bjc.2011.128.

Hansel, A., Jordan, A., Holzinger, R., Prazeller, P., Vogel, W., & Lindinger, W. (1995). Proton transfer reaction mass spectrometry: On-line trace gas analysis at the ppb level. *International Journal of Mass Spectrometry and Ion Processes, 149–150*(C), 609–619. doi:10.1016/0168-1176(95)04294-u.

Henderson, K. A., & Matthews, I. P. (2002). Biological monitoring of midwives' exposure to N2O using the Bio-VOC breath sampler. *Journal of Exposure Science & Environmental Epidemiology, 12*(5), 309–312. doi:10.1038/sj.jea.7500231.

Houben, R. M. G. J., & Dodd, P. J. (2016). The global burden of latent tuberculosis infection: A re-estimation using mathematical modelling. *PLoS Medicine, 13*(10). http://medicine.plosjournals.org/perlserv/?request=index-html&issn=1549-1676. doi:10.1371/journal.pmed.1002152.

James, M. (1952). Gas-liquid partition chromatography. A technique for the analysis of volatile materials. *The Analyst, 77*(921), 915–931. doi:10.1039/AN9527700915.

Jareño-Esteban, J. J., Muñoz-Lucas, M. A., Carrillo-Aranda, B., Maldonado-Sanz, J. A., de Granda-Orive, I., Aguilar-Ros, A., Civera-Tejuca, C., Gutiérrez-Ortega, C., & Callol-Sánchez, L. M. (2013). Volatile organic compounds in exhaled breath in a healthy population: Effect of tobacco smoking. *Archivos de Bronconeumologia, 49*(11), 457–461. http://www.elsevier.com/wps/find/journaldescription.cws_home/710346/description#description. doi:10.1016/j.arbr.2013.09.006.

Jones, A. W. (1983a). Role of rebreathing in determination of the blood-breath ratio of expired ethanol. *Journal of Applied Physiology, 55*(4), 1237–1241. doi:10.1152/jappl.1983.55.4.1237.

Jones, A.W., Mardh, G., & Anggard, E. (1983b). Determination of endogenous ethanol in blood and breath by gas chromatography-mass spectrometry. *Pharmacology Biochemistry and Behavior*, 18 Suppl 1, 267–272.

Kang, S., & Paul Thomas, C. L. (2016). How long may a breath sample be stored for at −80°C? A study of the stability of volatile organic compounds trapped onto a mixed Tenax: Carbograph trap adsorbent bed from exhaled breath. *Journal of Breath Research, 10*(2), 026011. doi:10.1088/1752-7155/10/2/026011.

Karl, T., Prazeller, P., Mayr, D., Jordan, A., Rieder, J., Fall, R., & Lindinger, W. (2001). Human breath isoprene and its relation to blood cholesterol levels: New measurements and modeling. *Journal of Applied Physiology, 91*(2), 762–770. https://www.physiology.org/journal/jappl. doi:10.1152/jappl.2001.91.2.762.

Kenfield, S. A., Wei, E. K., Stampfer, M. J., Rosner, B. A., & Colditz, G. A. (2008). Comparison of aspects of smoking among the four histological types of lung cancer. *Tobacco Control, 17*(3), 198–204. doi:10.1136/tc.2007.022582.

Khalid, T. Y., Costello, B. D. L., Ewen, R., White, P., Stevens, S., Gordon, F., Collins, P., McCune, A., Shenoy, A., Shetty, S., Ratcliffe, N. M., & Probert, C. S. (2013). Breath volatile analysis from patients diagnosed with harmful drinking, cirrhosis and hepatic encephalopathy: A pilot study. *Metabolomics, 9*(5), 938–948. doi:10.1007/s11306-013-0510-4.

Kharitonov, S. A. (2002). Dose-dependent onset and cessation of action of inhaled budesonide on exhaled nitric oxide and symptoms in mild asthma. *Thorax, 57*(10), 889–896. doi:10.1136/thorax.57.10.889.

Kharitonov, S. A. (2005). *Measurement of exhaled nitric oxide* (pp. 166–180). S. Karger AG. doi:10.1159/000083534.

Khoubnasabjafari, M., Mogaddam, M. R. A., Rahimpour, E., Soleymani, J., Saei, A. A., & Jouyban, A. (2022). Breathomics: Review of sample collection and analysis, data modeling and clinical applications. *Critical Reviews in Analytical Chemistry, 52*(7), 1461–1487. www.tandf.co.uk/journals/titles/10408347.asp. doi:10.1080/10408347.2021.1889961.

Kim, I. L.-D., Choi, S.-J., Kim, S.-J., & Jang, J. I.-S. U. (2015). Exhaled breath sensors. *Springer Science and Business Media LLC*, 19–49. doi:10.1007/978-94-017-9981-2_2.

Kim, K. H., Jahan, S. A., & Kabir, E. (2012). A review of breath analysis for diagnosis of human health. *TrAC – Trends in Analytical Chemistry, 33*, 1–8. doi:10.1016/j.trac.2011.09.013.

King, J., Mochalski, P., Kupferthaler, A., Unterkofler, K., Koc, H., Filipiak, W., Teschl, S., Hinterhuber, H., & Amann, A. (2010). Dynamic profiles of volatile organic compounds in exhaled breath as determined by a coupled PTR-MS/GC-MS study. *Physiological Measurement, 31*(9), 1169–1184. doi:10.1088/0967-3334/31/9/008.

Kischkel, S., Miekisch, W., Sawacki, A., Straker, E. M., Trefz, P., Amann, A., & Schubert, J. K. (2010). Breath biomarkers for lung cancer detection and assessment of smoking related effects – confounding variables, influence of normalization and statistical algorithms. *Clinica Chimica Acta, 411*(21–22), 1637–1644. doi:10.1016/j.cca.2010.06.005.

Kischkel, S., Miekisch, W., Fuchs, P., & Schubert, J. K. (2012). Breath analysis during one-lung ventilation in cancer patients. *European Respiratory Journal, 40,* 706–713.

Kramer, R., Sauer-Heilborn, A., Welte, T., Guzman, C. A., Höfle, M. G., & Abraham, W.-R. (2015). A rapid method for breath analysis in cystic fibrosis patients. *European Journal of Clinical Microbiology & Infectious Diseases, 34*(4), 745–751. doi:10.1007/s10096-014-2286-5.

Kwak, J., Fan, M., Harshman, S., Garrison, C., Dershem, V., Phillips, J., Grigsby, C., & Ott, D. (2014). Evaluation of Bio-VOC sampler for analysis of volatile organic compounds in exhaled breath. *Metabolites, 4*(4), 879–888. doi:10.3390/metabo4040879.

Lamote, K., Brinkman, P., Vandermeersch, L., Vynck, M., Sterk, P. J., Van Langenhove, H., Thas, O., Van Cleemput, J., Nackaerts, K., & van Meerbeeck, J. P. (2017). Breath analysis by gas chromatography-mass spectrometry and electronic nose to screen for pleural mesothelioma: A crosssectional case-control study. *Oncotarget, 8*(53), 91593–91602. http://www.impactjournals.com/oncotarget/index.php?journal=oncotarget&page=article&op=download&path%5B%5D=21335&path%5B%5D=67866. doi:10.18632/oncotarget.21335.

Lärstad, M., Söderling, A.-S., Caidahl, K., & Olin, A.-C. (2005). Selective quantification of free 3-nitrotyrosine in exhaled breath condensate in asthma using gas chromatography/tandem mass spectrometry. *Nitric Oxide, 13*(2), 134–144. doi:10.1016/j.niox.2005.05.009.

Lawal, O., Ahmed, W. M., Nijsen, T. M. E., Goodacre, R., & Fowler, S. J. (2017). Exhaled breath analysis: A review of "breath-taking" methods for off-line analysis. *Metabolomics, 13*(10), 110. http://www.kluweronline.com/issn/1573-3890/. doi:10.1007/s11306-017-1241-8.

Libardoni, M., Stevens, P. T., Waite, J. H., & Sacks, R. (2006). Analysis of human breath samples with a multi-bed sorption trap and comprehensive two-dimensional gas chromatography (GC × GC). *Journal of Chromatography B: Analytical Technologies in the Biomedical and Life Sciences, 842*(1), 13–21. doi:10.1016/j.jchromb.2006.05.008.

Lord, H., Yu, Y., Segal, A., & Pawliszyn, J. (2002). Breath analysis and monitoring by membrane extraction with sorbent interface. *Analytical Chemistry, 74*(21), 5650–5657. http://pubs.acs.org/journal/ancham. doi:10.1021/ac025863k.

Lourenço, C. É., & Turner, C. (2014). Breath analysis in disease diagnosis: Methodological considerations and applications. *Metabolites, 4*(2), 465–498. doi:10.3390/metabo4020465.

Lyon, A., Mincholé, A., Martínez, J. P., Laguna, P., & Rodriguez, B. (2018). Computational techniques for ECG analysis and interpretation in light of their contribution to medical advances. *Journal of the Royal Society Interface, 15*(138), 20170821. http://rsif.royalsocietypublishing.org/content/royinterface/15/138/20170821.full.pdf. doi:10.1098/rsif.2017.0821.

Machado, R. F., Laskowski, D., Deffenderfer, O., Burch, T., Zheng, S., Mazzone, P. J., Mekhail, T., Jennings, C., Stoller, J. K., Pyle, J., Duncan, J., Dweik, R. A., & Erzurum, S. C. (2005). Detection of lung cancer by sensor array analyses of exhaled breath. *American Journal of Respiratory and Critical Care Medicine, 171*(11), 1286–1291. doi:10.1164/rccm.200409-1184OC.

Mazzone, P. J., Hammel, J., Dweik, R., Na, J., Czich, C., Laskowski, D., & Mekhail, T. (2007). Diagnosis of lung cancer by the analysis of exhaled breath with a colorimetric sensor array. *Thorax, 62*(7), 565–568. doi:10.1136/thx.2006.072892.

Mazzone, P. J., Wang, X. F., Xu, Y., Mekhail, T., Beukemann, M. C., Na, J., Kemling, J. W., Suslick, K. S., & Sasidhar, M. (2012). Exhaled breath analysis with a colorimetric sensor array for the identification and characterization of lung cancer. *Journal of Thoracic Oncology,*

7(1), 137–142. https://www.journals.elsevier.com/journal-of-thoracic-oncology/. doi:10.1097/JTO.0b013e318233d80f.

McNerney, R., Wondafrash, B. A., Amena, K., Tesfaye, A., McCash, E. M., & Murray, N. J. (2010). Field test of a novel detection device for Mycobacterium tuberculosis antigen in cough. *BMC Infectious Diseases, 10*, 161. http://www.biomedcentral.com/1471-2334/10/161. doi:10.1186/1471-2334-10-161. United Kingdom.

Miekisch, W., Kischkel, S., Mieth, M., Schubert, J. K., Hengstenberg, A., & Beckmann, U. (2010). Construction and evaluation of a versatile CO2 controlled breath collection device. *IEEE Sensors Journal, 10*(1), 211–215. doi:10.1109/JSEN.2009.2035757.

Miekisch, W., Kischkel, S., Sawacki, A., Liebau, T., Mieth, M., & Schubert, J. K. (2008). Impact of sampling procedures on the results of breath analysis. *Journal of Breath Research, 2*(2), 026007. doi:10.1088/1752-7155/2/2/026007.

Minh, T. D. C., Oliver, S. R., Ngo, J., Flores, R., Midyett, J., Meinardi, S., Carlson, M. K., Rowland, F. S., Blake, D. R., & Galassetti, P. R. (2011). Noninvasive measurement of plasma glucose from exhaled breath in healthy and type 1 diabetic subjects. *American Journal of Physiology – Endocrinology and Metabolism, 300*(6), E1166. http://ajpendo.physiology.org/content/300/6/E1166.full.pdf+html. doi:10.1152/ajpendo.00634.2010. United States.

Mochalski, P., King, J., Haas, M., Unterkofler, K., Amann, A., & Mayer, G. (2014). Blood and breath profiles of volatile organic compounds in patients with end-stage renal disease. *BMC Nephrology, 15*(1), 1–14. http://www.biomedcentral.com/bmcnephrol/. doi:10.1186/1471-2369-15-43.

Montuschi, P., Santonico, M., Mondino, C., Pennazza, G., Mantini, G., Martinelli, E., Capuano, R., Ciabattoni, G., Paolesse, R., Natale, C. D., Barnes, P. J., & D'Amico, A. (2010). Diagnostic performance of an electronic nose, fractional exhaled nitric oxide, and lung function testing in asthma. *Chest, 137*(4), 790–796. doi:10.1378/chest.09-1836.

Mule, N. M., Patil, D. D., & Kaur, M. (2021). A comprehensive survey on investigation techniques of exhaled breath (EB) for diagnosis of diseases in human body. *Informatics in Medicine Unlocked, 26*, 100715. https://www.sciencedirect.com/science/journal/23529148. doi:10.1016/j.imu.2021.100715.

Nidheesh, V. R., Mohapatra, A. K., Unnikrishnan, V. K., Sinha, R. K., Nayak, R., Kartha, V. B., & Chidangil, S. (2021). Breath analysis for the screening and diagnosis of diseases. *Applied Spectroscopy Reviews, 56*(8–10), 702–732. doi:10.1080/05704928.2020.1848857.

O'Hara, M. E., O'Hehir, S., Green, S., & Mayhew, C. A. (2008). Development of a protocol to measure volatile organic compounds in human breath: A comparison of rebreathing and on-line single exhalations using proton transfer reaction mass spectrometry. *Physiological Measurement, 29*(3), 309–330. https://iopscience.iop.org/journal/0967-3334. doi:10.1088/0967-3334/29/3/003.

Ohlsson, J., Ralph, D. D., Mandelkorn, M. A., Babb, A. L., & Hlastala, M. P. (1990). Accurate measurement of blood alcohol concentration with isothermal rebreathing. *Journal of Studies on Alcohol, 51*(1), 6–13. doi:10.15288/jsa.1990.51.6.

Pabst, F., Miekisch, W., Fuchs, P., Kischkel, S., & Schubert, J. K. (2007). Monitoring of oxidative and metabolic stress during cardiac surgery by means of breath biomarkers: an observational study. *Journal of Cardiothoracic Surgery, 2*, 37.

Pathak, A., Swargiary, K., Kongsawang, N., Jitpratak, P., Ajchareeyasoontorn, N., Udomkittivorakul, J., & Viphavakit, C. (2023). Recent advances in sensing materials targeting clinical volatile organic compound (VOC) biomarkers: A review. *Biosensors, 13*(1), 114. doi:10.3390/bios13010114.

Peled, N., Hakim, M., Bunn, P. A., Miller, Y. E., Kennedy, T. C., Mattei, J., Mitchell, J. D., Hirsch, F. R., & Haick, H. (2012). Non-invasive breath analysis of pulmonary nodules. *Journal of Thoracic Oncology, 7*(10), 1528–1533. https://www.journals.elsevier.com/journal-of-thoracic-oncology/. doi:10.1097/JTO.0b013e3182637d5f.

Peng, G., Hakim, M., Broza, Y. Y., Billan, S., Abdah-Bortnyak, R., Kuten, A., Tisch, U., & Haick, H. (2010). Detection of lung, breast, colorectal, and prostate cancers from exhaled breath using a single array of nanosensors. *British Journal of Cancer, 103*(4), 542–551. doi:10.1038/sj.bjc.6605810.

Peng, G., Tisch, U., Adams, O., Hakim, M., Shehada, N., Broza, Y. Y., Billan, S., Abdah-Bortnyak, R., Kuten, A., & Haick, H. (2009). Diagnosing lung cancer in exhaled breath using gold nanoparticles. *Nature Nanotechnology, 4*(10), 669–673. http://www.nature.com/nnano/index.html. doi:10.1038/nnano.2009.235.

Pereira, J., Porto-Figueira, P., Cavaco, C., Taunk, K., Rapole, S., Dhakne, R., Nagarajaram, H., & Câmara, J. (2015). Breath analysis as a potential and non-invasive frontier in disease diagnosis: An overview. *Metabolites, 5*(1), 3–55. doi:10.3390/metabo5010003.

Pesesse, R., et al. (2019). Multimodal chemometric approach for the analysis of human exhaled breath in lung cancer patients by TD-GC × GC-TOFMS. *Journal of Chromatography B*, 146–153. doi:10.1016/j.jchromb.2019.01.029.

Phillips, C. O., Syed, Y., Parthaláin, N. M., Zwiggelaar, R., Claypole, T. C., & Lewis, K. E. (2012). Machine learning methods on exhaled volatile organic compounds for distinguishing COPD patients from healthy controls. *Journal of Breath Research, 6*(3), 036003. http://iopscience.iop.org/1752-7163/6/3/036003/pdf/1752-7163_6_3_036003.pdf. doi:10.1088/1752-7155/6/3/036003. United Kingdom.

Phillips, M. (1997). Method for the collection and assay of volatile organic compounds in breath. *Analytical Biochemistry, 247*(2), 272–278. doi:10.1006/abio.1997.2069.

Phillips, M., Basa-Dalay, V., Blais, J., Bothamley, G., Chaturvedi, A., Modi, K. D., Pandya, M., Natividad, M. P. R., Patel, U., Ramraje, N. N., Schmitt, P., & Udwadia, Z. F. (2012). Point-of-care breath test for biomarkers of active pulmonary tuberculosis. *Tuberculosis, 92*(4), 314–320. doi:10.1016/j.tube.2012.04.002.

Phillips, M., Basa-Dalay, V., Bothamley, G., Cataneo, R. N., Lam, P. K., Natividad, M. P. R., Schmitt, P., & Wai, J. (2010). Breath biomarkers of active pulmonary tuberculosis. *Tuberculosis, 90*(2), 145–151. doi:10.1016/j.tube.2010.01.003.

Phillips, M., Cataneo, R. N., Condos, R., Ring Erickson, G. A., Greenberg, J., La Bombardi, V., Munawar, M. I., & Tietje, O. (2007). Volatile biomarkers of pulmonary tuberculosis in the breath. *Tuberculosis, 87*(1), 44–52. doi:10.1016/j.tube.2006.03.004.

Phillips, M., Gleeson, K., Hughes, J. M. B., Greenberg, J., Cataneo, R. N., Baker, L., & McVay, W. P. (1999). Volatile organic compounds in breath as markers of lung cancer: A cross-sectional study. *Lancet, 353*(9168), 1930–1933. http://www.journals.elsevier.com/the-lancet/. doi:10.1016/S0140-6736(98)07552-7.

Poli, D., Carbognani, P., Corradi, M., Goldoni, M., Acampa, O., Balbi, B., Bianchi, L., Rusca, M., & Mutti, A. (2005). Exhaled volatile organic compounds in patients with non-small cell lung cancer: Cross sectional and nested short-term follow-up study. *Respiratory Research, 6*, 71. http://respiratory-research.com/content/6/1/71. doi:10.1186/1465-9921-6-t71Italy.

Prado, C., Marıń, P., & Periago, J. F. (2003). Application of solid-phase microextraction and gas chromatography–mass spectrometry to the determination of volatile organic compounds in end-exhaled breath samples. *Journal of Chromatography A, 1011*(1–2), 125–134. doi:10.1016/s0021-9673(03)01103-8.

Preti, G., Labows, J. N., Kostelc, J. G., Aldinger, S., & Daniele, R. (1988). Analysis of lung air from patients with bronchogenic carcinoma and controls using gas chromatography-mass spectrometry. *Journal of Chromatography B: Biomedical Sciences and Applications, 432*(C), 1–11. doi:10.1016/s0378-4347(00)80627-1.

Raninen, K. J., Lappi, J. E., Mukkala, M. L., Tuomainen, T. P., Mykkänen, H. M., Poutanen, K. S., & Raatikainen, O. J. (2016). Fiber content of diet affects exhaled breath volatiles in fasting and postprandial state in a pilot crossover study. *Nutrition Research, 36*(6), 612–619. www.elsevier.com/locate/nutres. doi:10.1016/j.nutres.2016.02.008.

Ratiu, I. A., Ligor, T., Bocos-Bintintan, V., Mayhew, C. A., & Buszewski, B. (2021). Volatile organic compounds in exhaled breath as fingerprints of lung cancer, asthma and COPD. *Journal of Clinical Medicine, 10*(1), 1–41. https://www.mdpi.com/2077-0383/10/1/32/pdf. doi:10.3390/jcm10010032.

Righettoni, M., Tricoli, A., Gass, S., Schmid, A., Amann, A., & Pratsinis, S. E. (2012). Breath acetone monitoring by portable Si:WO 3 gas sensors. *Analytica Chimica Acta, 738*, 69–75. doi:10.1016/j.aca.2012.06.002.

Righettoni, M., Schmid, A., Amann, A., & Pratsinis, S. E. (2013). Correlations between blood glucose and breath components from portable gas sensors and PTR-TOF-MS. *Journal of Breath Research, 7*(3), 037110. doi:10.1088/1752-7155/7/3/037110.

Rodríguez-Aguilar, M., Ramírez-García, S., Ilizaliturri-Hernández, C., Gómez-Gómez, A., Van-Brussel, E., Díaz-Barriga, F., Medellín-Garibay, S., & Flores-Ramírez, R. (2019). Ultrafast gas chromatography coupled to electronic nose to identify volatile biomarkers in exhaled breath from chronic obstructive pulmonary disease patients: A pilot study. *Biomedical Chromatography, 33*(12), e4684. http://onlinelibrary.wiley.com/journal/10.1002/. doi:10.1002/bmc.4684. (ISSN)1099-0801.

Rudnicka, Joanna, et al. (2011). Determination of volatile organic compounds as biomarkers of lung cancer by SPME–GC–TOF/MS and chemometrics. *Journal of Chromatography B, 879*, 3360–3366. doi:10.1016/j.jchromb.2011.09.001.

Sahota, A. S., Gowda, R., Arasaradnam, R. P., Daulton, E., Savage, R. S., Skinner, J. R., Adams, E., Ward, S. A., & Covington, J. A. (2016). A simple breath test for tuberculosis using ion mobility: A pilot study. *Tuberculosis, 99*, 143–146. http://www.elsevier.com/inca/publications/store/6/3/8/4/2/8/index.htt. doi:10.1016/j.tube.2016.05.005.

Sakumura, Y., Koyama, Y., Tokutake, H., Hida, T., Sato, K., Itoh, T., Akamatsu, T., & Shin, W. (2017). Diagnosis by volatile organic compounds in exhaled breath from lung cancer patients using support vector machine algorithm. *Sensors, 17*(2), 287. doi:10.3390/s17020287.

Salehi, S., Nikan, E., Khodadadi, A. A., & Mortazavi, Y. (2014). Highly sensitive carbon nanotubes-SnO2 nanocomposite sensor for acetone detection in diabetes mellitus breath. *Sensors and Actuators, B Chemical, 205*, 261–267. doi:10.1016/j.snb.2014.08.082.

Salvo, P., Ferrari, C., Persia, R., Ghimenti, S., Lomonaco, T., Bellagambi, F., & Di Francesco, F. (2015). A dual mode breath sampler for the collection of the end-tidal and dead space fractions. *Medical Engineering & Physics, 37*(6), 539–544. doi:10.1016/j.medengphy.2015.03.013.

Sanchez, J. M., & Sacks, R. D. (2003). GC analysis of human breath with a series-coupled column ensemble and a multibed sorption trap. *Analytical Chemistry, 75*(10), 2231–2236. doi:10.1021/ac020725g.

Sanchez, J. M., & Sacks, R. D. (2006). Development of a multibed sorption trap, comprehensive two-dimensional gas chromatography, and time-of-flight mass spectrometry system for the analysis of volatile organic compounds in human breath. *Analytical Chemistry, 78*(9), 3046–3054. doi:10.1021/ac060053k.

Santonico, M., Lucantoni, G., Pennazza, G., Capuano, R., Galluccio, G., Roscioni, C., La Delfa, G., Consoli, D., Martinelli, E., Paolesse, R., Di Natale, C., & D'Amico, A. (2012). In situ detection of lung cancer volatile fingerprints using bronchoscopic air-sampling. *Lung Cancer, 77*(1), 46–50. doi:10.1016/j.lungcan.2011.12.010.

Schmekel, B., Winquist, F., & Vikström, A. (2014). Analysis of breath samples for lung cancer survival. *Analytica Chimica Acta, 840*, 82–86. http://www.journals.elsevier.com/analytica-chimica-acta/. doi:10.1016/j.aca.2014.05.034.

Schubert, J. K., Miekisch, W., Birken, T., Geiger, K., & Nöldge-Schomburg, G. F. E. (2005). Impact of inspired substance concentrations on the results of breath analysis in mechanically ventilated patients. *Biomarkers, 10*(2–3), 138–152. doi:10.1080/13547500500050259.

Španěl, Patrik, & Smith, David (2020). Quantification of volatile metabolites in exhaled breath by selected ion flow tube mass spectrometry, SIFT-MS. *Clinical Mass Spectrometry, 16*, 18–24. doi:10.1016/j.clinms.2020.02.001.

Stefanuto, P. H., Zanella, D., Vercammen, J., Henket, M., Schleich, F., Louis, R., & Focant, J. F. (2020). Multimodal combination of GC × GC-HRTOFMS and SIFT-MS for asthma phenotyping using exhaled breath. *Scientific Reports, 10*(1). www.nature.com/srep/index.html. doi:10.1038/s41598-020-73408-2.

Sukul, P., Schubert, J. K., Oertel, P., Kamysek, S., Taunk, K., Trefz, P., & Miekisch, W. (2016). FEV manoeuvre induced changes in breath VOC compositions: An unconventional view on lung function tests. *Scientific Reports, 6*, 28029. www.nature.com/srep/index.html. doi:10.1038/srep28029.

Sukul, P., Trefz, P., Schubert, J. K., & Miekisch, W. (2014). Immediate effects of breath holding maneuvers onto composition of exhaled breath. *Journal of Breath Research, 8*(3), 037102. doi:10.1088/1752-7155/8/3/037102.

Svensson, S., Larstad, M., Broo, K., & Olin, A. (2007). Determination of aldehydes in human breath by on-fibre derivatization, solid-phase microextraction and GC-MS. *Journal of Chromatography B, 860*(1), 86–91. doi:10.1016/j.jchromb.2007.10.021.

Thekedar, B., Oeh, U., Szymczak, W., Hoeschen, C., & Paretzke, H. G. (2011). Influences of mixed expiratory sampling parameters on exhaled volatile organic compound concentrations. *Journal of Breath Research, 5*(1), 016001. doi:10.1088/1752-7155/5/1/016001.

Thun, M. J., Hannan, L. M., Adams-Campbell, L. L., Boffetta, P., Buring, J. E., Feskanich, D., Flanders, W. D., Sun, H. J., Katanoda, K., Kolonel, L. N., Lee, I. M., Marugame, T., Palmer, J. R., Riboli, E., Sobue, T., Avila-Tang, E., Wilkens, L. R., & Samet, J. M. (2008). Lung cancer occurrence in never-smokers: An analysis of 13 cohorts and 22 cancer registry studies. *PLoS Medicine, 5*(9), 1357–1371. http://medicine.plosjournals.org/archive/1549-1676/5/9/pdf/10.1371_journal.pmed.0050185-L.pdf. doi:10.1371/journal.pmed.0050185. United States.

Turner, C., Walton, C., Hoashi, S., & Evans, M. (2009). Breath acetone concentration decreases with blood glucose concentration in type I diabetes mellitus patients during hypoglycaemic clamps. *Journal of Breath Research, 3*(4), 046004. doi:10.1088/1752-7155/3/4/046004.

Ueta, I., Saito, Y., Hosoe, M., Okamoto, M., Ohkita, H., Shirai, S., Tamura, H., & Jinno, K. (2009). Breath acetone analysis with miniaturized sample preparation device: In-needle preconcentration and subsequent determination by gas chromatography-mass spectroscopy. *Journal of Chromatography B: Analytical Technologies in the Biomedical and Life Sciences, 877*(24), 2551–2556. doi:10.1016/j.jchromb.2009.06.039.

van den Velde, S., Quirynen, M., Van hee, P., & van Steenberghe, D. (2007). Halitosis associated volatiles in breath of healthy subjects. *Journal of Chromatography B: Analytical Technologies in the Biomedical and Life Sciences, 853*(1–2), 54–61. doi:10.1016/j.jchromb.2007.02.048.

Van Vliet, D., Smolinska, A., Jöbsis, Q., Rosias, P., Muris, J., Dallinga, J., Dompeling, E., & Van Schooten, F. J. (2017). Can exhaled volatile organic compounds predict asthma exacerbations in children? *Journal of Breath Research, 11*(1), 016016. http://iopscience.iop.org/article/10.1088/1752-7163/aa5a8b/pdf. doi:10.1088/1752-7163/aa5a8b.

Wang, C., Mbi, A., & Shepherd, M. (2010). A study on breath acetone in diabetic patients using a cavity ringdown breath analyzer: Exploring correlations of breath acetone with blood glucose and glycohemoglobin A1C. *IEEE Sensors Journal, 10*(1), 54–63. doi:10.1109/JSEN.2009.2035730.

Wang, Y., Hu, Y., Wang, D., Yu, K., Wang, L., Zou, Y., Zhao, C., Zhang, X., Wang, P., & Ying, K. (2012). The analysis of volatile organic compounds biomarkers for lung cancer in exhaled breath, tissues and cell lines. *Cancer Biomarkers, 11*(4), 129–137. doi:10.3233/CBM-2012-00270.

Wardencki, W., Chmiel, T., & Dymerski, T. (2013). Gas chromatography-olfactometry (GC-O), electronic noses (e-noses) and electronic tongues (e-tongues) for in vivo food flavour measurement. *Instrumental assessment of food sensory quality: A practical guide* (pp. 195–229). Elsevier. https://www.sciencedirect.com/book/9780857094391. doi:10.1533/9780857098856.2.195.

Wehinger, A., Schmid, A., Mechtcheriakov, S., Ledochowski, M., Grabmer, C., Gastl, G. A., & Amann, A. (2007). Lung cancer detection by proton transfer reaction mass-spectrometric analysis of human breath gas. *International Journal of Mass Spectrometry, 265*(1), 49–59. doi:10.1016/j.ijms.2007.05.012.

Wu, C., Siems, W. F., Asbury, G. R., & Hill, H. H. (1998). Electrospray ionization high-resolution ion mobility spectrometry-mass spectrometry. *Analytical Chemistry, 70*(23), 4929–4938. doi:10.1021/ac980414z.

Xiao, T., Wang, X.-Y., Zhao, Z.-H., Li, L., Zhang, L., Yao, H.-C., Wang, J.-S., & Li, Z.-J. (2014). Highly sensitive and selective acetone sensor based on C-doped WO3 for potential diagnosis of diabetes mellitus. *Sensors and Actuators B: Chemical, 199*, 210–219. doi:10.1016/j.snb.2014.04.015.

Yan, K., Zhang, D., Wu, D., Wei, H., & Lu, G. (2014). Design of a breath analysis system for diabetes screening and blood glucose level prediction. *IEEE Transactions on Biomedical Engineering, 61*(11), 2787–2795. http://ieeexplore.ieee.org/xpl/RecentIssue.jsp?reload=true&punumber=10. doi:10.1109/TBME.2014.2329753.

Zaric, B., Petrovic, S., Bjekic, M., Rajic, I., Popovic, A., & Djordjevic, D. (2014). Analysis of human exhaled breath in a population of young volunteers. *Arhiv za bioloske nauke, 66*(4), 1529–1538. doi:10.2298/abs1404529z.

Zeliger, H. I., Pan, Y., & Rea, W. J. (2012). Predicting co-morbidities in chemically sensitive individuals from exhaled breath analysis. *Interdisciplinary Toxicology, 5*(3), 123–126. doi:10.2478/v10102-012-0020-7.

Zhang, D., Guo, D., & Yan, K. (2017). *Breath analysis for medical applications* (pp. 1–309). Springer Singapore. http://www.springer.com/in/book/9789811043215. doi:10.1007/978-981-10-4322-2.

CHAPTER

3

Advances in photonic sensors for health and environmental monitoring

3.1 Introduction

Photonics can be defined as the technology of generating, controlling, manipulating, and detecting photons for a broad range of applications, including imaging, communication, and physical and chemical sensing. Generally, photonic sensors leverage light-matter interactions to accurately detect variations in the surrounding environment, providing high sensitivity for diverse applications such as industrial sensing, environmental monitoring, and the detection of gas and volatile organic compounds (VOCs). Optical waveguides, also commonly known as "dielectric waveguides," are structures designed to confine and guide light launched into integrated optical circuits and waveguide-based devices. Dielectric waveguides have been extensively covered in various textbooks and research articles, which readers may consult to understand their historical development (Doerr & Kogelnik, 2008; Snitzer, 1961). Optical waveguides are considered among the most promising tools in sensing, computing, and communication applications due to their unique characteristics, such as low cost, immunity to electromagnetic interference, and resistance to induced crosstalk. Optical waveguides can be categorized into two classes: (1) classical optical elements (such as diffractive gratings, lenses, or mirrors), arranged sequentially along the propagation of the wave, that confine the light through successive refocusing near the optical axis and (2) structures that allow the propagation of the light via total internal reflections (TIR).

With the continuous growth of the Internet of Things (IoT) and artificial intelligence, photonic sensors play a crucial role in the development of

Photonic Sensors for Detection of VOC Biomarkers: Technological Principles and Medical Applications
DOI: https://doi.org/10.1016/B978-0-443-29166-1.00012-3

Copyright © 2026 Elsevier Inc. All rights are reserved, including those for text and data mining, AI training, and similar technologies.

smart, digitally enhanced cities and interconnected systems (Weiss et al., 1995). Photonic sensors allow efficient monitoring of a wide range of parameters, including traffic flow, air quality, energy consumption, and structural health, which can significantly improve resource management and the quality of life. Moreover, these sensors exhibit great potential in medical applications, biomedicine, and disease detection (Ochoa et al., 2021). Photonic sensors can be utilized for continuous monitoring of various critical vital signs, including blood pressure, oxygen level, and heart rate, via non-invasive methods. In addition, the growing interest of researchers in photonic technology has led to breakthrough innovations in bioimaging, allowing more precise and accurate diagnostics and health monitoring (Butt, 2023; Kazanskiy et al., 2021). With the increasing concerns about environmental pollution and climate change, photonic sensors are considered to have significant potential to contribute to mitigating these issues through real-time monitoring. These photonic sensors can also be used for water and air quality monitoring by identifying heavy metal ions, toxic substances, and greenhouse gas emissions. The early detection and control of environmental risks enabled by these photonic devices can significantly improve strategies for environmental monitoring and management (Shahbaz et al., 2023).

Photonic resonant devices are employed in a broad range of label-free sensing applications due to their capability for multiplexed monitoring, potential for deployment in aquatic environments, and ability to focus electromagnetic energy into small mode volumes. A range of architectural designs, including microtoroids (Martin et al., 2004), photonic crystals (PCs) (Gangwar et al., 2023), ring/racetrack structures (Xu et al., 2022), and optical fibers (Gibson et al., 2008), have been exploited to fabricate optically resonant-based sensing devices. Due to their rapid and real-time detection capabilities, these devices have been increasingly applied in medical diagnostics as a non-invasive tool to identify and detect disease-related biomarkers present within the human body fluid, e.g., urine and blood (Panindre et al., 2018). This approach offers significant advantages over conventional tissue biopsies due to its portability, rapid response, and real-time performance that allows early identification of associated disease with minimal discomfort to the patient (Butt et al., 2023). The potential of photonic sensors has been demonstrated in liquid biopsy (i.e., body fluids) by utilizing fundamental light-matter interactions and sensing principles to detect and analyze biomarkers within these fluids (Dell'Olio et al., 2021). For example, plasmonic-based photonic sensors can be utilized to detect target-specific biomarkers by functionalizing the surface of the sensor with molecular recognition elements, such as aptamers or antibodies. When the target biomarkers attach to the functionalized receptors, a shift in the resonance signal occurs, allowing for accurate monitoring and quantification. Similarly, the surface of PC can be modified with target-specific receptors to accurately bind and identify target biomarkers present in biofluid samples. The interaction between the receptor and target biomarkers induces a change in the optical properties of the sensor, thereby enabling sensitive detection (Butt et al., 2023). The optical fiber based biosensor are produced by functionalizing the surface with target receptors for the selective detection of target biomarkers. Optical fiber-based sensors utilize a range of interrogation techniques that typically rely on demodulating variations in intensity, polarization, phase, or spectrum of the propagating light in the fiber, which are induced by variations in the refractive index caused by the presence of target biomarkers.

This chapter provides an extensive and in-depth information about the recent advancements in the field of photonic sensors based on optical fibers, optical waveguides, metasurfaces, wearable technologies, and plasmonics. These sensing devices are particularly attractive because of their portability and enhanced sensing performance. They are capable of identifying variations in the optical properties, including phase and polarization, induced by the surrounding environment. The chapter begins with the classification of photonic sensors. Subsequently provide an overview of the basic design, characteristics, and working principle of optical fiber sensors, optical waveguides, metasurfaces, wearable sensors, and plasmonic sensors, respectively. The chapter concluded with emphasis on future of the photonic sensor market, their application in VOC sensing, and future prospects.

3.2 Classification of photonic sensors

The field of photonic sensors can be categorized into several general types depending on their configuration, platform, and applications, as depicted in Fig. 3.1 (Butt et al., 2024).

(i) *Optical fiber-based sensors:* An optical fiber comprises a core and cladding with higher and lower refractive indices, respectively. It facilitates the detection of changes in various physical and chemical parameters such as pressure, temperature, and chemical characteristics (pH, water pollutants, biochemical metabolites, etc.). The working principle of an optical fiber-based sensor involves changes in the properties of launched light, such as intensity, polarization, phase, and amplitude. These optical properties are significantly influenced by variations in the surrounding environment, enabling accurate monitoring of physical and chemical conditions (Gangwar et al., 2024).

(ii) *Optical waveguide sensors*: These sensors involve the integration of miniaturized and complex waveguide components on a single chip, forming the basis of lab-on-a-chip technology (Pelucchi et al., 2022).

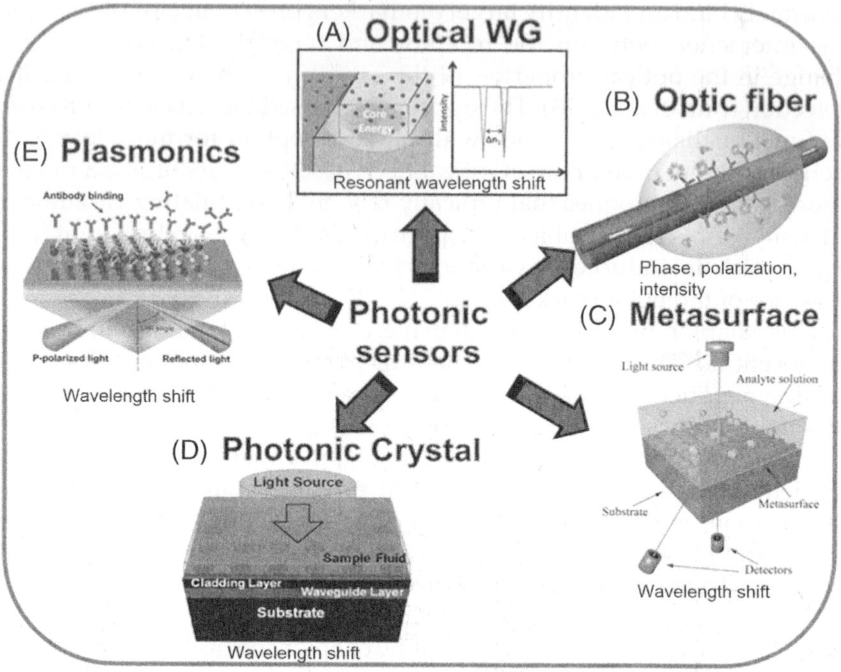

FIGURE 3.1 Photonic sensors overview. The photonic sensing technologies based on (A) optical waveguide, (B) optical fiber, (C) metasurface, (D) photonic crystal, and (E) plasmonics. *From Butt, M. A., Kazanskiy, N. L., Khonina, S. N., Voronkov, G. S., Grakhova, E. P., & Kutluyarov, R. V. (2023). A review on photonic sensing technologies: Status and outlook.* Biosensors, *13(5), 568. https://doi.org/10.3390/bios13050568.*

By exploiting advanced nanofabrication and lithography techniques, these sensing devices can attain high sensitivity, compact size, and robust performance. Optical waveguide sensors are generally utilized in applications that require compact design with enhanced functionality compared to other conventional sensing devices.

(iii) *Wearable sensors:* Wearable sensors are designed to be integrated into wearable technologies for real-time health monitoring, such as smartwatches, smart clothing, or other types of health-monitoring gadgets (Jha et al., 2024). These devices are considered potential tools for non-invasive monitoring of various vital signs, including oxygen levels and heart rate. These devices offer continuous collection of large datasets for personalized fitness and health applications.

(iv) *Metasurface-based sensors*: Metasurface-based sensing devices consist of artificially engineered surfaces made of subwavelength structures, which can alter the properties of the incident light in a specialized ways (Verma et al., 2024). The metasurface sensors utilize light manipulation capabilities to identify and analyze optical signals across a wide range of applications, from imaging to spectroscopy.

(v) *Plasmonic sensors:* The plasmonic sensors require the deposition of noble metals on top of the sensing head in order to introduce the surface plasmon resonance in the presence of propagated light to precisely and efficiently detect the target measurand (Viphavakit et al., 2018). The plasmonic sensors generally exhibit high sensitivity due to the significant enhancement of the electromagnetic field at the metal–dielectric interfaces, making them capable of detecting even small changes in the ambient environment.

(vi) *Photonic crystal-based sensors*: PCs are periodically arranged nanostructures that allow excellent confinement of the light field and are hence used to precisely control and manipulate the launched light (Gangwar et al., 2023). These tunable bandgap characteristics enable the accurate measurement of small variations in the surrounding environment (i.e., refractive index, temperature, humidity, etc.).

Each of the above aforementioned sensor types offers specific advantages and application domains, collectively contributing towards the progress and advancement of photonic sensing devices across a wide range of applications, including environmental monitoring, healthcare, and industrial sectors. The variety within photonic sensing reflects its capabilities and potential to address complex measurement demands. In the next section, an in-depth analysis is provided for each type of photonics sensor, including its working principle and limitations.

3.3 Optical fiber sensors

Optical fiber sensors are cylindrical waveguides that comprise a core and cladding, typically made of either glass or plastic. The fiber can be categorized into single-mode and multi-mode types based on the diameter of the core. Single-mode fibers have a core diameter of ~8 μm, while multi-mode fibers possess a core diameter of ~50 μm. In their standard form, optical fiber sensors undergo expansion or contraction in response to applied temperature and strains. However, when the core of the fiber is exposed to the surrounding environment, the sensor becomes significantly more sensitive to even small variations in the ambient environment. When light is launched through the optical fiber toward the sensing region, it is modulated by the changes in the surrounding medium. Subsequently, the output signal is recorded by an analytical device, which converts the optical signal into numerical values that can be analyzed to determine variations in the surrounding environment. These measurements precisely indicate the level of variation in the physical or chemical changes near the

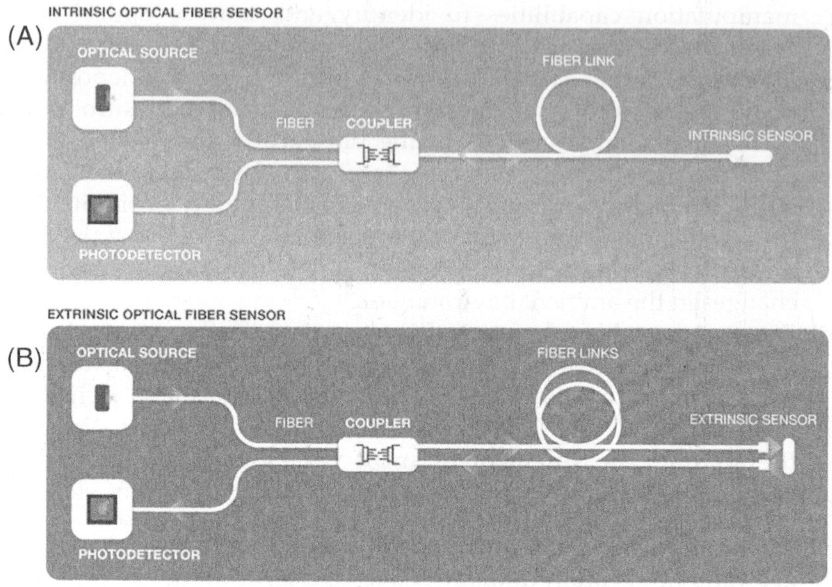

FIGURE 3.2 Schematic diagram illustrates (A) intrinsic and (B) extrinsic fiber sensors. Types of optical fiber sensor based on their location. *From Pendão, C., & Silva, I. (2022). Optical fiber sensors and sensing networks: Overview of the main principles and applications.* Sensors, 22(19), *7554. https://doi.org/10.3390/s22197554.*

sensing head. Fig. 3.2 depicts the schematic diagram of the optical fiber sensor configuration (Pendão & Silva, 2022). Fiber-based sensors typically offer several advantages over other sensing methods and hold significant promise for a broad range of applications. Notably, the major advantages of fiber sensors include compact size, low cost, multiplexing capability, and suitability for *in situ* monitoring, making them a potential candidate for use across various sectors (Culshaw, 2004). Furthermore, the fiber sensors are inherently immune to electromagnetic interference, making them especially valuable in hazardous environments where flammable material may be present, such as in hydrogen sensing (Pathak et al., 2023), fuel adulteration detection (Pathak et al., 2017), TNT explosive detection (Cennamo et al., 2013), or high-voltage electrical environments (Bohnert et al., 2005). The materials used in the fabrication of optical fibers also provide thermal stability, allowing them to operate in high-temperature conditions (Gangwar et al., 2023). Although the optical fiber-based sensing technology has gained significant advancements, the development in certain application areas, such as corrosion detection, remains largely underexplored and offers opportunities for future innovation.

Considering the type of applications, optical fiber sensors can be classified based on the type of measurand. These include physical parameters (e.g., pressure, temperature, strain), chemical parameters (e.g., pH level, refractive index, humidity, pesticides, etc.), and biomedical targets (e.g., carbon dioxide, oxygen, VOCs, proteins, and DNA). In addition, based on the location of the sensing region, optical fiber-based sensors can be categorized as intrinsic or extrinsic, as shown in Fig. 3.2 (Pendão & Silva, 2022). In the intrinsic type sensors (Fig. 3.2 (A)), the optical fiber itself serves as both the sensing head and the tranmission medium for the optical signal. These types of sensors operate based on the direct modulation of the launched light along the length of the fiber, and the light remains inside the optical fiber until it reaches the sensing head. In the intrinsic type of sensors, the surrounding environment can alter the properties of the optical signal passing through the fibers. After modulation, the signal is reflected or scattered back along the same optical fiber to reach the detector. Intrinsic fiber sensors are mostly suitable for distributed sensing over long distances to monitor multiple parameters. For instance, one method to monitor temperature variation is by studying Raman scattering in the optical fiber, whereas another method involves utilizing a fiber that exhibits temperature-dependent evanescent loss. Electrical voltage can be determined by examining the polarization of light as a function of electric field or voltage, considering nonlinear effects in specifically doped optical fibers. Angular variations can be monitored through the Sagnac effect, while directional sensing can be achieved using specialized long-period fiber gratings (Santos et al., 2009). The intrinsic sensors are also widely utilized as high-accuracy gyroscopes for navigation purposes in vehicles such as cars (Imamura et al., 2019) and airplanes (Zakirov & Umarov, 2020).

On the other hand, the extrinsic optical fiber sensor, also referred to as a hybrid sensor and typically utilizing multimode fiber, guides the launched light to and from the location of the sensing head, as shown in Fig. 3.2 (B). These types of sensors comprise an external sensing head separate from the used optical fibers and are based on compact-sized optical components, which are utilized to alter the characteristics of the transmitted light under various environmental conditions. The launched signal is transmitted to the sensing head via one arm of the optical fiber, and the reflected modulated signal is coupled back and sent to the detector through the other arm, as shown in Fig. 3.2 (B). In general, such types of sensing configurations are made of two fibers, one of which is used to transmit the signal to the sensing head, whereas the other fiber is used to collect and guide the modulated signal to the detector. Nowadays, more advanced techniques are being employed that use only one fiber to launch and collect the modulated signal for delivery to the detection system. A key advantage of an extrinsic optical fiber sensor lies in its ability to access environments that are inaccessible to conventional transmission methods, such as temperature monitoring within the jet engine or under intense electromagnetic conditions. These types of sensors are utilized to monitor

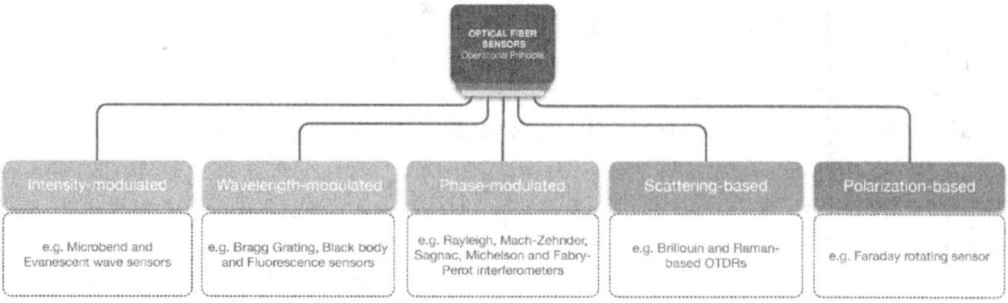

FIGURE 3.3 Principle of operation. Optical fiber sensor types categorized by principle of operation. *From Pendão, C., & Silva, I. (2022). Optical fiber sensors and sensing networks: Overview of the main principles and applications.* Sensors, *22(19), 7554. https://doi.org/10.3390/s22197554.*

rotation, vibration, temperature, velocity, acceleration, displacement, and torque (Kuang et al., 2004; Sabri et al., 2013). The additional advantage of extrinsic fiber sensors is their strong resistance to interference and signal distortion. However, combining extrinsic fiber sensors with other types of sensors can present challenges, as the conventional sensors typically produce electrical outputs that require conversion to an optical signal for compatibility (Pendão & Silva, 2022). Fabry-Perot interferometers can be considered as an example of such a sensing configuration, which consists of a cavity at the tip of an optical fiber that allows the light signal to exit the fiber and facilitate precise measurements, including acoustic wave (Wei et al., 2023), temperature (Chen et al., 2019), and pressure (Wei et al., 2020). The optical fiber sensors can be classified based on their working mechanisms into various types such as intensity, wavelength, phase, scattering, and polarization-based sensors, as shown in the following Fig. 3.3. A brief discussion of these sensing mechanisms is presented below, whereas a comprehensive discussion will be provided in the subsequent chapters.

- **Intensity-modulated sensors:** The intensity-modulated sensor is the earliest reported type of optical fiber sensor. This sensor can identify the changes in its characteristics or perturbations in the guided light due to attenuation, bending loss, and evanescent fields. Its simple configuration and cost-effectiveness make this device useful in various sectors. However, this sensor is also prone to considerable fluctuations in the output power, resulting in false detections, and hence requires an additional reference system in order to reduce fluctuations and provide more accuracy.
- **Wavelength-modulated sensors:** In wavelength-modulated sensors, the performance of the sensing devices is measured based on the shift in the output wavelength spectra corresponding to changes in the surrounding environment. For instance, fluorescence sensors, blackbody

sensors, and Bragg grating-based wavelength-modulated sensors are examples of this category.

- **Phase-modulated sensors:** Phase-modulated sensors utilize the interferometry principle to monitor the interference of the propagated optical signal. These types of sensors are popular in the sensing market due to their excellent sensing performance and detection accuracy. Nevertheless, such sensors are highly expensive. Mach–Zehnder, Michelson, Sagnac, and Fabry–Perot interferometers are considered among the most widely used phase-modulated sensors.
- **Polarization-based sensors:** Polarization-based sensors detect changes in the signal arising from variations in the state of polarization. These sensors utilize the birefringent properties of optical fibers, where the refractive index shifts based on polarization. When strain is applied to the fiber, birefringence is induced, resulting in a measurable phase difference.

3.4 Optical waveguide

Optical waveguides are considered as an especially engineered area within photonics, which relies on integrating and miniaturizing optical components onto a substrate to create chip-level devices (Thylén & Wosinski, 2014). This technological area includes the fabrication of photonic components, such as detectors, modulators, waveguides, and light sources, using III-V semiconductors or silicon. These optical waveguides allow precise control of light at micro- and nanoscale levels, facilitating the production of miniaturized, cost-effective, and efficient photonic circuits suitable for a broad range of applications. Currently, the fabrication of silicon (Si)-based optical waveguides can be performed efficiently and economically using silicon foundries, e-beam lithography (deep), UV photolithography, focused ion beam (FIB) micromachining, and several etching processes, such as inductively coupled plasma etching and reactive ion etching (RIE) (Belt et al., 2013; Corsetti et al., 2024; Somers et al., 2024).

Therefore, various approaches are available for optimizing complex photonic devices, ranging from prototyping for single-device fabrication to mass production schemes, with significant potential for low-cost device production. This domain can be categorised into different waveguide designs, including buried channel waveguides (Albert & Lit, 1990), slot waveguides (Butt et al., 2018b), and rib waveguides (Butt et al., 2018a). The waveguide configuration, along with optimized structural parameters, is one of the most critical factors contributing to improved sensing performance. The optimization of the structural parameters of the waveguide is performed numerically using the finite element method (FEM) (see

Chapter 4 for a detailed discussion of FEM), which offers time and cost efficiency while ensuring precise representation of the actual devices.

Choosing the appropriate optical waveguide is crucial for the intended application, especially in methods like optical sensing and spectroscopy. Biomolecules, such as DNA, proteins, and other biological components, depict reflection, absorption, and/or scattering at specific wavelengths. The selection of wavelength is critical to the sensitivity and selectivity of devices in sensing applications. Various biomolecules possess distinct absorption spectra, and by utilizing suitable wavelengths, the sensing device can effectively identify and differentiate target molecules within a complex environment.

Si-based biosensors utilize a silicon oxide substrate in order to determine light-matter interaction by utilizing near-infrared (NIR) light confined within the fabricated waveguide. The evanescent field, which extends from the waveguide into the surrounding medium, allows for the detection of changes in the surrounding refractive index (Butt et al., 2023). When the surrounding environment with different refractive indices interacts with the waveguide, it perturbs the evanescent field produced at the surface, which is then monitored by the appropriate detection tools. The generated evanescent field at the interface decays exponentially into the surrounding medium over a distance ranging from a few tens to hundreds of nanometers. The analytes located within the decay region alter the propagated signal significantly and hence exhibit a distinct spectrum compared to those present further from the surface (Butt et al., 2023). Therefore, the performance of the devices is determined based on variations in the evanescent field. An efficient detection tool can significantly discriminate between target molecules present in the bulk material and those absorbed onto the surface. Silicon oxide-based optical waveguides have been widely employed as alternative platforms for identifying toxic gases via evanescent field absorption sensing (Eranna et al., 2004). In case of gas sensing based on evanescent field absorption, the approach is effective only if the target gas demonstrates a well-defined absorption peak at the selected wavelength.

Optically resonant-based sensing devices have gained significant attention in the development of label-free biomolecular sensing due to their capability for multiplexed detection and their ability to confine the launched light into a compact waveguide structure. The major limitation of conventional biosensors is that the interaction with target molecules occurs primarily at the surface of the device, whereas the launched electromagnetic energy is tightly confined within the core. In 2012, Mancuso et al. demonstrated the fabrication of a nanoporous polymer ring resonator, which showed a 40% improvement in sensor sensitivity, attributed to the direct interaction of the guided optical field and the target analytes (Mancuso et al., 2012). In the reported work, the authors utilize an Ando

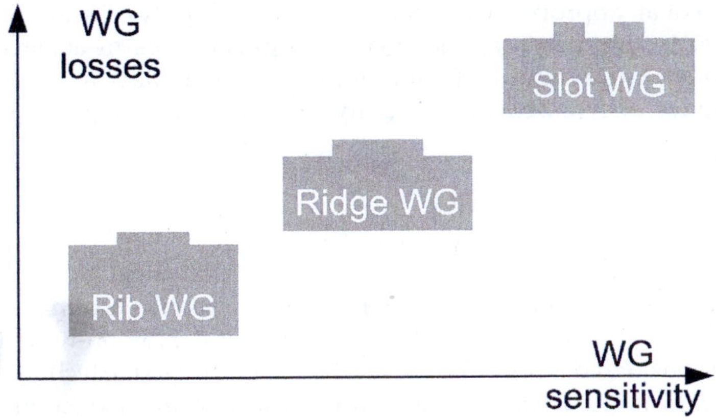

FIGURE 3.4 Optical waveguides. Comparison of various types of optical waveguides. *From Butt, M. A., Kazanskiy, N. L., Khonina, S. N., Voronkov, G. S., Grakhova, E. P., & Kutluyarov, R. V. (2023). A review on photonic sensing technologies: Status and outlook.* Biosensors, *13(5), 568. https://doi.org/10.3390/bios13050568.*

AQ4321D laser source to couple light into the polymeric waveguides. In order to ensure precise alignment, the chip and fiber were positioned on 3-axis stages, with the waveguide's input and output facets recessed by 3 mm to reduce the collection of the dispersed light.

In the past decade, extensive investigations have been conducted in order to improve the structural geometries of optical waveguides to achieve enhanced sensing performance (Wang et al., 2024). Typically, three different types of optical waveguides, namely rib waveguide, ridge waveguide, and slot waveguide, have been widely explored for sensing applications. The top cladding layer, which covers the measurand, interacts strongly with the evanescent field of the guided mode. The amount of light that reaches the upper cladding of each waveguide structure varies and is directly linked with increased optical losses. A higher degree of light penetration into the upper cladding results in greater losses due to absorption and scattering (Butt et al., 2023). The launched light is generally tightly confined within the Si core of waveguide structures, comprising high-refractive-index, e.g., strip and rib designs, whereas in the case of the slot waveguide design, the propagated light can be effectively confined within the sub-wavelength, low-refractive-index region formed between two parallel Si waveguides. Slot waveguides are considered to be extremely sensitive compared to ridge waveguides, as they enable significant spatial interaction between the surrounding environment and the evanescent field at the interface (Lim et al., 2024). Slot waveguides are therefore a preferable option that can be used for bulk index sensing. Fig. 3.4 represents the different optical waveguides along with their associated optical losses and sensitivities. The selection of an appropriate optical waveguide is highly critical and should be guided by the specific application. Low optical losses are achieved at the expense of the sensitivity offered by the rib waveguide. In contrast, slot waveguides offer excellent sensitivity but experience significant optical loss.

3.5 Wearable sensor

Wearable sensors have gained significant attention in health monitoring since the appearance of the COVID-19 pandemic. Human-wearable sensors are considered among the most valuable cutting-edge technologies introduced in recent decades, with a variety of applications including remote monitoring, health monitoring, ailment detection, home rehabilitation, detection of pesticide residues, and evaluation of medical treatment efficiency (Kaur et al., 2023). These wearable sensors enable continuous monitoring of various crucial physiological parameters, including heart rate, blood oxygen levels, wrist pulse, body temperature, respiration rate, as well as non-invasive biofluids such as perspiration and saliva. The major advantages of wearable sensors are their capability to provide real-time data for health management and early detection of health-related abnormalities. These sensors are commercially available in the market in the form of wrist bands, gloves, smart watches, tattoos, facemasks, patches, and glucometers (Nasiri & Khosravani, 2020; Promphet et al., 2021; Qiao et al., 2022). The fabrication of a wearable sensor requires a sensing element that responds to the target physical parameter. These selected sensing elements can rely on various principles, including capacitive, piezoelectric, magnetic, resistive, or optical (Shahbaz et al., 2023). The sensing element design is modified based on the nature of the sensor and its specific application. It is highly crucial to recognize that the sensing performance of the sensors can vary according to the sensor design, quality of manufacturing, environmental conditions, and selected application. The sensing performance of the device varies across different sensor types; therefore, the proper selection of a device for a target application requires careful consideration of various factors, including detection range, resolution, and the characteristics of the physical parameter being measured. The global wearable sensors market size was estimated at USD 1100 million in 2023 and is projected to surpass approximately USD 4900 million by 2033, growing at a compound annual growth rate (CAGR) of 15.77% from 2024 to 2033 (Research, 2024). A wearable sensor, e.g., a customized fitness monitoring system, is currently used by more than one in ten Americans, representing a threefold increase from 2012 (Shahbaz et al., 2023). The growth in the wearable sensor market has been driven by compactness, accessibility, and a user-friendly interface. For example, the integration of smartphones with portable devices such as smartwatches, glucometers, and other connected tools provides crucial information to physicians, enabling them to collect high-quality medical data from patients and accurately assess their health conditions (Nachiar et al., 2020). Despite the growth of the wearable sensor market, researchers continue to focus on enhancing these devices to collect additional and more meaningful physiological information from the body. The non-specific nature of many current wearable sensors (e.g., pulse monitoring and galvanic skin response) limits their accuracy and specificity, thereby reducing their effectiveness in reliable diagnostics. For example, heart rate can be affected by intense physical activity or even brisk walking, which complicates the identification of the underlying cause. Non-intrusive measurements are generally performed using an optical method, in which a specific wavelength of light interacts with the area of the skin to be assessed. The sensing device collects information from the skin by capturing the reflected, refracted, and absorbed light, and then interprets the obtained data, similar to the operation of a spectrophotometer (Kaur et al., 2023). A key factor in transmitting an optical signal through the skin is determining the penetration depth of the light. The device can use any light signal ranging from UV to IR, depending on the required depth of penetration and the absorption peaks relevant to the specific sensing application.

Wearable optical sensors developed for heart rate monitoring, based on photoplethysmography (PPG), have attracted significant attention. These sensors provide precise and real-time data, assisting clinicians in identifying the causes of excessive sweating and rapid heart rate (Colvonen, 2021). Several companies, such as Microsoft, Fitbit, Apple, Sony, Motorola, Masimo, Xiaomi, and MioGlobal, have developed various types of optical PPG sensing devices. These devices are often developed in the form of smartwatches, which can be worn on the wrist or placed on the chest, and they operate in reflection mode to acquire physiological data. Due to their small size, portability, and capability to generate stronger signals through transmission-based readings, particularly when worn on the finger, ring-type PPG devices have also become noteworthy. In recent days, numerous optical and electrical measurements can be performed using commercially available devices, including pulse tracking via PPG, blood oxygen level monitoring, step counting, respiration rate, and body temperature measurement, and even sleep quality assessment (Guillodo et al., 2020). Athletes are increasingly interested in adopting new technologies to track their fitness, improve physical performance, and maintain overall health. Nowadays, most athletes utilize wearable devices as part of their daily training and recovery routines. Sports teams currently employ wearable technology to monitor both the internal and external workloads on athletes. The next advancement for sports medicine practitioners involves unobtrusively tracking biomarkers from sweat and saliva, enabling the development of personalized hydration and recovery plans for each athlete (Seshadri et al., 2019). Wearable sensing technology has advanced remarkably over the past two decades from being regarded merely as tools to being recognized as strategic systems that integrate hardware with data analytics. However, there is still a need to standardize, evaluate, and validate these technologies from a clinical and sports medicine perspective. Earlier investigations have primarily been concerned with specific technical aspects of the wearable, such as materials (Yao et al., 2018), smart sensors (Choi et al., 2018; Heikenfeld et al., 2018), and soft interfaces (Someya et al., 2016), and have employed these components in wearable devices to monitor health conditions, including cystic fibrosis (Emaminejad et al., 2017), atrial fibrillation (Steinberg & Piccini, 2018), and diabetes (Wang, 2008). In digital health monitoring systems, wearable sensors serve as one of the most effective tools. For the rapid, real-time, and long-term monitoring and collection of health related data, the optical interrogator has been reported as the first fully integrated demonstration of a photonic wearable sensor (Li et al., 2021). The optical interrogator is known for its compactness, robustness, and cost-effectiveness due to its miniaturized integration of multiple photonic elements onto a silicon chip. In this design, the optical interrogator was fabricated within a bracelet, allowing it to perform real-time monitoring of heart rate and body temperature. Fig. 3.5 exhibits how the flexible integrated optical system comprises three fiber Bragg gratings (FBGs) mounted on skin-tight garments, along with a flexible electro-optical circuit board designed in the form of a wristband.

3.6 Metasurface based sensors

In the past two decades, metamaterials have gained significant attention in photonic sensing due to their exceptional electromagnetic properties (Butt et al., 2023; Kazanskiy et al., 2023; Qin et al., 2022). Metamaterials comprise an array of systematically arranged scattering components made of zero-index materials, negative-index media, and ultra-high-index materials. Metasurfaces are the two-dimensional equivalents of these artificial materials, consisting of an array of metamaterial structures. Metasurfaces have demonstrated a remarkable ability to manipulate electromagnetic wavefronts, primarily due to the interaction of electromagnetic waves and the meta-atom structures along with their functional configurations (Butt et al., 2023).

Metasurface technology has been widely exploited in chemical detection and biosensing, which have been studied in the optical and microwave domains. The sensitivity of the devices has also been reported to be enhanced by utilizing plasmon-induced transparency (Butt & Kazansky, 2020; Kazanskiy et al., 2021; Zhang et al., 2021). Modern advances in

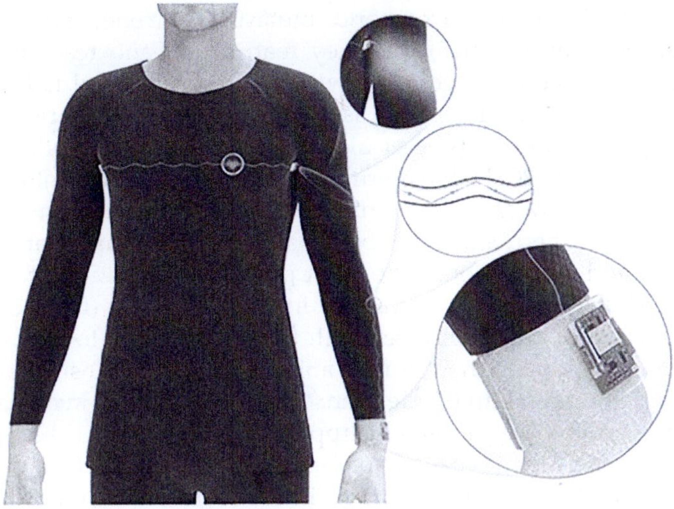

FIGURE 3.5 Photonics wearable sensor. Flexible optical sensors implemented on human body. *From Li, H., An, Z., Zhang, S., Zuo, S., Zhu, W., Zhang, S., Huang, B., Cao, L., Zhang, C., Zhang, Z., Song, W., Mao, Q., Mu, Y., Miao, C., Li, E., & Prades, J. D. (2021). Fully photonic integrated wearable optical interrogator.* ACS Photonics, *8(12), 3607–3618. https://doi.org/10.1021/acsphotonics.1c01236.*

science and technology require a transient response from metasurfaces; however, conventional metasurfaces are mostly limited in customization and tunability. Traditional metasurfaces provide fixed, pre-defined optical functionalities, mainly driven by shape, topology of meta-atoms, and configuration. Several researchers have already developed flexible and reconfigurable metasurfaces, in which the size, shape, and arrangement of meta-atoms can be adjusted or modified in response to external signals.

Metasurfaces are typically fabricated using dielectric materials, which potentially reduce optical loss due to their large bandgap energy that prevents optically produced interband transitions. Unlike metallic nanoparticles, which experience significant optical losses due to a large portion of the incident energy being converted into heat and the electric field being concentrated strongly near their external surfaces, dielectric materials exhibit minimal optical losses and therefore achieve exceptional electric field concentration within their interiors. Metasurface-based devices can be divided into three categories: (1) narrowband (Butt & Kazansky, 2020), (2) broadband (Azad et al., 2016), and (3) wideband (Lalbakhsh et al., 2021), based on the utilized materials, surface morphology, and their frequency responses.

One example of a narrowband metasurface relies on Fano resonance(Limonov et al., 2017). The key feature of Fano resonance lies in its sharp resonance peak, coupled with a significant local field enhancement. This characteristic serves as the foundation for various biological sensing applications (Lee et al., 2015) and nonlinear photonics systems (Günay et al., 2021). Furthermore, high-performance color filters (Z. Xu et al., 2021), broadband absorbers (Z. Xu et al., 2021), and polarizers (Intaravanne & Chen, 2020), can be considered as examples of multiband and broadband applications. The most popular application of metamaterials and metasurfaces is in refractive index biosensing (Lan et al., 2020). The interaction of target analytes with the device leads to changes in the refractive index. Due to their unique ability to conduct sensitive, label-free biosensing, these refractive index sensors can be utilized in a broad range of chemical and biological sensing applications.

The following four key factors determine whether a metasurface sensor can effectively detect small variations in the surrounding environment:

1. The operating frequency of the sensor must be tuned to be as low as possible in order to avoid substrate and background absorption. Conventional sensing devices pose a significant challenge due to their limited surface area and unique characteristics to operate at higher frequencies. Therefore, it is necessary to maintain a compact sensor design while reducing its operating frequency.
2. The fabricated sensor must produce a detectable readout signal along with a sharp resonant peak to maintain accuracy in tracking the change in the transmission spectra.
3. The sensor must exhibit a linear response, which is related to the quality factor (Q-factor) of the sensor. A high Q-factor in metasurface sensors enhances sensitivity by producing sharp resonance peaks, allowing monitoring of small changes in the ambient environment.
4. Sensitivity is the final and most critical requirement of metasurface sensors. High sensitivity enables the sensor to distinguish even small shifts in the output spectra resulting from variation in the ambient environment.

Resonant micro- and nano-photonic refractive index sensors have been employed over the past decades for real-time monitoring of chemical and biological samples without labelling them, e.g., finding organic liquid components or detecting target biomolecules in a biological fluid (Butt et al., 2020; Zubaidah binti Haji Jumat et al., 2021). The basic working mechanism of such devices exploits various properties of light, such as phase, intensity, and wavelength, to monitor the performance of the sensing device by analyzing the reflection or transmission spectra obtained from the sensor. The metasurface sensor can be characterized using the wavelength interrogation method. This approach, typically utilized in resonant

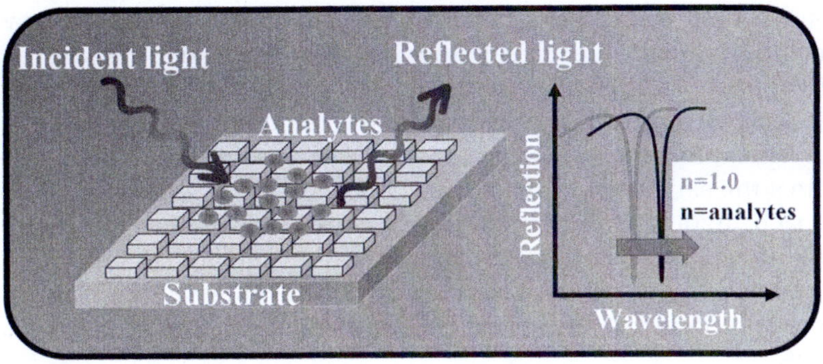

FIGURE 3.6 Metasurface refractive index sensor. Working principle of a metasurface sensor based on the wavelength interrogation technique. *From Kazanskiy, N. L., Khonina, S. N., & Butt, M. A. (2023). Recent development in metasurfaces: A focus on sensing applications.* Nanomaterials, 13*(1), 118. https://doi.org/10.3390/nano13010118.*

structures, involves monitoring shifts in the resonant wavelength, which represents the most popular interrogation technique adopted in photonic sensors (Butt et al., 2020). The variation in the refractive index of the ambient environment greatly influences the effective refractive index of the resonant mode, resulting in a shift in the resonant wavelength. Optical systems utilizing wavelength interrogation typically rely on broadband light sources, such as superluminescent diodes or halogen lamps, which provide a broad spectral range where the resonant wavelength is expected to occur (Kazanskiy et al., 2020, 2023). Fig. 3.6 (left) exhibits a simple schematic diagram of a metasurface-based refractive index sensor, where a broadband light interacts with the metasurfaces, launched from a specific angle, while the reflected beam is collected via a spectrum analyzer. The right figure shows the reflection spectrum for the same device, demonstrating a resonance dip at a certain wavelength, which undergoes a redshift when the ambient environment changes from air to certain analytes.

3.7 Plasmonic sensors

The integration of noble metals such as silver, gold, copper, and aluminium has led to the development of surface plasmon resonance (SPR), a phenomenon that has been extensively exploited over the past decades to achieve enhanced detection accuracy and high sensitivity (Liu et al., 2024; Vaidyanathan et al., 2024). SPR is known as the resonant oscillation of conduction electrons at the metal–dielectric interface, which is stimulated by incident light (Pathak et al., 2019). The SPR phenomenon can be categorized into two main classes: conventional SPR, which is based on the thin films and bulk-metal dielectric interfaces, and localized surface plasmon resonance (LSPR), which arises from metallic nanostructures such as gold or silver nanoparticles. Two different basic approaches have been reported in optical sensing to integrate plasmonic materials with optical waveguides by exploiting surface plasmon polaritons (SPP). The first method requires the activation of the surface plasmon wave, whereas the second approach involves the stimulation of a pure plasmonic mode. In both cases, the excited plasmon wave, or SPP at the metal-dielectric interface, is affected even by small variations in the localized refractive index, which is generally detected and analyzed to evaluate the performance of the sensors.

Plasmonic-based photonic sensors are utilized in a broad range of applications, from basic measurement to advanced artificial vision systems for object identification (Butt et al., 2023). One of the major challenges faced by the sensor industry is the development of innovative nanostructures with tailored functionalities for target-specific detection (Brunetti et al., 2022; Koya et al., 2020). The concept of utilizing SPPs, among other reported nanotechnologies, distinguishes this approach from its potential competitors. Metallic nanostructures have shown great potential to produce and propagate electromagnetic radiation in fundamentally new and unprecedented ways. Due to the notable advancements in micro- and nanofabrication techniques, plasmonic-based photonic sensors have shown significant performance in a broad range of applications, including biosensing, chemical sensing, gas sensing, food safety, environmental monitoring, and medical diagnostics. These reported sensors have shown exceptional performance in detecting various measurable parameters, particularly in biosensing. In 2020, a group of researchers from Japan and Italy reported a nanoplasmonic-based sensor for COVID-19 detection, utilizing surface patterning and biotin tag-based protein functionalization methods (Funari et al., 2022). The fabricated nanoplasmonic sensor relies on the shift in the wavelength spectra in the presence of antibodies, which occurs due to the LSPR phenomenon generated by gold nanoparticles.

Comparing the performance of plasmonic sensors to those based on conventional sensing platforms, e.g., silicon photonic or optical fibers, reveals that the plasmonic-based sensors show significantly smaller footprints and demonstrate extremely high sensing capabilities, making them highly disirable for modern sensing applications. Plasmonic waveguide structures, particularly those based on metal-insulator-metal configurations, have gained significant attention due to their small footprint, ease of integration, high detection accuracy, and effective balance between localized light and transmission loss, as well as their ability to overcome the diffraction limit of light. In 2017, Wang et al. reported a refractive index sensor based on leaky resonant scattering from a single semiconductor nanowire and achieved a sensitivity of 235 nm/RIU (Wang et al., 2017). In the same year, Teng et al. reported the application of a standard optical fiber combined with long-period gratings as the basis for a plasmonic-based refractive index optical fiber sensor (Teng et al., 2017). It was observed that bending the standard optical fiber with a long-period grating into a U-shape significantly improves the sensing performance of the sensor. With optimized structural parameters, the sensor exhibited a maximum sensitivity of 1130%/RIU, along with a resolution of 8.44×10^{-4}, measured in the refractive index range from 1.33 to 1.41.

The plasmonic-based sensor has been widely investigated for temperature and refractive index monitoring (Amirjani & Sadrnezhaad, 2021; Gallinet et al., 2015). Although in some cases the designed sensors exhibited extremely high sensitivity (e.g., 90,500 nm/RIU) (Pathak et al., 2022), the proposed sensing configurations were mainly analyzed for a single parameter at a time, either refractive index or temperature. Moreover, the geometric complexity, including the intricate design, precise thickness, and density of the metallic nanostructures or thin films, makes it challenging to fabricate these devices without compromising the structural integrity and sensing performance. In 2020, Mahdiye et al. reported a metal-insulator-metal-coated ring resonator for refractive index sensing (Rahmatiyar et al., 2020). The reported sensor exhibited a maximum sensitivity of 1295 nm/RIU and a high figure of merit of 159.6 RIU^{-1}. Later in 2019, Yang et al. reported another complex design using a metal-insulator-metal waveguide combined with symmetric triangular stubs coupled to a circular split-ring resonator for refractive index sensing. The fabricated device exhibited a high sensitivity of 1500 nm/RIU and a high figure of merit of 65.2 (Yang et al., 2019). All the reported designs involve rigorous optimization of various structural parameters to attain high sensing performance. While the findings of the reported research appear promising, significant challenges emerge during the actual fabrication of these devices, where several important factors require precise adjustment at the nanoscale.

Recent developments in this field have demonstrated the significant potential of metal-insulator-metal plasmonic waveguides in temperature sensing applications when integrated with thermally sensitive materials such as polydimethylsiloxane (PDMS) or ethanol (Kazanskiy et al., 2021). In 2018, Zhu and Lou reported a highly sensitive and multifunctional plasmonic temperature sensor utilizing ethanol-sealed elliptical resonators (Zhu & Lou, 2018). The reported design exhibited an exceptional sensitivity of −3.64 nm/°C compared to the conventional sensor. Even after achieving high sensitivity, the capability of the proposed sensor was limited to temperature detection only. Sensors based on PDMS are highly sensitive to variations in temperature due to their high thermo-optic coefficient. These PDMS-based sensors are generally utilized in areas that require a high level of precision in temperature monitoring. PDMS, poly (methyl methacrylate) (PMMA), and other types of functional polymers also offer high permeability to various gases and are therefore known to be suitable candidates for various gas sensing applications when integrated with plasmonic waveguides (Pathak et al., 2023). A functional polymer, namely polyhexamethylene biguanide, shows a linear relationship in the presence of various CO_2 concentrations. These polymers can potentially be used to design a new class of CO_2 gas sensors. In 2022, Butt et al. reported an on-chip plasmonic Bragg grating-based CO_2 sensor (Butt et al., 2022). The sensor design comprises the coating of gold nanoblocks with polyhexamethylene biguanide, which are periodically stacked in the MIM waveguide, resulting in a broadband Bragg reflection. The designed sensor exhibited a sensitivity of 226 pm/ppm for a broad range of CO_2 concentrations, varying from 0 ppm to 524 ppm. Overall, the plasmonic-based photonic sensor opens new opportunities for further exploration to improve performance, detection accuracy, and response time of the sensing devices.

3.8 Photonics sensors market

The revenue generated by photonic sensors and detectors from the worldwide market was valued at USD 17.8 billion in 2022 and is projected to reach USD 70.20 billion by 2031, growing at a (CAGR) of 16.93%. This global photonic market is expanding due to its increasing applications in consumer electronics, chemicals, healthcare, medical sectors, and industrial automation ("Photonic Sensors & Detectors Market Outlook," no date).

In order to maintain human safety, enable *in-situ* monitoring, and improve interactions at machine-environment interfaces, photonic sensors and detectors are widely employed in various hazardous environments. Nowadays, photonic sensors, which offer lightweight, compact, and low-noise alternatives to traditional bulky sensors, are greatly influencing the image sensing sector. Science now increasingly recognizes the value of photonics sensors, specifically for revealing information that is not immediately observable. Over the forecast period, the photonic sensor market is expected to grow due to the availability of faster, lighter, more cost-effective, miniaturized, and highly functional components and systems that offer enhanced energy efficiency. Recent advancements in photonic sensing, including miniaturization, improved sensitivity, and target specificity, are expected to drive the growth of the photonic sensor market and increase the commercial value of these devices. Ongoing research and development remain key contributors to the advancement of photonic sensors as well as their growing market value. The emergence of internet of things (IoT), autonomous vehicles, and smart cities, is anticipated to create additional market opportunities for photonic sensors (Shahbaz et al., 2023).

FIGURE 3.7 Fabrication of photonic devices. Standard and advanced nano/micro fabrication techniques utilized in the development of photonic devices. *From Butt, M. A., Mateos, X., & Piramidowicz, R. (2024). Photonics sensors: A perspective on current advancements, emerging challenges, and potential solutions (Invited).* Physics Letters A, 516, *129633. https://doi.org/10.1016/j.physleta.2024.129633.*

3.9 Fabrication method of photonic sensors

The fabrication of miniaturized photonic sensors involves various complex processes to fabricate accurate and precise optical components on various substrates. Several techniques have been reported over the past decades to fabricate such photonic devices, as shown in Fig. 3.7. Among these approaches, photolithography is one of the most common techniques used for fabricating miniaturized devices. This process involves patterning a photoresist layer by exposing it to ultraviolet (UV) or deep ultraviolet (DUV) light through a mask, followed by etching to transfer the pattern onto the chosen substrate. This technique is utilized to fabricate complex waveguide structures, including grating, interferometry, microfluidic, and other optical modules, with high precision and accuracy. Another approach is electron-beam lithography (EBL), which offers finer features at the sub-10 nm scale by utilizing a focused electron beam to directly write desired patterns on a substrate. Nanoimprint lithography (NIL) is another widely recognized nanofabrication technology utilized to fabricate nanoscale structures. In NIL, a template with the required pattern is pressed into a polymer film, thereby transferring the pattern onto the substrate. This process allows the fabrication of high-resolution structures with potential applications in photonic sensors, including waveguides, optical gratings, and PC (Grego et al., 2010). Although NIL offers several advantages, it also presents some limitations in achieving precise alignment and registration of multiple layers with nanoscale features. Furthermore, the selection of appropriate imprint materials and the risk of defects during the fabrication process may significantly affect the printing quality and reliability of optical components developed by NIL.

The patterning of the device requires the deposition of appropriate material on the substrate. Several techniques have been reported over-the past decades for material deposition, such as physical vapor deposition (PVD), chemical vapor deposition (CVD), molecular beam epitaxy (MBE), and atomic layer deposition (ALD), which are utilized to create optical coatings, metal layers for plasmonic sensors, and semiconductor layers for integrated photonic devices. Alternatively, the sol-gel and dip-coating methods offer a cost-effective way to produce high-quality and uniform thin films for various applications. This process includes repeatedly immersing and withdrawing a substrate from a sol solution, forming a uniform thin film upon drying and heat treatment. Its simplicity, precise thickness control, and ability to cover large areas make it ideal for achieving complex compositions and customized optical properties through doping. However, the major drawbacks include cracking of the deposited films during the drying process, difficulty in achieving uniformity, specifically on non-planar substrates, and constraints in scaling for mass production. During material deposition, controlling the refractive index and film porosity is a key factor carefully considered for optimizing the optical performance. Regardless of these limitations, the dip-coating and sol-gel methods remain valuable for the deposition of functional thin films in photonic devices. Ongoing efforts aim to increase the versatility of this approach and improve performance through process optimization and material design (Butt, 2022). Continued advancements in fabrication techniques will further improve innovation and facilitate the fabrication of more advanced and efficient photonic sensor devices.

Etching techniques are crucial for fabricating precise and finely resolved structures and patterns on selected substrates. Reactive ion etching (RIE), a well-known form of dry etching that utilizes a plasma approach, is widely utilized due to its ability to create deep, high-resolution features, as well as its high etch rate, which is beneficial for semiconductors, silicon, and dielectrics. The wet chemical etching technique allows greater flexibility in selectively dissolving materials with chemical solutions, enabling versatile structuring on various substrates. This technique provides a more accurate definition of intricate patterns when combined with photolithography. It plays a key role in the advancement of micro-resonators, optical waveguides, and gratings in micro-photonic devices used across a broad range of applications, including telecommunications, biomedical fields, and sensing. Direct laser writing and laser micromachining enable accurate patterning and modification of optical materials. Additionally, 3D printing techniques, including two-photon polymerization and soft lithography, are gaining significant attention for the rapid prototyping and mass production of precise, finely resolved complex photonic sensors. These fabrication approaches are essential for achieving the required features, along with the optical properties and functionalities that are vital for photonic sensors used in various applications.

3.10 Application of photonic devices in volatile organic compound monitoring

Photonic devices have gained significant attention for the real-time monitoring of VOC gases due to their remarkable characteristics, as discussed in the above sections. These include fast response time, compact size, low cost, excellent immunity to electromagnetic interference, operation in harsh environments, absence of electrical hazards, and multiplexing capability (Gangwar et al., 2024). In order to determine various chemical parameters, including toxic gases, proteins, and viruses in both industrial and medical sectors, several types of optical sensors have been reported and commercialized over the past decades (Prabowo et al., 2018). Among the reported optical sensing approaches, colorimetry, SPR, interferometry, fluorescence, optical fibers, planar waveguides, and PC -based sensors have witnessed notable progress due to their rapid sensing response (Baird & Myszka, 2001). Given the low concentrations of VOCs in samples and the presence of substantial interferences in mixed-gas environments, improving device sensitivity and target selectivity has remained a long-term goal. Recent advancements in functional sensing materials have enabled the development of more efficient VOC sensors with significantly improved sensing performance (Pathak et al., 2023; Toma et al., 2021). Additionally, integrating such sensing materials with various types of photonic platforms, such as optical fibers, planar waveguides, and PC, along with machine learning (ML) and the IoT, allows for more efficient signal processing and analysis. The measurements usually generate a large volume of data due to continuous and real-time monitoring. Therefore, the integration of ML and IoT with a smart sensor system leverages previously generated data to predict future responses, thereby enabling the development of intelligent, wearable, and portable devices for point-of-care testing (Mazzone et al., 2007). Based on the above advancements in photonics technologies, a range of major diseases has been included in VOC-based optical sensing applications, such as COVID-19 (Soler et al., 2020), lung diseases (Cognetti et al., 2022), liver diseases (Soltani et al., 2023), inflammation (Leanse et al., 2023), diabetes (Doulamis et al., 2021), and bacterial infections (King et al., 2015; Nath et al., 2020).

Real-time monitoring of VOC concentrations is crucial for the identification of specific diseases. As discussed in Chapter 1, these VOCs serve as biomarkers for non-invasive diagnosis of cancerous, metabolic, infectious, genetic, and other diseases. Over the past decades, various sensor technologies, including metal oxide semiconductor sensors (MOSs), photoionization detectors (PIDs), chemiresistive sensors, and optical sensors, have been successfully explored for VOC detection at concentrations as low as the ppt level (Majhi et al., 2021; Wang et al., 2020a; Wei Xu et al., 2021; Yamazoe & Shimanoe, 2009). Among all these methods, the optical fiber-based VOC sensors have made significant progress in VOC detection, demonstrating superior sensing performance compared to conventional gas sensors. The integration of advanced sensing materials with optical fiber makes these devices promising candidates for medical applications and industrial purposes. These sensing materials exhibit variations in their chemical and physical properties, such as refractive index (RI), absorption, conductivity, stress, and strain, in the presence of VOC gases (Jamila et al., 2024). Evanescent wave-based intrinsic sensors are considered the most advanced sensing technique, as they monitor changes in the intensity of transmitted light due to variations in the refractive index of the sensing materials with the ambient environment. Recently, researchers have reported advanced sensing strategies by functionalizing the surface of the fiber, typically over the core, with high refractive index materials and/or nanomaterials for the selective detection of target molecules.

In general, optical fiber-based VOC sensors composed of various materials and nanocomposites, as these materials exhibit quantum effects at the nanoscale and are thus considered promising candidates for VOC sensing (Elosua et al., 2017). For example, in 2018, Narasimman et al. reported an optical fiber sensor coated with zinc oxide (ZnO) nanoflakes and achieved improved selectivity for acetone and ammonia (Narasimman et al., 2018). Later in 2021, Prasanth et al. fabricated a tin oxide (SnO_2) coated plastic-clad silica multimode fiber and achieved a good sensitivity of 3.40 nm/vol% and stability towards 80% ethanol (Prasanth et al., 2021). Recently, Nirmala and Vadivel synthesized a $Zn_3(VO_4)_2$/PMMA composite using the hydrothermal method and deposited it onto an optical fiber for the detection of various toxic gases (Nirmala Devi & Vadivel, 2024). The fabricated sensor exhibited an average response time of 22.4 minutes along with a maximum recovery time of 31.5 minutes for 127 counts/kPa of ammonia. Given the potential of optical fiber, there is increasing demand for the development of optical fiber-based VOC sensors that utilize various nanomaterials and nanocomposites for environmental monitoring and medical diagnosis (Velumani et al., 2022). 2D sensing materials with thickness ranging from a monolayer to a few nanometers demonstrate unique and exciting sensing capabilities. Since 2004, graphene has emerged as a leading sensing material, demonstrating remarkable properties utilized in energy storage, nanoelectronics, biological applications, and sensors (Urade et al., 2023). Similar to graphene, several other 2D materials have been reported in recent decades, such as hexagonal boron nitride (h-BN), molybdenum disulfide (MoS_2), and tungsten disulfide (WS_2), which hold great potential in sensing applications due to their unique optical, electronic, and morphological properties (Donarelli & Ottaviano, 2018; Shim et al., 2017).

3.11 Advantages, limitations, and challenges of photonic devices

The increasing demand for sensors across a broad range of sectors, including security, defense, aerospace, automotive, environmental monitoring, food quality monitoring, and healthcare, has led to significant advancements in the past decades. Concerning complementary metal-oxide-semiconductor (CMOS)-compatible silicon-on-insulator (SOI) technology, the focus is on advancing and integrating photonic innovation with microfluidic technology in order to improve sensing performance, specifically in terms of portability, compactness, and sensitivity. The compactness of photonic devices allows them to precisely integrate multiple functionality on a single platform, known as lab-on-a-chip systems. These lab-on-a-chip enable cutting-edge photonic sensing due to their high sensitivity, target selectivity, along with the excellent stability, immunity to electromagnetic interference, and technological advancements at lower costs.

There are various applications of waveguide-based optical sensors that have been reported in the past decades, including label-free and real-time monitoring of chemical or biological targets that precisely interact with the functionalized waveguide surfaces. By exploiting well-known photonic integration substrates such as silicon nitride or silicon substrates, these devices exhibit significant prospects for small and low-cost mass production. The most common sensor configurations are based on interferometric techniques, such as the Mach–Zehnder (Doughan et al., 2024) and Michelson (Ghoname et al., 2024), or involve resonant components, including disk, ring, and Bragg resonators, which may significantly improve performance through using the Vernier effect (Claes et al., 2010).

These sensor designs allow extended interaction with the target analyte while maintaining compactness and high sensitivity, making them suitable for high-density integration into massive parallel arrays (Butt et al., 2023). Different optimization methods have been reported in the past for certain types of waveguides for both (1) surface sensing, when the measurand is attached to the top of the waveguide, and (2) homogeneous sensing, which involves monitoring the variation of refractive index in the waveguide cladding under different ambient conditions. These investigations, however, mostly focused on a few types of waveguides made of silicon (Debackere et al., 2007; Dell'Olio & Passaro, 2007), silicon nitride (Milvich et al., 2018), and polymers (Chen et al., 2021; Zhang et al., 2019). Therefore, surface sensitivities among different types of waveguides and integration platforms cannot be directed compared. In addition, most of the sensitivity assessments consider only specific surface layers with fixed refractive indices. Another interesting form of waveguide is PC, which offer an exciting way to achieve improved sensing performance by confining and manipulating the behavior of launched light. Over the past few decades, numerous configurations of PC have been widely used in photonic sensing. These crystals exhibit strong confinement of light in a very small region, enabling chemical and biological sensing at the nanoscale. From the technical point of view, the photonic crystals-based optical sensors include integrated planar PC and photonic crystal fibers, which are considered among the most appropriate forms of devices used in multiplexing and label-free detection. PC planar waveguides are generally fabricated in similar ways as planar waveguides, using conventional E-beam lithography, ICP etching, and PECVD, whereas the photonic crystal fibers are fabricated using the stack-and-draw method. The major limitations of the photonic crystal-based optical sensors are imperfections that occur during the fabrication, which can significantly affect the sensing performance. They may also require complex integration with other components, limiting scalability. Additionally, their performance can be impacted by environmental conditions, leading to variability in measurements.

It has been evident that the optical fiber-based sensors have undergone great advancements in the past decades in terms of sensing performance and robust design. The deposition of novel sensing materials is widely employed in order to achieve target selectivity and rapid response time (Pendão & Silva, 2022). However, maintaining the uniform deposition and thickness of sensing materials over the cylindrical fiber surface requires full control and precise repeatability. Repeatability is specifically important in the case of 2D material deposition to improve sensitivity. Another challenge associated with optical fiber sensors is the cross-sensitivity towards temperature and other toxic compounds present in the measurand, which may lead to poor selectivity as well as reduced accuracy (Memon et al., 2022).

Metasurface-based sensors are an emerging area of investigation due to their unique characteristics that enable control over light guiding and scattering in extremely miniature devices. The geometric dimensions of each meta-atom, which are crucial components of metamaterials, along with their subwavelength spacing, play a vital role in determining the performance of metamaterials (Butt et al., 2023). The meta-atoms are subwavelength nanostructures made of noble metals or high-index dielectrics and are designed to demonstrate effective local optical responses, which are generally defined in terms of amplitude, phase, and electric and magnetic polarizabilities. Dielectric materials can exhibit a positive dielectric constant in the optical range while remaining loss-free. However, these losses are generally associated with the metallic components, which can significantly restrict the performance of the metamaterial system. Various noble metals, such as silver and gold, are utilized as metamaterials as they exhibit relatively lower losses. For effective sensing applications in the optical range, the losses arising from these metals are still considered high. Additionally, the inherently high and non-tunable negative permittivity values of metals present another challenge (Kazanskiy et al., 2023).

As the market for plasmonic optical sensors continues to grow and technological advancements progress, there remains significant potential to explore this area within the domain of classical optics. In addition to planar waveguides, optical fibers, whispering-gallery mode, PC, and wearable sensors have gained significant attention because of their high-Q factors and moderate mode volumes (Xu et al., 2019). Several researchers have already achieved extremely high sensitivity and single-molecule detection by integrating the properties of plasmonic materials with optical fibers, waveguides, and PC (Baaske et al., 2014; Vincent et al., 2020). Although gold and silver are considered the most commonly used plasmonic materials to improve sensing performance, they possess great limitations in terms of their high cost (for gold), inherent optical loss, and oxidation issue (for silver). In order to resolve these limitations, several researchers have reported alternative materials, namely aluminum and sodium, which offer relatively low losses and are more cost-effective (Jiao et al., 2021; Wang et al., 2020b). Despite the low loss of sodium, its instability in air appears to be a serious challenge for sodium-based devices; however, encasing the device within quartz or epoxy helps to achieve stable performance (Wang et al., 2020b). Furthermore, the integration of plasmonic nanoparticles under controlled deposition has also been reported for quantitative surface-enhanced Raman spectroscopy (SERS) (Gangwar et al., 2024).

3.12 Conclusion

Recent years have witnessed immense growth in photonic sensing technologies due to the increasing demand for highly sensitive, accurate, and portable solutions with extensive applications, including industrial automation, environmental monitoring, and medical diagnosis. The integration of photonic devices with microelectronic components has led to notable progress in the development of compact and cost-effective sensing systems, which can easily be integrated with other waveguides or electronic devices. This synergy has opened new avenues in sensing applications, specifically in healthcare, where such integrated systems allow continuous and real-time monitoring of various physiological parameters. For example, optical fiber-based sensors measure the changes in the properties of the propagated light due to variations in the surrounding parameters such as temperature, strain, pressure, and refractive index. Compared to traditional electronic sensors, they offer several advantages, including improved sensitivity, immunity to electromagnetic interference, the ability to multiplex, and a rapid response time. SPR-based technology is also considered a potential tool to improve the sensing performance of photonic devices in terms of sensitivity and response time. This sensing principle is based on light-matter interactions taking place between the light and the metallic surface coating, enabling the detection of variations in target molecule properties, making them an ideal tool for label-free detection of various biomolecules and other important analytes. Plasmonic sensors, in particular, are being explored for monitoring various environmental pollutants, including heavy metals, organic contaminants, and gaseous compounds. Similarly, photonic crystal sensors utilize shifts in their optical properties to sense changes in the local environment, making them suitable for applications in chemical and biological detection. Their inherent high sensitivity and ability to selectively detect target molecules without labels make them an appropriate tool for real-time environmental monitoring. The sensors can selectively identify and monitor gases, chemical pH values, salinity, dissolved oxygen, and turbidity when functionalized with target-selective materials. With continuous advancements in materials science, fabrication techniques, and device integration, photonic sensors are poised to become crucial tools for next-generation diagnostics, environmental surveillance, and industrial monitoring, marking a new era of precise, real-time, and intelligent sensing technologies.

References

Albert, J., & Lit, J. W. Y. (1990). Full modeling of field-assisted ion exchange for graded index buried channel optical waveguides. *Applied Optics*, *29*(18), 2798–2804. doi:10.1364/AO.29.002798.

Amirjani, A., & Sadrnezhaad, S. K. (2021). Computational electromagnetics in plasmonic nanostructures. *Journal of Materials Chemistry C, 9*(31), 9791–9819. http://pubs.rsc.org/en/journals/journal/tc. doi:10.1039/d1tc01742j.

Azad, A. K., Kort-Kamp, W. J. M., Sykora, M., Weisse-Bernstein, N. R., Luk, T. S., Taylor, A. J., Dalvit, D. A. R., & Chen, H. T. (2016). Metasurface broadband solar absorber. *Scientific Reports, 6*. www.nature.com/srep/index.html. doi:10.1038/srep20347.

Baaske, M. D., Foreman, M. R., & Vollmer, F. (2014). Single-molecule nucleic acid interactions monitored on a label-free microcavity biosensor platform. *Nature Nanotechnology, 9*(11), 933–939. http://www.nature.com/nnano/index.html. doi:10.1038/nnano.2014.180.

Baird, C. L., & Myszka, D. G. (2001). Current and emerging commercial optical biosensors. *Journal of Molecular Recognition, 14*(5), 261–268. doi:10.1002/jmr.544.

Belt, M., Bovington, J., Moreira, R., Bauters, J. F., Heck, M. J. R., Barton, J. S., Bowers, J. E., & Blumenthal, D. J. (2013). Sidewall gratings in ultra-low-loss Si3N4 planar waveguides. *Optics Express, 21*(1), 1181–1188. http://www.opticsinfobase.org/view_article.cfm?gotourl=http%3A%2F%2Fwww.opticsinfobase.org%2FDirectPDFAccess%2FEE48134B-F90F-3323-4685CAEFE31704D1_248310%2Foe-21-1-1181.pdf%3Fda%3D1%26id%3D248310%26seq%3D0%26mobile%3Dno&org=. doi:10.1364/OE.21.001181.

Bohnert, K., Gabus, P., Kostovic, J., & Brändle, H. (2005). Optical fiber sensors for the electric power industry. *Optics and Lasers in Engineering, 43*(3–5), 511–526. doi:10.1016/j.optlaseng.2004.02.008.

Brunetti, G., Sasanelli, N., Armenise, M. N., & Ciminelli, C. (2022). Nanoscale optical trapping by means of dielectric bowtie. *Photonics, 9*(6), 425. https://www.mdpi.com/2304-6732/9/6/425/pdf?version=1655384630. doi:10.3390/photonics9060425.

Butt, M. A., Khonina, S. N., & Kazanskiy, N. L. (2020). A plasmonic colour filter and refractive index sensor applications based on metal–insulator–metal square µ -ring cavities. *Laser Physics, 30*(1), 016205. doi:10.1088/1555-6611/ab5578.

Butt, M. A., Kazanskiy, N. L., Khonina, S. N., Voronkov, G. S., Grakhova, E. P., & Kutluyarov, R. V. (2023). A review on photonic sensing technologies: Status and outlook. *Biosensors, 13*(5), 568. http://www.mdpi.com/journal/biosensors/. doi:10.3390/bios13050568.

Butt, M. A. (2023). Numerical assessment of a metal-insulator-metal waveguide-based plasmonic sensor system for the recognition of tuberculosis in blood plasma. *Micromachines, 14*(4), 729. http://www.mdpi.com/journal/micromachines. doi:10.3390/mi14040729.

Butt, M. A., Kazanskiy, N. L., & Khonina, S. N. (2020). Highly integrated plasmonic sensor design for the simultaneous detection of multiple analytes. *Current Applied Physics, 20*(11), 1274–1280. doi:10.1016/j.cap.2020.08.020.

Butt, M. A., Kazanskiy, N. L., & Khonina, S. N. (2022). On-chip symmetrically and asymmetrically transformed plasmonic Bragg grating formation loaded with a functional polymer for filtering and CO_2 gas sensing applications. *Measurement, 201*, 111694. doi:10.1016/j.measurement.2022.111694.

Butt, M. A., Khonina, S. N., & Kazanskiy, N. L. (2018). Modelling of Rib channel waveguides based on silicon-on-sapphire at 4.67 µm wavelength for evanescent field gas absorption sensor. *Optik, 168*, 692–697. doi:10.1016/j.ijleo.2018.04.134.

Butt, M. A., Khonina, S. N., & Kazanskiy, N. L. (2018). Silicon on silicon dioxide slot waveguide evanescent field gas absorption sensor. *Journal of Modern Optics, 65*(2), 174–178. www.tandf.co.uk/journals/titles/09500340.asp. doi:10.1080/09500340.2017.1382596.

Butt, M. A. (2022). Thin-film coating methods: A successful marriage of high-quality and cost-effectiveness—a brief exploration. *Coatings, 12*(8), 1115. doi:10.3390/coatings12081115.

Butt, M. A., & Kazansky, N. L. (2020). Narrowband perfect metasurface absorber based on impedance matching. *Photonics Letters of Poland, 12*(3), 88. doi:10.4302/plp.v12i3.1041.

Butt, M. A., Mateos, X., & Piramidowicz, R. (2024). Photonics sensors: A perspective on current advancements, emerging challenges, and potential solutions (Invited). *Physics Letters A, 516*, 129633. doi:10.1016/j.physleta.2024.129633.

Cennamo, N., D'Agostino, G., Galatus, R., Bibbò, L., Pesavento, M., & Zeni, L. (2013). Sensors based on surface plasmon resonance in a plastic optical fiber for the detection of trinitrotoluene. *Sensors and Actuators B: Chemical, 188*, 221–226. doi:10.1016/j.snb.2013.07.005.

Chen, M., Zhao, Y., Wei, H. E., Zhu, C., & Krishnaswamy, S. (2021). 3D printed castle style Fabry-Perot microcavity on optical fiber tip as a highly sensitive humidity sensor. *Sensors and Actuators B: Chemical, 328*, 128981. doi:10.1016/j.snb.2020.128981.

Chen, M., Zhao, Y., Wei, H. E., & Krishnaswamy, S. (2019). Cascaded FPI/LPFG interferometer for high-precision simultaneous measurement of strain and temperature. *Optical Fiber Technology, 53*, 102025. doi:10.1016/j.yofte.2019.102025.

Choi, J., Ghaffari, R., Baker, L. B., & Rogers, J. A. (2018). Skin-interfaced systems for sweat collection and analytics. *Science Advances, 4*(2), eaar3921. http://advances.sciencemag.org/content/advances/4/2/eaar3921.full.pdf. doi:10.1126/sciadv.aar3921.

Claes, T., Bogaerts, W., & Bienstman, P. (2010). Experimental characterization of a silicon photonic biosensor consisting of two cascaded ring resonators based on the Vernier-effect and introduction of a curve fitting method for an improved detection limit. *Optics Express, 18*(22), 22747. doi:10.1364/oe.18.022747.

Cognetti, J. S., Moen, M. T., Brewer, M. G., Bryan, M. R., Tice, J. D., McGrath, J. L., & Miller, B. L. (2022). A photonic biosensor-integrated tissue chip platform for real-time sensing of lung epithelial inflammatory markers. *Lab on a Chip, 23*(2), 239–250. http://pubs.rsc.org/en/journals/journal/lc. doi:10.1039/d2lc00864e.

Colvonen, P. J. (2021). Response to: Investigating sources of inaccuracy in wearable optical heart rate sensors. *NPJ Digital Medicine, 4*(1), 38. https://www.nature.com/npjdigitalmed/. doi:10.1038/s41746-021-00408-5.

Corsetti, S., Notaros, M., Sneh, T., Stafford, A., Page, Z. A., & Notaros, J. (2024). Silicon-photonics-enabled chip-based 3D printer. *Light: Science and Applications, 13*(1), 78. https://www.nature.com/lsa/. doi:10.1038/s41377-024-01478-2.

Culshaw, B. (2004). Optical fiber sensor technologies: Opportunities and—perhaps—pitfalls. *Journal of Lightwave Technology, 22*(1), 39–50. doi:10.1109/JLT.2003.822139.

Debackere P., Taillaert D., De Vos K., Scheerlinck S., Bienstman P., Baets R. 2007. Si based waveguide and surface plasmon sensors. In Proceedings of SPIE – The International Society for Optical Engineering (Vol. 6477). doi: 10.1117/12.702040.

Dell'Olio, F., Su, J., Huser, T., Sottile, V., Cortés-Hernández, L. E., & Alix-Panabières, C. (2021). Photonic technologies for liquid biopsies: Recent advances and open research challenges. *Laser and Photonics Reviews, 15*(1), 2000255. http://onlinelibrary.wiley.com/journal/10.1002/. doi:10.1002/lpor.202000255. (ISSN)1863-8899.

Dell'Olio, F., & Passaro, V. M. (2007). Optical sensing by optimized silicon slot waveguides. *Optics Express, 15*(8), 4977. doi:10.1364/OE.15.004977.

Doerr, Christopher R., & Kogelnik, Herwig (2008). Dielectric Waveguide Theory. *Journal of Lightwave Technology, 26*(9), 1176–1187. doi:10.1109/JLT.2008.923632.

Donarelli, M., & Ottaviano, L. (2018). 2D materials for gas sensing applications: A review on graphene oxide, MoS2, WS2 and phosphorene. *Sensors, 18*(11), 3638. doi:10.3390/s18113638.

Doughan, I., Oyemakinwa, K., Ovaskainen, O., & Roussey, M. (2024). Strip-loaded Mach–Zehnder interferometer for absolute refractive index sensing. *Scientific Reports, 14*(1), 53326. doi:10.1038/s41598-024-53326-3.

Doulamis, A., Doulamis, N., Angeli, A., Lazaris, A., Luthman, S., Jayapala, M., Silbernagel, G. Ü., Napp, A., Lazarou, I., Karalis, A., Hoveling, R., Terzopoulos, P., Yamas, A., Georgiadis, P., Maulini, R., & Muller, A. (2021). A non-invasive photonics-based device for monitoring of diabetic foot ulcers: Architectural/sensorial components & technical specifications. *Inventions, 6*(2), 27. doi:10.3390/inventions6020027.

Elosua, C., Arregui, F. J., Del Villar, I., Ruiz-Zamarreño, C., Corres, J. M., Bariain, C., Goicoechea, J., Hernaez, M., Rivero, P. J., Socorro, A. B., Urrutia, A., Sanchez, P., Zubiate, P., Lopez-Torres, D., De Acha, N., Ascorbe, J., Ozcariz, A., & Matias, I. R. (2017).

Micro and nanostructured materials for the development of optical fibre sensors. *Sensors (Switzerland), 17*(10), 2312. http://www.mdpi.com/1424-8220/17/10/2312/pdf. doi:10.3390/s17102312.

Emaminejad, S., Gao, W., Wu, E., Davies, Z. A., Nyein, H. Y. Y., Challa, S., Ryan, S. P., Fahad, H. M., Chen, K., Shahpar, Z., Talebi, S., Milla, C., Javey, A., & Davis, R. W. (2017). Autonomous sweat extraction and analysis applied to cystic fibrosis and glucose monitoring using a fully integrated wearable platform. *Proceedings of the National Academy of Sciences of the United States of America, 114*(18), 4625–4630. http://www.pnas.org/content/114/18/4625.full.pdf. doi:10.1073/pnas.1701740114.

Eranna, G., Joshi, B. C., Runthala, D. P., & Gupta, R. P. (2004). Oxide materials for development of integrated gas sensors – a comprehensive review. *Critical Reviews in Solid State and Materials Sciences, 29*(3–4), 111–188. doi:10.1080/10408430490888977.

Funari, R., Fukuyama, H., & Shen, A. Q. (2022). Nanoplasmonic multiplex biosensing for COVID-19 vaccines. *Biosensors and Bioelectronics, 208,* 114193. doi:10.1016/j.bios.2022.114193.

Gallinet, B., Butet, J., & Martin, O. J. F. (2015). Numerical methods for nanophotonics: Standard problems and future challenges. *Laser and Photonics Reviews, 9*(6), 577–603. http://www3.interscience.wiley.com/journal/113511747/home. doi:10.1002/lpor.201500122.

Gangwar, R. K., Kumari, S., Pathak, A. K., Gutlapalli, S. D., & Meena, M. C. (2023). Optical fiber based temperature sensors: A review. *Optics, 4*(1), 171–197. www.mdpi.com/journal/optics. doi:10.3390/opt4010013.

Gangwar, R. K., Pathak, A. K., Chiavaioli, F., Abu Bakar, M. H., Kamil, Y. M., Mahdi, M. A., & Singh, V. K. (2024). Optical fiber SERS sensors: Unveiling advances, challenges, and applications in a miniaturized technology. *Coordination Chemistry Reviews, 510,* 215861. doi:10.1016/j.ccr.2024.215861.

Gangwar, R. K., Pathak, A. K., & Kumar, S. (2023). Recent progress in photonic crystal devices and their applications: A review. *Photonics, 10*(11), 1199. http://www.mdpi.com/journal/photonics. doi:10.3390/photonics10111199.

Ghoname, A. O., Hassanien, A. E., Goddard, L. L., & Gong, S. (2024). Compact lithium niobate Michelson interferometer modulators based on spiral waveguides. *IEEE Journal of Selected Topics in Quantum Electronics, 30*(4), 1–8. http://ieeexplore.ieee.org/xpl/RecentIssue.jsp?punumber=2944. doi:10.1109/JSTQE.2024.3391693.

Gibson, R., Selfridge, R., Schultz, S., Wang, W., & Forber, R. (2008). Electro-optic sensor from high Q resonance between optical D-fiber and slab waveguide. *Applied Optics, 47*(13), 2234–2240. http://ao.osa.org/Issue.cfm. doi:10.1364/AO.47.002234.

Grego, S., Huffman, A., Lueck, M., Stoner, B. R., & Lannon, J. (2010). Nanoimprint lithography fabrication of waveguide-integrated optical gratings with inexpensive stamps. *Microelectronic Engineering, 87*(10), 1846–1851. http://www.journals.elsevier.com/microelectronic-engineering/. doi:10.1016/j.mee.2009.11.003.

Guillodo, E., Lemey, C., Simonnet, M., Walter, M., Baca-García, E., Masetti, V., Moga, S., Larsen, M., Ropars, J., & Berrouiguet, S. (2020). Clinical applications of mobile health wearable–based sleep monitoring: Systematic review. *JMIR mHealth and uHealth, 8*(4), e10733. doi:10.2196/10733.

Günay, M., Cicek, A., Korozlu, N., Bek, A., & Tasgin, M. E. (2021). Fano enhancement of unlocalized nonlinear optical processes. *Physical Review B, 104*(23), 235407. https://journals.aps.org/prb/abstract/10.1103/PhysRevB.104.235407. doi:10.1103/PhysRevB.104.235407.

Heikenfeld, J., Jajack, A., Rogers, J., Gutruf, P., Tian, L., Pan, T., Li, R., Khine, M., Kim, J., Wang, J., & Kim, J. (2018). Wearable sensors: Modalities, challenges, and prospects. *Lab on a Chip, 18*(2), 217–248. doi:10.1039/c7lc00914c.

Imamura T., Matsui T., Yachi M., Kumagai H. 2019. A low-cost interferometric fiber optic gyro for autonomous driving. In Proceedings of the 32nd International Technical Meeting of the Satellite Division of the Institute of Navigation, ION GNSS+ (pp. 1685–1695). Institute of Navigation Japan.

Intaravanne, Y., & Chen, X. (2020). Recent advances in optical metasurfaces for polarization detection and engineered polarization profiles. *Nanophotonics, 9*(5), 1003–1014. http://www.degruyter.com/view/j/nanoph?rskey=PZEKdQ&result=1&q=Nanophotonics. doi:10.1515/nanoph-2019-0479.

Jamila, R. M., Narasimman, S., Prasanth, A., Muthukumar, M., Alex, Z. C., & Anand, G. T. (2024). Fiber optic sensor coated with multiple layers of hexagonal boron nitride nanosheets (BNNS) for the detection of volatile organic compounds. *ACS Applied Materials and Interfaces, 16*(27), 35525–35540. http://pubs.acs.org/journal/aamick. doi:10.1021/acsami.4c05230.

Jha, R., Mishra, P., & Kumar, S. (2024). Advancements in optical fiber-based wearable sensors for smart health monitoring. *Biosensors and Bioelectronics, 254*, 116232. doi:10.1016/j.bios.2024.116232.

Jiao, F., Li, F., Shen, J., Guan, C., Khan, S. A., Wang, J., Yang, Z., & Zhu, J. (2021). Wafer-scale flexible plasmonic metasurface with passivated aluminum nanopillars for high-sensitivity immunosensors. *Sensors and Actuators B: Chemical, 344*, 130170. doi:10.1016/j.snb.2021.130170.

Kaur, B., Kumar, S., & Kaushik, B. K. (2023). Novel wearable optical sensors for vital health monitoring systems—a review. *Biosensors, 13*(2), 181. http://www.mdpi.com/journal/biosensors/. doi:10.3390/bios13020181.

Kazanskiy, N. L., Khonina, S. N., Butt, M. A., Kaźmierczak, A., & Piramidowicz, R. (2021). A numerical investigation of a plasmonic sensor based on a metal-insulator-metal waveguide for simultaneous detection of biological analytes and ambient temperature. *Nanomaterials, 11*(10), 2551. https://www.mdpi.com/2079-4991/11/10/2551/pdf. doi:10.3390/nano11102551.

Kazanskiy, N. L., Butt, M. A., & Khonina, S. N. (2021). Carbon dioxide gas sensor based on polyhexamethylene biguanide polymer deposited on silicon nano-cylinders metasurface. *Sensors (Switzerland), 21*(2), 1–14. https://www.mdpi.com/1424-8220/21/2/378/pdf. doi:10.3390/s21020378.

Kazanskiy, N. L., Khonina, S. N., & Butt, M. A. (2020). Plasmonic sensors based on Metal-insulator-metal waveguides for refractive index sensing applications: A brief review. *Physica E: Low-dimensional Systems and Nanostructures, 117*, 113798. doi:10.1016/j.physe.2019.113798.

Kazanskiy, N. L., Khonina, S. N., & Butt, M. A. (2021). Polarization-insensitive hybrid plasmonic waveguide design for evanescent field absorption gas sensor. *Photonic Sensors, 11*(3), 279–290. http://www.springer.com/physics/optics+%26+lasers/journal/13320. doi:10.1007/s13320-020-0601-6.

Kazanskiy, N. L., Khonina, S. N., & Butt, M. A. (2023). Recent development in metasurfaces: A focus on sensing applications. *Nanomaterials, 13*(1), 118. http://www.mdpi.com/journal/nanomaterials. doi:10.3390/nano13010118.

King D.R.Ø., Hedegaard E., Ackermann T.N., Rius C., Muñoz-Berbel X., Knudsen H., Rodríguez-Rodríguez I., Dorado B., Alvarez E., Ducrée J., Andrés V., Simonsen U., Llobera A. 2015. Living photonics: Monitoring light propagation through cells (LiPhos). In Progress in Biomedical Optics and Imaging – Proceedings of SPIE (Vol. 9518). SPIE. doi:10.1117/12.2181306. http://spie.org/x1848.xml

Koya, A. N., Cunha, J., Guo, T. L., Toma, A., Garoli, D., Wang, T., Juodkazis, S., Cojoc, D., & Zaccaria, R. P. (2020). Novel plasmonic nanocavities for optical trapping-assisted biosensing applications. *Advanced Optical Materials, 8*(7), 1901481. http://onlinelibrary.wiley.com/journal/10.1002/. doi:10.1002/adom.201901481. (ISSN)2195-1071.

Kuang, K. S. C., Quek, S. T., & Maalej, M. (2004). Assessment of an extrinsic polymer-based optical fibre sensor for structural health monitoring. *Measurement Science and Technology, 15*(10), 2133–2141. http://www.iop.org/EJ/journal/0957-0233. doi:10.1088/0957-0233/15/10/024.

Lalbakhsh, A., Afzal, M. U., Hayat, T., Esselle, K. P., & Mandal, K. (2021). All-metal wide-band metasurface for near-field transformation of medium-to-high gain electromagnetic sources. *Scientific Reports, 11*(1), 9421. www.nature.com/srep/index.html. doi:10.1038/s41598-021-88547-3.

Lan, G., Jin, Z., Nong, J., Luo, P., Guo, C., Sang, Z., Dong, L., & Wei, W. (2020). Narrowband perfect absorber based on dielectric-metal metasurface for surface-enhanced infrared sensing. *Applied Sciences (Switzerland), 10*(7), 2295. https://res.mdpi.com/d_attachment/applsci/applsci-10-02295/article_deploy/applsci-10-02295.pdf. doi:10.3390/app10072295.

Leanse, L. G., Hong, W., & de Souza Rastelli, A. N. (2023). Editorial: Photonics-based diagnosis and treatment of infectious and inflammatory diseases. *Frontiers in Microbiology, 14,* 1205032. https://www.frontiersin.org/journals/microbiology#. doi:10.3389/fmicb.2023.1205032.

Lee, K. L., Huang, J. B., Chang, J. W., Wu, S. H., & Wei, P. K. (2015). Ultrasensitive biosensors using enhanced fano resonances in capped gold nanoslit arrays. *Scientific Reports, 5,* 8547. www.nature.com/srep/index.html. doi:10.1038/srep08547.

Li, H., An, Z., Zhang, S., Zuo, S., Zhu, W., Zhang, S., Huang, B., Cao, L., Zhang, C., Zhang, Z., Song, W., Mao, Q., Mu, Y., Miao, C., Li, E., & Prades, J. D. (2021). Fully photonic integrated wearable optical interrogator. *ACS Photonics, 8*(12), 3607–3618. http://pubs.acs.org/journal/apchd5. doi:10.1021/acsphotonics.1c01236.

Lim, J., Shim, J., Kim, I., Kim, S. K., Lim, H., Ahn, S. Y., Park, J., Geum, D. M., & Kim, S. H. (2024). Ultrasensitive mid-infrared optical gas sensor based on germanium-on-insulator photonic circuits with limit-of-detection at sub-ppm level. *ACS Photonics.* http://pubs.acs.org/journal/apchd5. doi:10.1021/acsphotonics.4c01185.

Limonov, M. F., Rybin, M. V., Poddubny, A. N., & Kivshar, Y. S. (2017). Fano resonances in photonics. *Nature Photonics, 11*(9), 543–554. http://www.nature.com/nphoton/index.html. doi:10.1038/NPHOTON.2017.142.

Liu, W., Han, H., & Wang, J. (2024). Recent advances in the 3D chiral plasmonic nanomaterials. *Small, 20*(8), 2305725. http://onlinelibrary.wiley.com/journal/10.1002/(ISSN)1613-6829. doi:10.1002/smll.202305725.

Majhi, S. M., Mirzaei, A., Kim, H. W., Kim, S. S., & Kim, T. W. (2021). Recent advances in energy-saving chemiresistive gas sensors: A review. *Nano Energy, 79,* 105369. http://www.journals.elsevier.com/nano-energy/. doi:10.1016/j.nanoen.2020.105369.

Mancuso, M., Goddard, J. M., & Erickson, D. (2012). Nanoporous polymer ring resonators for biosensing. *Optics Express, 20*(1), 245–255. http://www.opticsinfobase.org/view_article.cfm?gotourl=http%3A%2F%2Fwww%2Eopticsinfobase%2Eorg%2FDirectPDFAccess%2FDA392047%2DD615%2D7CF3%2DD6B40AF09A6C9E68%5F225973%2Epdf%3Fda%3D1%26id%3D225973%26seq%3D0%26mobile%3Dno&org=. doi:10.1364/OE.20.000245.

Martin, A. L., Armani, D. K., Yang, L., & Vahala, K. J. (2004). Replica-molded high-Q polymer microresonators. *Optics Letters, 29*(6), 533–535. http://ol.osa.org/Issue.cfm. doi:10.1364/OL.29.000533.

Mazzone, P. J., Hammel, J., Dweik, R., Na, J., Czich, C., Laskowski, D., & Mekhail, T. (2007). Diagnosis of lung cancer by the analysis of exhaled breath with a colorimetric sensor array. *Thorax, 62*(7), 565–568. doi:10.1136/thx.2006.072892.

Memon, S. F., Wang, R., Strunz, B., Chowdhry, B. S., Pembroke, J. T., & Lewis, E. (2022). A review of optical fibre ethanol sensors: Current state and future prospects. *Sensors, 22*(3), 950. doi:10.3390/s22030950.

Milvich, J., Kohler, D., Freude, W., & Koos, C. (2018). Surface sensing with integrated optical waveguides: A design guideline. *Optics Express, 26*(16), 19885. doi:10.1364/OE.26.019885.

Nachiar C.C., Ambika N., Moulika R., Poovendran R. 2020. Design of cost-effective wearable sensors with integrated health monitoring system. In Proceedings of the 4th International Conference on IoT in Social, Mobile, Analytics and Cloud, ISMAC 2020 (pp. 1289–1292). IEEE. doi: 10.1109/I-SMAC49090.2020.9243462. http://ieeexplore.ieee.org/xpl/mostRecentIssue.jsp?punumber=9243168

Narasimman, S., Balakrishnan, L., & Alex, Z. C. (2018). Fiber-optic ammonia sensor based on amine functionalized ZnO nanoflakes. *IEEE Sensors Journal, 18*(1), 201–208. doi:10.1109/jsen.2017.2769718.

Nasiri, S., & Khosravani, M. R. (2020). Progress and challenges in fabrication of wearable sensors for health monitoring. *Sensors and Actuators A: Physical, 312*, 112105. doi:10.1016/j.sna.2020.112105.

Nath, P., Kabir, A., Doust, S. K., Kreais, Z. J., & Ray, A. (2020). Detection of bacterial and viral pathogens using photonic point-of-care devices. *Diagnostics, 10*(10), 841. doi:10.3390/diagnostics10100841.

Nirmala Devi, K., & Vadivel, S. (2024). Room temperature fiber optic gas sensor technology based Zn3(VO4)2/PMMA hybrid composites via facile hydrothermal route. *Optical Fiber Technology, 82*, 103642. doi:10.1016/j.yofte.2023.103642.

Ochoa, M., Francisco Algorri, J. É., Roldán-Varona, P., Rodríguez-Cobo, L., & Miguel López-Higuera, J. É. (2021). Recent advances in biomedical photonic sensors: A focus on optical-fibre-based sensing. *Sensors, 21*(19), 6469. doi:10.3390/s21196469.

Panindre P., Susan Mousavi N., Kumar S. 2018. Design optimization of a single-mode microring resonator for label-free detection of biomarkers within a tunable spectral range of 2 nm. In Proceedings of SPIE – The International Society for Optical Engineering (Vol. 10711). SPIE. doi: 10.1117/12.2318934. http://spie.org/x1848.xml

Pathak, A. K., Rahman, B. M. A., Singh, V. K., & Kumari, S. (2019). Sensitivity enhancement of a concave shaped optical fiber refractive index sensor covered with multiple Au nanowires. *Sensors, 19*(19), 4210. doi:10.3390/s19194210.

Pathak, A. K., Rahman, B. M. A., & Viphavakit, C. (2022). Nanowire embedded micro-drilled dual-channel approach to develop highly sensitive biosensor. *IEEE Photonics Technology Letters, 34*(13), 707–710. doi:10.1109/lpt.2022.3182783.

Pathak, A. K., Gangwar, R. K., Priyadarshini, P., & Singh, V. K. (2017). A robust optical fiber sensor for the detection of petrol adulteration. *Optik, 149*, 43–48. doi:10.1016/j.ijleo.2017.09.036.

Pathak, A. K., Verma, S., Sakda, N., Viphavakit, C., Chitaree, R., & Rahman, B. M. A. (2023). Recent advances in optical hydrogen sensor including use of metal and metal alloys: A review. *Photonics, 10*(2), 122. http://www.mdpi.com/journal/photonics. doi:10.3390/photonics10020122.

Pathak, A., Swargiary, K., Kongsawang, N., Jitpratak, P., Ajchareeyasoontorn, N., Udomkittivorakul, J., & Viphavakit, C. (2023). Recent advances in sensing materials targeting clinical volatile organic compound (VOC) biomarkers: A review. *Biosensors, 13*(1), 114. doi:10.3390/bios13010114.

Pelucchi, E., Fagas, G., Aharonovich, I., Englund, D., Figueroa, E., Gong, Q., Hannes, H., Liu, J., Lu, C. Y., Matsuda, N., Pan, J. W., Schreck, F., Sciarrino, F., Silberhorn, C., Wang, J., & Jöns, K. D. (2022). The potential and global outlook of integrated photonics for quantum technologies. *Nature Reviews Physics, 4*(3), 194–208. nature.com/natrevphys/. doi:10.1038/s42254-021-00398-z.

Pendão, C., & Silva, I. (2022). Optical fiber sensors and sensing networks: Overview of the main principles and applications. *Sensors, 22*(19), 7554. doi:10.3390/s22197554.

Photonic Sensors & Detectors Market Outlook. n.d. https://www.futuremarketinsights.com/reports/photonics-sensor-and-detector-market.

Prabowo, B. A., Purwidyantri, A., & Liu, K. C. (2018). Surface plasmon resonance optical sensor: A review on light source technology. *Biosensors, 8*(3), 80. http://www.mdpi.com/journal/biosensors/. doi:10.3390/bios8030080.

Prasanth, A., Meher, S. R., & Alex, Z. C. (2021). Experimental analysis of SnO_2 coated LMR based fiber optic sensor for ethanol detection. *Optical Fiber Technology, 65*, 102618. doi:10.1016/j.yofte.2021.102618.

Promphet, N., Ummartyotin, S., Ngeontae, W., Puthongkham, P., & Rodthongkum, N. (2021). Non-invasive wearable chemical sensors in real-life applications. *Analytica Chimica Acta, 1179*, 338643. doi:10.1016/j.aca.2021.338643.

Qiao, Y., Qiao, L., Chen, Z., Liu, B., Gao, L. I., & Zhang, L. (2022). Wearable sensor for continuous sweat biomarker monitoring. *Chemosensors, 10*(7), 273. doi:10.3390/chemosensors10070273.

Qin, J., Jiang, S., Wang, Z., Cheng, X., Li, B., Shi, Y., Tsai, D. P., Liu, A. Q., Huang, W., & Zhu, W. (2022). Metasurface micro/nano-optical sensors: Principles and applications. *ACS Nano, 16*(8), 11598–11618. http://pubs.acs.org/journal/ancac3. doi:10.1021/acsnano.2c03310.

Rahmatiyar, M., Afsahi, M., & Danaie, M. (2020). Design of a refractive index plasmonic sensor based on a ring resonator coupled to a MIM waveguide containing tapered defects. *Plasmonics, 15*(6), 2169–2176. http://www.springerlink.com/content/1557-1955. doi:10.1007/s11468-020-01238-z.

Research, P. (2024). *Wearable sensors market size, share, and trends*. https://www.grandviewresearch.com/industry-analysis/global-wearable-sensor-market.

Sabri, N., Aljunid, S. A., Salim, M. S., Ahmad, R. B., & Kamaruddin, R. (2013). Toward optical sensors: Review and applications. *Journal of Physics: Conference Series, 423*(1), 012064. doi:10.1088/1742-6596/423/1/012064.

Santos, J. L., Frazãoa, O., Baptista, J. M., Jorge, P. A. S., Dias, I., Araújo, F. M., & Ferreira, L. A. (2009). Optical fibre sensing networks. *SBMO/IEEE MTT-S International Microwave and Optoelectronics Conference Proceedings* (pp. 290–298). IEEE. doi:10.1109/IMOC.2009.5427580.

Seshadri, D. R., Li, R. T., Voos, J. E., Rowbottom, J. R., Alfes, C. M., Zorman, C. A., & Drummond, C. K. (2019). Wearable sensors for monitoring the internal and external workload of the athlete. *NPJ Digital Medicine, 2*(1). https://www.nature.com/npjdigitalmed/. doi:10.1038/s41746-019-0149-2.

Shahbaz, M., Butt, M. A., & Piramidowicz, R. (2023). A concise review of the progress in photonic sensing devices. *Photonics, 10*(6), 698. http://www.mdpi.com/journal/photonics. doi:10.3390/photonics10060698.

Shim, J., Park, H.-Y., Kang, D.-H. O., Kim, J.-O. K., Jo, S.-H., Park, Y., & Park, J.-H. (2017). Electronic and optoelectronic devices based on two-dimensional materials: From fabrication to application. *Advanced Electronic Materials, 3*(4). doi:10.1002/aelm.201600364.

Snitzer, E. (1961). Cylindrical Dielectric Waveguide Modes. *Journal of the Optical Society of America, 51*(5), 491–498. doi:10.1364/JOSA.51.000491.

Soler, M., Scholtz, A., Zeto, R., & Armani, A. M. (2020). Engineering photonics solutions for COVID-19. *APL Photonics, 5*(9). http://aip.scitation.org/journal/app. doi:10.1063/5.0021270.

Soltani, O., Francoeur, S., & Kanzari, M. (2023). Detecting cancerous human liver cells with high performances using photonic crystals. *Physica B: Condensed Matter, 650*, 414557. doi:10.1016/j.physb.2022.414557.

Somers, P., Münchinger, A., Maruo, S., Moser, C., Xu, X., & Wegener, M. (2024). The physics of 3D printing with light. *Nature Reviews Physics, 6*(2), 99–113. https://www.nature.com/natrevphys/. doi:10.1038/s42254-023-00671-3.

Someya, T., Bao, Z., & Malliaras, G. G. (2016). The rise of plastic bioelectronics. *Nature, 540*(7633), 379–385. http://www.nature.com/nature/index.html. doi:10.1038/nature21004.

Steinberg, B. A., & Piccini, J. P. (2018). Screening for atrial fibrillation with a wearable device. *JAMA, 320*(2), 139. doi:10.1001/jama.2018.8513.

Teng, C., Yu, F., Ding, Y., & Zheng, J. (2017). Refractive index sensor based on multi-mode plastic optical fiber with long period grating. In *Proceedings of SPIE – The International Society for Optical Engineering: 10231*. SPIE http://spie.org/x1848.xml. doi:10.1117/12.2265305.

Thylén, L., & Wosinski, L. (2014). Integrated photonics in the 21st century. *Photonics Research, 2*(2), 75. doi:10.1364/PRJ.2.000075.

Toma, K., Tsujii, M., Arakawa, T., Iwasaki, Y., & Mitsubayashi, K. (2021). Dual-target gas-phase biosensor (bio-sniffer) for assessment of lipid metabolism from breath acetone and isopropanol. *Sensors and Actuators B: Chemical, 329*, 129260. doi:10.1016/j.snb.2020.129260.

Urade, A. R., Lahiri, I., & Suresh, K. S. (2023). Graphene properties, synthesis and applications: A review. *JOM, 75*(3), 614–630. https://www.springer.com/journal/11837. doi:10.1007/s11837-022-05505-8.

Vaidyanathan, A., Mondal, B., Sekhar Rout, C., & Chakraborty, B. (2024). Plasmonic gas sensors based on nanomaterials: Mechanisms and recent developments. *Journal of Physics D: Applied Physics, 57*(26), 263002. doi:10.1088/1361-6463/ad32a7.

Velumani, M., Prasanth, A., Narasimman, S., Chandrasekhar, A., Sampson, A., Meher, S. R., Rajalingam, S., Rufus, E., & Alex, Z. C. (2022). Nanomaterial-based sensors for exhaled breath analysis: A review. *Coatings, 12*(12), 1989. www.mdpi.com/journal/coatings. doi:10.3390/coatings12121989.

Verma, S., Pathak, A. K., & Rahman, B. M. A. (2024). Review of biosensors based on plasmonic-enhanced processes in the metallic and meta-material-supported nanostructures. *Micromachines, 15*(4), 502. http://www.mdpi.com/journal/micromachines. doi:10.3390/mi15040502.

Vincent, S., Subramanian, S., & Vollmer, F. (2020). Optoplasmonic characterisation of reversible disulfide interactions at single thiol sites in the attomolar regime. *Nature Communications, 11*(1). http://www.nature.com/ncomms/index.html. doi:10.1038/s41467-020-15822-8.

Viphavakit, C., Keeffe, S. O, Yang, M., Andersson-Engels, S., & Lewis, E. (2018). Gold enhanced hemoglobin interaction in a Fabry-Pérot based optical fiber sensor for measurement of blood refractive index. *Journal of Lightwave Technology, 36*(4), 1118–1124. doi:10.1109/JLT.2017.2782078.

Wang, J. (2008). Electrochemical glucose biosensors. *Chemical Reviews, 108*(2), 814–825. doi:10.1021/cr068123a.

Wang, X., Li, Z., & Su, L. (2024). Soft optical waveguides for biomedical applications, wearable devices, and soft robotics: A review. *Advanced Intelligent Systems, 6*(1). https://onlinelibrary.wiley.com/journal/26404567. doi:10.1002/aisy.202300482.

Wang, Y., Tan, B., Liu, S., Tian, T., Xu, G., & Cai, B. (2020a). An optical fiber-waveguide-fiber platform for ppt level evanescent field-based sensing. *Sensors and Actuators, B: Chemical, 306*. https://www.journals.elsevier.com/sensors-and-actuators-b-chemical. doi:10.1016/j.snb.2019.127548.

Wang, Y., Gao, B., Zhang, K., Yuan, K., Wan, Y., Xie, Z., Xu, X., Zhang, H., Song, Q., Yao, L., Fang, X., Li, Y., Xu, W., Zhang, J., & Dai, L. (2017). Refractive index sensor based on leaky resonant scattering of single semiconductor nanowire. *ACS Photonics, 4*(3), 688–694. http://pubs.acs.org/journal/apchd5. doi:10.1021/acsphotonics.7b00064.

Wang, Y., Yu, J., Mao, Y. F., Chen, J., Wang, S., Chen, H. Z., Zhang, Y., Wang, S. Y., Chen, X., Li, T., Zhou, L., Ma, R. M., Zhu, S., Cai, W., & Zhu, J. (2020b). Stable, high-performance sodium-based plasmonic devices in the near infrared. *Nature, 581*(7809), 401–405. http://www.nature.com/nature/index.html. doi:10.1038/s41586-020-2306-9.

Wei, H., Chen, M., & Krishnaswamy, S. (2020). Three-dimensional-printed Fabry-Perot interferometer on an optical fiber tip for a gas pressure sensor. *Applied Optics, 59*(7), 2173–2178. https://www.osapublishing.org/ao/abstract.cfm?uri=ao-59-7-2173. doi:10.1364/AO.385573.

Wei, H., Wu, Z., Sun, K., Zhang, H., Wang, C., Wang, K., Yang, T., Pang, F., Zhang, X., Wang, T., & Krishnaswamy, S. (2023). Two-photon 3D printed spring-based Fabry–Pérot cavity resonator for acoustic wave detection and imaging. *Photonics Research, 11*(5), 780. doi:10.1364/prj.481858.

Weiss, M. N., Srivastava, R., Groger, H., Lo, P., & Luo, S. F. (1995). A theoretical investigation of environmental monitoring using surface plasmon resonance waveguide sensors. *Sensors and Actuators: A Physical, 51*(2–3), 211–217. doi:10.1016/0924-4247(95)01208-7.

Xu, W., Cai, Y., Gao, S., Hou, S., Yang, Y., Duan, Y., Fu, Q., Chen, F., & Wu, J. (2021). New understanding of miniaturized VOCs monitoring device: PID-type sensors performance evaluations in ambient air. *Sensors and Actuators B: Chemical, 330*, 129285. doi:10.1016/j.snb.2020.129285.

Xu, Y., Liu, T., Liu, S., Sun, X., & Zhang, D. (2022). Mutual-coupling in high-Q silicon dual-concentric micro-ring/racetrack resonator. *IEEE Photonics Journal, 14*(4). http://www.ieee.org. doi:10.1109/JPHOT.2022.3186914.

Xu, Y., Bai, P., Zhou, X., Akimov, Y., Png, C. E., Ang, L. K., Knoll, W., & Wu, L. (2019). Optical refractive index sensors with plasmonic and photonic structures: Promising and inconvenient truth. *Advanced Optical Materials, 7*(9), 1801433. http://onlinelibrary.wiley.com/journal/10.1002/. doi:10.1002/adom.201801433. (ISSN)2195-1071.

Xu, Z., Li, N., Dong, Y., Fu, Y. H., Hu, T., Zhong, Q., Zhou, Y., Li, D., Zhu, S., & Singh, N. (2021). Metasurface-based subtractive color filter fabricated on a 12-inch glass wafer using a CMOS platform. *Photonics Research, 9*(1), 13–20. https://www.osapublishing.org/DirectPDFAccess/4418951C-9091-A414-30C633B0E5A9208E_444742/prj-9-1-13.pdf?da=1&id=444742&seq=0&mobile=no. doi:10.1364/PRJ.404124.

Yamazoe, N., & Shimanoe, K. (2009). New perspectives of gas sensor technology. *Sensors and Actuators, B: Chemical, 138*(1), 100–107. doi:10.1016/j.snb.2009.01.023.

Yang, X., Hua, E., Wang, M., Wang, Y., Wen, F., & Yan, S. (2019). Fano resonance in a MIM waveguide with two triangle stubs coupled with a split-ring nanocavity for sensing application. *Sensors, 19*(22), 4972. doi:10.3390/s19224972.

Yao, S., Swetha, P., & Zhu, Y. (2018). Nanomaterial-enabled wearable sensors for healthcare. *Advanced Healthcare Materials, 7*(1). http://onlinelibrary.wiley.com/journal/10.1002/(ISSN)2192-2659. doi:10.1002/adhm.201700889.

Zakirov, R., & Umarov, A. (2020). Uzbekistan Fiber optic gyroscope and accelerometer application in aircraft inertial system. *International Conference on Information Science and Communications Technologies, ICISCT 2020*. Institute of Electrical and Electronics Engineers Inc http://ieeexplore.ieee.org/xpl/mostRecentIssue.jsp?punumber=9351351. doi:10.1109/ICISCT50599.2020.9351385.

Zhang, D., Wei, H., & Krishnaswamy, S. (2019). 3D printing optofluidic mach-zehnder interferometer on a fiber tip for refractive index sensing. *IEEE Photonics Technology Letters, 31*(21), 1725–1728. https://ieeexplore.ieee.org/servlet/opac?punumber=68. doi:10.1109/LPT.2019.2943897.

Zhang, S., Wong, C. L., Zeng, S., Bi, R., Tai, K., Dholakia, K., & Olivo, M. (2021). Metasurfaces for biomedical applications: Imaging and sensing from a nanophotonics perspective. *Nanophotonics, 10*(1), 259–293. http://www.degruyter.com/view/j/nanoph?rskey=PZEKdQ&result=1&q=Nanophotonics. doi:10.1515/nanoph-2020-0373.

Zhu, J., & Lou, J. (2018). Ultrasensitive and multifunction plasmonic temperature sensor with ethanol-sealed asymmetric ellipse resonators. *Molecules, 23*(10), 2700. doi:10.3390/molecules23102700.

Zubaidah binti Haji Jumat, S., Chou Chao, C. T., Chou Chau, Y. F., Mahadi, A. H., Kooh, M. R. R., Kumara, N. T. R. N., & Chiang, H. P. (2021). Plasmonic refractive index sensor based on the combination of rectangular and circular resonators including baffles. *Chinese Journal of Physics, 71*, 286–299. http://www.journals.elsevier.com/chinese-journal-of-physics/. doi:10.1016/j.cjph.2021.02.006.

CHAPTER

4

Simulation and design optimization of optical VOC sensors using the finite element method

4.1 Introduction

Recent developments in guided-wave photonics, including optical fiber and nanophotonic devices, have introduced various arbitrarily shaped waveguides, commonly demonstrating anisotropic, inhomogeneous, dispersive, and nonlinear behaviors. Most optical waveguides of arbitrary geometries do not lend themselves to analytical solutions, and hence, computational tools for the modeling and simulation are considered to be essential for successful design, optimization, practical realization, and minimizing the fabrication tolerance. To accomplish this purpose, several numerical approaches have been reported, e.g., the finite element method (FEM), finite difference time domain (FDTD), and machine learning (ML). In particular, FEM is a numerical tool utilized to solve partial differential equations in two or three-dimensional space variables that arise in mathematical and engineering modeling (Rylander et al., 2013). Generally, in the finite-element technique, the domain over which a mathematical problem is to be modeled is subdivided into several smaller domains known as finite elements, over which simple functions are used to approximate the relevant field quantities. Proper mesh refinement ensures that the solutions converge towards the true values. The FEM is today extensively used in industry, the healthcare sector, as well as by the research community, and indeed, it could be said that without it many optical waveguide problems would be impossible to solve (Koshiba, 2014).

Photonic Sensors for Detection of VOC Biomarkers: Technological Principles and Medical Applications
DOI: https://doi.org/10.1016/B978-0-443-29166-1.00005-6

Copyright © 2026 Elsevier Inc. All rights are reserved, including those for text and data mining, AI training, and similar technologies.

This chapter provides a comprehensive discussion on the fundamentals of FEM, its associated tools, and recent advancements in FEM research, with a particular emphasis on optical waveguides and their applications in the development of VOC sensors. Beginning with a concise overview of FEM basics, including its theoretical underpinnings and traditional nodal elements, the chapter discusses innovative developments such as the introduction of edge elements aimed at eliminating nonphysical solutions. Additionally, the chapter discusses the implementation of a perfectly matched layer (PML) in order to mitigate the unwanted reflections from the boundaries. Throughout the chapter, we highlight various FEM methodologies tailored for beam propagation analysis, guided-mode analysis, and more. Finally, the chapter also demonstrates various types of optical waveguides (i.e., planar waveguide, photonic crystal waveguide, and optical fiber), their numerical analysis, optimization, and their application in VOC sensing.

4.2 Fundamentals of the finite element method

4.2.1 The procedure of the finite element method

The primary steps involved in the development of a photonic modal analysis by utilizing FEM are summarized below (Atia et al., 2018):

- Subdividing the large and complex structures into several smaller domains
- Setting up an interpolation equation based on values at the elemental boundaries or the nodes to compute the unknown parameters, such as a field or a potential
- Constructing the global eigenvalue equations by using a variational formalism that is equivalent to the governing differential equations
- After assembling the equation, the global equations are solved by utilizing an eigenvalue matrix solver
- Lastly, interpreting the eigenvalues and eigenvectors to determine the desired parameters

4.2.2 Domain discretization

In order to apply the FEM, the selected domains must be subdivided into several smaller regions, also known as elements, as shown in Fig. 4.1. To ensure accurate modeling, the size and shape of each element need to be adjusted so as to closely match the domain's shape and curvature of the boundaries. These discretizations of the domain form a mesh (can be fine, extremely fine, or coarse, based on the geometry) that may take regular or irregular forms based on the domain. Irregular mesh distribution is

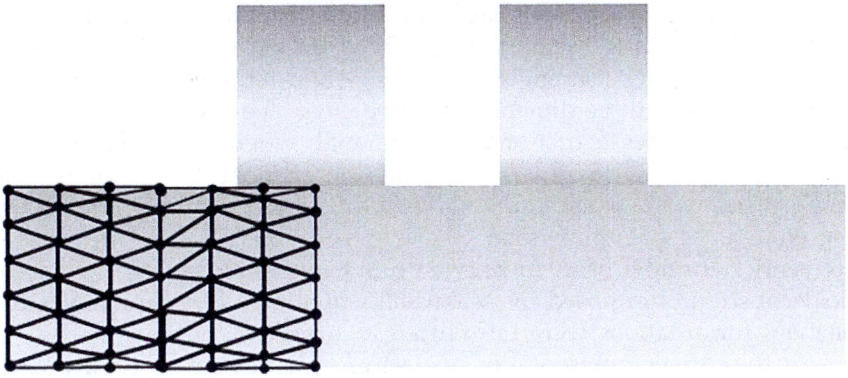

FIGURE 4.1 Domain discretization. This schematic illustrates triangular finite element discretization on the surface of a complex body.

generally considered more appropriate and well arranged to represent the domain of interest, as compared to regular mesh distribution. Finer meshing is generally utilized for domains of particular interest with extremely thin layers or small areas, while the coarse meshing is used for large or the rest of the domains. In order to perform photonic computations, the irregular mesh elements are considered for narrow, thin, and pointed structures where the change in the electromagnetic (EM) fields occurs rapidly. In regions where the EM field distribution is expected to vary gradually, coarser, larger domains may be used. It is also crucial to carefully choose the shape and the size of the boundary elements in order to match the exact boundary as closely as possible. To achieve this, various fundamental geometries can be utilized, including isoparametric elements, straight edge elements, edge elements, and infinite elements. Isoparametric elements, also known as curvilinear elements, are characterized by curved edges, with nodal positional coordinates described by the mapping between the local and global Cartesian coordinate, which are generally useful in solving the curved boundaries of any given geometry. Whereas the straight edge elements involve the rectangles, triangles, and quadrilaterals for two-dimensional domains, while prisms, tetrahedra, bricks, and pyramids are used for three-dimensional domains. The infinite elements are generally used for unbounded or open-space photonic problems. The form of the infinite element's shape function diminishes exponentially along the directions where the field continues infinitely (Rahman & Davies, 1984a, 1984b). PML effectively truncates the computational region via coordinate stretching, and are useful in regions where it is important to avoid spurious reflections arising from the finiteness of the computational domain.

4.2.3 Node and edge element

Several types of elements are available based on their shape and interpolation functions (Koshiba, 2014). Two-dimensional problems are typically solved using triangular elements, while tetrahedral elements are utilized to solve three-dimensional problems. The linear or the lowest-order element uses a first-order polynomial, whereas the second-order polynomial problem requires the quadratic element. The primary challenge encountered when applying FEM to EM eigenvalue problems was the emergence of nonphysical, spurious solutions. These arise because conventional nodal basis functions do not inherently satisfy the vector-field constraints imposed by Maxwell's equations. Consequently, edge-element formulations were introduced to suppress spurious modes and ensure the correct representation of electromagnetic field continuity across material interfaces, resulting in more accurate and physically meaningful solutions (Koshiba, 2014).

4.2.4 Maxwell's equations

Maxwell's equations are a set of 4 differential equations that describe the behaviour of electromagnetic fields in optical waveguides and form the basics of FEM. In the FEM, these Maxwell equations are solved to determine the field distribution, light-matter interactions, as well as propagation characteristics. The four foundational Maxwell's equations are given below:

$$\frac{\partial \vec{B}}{\partial t} = -\nabla \times E \ (Faraday's\ Law) \tag{4.1}$$

$$\frac{\partial \vec{D}}{\partial t} = \nabla \times \vec{H} - \vec{J} \ (Maxwell - Ampere\ Law) \tag{4.2}$$

$$\nabla . \vec{B} = 0 \ (Gauss's\ Law\ for\ Magnetism) \tag{4.3}$$

$$\nabla . \vec{D} = \rho \ (Gauss's\ Law\ for\ electricity) \tag{4.4}$$

where $\vec{D}$ and $\vec{E}$ are the electric flux density and electric field, respectively, whereas $\vec{B}$ and $\vec{H}$ represent the magnetic flux density and magnetic field, respectively. All field variables are functions of time t and space $\vec{x} = (x, y, z) \in \mathbb{R}^3$. The current density $\vec{J}$ and the charge density (ρ) are related by the continuity equation as given below:

$$\frac{d\rho}{dt} + \nabla \cdot \vec{J} = 0 \tag{4.5}$$

Readers interested in detailed mathematical derivations and FEM formulations are referred to standard texts such as (Monk, 2003; Polycarpou, 2022).

4.2.5 Variational formulations

In FEM, a variational formulation transforms a differential equation into its weak, or integral, form, making it suitable for numerical computation. The variational formulation forms the basis for constructing the global matrix system from individual finite elements and is widely used in electromagnetic waveguide analysis. Detailed mathematical derivations can be found in other literatures (Koshiba, 2014; Stallcup, 2018).

4.2.6 Perfectly matched layer

To construct a reliable FEM solver, the role of simulation boundaries must be addressed properly. As described above, conventional FEM boundaries assume that both the field and its normal derivative become zero at the simulation boundary, similar to enclosing the computational region within a magnetic or metallic box. However, this approximation is generally invalid, specifically when one needs to simulate the design with open boundaries. To address this complication, an artificial boundary condition known as the PML has been developed (Koshiba, et al., 1999). The PML is defined as a non-physical layer utilized for wave equations, generally employed to truncate the analysis region and simulate open boundary conditions. As its name suggests, the PML is designed to seamlessly match the adjacent domain, preventing the reflection of any incident waves from the PML layer. This feature provides a significant advantage to PML compared to other absorbing materials. In 1994, Berenger et al. reported the first PML, also known as "split-field PML" which can be used with Maxwell's equations (Berenger, 1994). Since then, the field has gained significant interest, and various modifications have been introduced to PML for Maxwell's equations as well as other wave equations, including Schrodinger's equation. During the simulation of photonic devices, optical waveguides, and circuits, the spurious reflections from different sizes of elements can be significantly reduced by considering the appropriate absorbing boundary conditions (Berenger, 1994). Later in 1998, Teixeira and Chew reported an anisotropic PML condition (Teixeira & Chew, 1998), without requiring the division of field components, this method is widely applied in mesh truncation for FEM studies.

4.2.7 Variations of the conventional finite element method

Mostly, the FEM simulations rely on second-order triangular elements, as first-order ones often result in poor accuracy (Rodriguez-Esquerre et al., 2005). The major drawback of conventional second-order elements is that their extra nodes between triangle corners cannot be tuned to fit the curved boundary. To address this issue, curvilinear elements can be utilized (Wang, 2019). Unlike conventional elements, curvilinear elements allow improved convergence rates while modeling curved regions. However, the first-order elements are generally preferred due to their simplicity and easy mesh refinement. Additionally, most of the mesh generation software depends on the first-order triangular elements. Consequently, various efforts have been devoted toward the improvement of the accuracy of the first-order element. Recently, a new method termed "smoothed Galerkin" has been reported (Liu, 2008). This new method relies on weakening the weak form (W2) as in mesh-free methods (Atia et al., 2018; Liu, 2009). Other than approximating the field variable, the weak form also approximates the field derivatives by utilizing the gradient smoothing technique (LeVeque, 2002). The weak form is distinguished by its lower sensitivity towards the mesh distortion and high accuracy. To improve the accuracy of Galerkin-based methods, the smoothed Galerkin form is utilized, which introduces the smoothed finite element method (SFEM), widely used to solve various mechanical problems. In conventional FEM, the shape function derivatives at element boundaries are numerically discontinuous, resulting in poor performance. SFEM overcomes this drawback by ensuring that the derivatives at element boundaries are approximated to be continuous. SFEM has recently been introduced for time-domain analysis of optical and photonic waveguides (Atia et al., 2015).

4.2.8 Time domain methods

The advancement in the field of nanophotonics has widely initiated significant research efforts focused on developing numerical techniques that could potentially keep this rapid growth continuing. These advanced numerical tools have been employed to solve Maxwell's equations in both time and frequency domains. The time domain formulation of Maxwell's equations offers significant advantages compared to frequency domain methods. This advantage includes the capability to cover a wide range of frequencies within a single simulation. Furthermore, time-domain methods provide detailed temporal information such as transient responses, pulse propagation, and time-resolved scattering or reflection, making them crucial tools for real-time design and performance prediction.

In time-domain methods, the finite-difference time-domain (FDTD) method has become extremely popular due to its simplicity and versatility.

FDTD discretizes Maxwell's equations in both space and time using finite differences (Atia et al., 2018). FDTD has been widely utilized in simulating various photonic devices, and the application was extended to integrate various materials, including dispersive, anisotropic, and nonlinear types. Besides several advancements, the approach suffers from inefficient discretization of irregular or complex geometries. Consequently, it becomes challenging to model complex photonic designs and curved geometries with high precision, including structures like photonic crystal fibers. This problem is generally referred to as the "staircasing problem". Besides this, the traditional FDTD approach suffers poor stability, where the time step size must satisfy the Courant–Friedrichs–Lewy condition, making the method highly sensitive to the selection of step size (Atia et al., 2018).

4.3 Guided-mode analysis

4.3.1 Straight waveguide analysis

While the general guided-mode analysis cannot be well addressed by a formulation based on a single scalar quantity, a suitable approximation can be discovered for weakly guiding structures. This has led to the development of what is known as the approximate scalar FEM (Koshiba et al., 1984a). In the approximate scalar FEM, the traditional nodal elements can be utilized without getting any non-physical or spurious solutions. However, the rigorous evaluation of guided modes in high-confinement structures, such as photonic crystal waveguides, high-index contrast waveguides, microstructured specialized optical fibers, and photonic crystal fibers based on the photonic band-gap phenomenon, full vector analysis is essential, which requires the development of different types of full vectorial FEM (FV-FEM).

The FV-FEM are extremely useful and suitable for the formulation of various complex waveguide problems. The most crucial problem associated with this technique is the appearance of spurious solutions, see detailed discussion in (Koshiba et al., 1984b, 1986; Koshiba & Inoue, 1992; Koshiba & Tsuji, 2000).

4.3.2 Curved waveguide analysis

As discussed in Section 4.3.1, curvilinear hybrid elements are used to analyze straight optical waveguides and are considered to be equally effective in resolving issues associated with the curved waveguide structures. In 2006, Kakihara et al. utilized FV-FEM in a cylindrical coordinate system to study the bending loss in a photonic waveguide (Kakihara et al., 2006). The discretization of the structure was performed using hybrid edge/nodal elements for the cross-section of a 3D curved waveguide. The solution region was bounded by anisotropic PML within the cylindrical coordinate system, effectively addressing the leaky modes of curved waveguides. Later, in 2009, the same approach was utilized by Tadashi et al. in order to design and fabricate bend-insensitive all-solid photonic band gap fibers comprising heterostructured cladding (Tadashi et al., 2009).

4.4 Beam propagation methods

Nowadays, the beam propagation approach is widely used to investigate the propagation of light in longitudinally varying optical waveguides (Koshiba, 2014; Koshiba, et al., 1999). In the FEM-beam propagation method (FE-BPM), the EM field in the transverse (xy) plane is discretized using FEM, while the Crank-Nicholson method facilitates the propagation analysis. It is possible to utilize non-uniform element meshes and dynamically tune them along the direction of propagation, improving the computational efficiency while maintaining numerical accuracy. In 1998, Yasuhide and Masanori introduced a straightforward and effective mesh-generation technique for beam-propagation analysis and approximate scalar-guided mode solutions. Subsequently, in 2000, the same group presented full-vectorial guided-mode and beam-propagation solutions (Tsuji & Koshiba, 2000).

The primary concern when employing FE-BPM to analyze light propagation within a finite spatial domain is identifying the boundary conditions at the edges of the computational window. Recently, an anisotropic PML has been successfully integrated into both the two-dimensional FE-BPM for planar optical waveguides (Tadashi et al., 2009) and an approximate scalar three-dimensional FE-BPM (Saitoh & Koshiba, 2001a), as well as a complete vector 3D-FE-BPM for arbitrarily shaped optical waveguides (Saitoh & Koshiba, 2001b). We limited this chapter mainly to 2 widely known time discretization methods, namely (1) the Newmark-beta and (2) the Crank–Nicolson. A detailed discussion on these methods is covered by Atia (2019).

4.5 Types of finite element method used for analysis

4.5.1 Linear finite element method

In this part, we explore the FEM, employing linear components to tackle the scalar wave equation concerning optical fibers with varying refractive index profiles. In the case of linear FEM, the solution varies linearly between the nodes, which utilizes the elements specified by linear shape functions. These components can be simple line segments shown in one dimension, whereas, for two dimensions, they can be shown either in the form of triangles or quadrilaterals; and for three dimensions, they are more complex and can be represented in the form of tetrahedra or hexahedra. Linear FEM is computationally efficient and straightforward to implement, making it suitable for problems where the solution is expected to vary linearly or when computational resources are limited. The methodology is tailored to accommodate various refractive index structures, with convergence analyses duly conducted. The key in the process lies in carefully determining the count and positioning of nodes within the component to ensure its precise geometric definition. The node count must be adequate to allow the assumed degree of interpolation corresponding to the primary parameters. For a linear polynomial approximation, two nodes suffice, each housing a primary unknown, effectively delineating the element's geometry. Although the linear FEM is simple, its accuracy is lower compared to the higher-order elements, specifically the problems involving curved geometries or complex feature distributions.

4.5.2 Quadratic finite element method

Quadratic FEM is another FEM technique generally utilized for the numerical analysis of complex waveguides (Nagarkar & Pawar, 2014). Quadratic FEM utilizes elements with quadratic shape functions, incorporating mid-side nodes in addition to the corner nodes, which allows for a quadratic variation within each element. This method provides higher accuracy as it can better represent curved geometries and capture more complex variations in the solution. Although it requires more computational resources and is more complex to implement, quadratic FEM is ideal for problems requiring greater precision, such as in the stress analysis of complex structures or when modeling phenomena with sharp gradients. This method is preferred when the accuracy of the solution is paramount.

4.6 Applications

Among several reported works, this section aims to cover some examples of the application of the FEM approach in solving various complex designs and optimizing the structure to detect various VOC concentrations. A detailed discussion of various optical waveguide-based VOC sensors is presented in the following chapters. Therefore, rather than providing a comprehensive review, this section briefly highlights the role of the FEM in the design, analysis, and optimisation of devices for VOC sensing.

4.6.1 Planar waveguide volatile organic compounds sensor

In 2017, Ghosh and Rahman reported a design and numerical optimization of a miniaturized Mach–Zehnder interferometer based plasmonic

VOC sensor for the detection of ethanol vapor (Ghosh & Rahman, 2017). The reported work demonstrates the improved version of the H-field-based variational equation for designing and optimizing the horizontal slot waveguide VOC sensors. Designing a composite plasmonic (Alam et al., 2014) and pure optical waveguide (Berini, 2000), an appropriate analysis of real and complex effective indices (n_{eff}), hybrid quasi-TE and -TM modes, and dominating and non-dominating field components is considered to be highly crucial. Moreover, for designing a composite plasmonic planar waveguide, a thorough calculation is required in order to analyze the confinement factors (Γ) within the specific area, along with the variation in the modal n_{eff} due to the presence of the sensing medium. All the parameters mentioned above can be determined by solving Maxwell's equations for the structure. However, in several cases, these equations turn out to be too complex to be solved analytically or semi-analytically. Therefore, a fully vectorial, numerically efficient FV-FEM can be utilized to accurately compute the modes of such waveguides. A 2D FV-FEM approach is employed to analyze the modal analysis of the waveguide problem, whereas a 3D FV-FEM is employed to ascertain the resonating characteristics of the structure. In the reported work, the authors utilized the H field-based variational formulation. This technique comprises advantages over other approaches, i.e., E and combined E + H field formulations reported by Berk et al. (Berk, 1956). In this work, the H field components were considered continuous at the dielectric interfaces as well as in the condition where the index contrast can be high, e.g., the metal-dielectric interface. The formulation involves a natural boundary condition termed a perfect electric wall (PEW). Depending on the modal requirements of the waveguide, suitable boundary conditions like PEW and/or perfect magnetic wall (PMW) are applied as needed. This approach offers one-fold and two-fold symmetrical advantages, enhancing accuracy while reducing computational resources. Properly applying PEW and PMW boundaries prevents significant degeneration of different polarization states, leading to improved accuracy with significantly reduced computational demand.

The variational approach is based on Maxwell's curl equations, resulting in Euler equations that lead to Helmholtz's equation, though they may not fulfill Maxwell's divergence equations. Consequently, spurious modes emerge throughout the eigenvalue matrix alongside real modes. This shortfall in formulation can be rectified by incorporating an additional component, ensuring that the Euler equation of the modified functional J_e complies with all 4 of Maxwell's equations.

$$J_e = \left[\int\int (\nabla \times H)^* . \widehat{\varepsilon_r^{-1}} (\nabla \times H) dxdy + p \int\int (\nabla . H)^* . (\nabla . H) dxdy\right] - k_0^2 \left[\int\int H^* . \hat{\mu}_r H dxdy\right] \tag{4.6}$$

where J_e denotes the numerical errors that occur because of the discretization of the desired domain. Rahman and Davis discussed PFM in 1984 (Rahman & Davies, 1984c), where p represents a weighting factor which is included in order to balance the divergence and curl parts of the equation externally. This penalty term is typically defined as the inverse cube of the waveguide's effective refractive index, i.e., $1/n_{eff}^3$ of the specific waveguide, which is considered to be very useful in order to provide precise modal solutions for any waveguides consisting of only positive dielectric constants. In contrast to the conventional penalty method previously discussed, Souvik and Rahman reported a more straightforward approach to obtain the modal solution for both pure and composite waveguides, considering that their local permittivity may include both positive and negative values (Ghosh & Rahman, 2017). The variation in the plasmonic field takes place in small dimensions. To address this, the divergence integral is tailored by utilizing the local permittivity rather than relying on the global penalty parameter (p), as reported earlier. In this work, the authors considered utilizing the dielectric constant of each discretized element within the divergence part during the design of the global eigenvalue equation. The modified version of the H-field-based variational equation is shown below:

$$k_0^2 = \left(\frac{\omega}{c}\right)^2 \frac{\left[\int\int (\nabla \times H)^* . \widehat{\varepsilon_r^{-1}} (\nabla \times H) dxdy + \int\int (\nabla . H)^* \varepsilon_r^{-1} . (\nabla . H) dxdy\right]}{\int\int H^* . \hat{\mu}_r H dxdy} \tag{4.7}$$

where k_0, ε, ω, and μ exhibit wavenumber, relative permittivity, angular frequency, and permeability tensor, respectively. The quasi-TE and -TM modes can be obtained through post-processing of the eigenvectors corresponding to the selected eigenvalues. In order to evaluate the power confinement in the selected region of the planar VOC sensor, the sensor design underwent rigorous simulation and optimization. The confinement factor in such a planar waveguide can be determined by utilizing the eigenvectors calculated from the following relation:

$$\Gamma_{slot} = \frac{\int\int_{slot}^{1} Re(E \times H^*) . zdxdy}{\int\int_{\infty}^{1} Re(E \times H^*) . zdxdy} \tag{4.8}$$

Where *Re* denotes the real part of complex magnetic (H) and electric (E) fields. The symbol $*$ indicates the complex conjugate, and integrations are conducted across the entire computational domain for the denominator, while only in the slot region for the numerator.

In this work, the authors considered a silicon-on-insulator (SOI)-based composite plasmonic waveguide (CPWG), which is numerically investigated and comprises a thin layer of porous ZnO as a horizontal slot used for the detection of ethanol vapor. These numerical analyses revealed that the CPWG is capable of supporting a hybrid plasmonic supermode formed by the coupling of surface plasmon polaritons and a dielectric mode over a considerable distance. A nanoscale field confinement is achieved and exploited to monitor the variation in the refractive index due to the absorption and condensation of ethanol vapor. All the structural parameters are optimized using FV-FEM to achieve maximum sensing performance. The reported CPWG exhibits a high sensitivity of 0.7 per RIU for a 40% porous ZnO layer as a horizontal slot. Various phase sensitivities of 0.30, 0.34, 0.38, and 0.40 were observed under the 51.57%, 52.05%, 52.59%, and 53.16% condensation of ethanol within the horizontally slotted ZnO layer comprising porosity of 30%, 40%, 50%, and 60%, respectively.

Following the same approach, Chandra and Ranjan performed a numerical investigation over multiple types of slot waveguides and optimized their gas sensing response (Chandra & Ranjan, 2021). In the reported work, the authors designed three different types of slot waveguides, namely, partial-strip-loaded slot, conventional slot, and full-strip-loaded slot waveguides, and analyzed their sensing performance for different concentrations of methane gas. In order to improve the evanescent field ratio within the slot region, and thereby enhance the sensitivity of the sensor, the slot waveguides were considered to be made of germanium on calcium fluoride substrates. Various structural parameters of the waveguide, including propagation loss, sensing response, and evanescent field ratio, were considered for the analysis and comparison among the designed structures by varying the structural parameters such as width, thickness, and height of the germanium slots. The simulation results exhibited that the full-strip-loaded slot waveguide demonstrates excellent and higher performance in terms of sensitivity (73.76 L/mol) and evanescent field (28%), followed by the partial-strip-loaded slot waveguide, for a target propagation loss of 3 dB/cm. Interestingly, the conventional slot waveguide exhibited a marginally longer optimal length of 0.86 cm compared to the other structures. In 2022, Reddy and Singh proposed a porous-silicon-assisted hybrid plasmonic slot waveguide-based sensor to detect various concentrations of ethanol (Reddy & Singh, 2022). The sensing configuration is made of a metal-insulator-semiconductor-insulator-metal comprising porous silicon as the target sensing material embedded within the configuration. The device performance was optimized and analyzed through modeling and numerical simulations using the FEM in COMSOL Multiphysics, showing improved electric field confinement in a low-index slot and demonstrating a good red shift in resonance wavelength in the presence of different ethanol concentrations. The red shift observed in the output wavelength can be precisely tuned by varying the porosity in silicon and the structural parameters of the hybrid plasmonic waveguide. By optimizing the structural parameter of the design, the proposed sensor exhibited a high sensitivity of 400.43–612.43 nm/RIU along with a remarkable figure of merit (FoM) of 12.42–19.46/RIU, and an excellent Q-factor of 46.8–52.9, for 10% to 25% porosity of silicon.

4.6.2 Photonic crystal waveguide

The research related to photonic crystals (PhCs) is growing on a large scale in various sectors. PhC waveguides are considered as a fascinating sensing device utilized in a wide range of applications, including physical sensing, chemical sensing, biosensing, gas sensing, etc. (Gangwar et al., 2023). PhC waveguides can be categorized into 3 types, namely (1) One-dimensional PhCs (1D-PhCs), (2) two-dimensional PhCs (2D-PhCs), and (3) three-dimensional PhCs (3D-PhCs), representing different dimensions of periodic designs in order to manipulate and control the propagation of light. The variation of refractive index occurs only in one direction in 1D-PhCs, allowing control of light propagation within the structure with high accuracy. In the 2D-PhCs, the control of light was more precise in 2D, allowing more advanced control of light and supporting the development of planar devices like waveguides and microcavities (Gangwar et al., 2023). In contrast, the 3D-PhCs exhibit refractive index variations across the 3D, allowing precise manipulation of light within volumetric architectures. These types of waveguides have been widely exploited numerically to design and optimize various gas sensors (Goyal et al., 2017). For VOC sensing, the presence of periodic air holes in PhCs works as a natural candidate to accommodate VOC molecules in these air holes. The interaction of gaseous molecules with air holes results in the variation in the refractive index as well as the resonant wavelength of PhCs corresponding to various concentrations of gaseous molecules, which can be detected for gas sensing. In the past decades, several types of PhC fiber-based sensors for VOC detection have been developed and reported (Lopez-Torres et al., 2020). Out of several published articles, this chapter provides an example of such PhC based VOC sensor and the application of FEM analysis in optimizing such a structure.

In 2020, Paul et al. reported an oligoporous-core quasi-cladded PhC fiber based VOC sensor to monitor various VOC molecules, including ethanol, methanol, isopropanol (IPA), and butanol (Paul et al., 2020). Based on crystallography, the quasi-PhC fiber comprises ordered air holes with a periodicity, as shown in Fig. 4.2(A). The available spaces are considered to be filled with quasi-crystalline patterns. In the reported work, the diameter of the circular air hole was considered to be d = 0.75 μm, whereas the horizontal and vertical pitch sizes were considered to be Λ_X = 2 μm and Λ_Y = 2 μm, respectively. Effective sensing performance typically depends on careful optimization of the design for fabrication tolerances and practical realization. The sensor design proposed by the authors can be easily manufactured via the

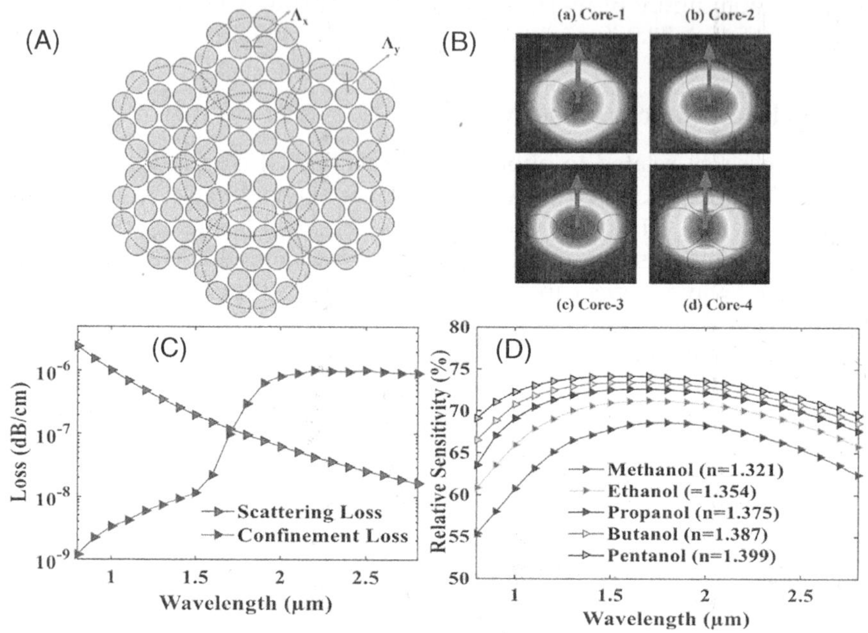

FIGURE 4.2 Schematic of quasi PCF sensor. Schematic diagram of (A) cross-section of the sensor (B) light confinement and field distribution in different region of PCF (C) scattering and confinement loss obtained from designed sensor, and (D) sensitivity of the sensor in presence of different VOCs. *From Paul, B. K., Ahmed, K., Dhasarathan, V., & Nguyen, T. K. (2020). Oligoporous-core Quasi cladding photonic crystal fiber based micro-sensor for alcohol detection.* Physica B: Condensed Matter, 584. *https://doi.org/10.1016/j.physb.2020.412104.*

sol-gel method. The reported design comprises a combination of 6 circular crystalline structures in order to form a larger hexagonal structure, as shown in Fig. 4.2(A). In order to determine the overall sensing performance of the designed sensor, power fraction, relative sensitivity, numerical aperture (NA), modal loss, and chromatic dispersion were evaluated using FEM based on commercially available software COMSOL Multiphysics. The relative sensitivity (R_s) of the design can be calculated using the following relation (Islam et al., 2018):

$$R_s\% = \frac{n_{ch}}{R_e|n_{eff}|} \times P_t \tag{4.9}$$

where n_{ch} represents the refractive index of different VOC analytes $R_e\,|\,n_{eff}\,|$ and represents the real part of the effective index, whereas P_t shows the overall fraction of the light beam. However, the overall power loss in the designed sensor depends on the power fraction of the designed model. The total power fraction can be defined as the ratio of power within the sample to the total power, as given by the following relation:

$$P_t = \frac{\int R_e(E_x H_y - E_x H_y)dxdy_{Sample}}{\int R_e(E_x H_y - E_x H_y)dxdy_{total}} \tag{4.10}$$

Here, E_x and $H_y$represent the electric and magnetic field components corresponding to x- and y-polarizations, respectively. However, the total power fraction of the designed sensor was estimated using confinement loss, which can be determined using the imaginary part of the refractive index, as given below (Pathak et al., 2019):

$$\alpha_c\left(\frac{dB}{m}\right) = 8.8686 \times 2\pi/\lambda \times Im[n_{eff}] \tag{4.11}$$

The NA can be defined as the internal geometrical compatibility of the structure, which depends on the effective area of total confinement (Paul et al., 2020). The NA of the design was estimated using the following relation (Islam et al., 2018):

$$NA = \frac{1}{\sqrt{1 + \frac{\pi A_{eff}}{\lambda^2}}} \tag{4.12}$$

where A_{eff}, indicates the area where the modes are effectively confined within the design structure and can be calculated using the following relation (Islam et al., 2018):

$$A_{eff}(\mu m) = \frac{\left[\int I(r)rdr\right]^2}{\left[\int I^2(r)rdr\right]^2} \tag{4.13}$$

where, I(r) can be defined as $| E_t |^2$. Using these equations, the authors optimized the structural parameter of the designed sensor.

Fig. 4.2(B) exhibits the mode profile at 1.5 μm wavelength for the designed structure. A tight confinement via the design core is evidently observed in Fig. 4.2(B) with minimal confinement loss. In contrast, the refractive index difference between the core and cladding of the designed configuration exhibits a significantly low scattering loss. Fig. 4.2(C) illustrates the confinement and associated scattering loss of the designed sensor at 1.5 μm wavelength in the presence of ethanol. The maximum confinement and scattering losses were obtained to be 1.15×10^{-8} dB/cm and 1.53×10^{-7} dB/cm, respectively. The convergence error of 7.953×10^{-9} has been found at the operating wavelength $\lambda = 1.55$ μm. The value of NA increases when light is more tightly confined within the smaller core region. However, an NA larger than 0.4 is generally rare for the glass optical fibers (Paul et al., 2020). Fig. 4.2(C) shows that the scattering loss of the designed sensor shows an inverse trend compared to the confinement loss.

While the confinement loss demonstrates an upward trend, the scattering loss displays a downward trend. Additionally, it is evident that the scattering loss and confinement loss intersect at 1.75 μm, both of which are equal to 10^{-7} dB/cm. Fig. 4.2(D) illustrates the sensitivity of various VOCs, including methanol, propanol, ethanol, pentanal, and butanol, with respect to wavelength. The obtained result reveals that the sensitivity of the sensor improves with the increase in the refractive index of the analytes, which generally depends on the chemical properties. Therefore, the sensitivities toward the alcohol analytes can be ranked as methanol < ethanol < propanol < butanol < pentanol, respectively. Relative sensitivities of methanol, ethanol, propanol, butanol, and pentanol were analyzed to be 66.8%, 71.2%, 72.6%, 73.4%, and 74.8%, respectively, at 1.5 μm (Paul et al., 2020). In 2023, Jokar et al. reported a novel PhC waveguide design comprising a cavity to identify multiple VOCs in water utilizing FDTD simulations (Houssein Jokar et al., 2023). The design of the sensor was made of a simple PhC comprising two waveguide couplers and a micro-cavity surrounded by multiple air holes, as shown in Fig. 4.3(A). A micro-cavity is considered in the center of this PhC structure, which enhances a specific wavelength and transfers it to the output end. Later, the sensitivity of the device was improved by considering two additional cavities positioned above and below the central cavity [as shown in Fig. 4.3(B)], and the performance of the device was compared with a single cavity, and an enhanced output intensity at the resonant wavelength was obtained with the latter design. The sensing configuration was tested in the presence of ethanol, methanol, and acetone with a 30% concentration in water. The proposed design exhibited a maximum sensitivity of 356.43 nm/RIU for ethanol, whereas lower sensitivities of 210 nm/RIU and 273.33 nm/RIU were recorded for methanol and acetone, respectively.

4.6.3 Fiber optic waveguides

Several standard optical fiber-based VOC sensors have been designed and optimized using FEM. However, the actual materials analysis hasn't been well exploited for actual gas sensing (Pathak & Singh, 2019; Pathak et al., 2022). In 2021, Pathak and Viphavakit reported, designed and analyzed the potential of D-shaped polished standard optical fiber for early detection of diabetes based on composite plasmonic material (Pathak & Viphavakit, 2021). The reported work focused on the detection of a diabetes biomarker. The exhaled breath of a diabetic patient generally comprises several VOCs, including acetone, ethanol, and IPA. Among the various VOCs, IPA has been commonly detected in both Type I and Type II diabetic patients, as discussed in Chapter 1. In the present work, the authors reported a simple design of optical fiber-based composite plasmonic sensor comprising a D-shaped single-mode fiber (SMF) coated with a thin layer of silver, followed by porous-titanium oxide ($Ag/P\text{-}TiO_2$) composite to

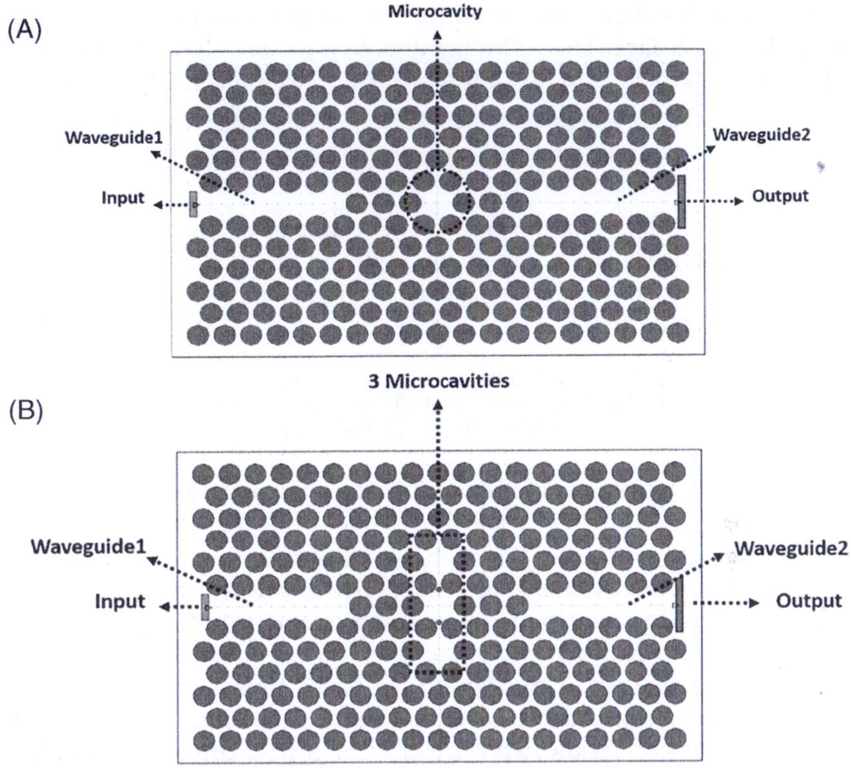

FIGURE 4.3 2D Photonic crystal. (A) The layout of the primarily designed biochemical sensor based on 2D PhC using a triangular lattice of air holes and a micro-cavity in the middle (B) the improved design of the sensor by introducing two micro-cavities above and below the central cavity. *From Houssein Jokar, M., Naraghi, A., Seifouri, M., & Olyaee, S. (2023). Design of bio-alcohol sensor based on waveguide-coupled photonic crystal cavity.* Results in Optics, 13. *https://doi.org/10.1016/j.rio.2023.100563.*

identify various concentrations of IPA molecules. In the current work, the authors utilized TiO_2 due to its excellent selectivity towards IPA (Alev et al., 2015; Shao et al., 2017; Vishwakarma & Yadav, 2018). The proposed sensor utilizes a wavelength interrogation method, and the sensor applies modal loss analysis in order to optimize the structural parameters of the sensor and investigate its performance corresponding to various volume fractions of IPA trapped within the porous TiO_2 air voids. For the sensor design, a standard SMF with core and cladding diameters of 9 μm and 125 μm, respectively, was utilized. The cladding material of the fiber was considered to be made of fused silica, whereas the core of the fiber was considered to be silica with germanium doping to maintain the refractive index difference between the core and the cladding and facilitate the total internal reflection (TIR) phenomenon. Later, the fiber was considered polished to form the

D-shape, and the Ag/P-TiO_2 composite material was considered on top of this polished surface. The Sellmeier equation given below was utilized to calculate the refractive index of the fused silica (Pathak et al., 2022):

$$n_s = \sqrt{1 + \frac{a_1 \lambda^2}{\lambda^2 - b_1^2} + + \frac{a_2 \lambda^2}{\lambda^2 - b_2^2} + \frac{a_3 \lambda^2}{\lambda^2 - b_3^2}} \tag{4.14}$$

where, a_1, a_2, a_3, b_1, b_2, and b_3 represent Sellmeier coefficients, the values of which are taken from a previously reported article (Ghosh et al., 1994). The dielectric constant of the Ag film was determined using Lorentz-Drude mode (Rakić et al., 1998):

$$\varepsilon_m = 1 - \frac{\Omega_p^2}{\omega(\omega - i\Gamma_0} + \sum_{j=1}^{k} \frac{f_j \omega_p^2}{\left(\omega_j^2 - \omega^2\right) + j\omega\Gamma_i} \tag{4.15}$$

where ε_m and ω_p represent the dielectric constant and the plasma frequency, respectively, whereas k is the number of oscillators with frequency ω_p and strength f_j. $\Omega_p = \sqrt{(f_0.\omega_p)}$, shows the plasma frequency related to intraband transition (Rakić et al., 1998). The sensor was developed using pure and hybrid plasmonic materials, which typically required accurate evaluation of the real and imaginary components of the modal effective index (n_{eff}). In the case of an optical fiber-based sensor design, a rigorous modal analysis is essential in order to determine the modal loss with the change in the n_{eff} under the exposure to various VOC concentrations. All of the above parameters can be determined by solving Maxwell's equations for a cylindrical waveguide. Here, the authors utilized the FV-FEM based on commercially available software COMSOL Multiphysics to evaluate the exact propagated modes through such waveguides.

The proposed IPA sensor consists of a thick layer of P-TiO_2 on top of a thin plasmonic layer of Ag. P-TiO_2 has a lower refractive index compared to the bulk TiO_2 due to the presence of air voids throughout the surface ($n = 1$). As the porosity of P-TiO_2 increases, its equivalent refractive index decreases. The increased pore volume resulting from the porosity results in more absorption and condensation of IPA in the air voids. The P-TiO_2 exhibits several advantages over the bulk TiO_2 due to its uniform pore size and high specific surface area (Guo et al., 2009; Su et al., 2006). When the IPA is considered as an ambient environment surrounding the sensor, it is expected to be absorbed within the air voids of P-TiO_2, and the available air from the air voids is considered to be occupied by the condensed form of IPA. Depending on the multiple volume fractions of IPA absorbed by P-TiO_2, both the dielectric constant and the n_{eff} change. Various quantitative models have been reported in past decades in order to estimate the permittivity of such heterogeneous media, e.g., the Maxwell-Garnet theory, Bruggeman model, and the Lorentz-Lorenz model. Here, the authors have used the Lorentz-Lorenz model to evaluate the equivalent refractive index and n_{eff} of P-TiO_2 as given in the equation below: (Garahan et al., 2007; Ghosh and Rahman, 2017):

$$\frac{n_e^2 - 1}{n_e^2 + 2} = (1 - P)\left(\frac{n_c^2 - 1}{n_c^2 + 2}\right) + (P - V)\left(\frac{n_a^2 - 1}{n_a^2 + 2}\right) + V\left(\frac{n_d^2 - 1}{n_d^2 + 2}\right) \tag{4.16}$$

Here, n_e, n_c, n_a, and n_d represent equivalent refractive index, refractive index of bulk TiO_2 (DeVore, 1951), refractive index of air, and dispersed media (condensed IPA) (Sani & Dell'Oro, 2016), respectively. Similarly, V represents the volume fraction of IPA, whereas P depicts the porosity of the TiO_2. Using the same FV-FEM, the authors optimized all the structural parameters of the designed sensor, including the thickness of Ag, P-TiO_2, porosity volume, as well as the polishing depth. The sensing performance of the device was investigated for IPA condensed in P-TiO_2, varied between 0% and 100%. The designed sensor demonstrated a maximum sensitivity of 0.25 nm/%, 0.35 nm/%, 0.45 nm/%, and 1.35 nm/% for 20%, 40%, 60%, and 80% volume fractions of IPA condensed in P-TiO_2.

Following a similar approach, Chauhan and Singh reported a numerical investigation over TiO_2 deposited tapered SMF for alcohol sensing (Chauhan & Singh, 2023). In the reported work, the authors utilized FEM to optimize the structural parameters (i.e., metal thickness and tapering diameter) and analyze the sensing performance of the designed sensor based on the wavelength interrogation technique. The sensor comprises a tapered SMF coated with silver and a thin layer of TiO_2 to achieve SPR and improve the sensitivity of the device. In addition to improving the sensitivity of the sensor, the thin layer of TiO_2 over silver is utilized to protect the silver from oxidation and avoid toxicity. The authors performed numerical analysis using COMSOL Multiphysics-based FEM simulations. The obtained result exhibited a maximum wavelength sensitivity of 3250 nm/RIU and amplitude sensitivity of 1891 RIU^{-1}. The sensor also achieved an average refractive index and alcohol sensitivity of 2350 nm/RIU, and 0.886 nm/%, respectively, along with high linearity over index values ranging from 1.33 to 1.38 and alcohol concentration levels from 0 to 60%. Additionally, the authors also observed a reasonable resolution of 3.07×10^{-5} RIU and a FoM and detection accuracy of 195 RIU^{-1} and 0.008, respectively.

4.7 Conclusion

This chapter provides an overview of the FEM analysis method, along with its basic introduction and crucial aspects used to analyze and optimize various complex optical waveguides. After a brief discussion of the fundamentals of FEM (procedure, discretization, node and edge element, PML), guided beam analysis, beam propagation, and types of FEM analysis, three different types of waveguides have been discussed, based on the FEM analysis used in VOC sensing. FEM is not limited to gas sensing applications, they are also widely utilized in designing and optimizing optical waveguides exhibiting effects like electro-optic, acousto-optic, magneto-optic, and Kerr nonlinearity.

References

Alam, M. Z., Bahrami, F., Aitchison, J. S., & Mojahedi, M. (2014). Analysis and optimization of hybrid plasmonic waveguide as a platform for biosensing. *IEEE Photonics Journal, 6*(4), 1–10. doi:10.1109/JPHOT.2014.2331232.

Alev, O., Şennik, E., Kılınç, N., & Öztürk, Z. Z. (2015). Gas sensor application of hydrothermally growth TiO_2 nanorods. *Procedia Engineering, 120*, 1162–1165. doi:10.1016/j.proeng.2015.08.747.

Atia, K. S. R., et al. (2019). Finite element method for sensing applications. *Computational Photonic Sensors* (pp. 109–151). Springer.

Atia, K. S. R., Ghosh, S., Heikal, A. M., Hameed, M. F. O., Rahman, B. M. A., & Obayya, S. S. A. (2018). Finite element method for sensing applications. *Computational photonic sensors* (pp. 109–151). Springer International Publishing. http://www.springer.com/in/book/9783319765556. doi:10.1007/978-3-319-76556-3_6.

Atia, K. S. R., Heikal, A. M., & Obayya, S. S. A. (2015). Efficient smoothed finite element time domain analysis for photonic devices. *Optics Express, 23*(17), 22199–22213. http://www.opticsexpress.org/. doi:10.1364/OE.23.022199.

Berenger, J. P. (1994). A perfectly matched layer for the absorption of electromagnetic waves. *Journal of Computational Physics, 114*(2), 185–200. doi:10.1006/jcph.1994.1159.

Berini, P. (2000). Plasmon-polariton waves guided by thin lossy metal films of finite width: Bound modes of symmetric structures. *Physical Review B – Condensed Matter and Materials Physics, 61*(15), 10484–10503. doi:10.1103/PhysRevB.61.10484.

Berk, A. (1956). Variational principles for electromagnetic resonators and waveguides. *IRE Transactions on Antennas and Propagation, 4*(2), 104–111. doi:10.1109/TAP.1956.1144365.

Chandra, V., & Ranjan, R. (2021). Performance analysis of different slot waveguide structures for evanescent field based gas sensor applications. *Optical and Quantum Electronics, 53*(8). doi:10.1007/s11082-021-03102-8.

Chauhan, M., & Singh, V. K. (2023). TiO_2 coated tapered optical fiber SPR sensor for alcohol sensing application. *Journal of Optics (India), 52*(4), 2285–2295. doi:10.1007/s12596-023-01131-y.

DeVore, J. R. (1951). Refractive indices of rutile and sphalerite. *Journal of the Optical Society of America, 41*(6), 416. doi:10.1364/JOSA.41.000416.

Gangwar, R. K., Pathak, A. K., & Kumar, S. (2023). Recent progress in photonic crystal devices and their applications: A review. *Photonics, 10*(11). http://www.mdpi.com/journal/photonics. doi:10.3390/photonics10111199.

Garahan, A., Pilon, L., Yin, J., & Saxena, I. (2007). Effective optical properties of absorbing nanoporous and nanocomposite thin films. *Journal of Applied Physics, 101*(1). doi:10.1063/1.2402327.

Ghosh, G., Endo, M., & Iwasaki, T. (1994). Temperature-dependent sellmeier coefficients and chromatic dispersions for some optical fiber glasses. *Journal of Lightwave Technology, 12*(8), 1338–1342. doi:10.1109/50.317500.

Ghosh, S., & Rahman, B. M. A. (2017). A compact mach-zehnder interferometer using composite plasmonic waveguide for ethanol vapor sensing. *Journal of Lightwave Technology, 35*(14), 3003–3011. doi:10.1109/JLT.2017.2703827.

Goyal, A. K., Dutta, H. S., & Pal, S. (2017). Recent advances and progress in photonic crystal-based gas sensors. *Journal of Physics D: Applied Physics, 50*(20). http://iopscience.iop.org/article/10.1088/1361-6463/aa68d3/pdf. doi:10.1088/1361-6463/aa68d3.

Guo, B., Shen, H., Shu, K., Zeng, Y., & Ning, W. (2009). The study of the relationship between pore structure and photocatalysis of mesoporous TiO_2. *Journal of Chemical Sciences, 121*(3), 317–321. http://www.ias.ac.in/chemsci/Pdf-May2009/317.pdf. doi:10.1007/s12039-009-0036-5.

Houssein Jokar, M., Naraghi, A., Seifouri, M., & Olyaee, S. (2023). Design of bio-alcohol sensor based on waveguide-coupled photonic crystal cavity. *Results in Optics, 13*. doi:10.1016/j.rio.2023.100563.

Islam, M. S., Sultana, J., Ahmed, K., Islam, M. R., Dinovitser, A., Ng, B. W. H., & Abbott, D. (2018). A novel approach for spectroscopic chemical identification using photonic crystal fiber in the terahertz regime. *IEEE Sensors Journal, 18*(2), 575–582. http://ieeexplore.ieee.org/xpl/RecentIssue.jsp?punumber=7361. doi:10.1109/JSEN.2017.2775642.

Kakihara, K., Kono, N., Saitoh, K., & Koshiba, M. (2006). Full-vectorial finite element method in a cylindrical coordinate system for loss analysis of photonic wire bends. *Optics Express, 14*(23), 11128. doi:10.1364/OE.14.011128.

Koshiba, M. (2014). Optical waveguide theory by the finite element method. *IEICE Transactions on Electronics, E97-C*(7), 625–635. https://www.jstage.jst.go.jp/article/transele/E97.C/7/E97.C_625/_pdf. doi:10.1587/transele.E97.C.625.

Koshiba, M., Hayata, K., & Suzuki, M. (1984a). Approximate scalar finite-element analysis of anisotropic optical waveguides with off-diagonal elements in a permittivity tensor. *IEEE Transactions on Microwave Theory and Techniques, 32*(6), 587–593. doi:10.1109/TMTT.1984.1132733.

Koshiba, M., Hayata, K., & Suzuki, M. (1984b). Vectorial finite-element method without spurious solutions for dielectric waveguide problems. *Electronics Letters, 20*(10), 409. doi:10.1049/el:19840283.

Koshiba, M., Hayata, K., & Suzuki, M. (1986). Finite-element solution of anisotropic waveguides with arbitrary tensor permittivity. *Journal of Lightwave Technology, 4*(2), 121–126. doi:10.1109/JLT.1986.1074687.

Koshiba, M., & Inoue, K. (1992). Simple and efficient finite-element analysis of microwave and optical waveguides. *IEEE Transactions on Microwave Theory and Techniques, 40*(2), 371–377. doi:10.1109/22.120111.

Koshiba, M., & Tsuji, Y. (2000). Curvilinear hybrid edge/nodal elements with triangular shape for guided-wave problems. *Journal of Lightwave Technology, 18*(5), 737–743. doi:10.1109/50.842091.

Koshiba, M., Tsuji, Y., & Hikari, M. (1999). Finite element beam propagation method with perfectly matched layer boundary conditions. *IEEE Transactions on Magnetics, 35*(3), 1482–1485. doi:10.1109/20.767247.

LeVeque R.J., Finite volume methods. Cambridge University Press, (2002), 64–86, doi:10.1017/cbo9780511791253.005.

Liu, G. R. (2008). A generalized gradient smoothing technique and the smoothed bilinear form for Galerkin formulation of a wide class of computational methods. *International Journal of Computational Methods, 5*(2), 199–236. doi:10.1142/S0219876208001510.

Liu, G. R. (2009). *Meshfree methods: Moving beyond the finite element method* (2nd ed., pp. 1–749). CRC Press. http://www.tandfebooks.com/doi/book/9781420082104. doi:10.1201/9781420082104.

Lopez-Torres, D., Elosua, C., & Arregui, F. J. (2020). Optical fiber sensors based on microstructured optical fibers to detect gases and volatile organic compounds—A review. *Sensors, 20*(9), 2555. doi:10.3390/s20092555.

Monk, Peter (2003). *Finite Element Methods for Maxwell's Equations*. Oxford Academic.
Nagarkar, R. P., & Pawar, M. M. (2014). Modal analysis of optical waveguide using finite element method. *Emerging Trends in Electronics and Telecommunication Engineering 2013*. https://www.ijcaonline.org/proceedings/ncet/number1/15646-1403/.
Pathak, A. K., Rahman, B. M. A., Singh, V., & Kumari, S. (2019). Sensitivity enhancement of a concave shaped optical fiber refractive index sensor covered with multiple Au nanowires. *Sensors, 19*(19), 4210. doi:10.3390/s19194210.
Pathak, A. K., Rahman, B. M. A., & Viphavakit, C. (2022). Nanowire embedded micro-drilled dual-channel approach to develop highly sensitive biosensor. *IEEE Photonics Technology Letters, 34*(13), 707–710. doi:10.1109/lpt.2022.3182783.
Pathak, A. K., & Viphavakit, C. (2021). VOC biomarker monitoring for diabetes through exhaled breath using Ag/P-TiO_2 composite plasmonic sensor. *IEEE Sensors Journal, 21*(20), 22631–22637. doi:10.1109/JSEN.2021.3104766.
Paul, B. K., Ahmed, K., Dhasarathan, V., & Nguyen, T. K. (2020). Oligoporous-core Quasi cladding photonic crystal fiber based micro-sensor for alcohol detection. *Physica B: Condensed Matter, 584*. http://www.journals.elsevier.com/physica-b-condensed-matter/. doi:10.1016/j.physb.2020.412104.
Polycarpou, Anastasis C. (2022). *Introduction to the Finite Element Method in Electromagnetics*. Springer.
Rahman, B., & Davies, J. (1984a). Finite-element solution of integrated optical waveguides. *Journal of Lightwave Technology*, 2(5), 682–688. doi:10.1109/JLT.1984.1073669.
Rahman, B. M. A., & Davies, J. B. (1984b). Finite-element analysis of optical and microwave waveguide problems. *IEEE Transactions on Microwave Theory and Techniques, 32*(1), 20–28. doi:10.1109/TMTT.1984.1132606.
Rahman, B. M. A., & Davies, J. B. (1984c). Penalty function improvement of waveguide solution by finite elements. *IEEE Transactions on Microwave Theory and Techniques, 32*(8), 922–928. doi:10.1109/TMTT.1984.1132789.
Rakić, A. D., Djurišić, A. B., Elazar, J. M., & Majewski, M. L. (1998). Optical properties of metallic films for vertical-cavity optoelectronic devices. *Applied Optics, 37*(22), 5271. doi:10.1364/ao.37.005271.
Reddy, S. K., & Singh, M. (2022). Porous-silicon assisted hybrid plasmonic slot waveguide based on-chip ethanol sensor. *IEEE Sensors Journal, 22*(3), 2062–2069. doi:10.1109/JSEN.2021.3135762.
Rodriguez-Esquerre, V. F., Koshiba, M., & Hernandez-Figueroa, H. E. (2005). Finite-element analysis of photonic crystal cavities: Time and frequency domains. *Journal of Lightwave Technology, 23*(3), 1514–1521. doi:10.1109/jlt.2005.843441.
Rylander, T., Ingelström, P., & Bondeson, A. (2013). *Computational electromagnetics texts in applied mathematics*: Vol. 2. Springer https://link.springer.com/book/10.1007/978-1-4614-5351-2. doi:10.1007/978-1-4614-5351-2.
Saitoh, K., & Koshiba, M. (2001a). Approximate scalar finite-element beam-propagation method with perfectly matched layers for anisotropic optical waveguides. *Journal of Lightwave Technology, 19*(5), 786–792. doi:10.1109/50.923493.
Saitoh, K., & Koshiba, M. (2001b). Full-vectorial finite element beam propagation method with perfectly matched layers for anisotropic optical waveguides. *Journal of Lightwave Technology, 19*(3), 405–413. doi:10.1109/50.918895.
Sani, E., & Dell'Oro, A. (2016). Spectral optical constants of ethanol and isopropanol from ultraviolet to far infrared. *Optical Materials, 60*, 137–141. doi:10.1016/j.optmat.2016.06.041.
Shao, S., Wang, W., Zhou, K., Jiang, F., Wu, H., & Koehn, R. (2017). GQDs-TiO_2 heterojunction based thin films for volatile organic compounds sensor with excellent performance at room temperature. *Materials Letters, 186*, 193–197. http://www.journals.elsevier.com/materials-letters/. doi:10.1016/j.matlet.2016.10.012.
Stallcup, I. A. (2018). *Finite Element and Finite Difference Methods for Maxwell's Equations in Metamaterials*.

Su, C., Lin, K. F., Lin, Y. H., & You, B. H. (2006). Preparation and characterization of high-surface-area titanium dioxide by sol-gel process. *Journal of Porous Materials, 13*(3), 251–258. doi:10.1007/s10934-006-8012-7.

Tadashi, M., Saitoh, K., Taru, T., Nagashima, T., Maeda, K., & Sasaki, T. (2009). Bend-insensitive and effectively single-moded all-solid photonic bandgap fibers with heterostructured cladding. *Proceedings of the 35th European Conference on Optical Communication*. IEEE https://ieeexplore.ieee.org/document/5287028.

Teixeira, F. L., & Chew, W. C. (1998). General closed-form PML constitutive tensors to match arbitrary bianisotropic and dispersive linear media. *IEEE Microwave and Guided Wave Letters, 8*(6), 223–225. doi:10.1109/75.678571.

Tsuji, Y., & Koshiba, M. (2000). Adaptive mesh generation for full-vectorial guided-mode and beam-propagation solutions. *IEEE Journal on Selected Topics in Quantum Electronics, 6*(1), 163–169. doi:10.1109/2944.826884.

Vishwakarma, A. K., & Yadav, L. (2018). Detection of propanol gas using titanium dioxide based thick film. *IOP Conference Series: Materials Science and Engineering, 404*(1), 012020. doi:10.1088/1757-899x/404/1/012020.

Wang X.-H., (2019). *Finite element methods for nonlinear optical waveguides.*, doi:10.1201/9780203756027 2019.

CHAPTER

5

VOC sensing materials: Principles and recent advances

5.1 Introduction

Volatile organic compounds (VOCs) are known for their low molecular weight, with high vapor pressure under ambient conditions and at room temperature (Lai et al., 2021). Based on their volatility, VOCs can be categorized into three classes: (1) volatile, (2) semi-volatile, and (3) very-volatile. VOCs can also be classified based on the chemical classes, e.g., pyrazine, terpenes, benzenoids, organic acids, alcohols, ketones, and alkenes (Almeida et al., 2023). Typically, these VOCs can be produced from both natural processes and human activities. Major anthropogenic sources of VOC production involve emissions from transportation, industrial operations, agricultural activities, and metabolic processes in humans. The primary natural sources of VOC emission include geological activities, microbial activity in soil, forest fires, animal emissions, and volcanic activity (Han et al., 2024). VOCs play a crucial role in various photochemical processes, which produce harmful and potentially lethal byproducts. For instance, it has been reported that in northern China, VOCs released from oil production are the primary cause of increased atmospheric oxidation capacity, leading to severe ozone air pollution (Chen et al., 2020a, 2020b; Han et al., 2024). These VOC vapors are generally teratogenic, carcinogenic, toxic, and hazardous, leading to serious health problems when exposure exceeds the threshold limit set by various government agencies (Pathak & Viphavakit, 2022). In addition, several VOCs have an extremely unpleasant odor, which can lower the air quality of the atmosphere. These VOCs have unique physical and chemical properties, i.e., small molecules

Photonic Sensors for Detection of VOC Biomarkers: Technological Principles and Medical Applications
DOI: https://doi.org/10.1016/B978-0-443-29166-1.00006-8

Copyright © 2026 Elsevier Inc. All rights are reserved, including those for text and data mining, AI training, and similar technologies.

(< 300 da), typically contain two functional groups, and have the ability to quickly disperse in air and water (Weisskopf et al., 2021).

Long-term exposure to VOCs can cause irreversible damage to the immune system as well as the respiratory system, and can be life-threatening (Zhou, 2023). Considering the toxicity and threat of VOCs, the World Health Organization (WHO) has set a threshold limit for each VOC to prevent excessive exposure to human beings (Szulczyński & Gębicki, 2017). Additionally, several VOC vapors are released from the human body on a daily basis, serving as a potential biomarker to understand the state of human health, the presence of disease, and the occurance of lesions. For example, acetone is popularly known as a potential biomarker to identify diabetes, whereas variation in the concentration level of benzene in the exhaled breath indicates lung cancer (Yin et al., 2021). Therefore, VOC monitoring is considered highly crucial for disease diagnosis, human health care, environmental monitoring, and other applications. While their low concentrations, predominantly nonpolar nature, and limited reactivity make it challenging to produce portable and real-time sensing configurations for these VOCs, this field reminas an active area of research aimed to developing target-selective materials (Pathak and Viphavakit, 2022) that can lead to the commercialization of sensors (Dai et al., 2022).

Several reviews have appeared in the recent literature exploring advancements in VOC sensing materials. For instance, Pathak et al. published a recent review on VOC sensing materials for clinical applications, providing in-depth insights into the types and subtypes of these materials, along with their advantages and limitations (Pathak et al., 2023). Han et al., 2024 reported a detailed overview of the basics of VOC sensing and applicable materials, along with their development strategies, major challenges, and opportunities. Previously published reports have shown that nanoscale sensing materials, ranging in size from 2 to 100 nm, are widely used in the development of highly sensitive vapor/gas sensors. This is primarily due to their large surface area and unique material characteristics (i.e., chemical, optical, and electrical) compared to bulk materials. These sensing materials typically include metal oxides (e.g., zinc oxide, nickel oxide, molybdenum oxide, and titanium dioxide), carbon-based materials (e.g., graphene, multi-walled carbon nanotubes, and activated carbon [AC]), and polymers (e.g., polydimethylsiloxane, polypyrrole, polythiophene, and molecularly imprinted polymers [MIP]) with various morphological structures, deposited onto traditional transducers such as capacitive, gravimetric, resistive, and optical sensors. The sensing ability of these materials is highly influenced by the chemisorption of negatively charged oxygen from the ambient air environment (Tricoli et al., 2010). Gas chemisorption is an energy-dependent process, requiring various surface temperatures for oxygen species to adsorb onto the surface of the sensing material. In ambient conditions, oxygen molecules capture electrons from the sensing material and exist in several forms depending on the temperature, i.e., O_2^- at <100°C, O^- between 100°C and 300°C, and O^{2-} at >300°C. Typically, relatively high temperatures may enhance the activation of adsorbed VOC molecules and speed up the rate of gas diffusion. It is conducive to the occurrence of chemical reactions between the oxygen species and the target VOC molecules (Pathak et al., 2023). It is also influenced by the chemical reaction taking place between gaseous molecules and sensing material, as well as the diffusion of VOCs through the bulk sensing material.

From the above discussion, it is evident that the choice of sensing material plays a crucial role in VOC detection, as it directly influences the sensitivity, selectivity towards target analytes, response time, and limit of detection (LOD) of the sensor. This chapter aims to provide detailed information on each sensing material along with its surface morphology, enabling researchers to understand the advantages and limitations of each materials. Additionally, it will help them in selecting the most appropriate material for their specific applications.

5.2 Metal oxides

Oxides are chemical compounds that contain at least one or more oxygen atoms along with other composite elements, e.g., CO_2 and H_2O. Metal oxides (MOs) are widely popular chemical compounds that are produced by combining a metal and oxygen. In general, the MOs are recognized for their ability to selectively detect target analytes using various techniques such as electrical, optical, and chemical.

5.2.1 Zinc oxide

ZnO is a MO belonging to the class of n-type semiconductors that possess a large binding energy of 60 meV at room temperature (RT) (Zhu et al., 2018) and have a wide band-gap energy (3.37 ev at 300 K). Nanostructured ZnO, with its high surface-to-volume ratio, is a promising candidate for VOC and gas-sensing detection (Al-Hardan et al., 2010). ZnO has been reported as one of the most popular gas sensing materials as it provides several advantages, including simple synthesis, biocompatibility, good mechanical stability, and offers precise and selective response toward various VOCs and semi-VOCs (Gakhar et al., 2022). A large number of research and investigations have been reported in the past decades for gas sensing using various types of ZnO nanostructures e.g., nanotubes , nanorods, nanosheets, nanowires, nanoparticles, thin films, and 3D hierarchical structures (Pathak et al., 2018;

Swargiary et al., 2022; Zhu et al., 2018,(Saeed, Marwani, Shahzad, Asiri, & Rahman, 2024)). These nanostructures can be readily synthesized by widely-known approaches, e.g., radio frequency (RF) sputtering (Al-Hardan et al., 2010; Teimoori et al., 2017), sol-gel coating (Muthukrishnan et al., 2016), and hydrothermal (Peng et al., 2016; Renganathan et al., 2011; Swargiary et al., 2022), solvothermal, combined with calcination (Zhu et al., 2018). Among various classes of VOCs, ethanol and acetone are among the most reported and investigated VOCs to be detected by various ZnO nanostructures, while significant research exists that shows their promising response toward other VOCs such as n-butanol, acetylene, benzene, and cyclohexane based on their different structures. Hardan et al. reported the integration of ZnO thin films on a thermally oxidized silicon substrate in order to identify acetone, isopropanol, and ethanol, and report the highest sensitivity for acetone monitoring (~500 ppm) at high temperature of 400°C (Al-Hardan et al., 2010). The operating mechanism is based on the ionosorption model, where the resistance of the MO semiconductor changes in the presence of the test gas. Bora et al. reported ZnO nanorods coated on a multimode optical fiber for the detection of various volatile chemical vapors including methanol, ethanol, benzene, and toluene and obtained an average sensitivity of 22.2% at 50 ppm for ethanol vapors (Bora et al., 2014). The investigation offered a compact, simple and low-cost optical fiber sensor using light scattering from the coated ZnO nanorods of a specific length, diameter, and surface-to-volume ratio of nanorods. Later in 2016, Muthukrishnan et al. proposed an alternative ZnO deposition using sol-gel dip coating approach and achieved thin ZnO layer in order to lower the cost per production by reducing processing complications (Muthukrishnan et al., 2016). The sensing performance of the device was investigated for acetone, acetaldehyde, and ethanol by chemiresistive technique, and they report a rapid response/recovery time on the order of seconds. The sensor exhibited the highest selectivity for acetone at 71.32% compared to other test gases, with a minimum detection limit of 2 ppm for acetone at RT. The sensor exhibited a high sensitivity of 1.08, which was calculated by utilizing the ratios between the changes in the electrical resistance of the sensor in air to the test gases. In 2018, Zhu et al. synthesized porous coral ZnO for the detection of methylbenzene, formaldehyde, acetone, and methanol for low concentrations of 600 ppm at a high temperature 360°C (Zhu et al., 2018). Due to the porosity of ZnO, the fabricated device shows significant performance toward these VOCs with rapid response/recovery for all four VOC gases. Most recently, Swargiary et al., 2022 also reported the potential of ZnO nanorods coated multimode fiber in order to identify various concentrations of isopropanol vapor exhaled from diabetic patients. The reported sensor showed a high sensitivity of 0.053 nm/% for isopropanol vapor detected at RT. Fig. 5.1 illustrates

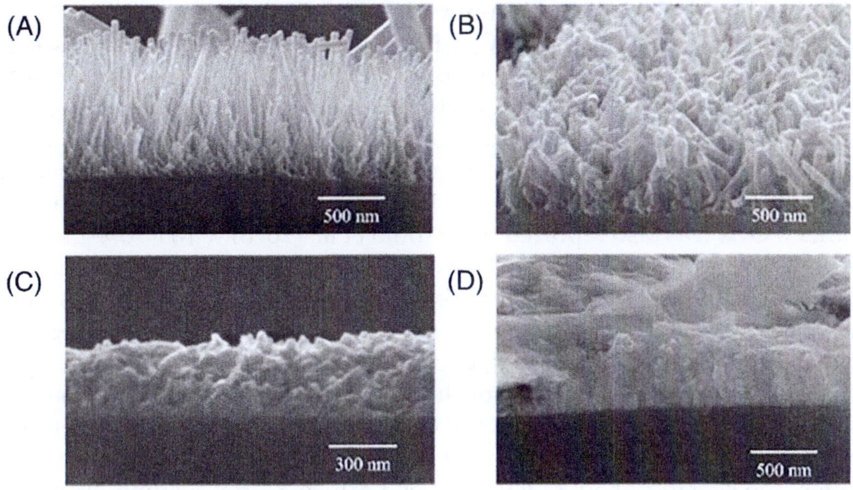

FIGURE 5.1 SEM image of ZnO nanorods.SEM image of ZnO nanorods decorated over CSF under different growth times of (A) 3 hours, (B) 4 hours, (C) 5 hours, and (D) 7 hours. *From Swargiary, K., Metem, P., Kulatumyotin, C., Thaneerat, S., Ajchareeyasoontorn, N., Jitpratak, P., Bora, T., Mohammed, W. S., Dutta, J., & Viphavakit, C. (2022). ZnO nanorods coated single-mode–multimode–single-mode optical fiber sensor for VOC biomarker detection.* Sensors, *22(16). https://doi.org/10.3390/s22166273.*

the surface morphology of ZnO nanorods grown over the multimode fiber using the spray coating technique for VOC sensing.

Although various ZnO nanostructure-decorated sensors have been shown to have great potential with excellent sensing performance for various types of VOCs, achieving high sensitivity and target selectivity to effectively identify *multiple* VOC vapors in a mixed volatile gaseous environment is still challenging. The primary drawback of ZnO structures as gas sensing materials is their poor selectivity (Wang et al., 2017a). Various researchers have reported that the property of ZnO nanostructures can be tuned by composite structuring, doping, and/or utilizing surface modification by other noble metals, thereby providing significant improvement to the target selectivity. Ma et al. (2017) proposed Ag-doped ZnO/SnO_2 hollow nanofiber structured sensors for quick responsivity (~5 seconds) and excellent selectivity with a sensitivity of 1 ppm concentration of ethanol at 200°C. The improved sensing response was attributed to the fact that the doping of a noble metal along with ZnO creates sites for adsorbates, promotes surface reactions, and acts as a catalyst, thereby, improving the thermal stability of the nanostructure. The major advantage of this structure is its high surface-to-volume ratio, its surface roughness and porosity, which makes it a promising candidate for the efficient detection of ethanol vapor in the surrounding environment. That same year,

Wang et al. (2017a) utilized Mn-doped ZnO (MZO) for the selective detection of 20 ppm acetone at 340°C, while CdO-activated MZO was used for the selective detection of 20 ppm ethanol at 240°C. These gas sensors were also employed as a sensor array to differentiate among ethanol and acetone mixtures at varying concentration ratios, demonstrating their potential as high-performance gas sensors. Similarly high sensitivity and selectivity with stable response/recovery characteristics toward methanol among other VOCs was also reported by Mandal et al. (2018) with a rose-like ZnO microcube/MoO_3 micrograss-based composite structure. More recently, Dong et al., 2023 synthesized a novel composite of hollow cubes made of SnO_2/ZnO using a self-template hydrothermal method followed by a calcination process. Dong et al. explored the synthesis mechanism of the composite structure and its morphology, as shown in Fig. 5.2. The ZnO nanostructures were grown on the thin walls of hollow cubes, along with the SnO_2 nanorods attached to the hollow cube walls. Using these structures, the sensing of VOC vapors was performed for acetone, formaldehyde, isopropanol, n-butanol, methanol, ethanol, benzene, as well as three other reductive gases, i.e., hydrogen, ammonia, and carbon monoxide, at 220°C. Among all the VOC vapors, the specially grown hollow SnO_2/ZnO cube exhibited superior sensitivity, target selectivity, and reproducibility toward 100 ppm formaldehyde, along with long-term stability and rapid response time of 148 seconds. The sensing performance of the device was analyzed from the ratio between the baseline and balanced resistance of the composite SnO_2/ZnO nanostructure in the air and the target gases.

From the discussions above, it is evident that the selectivity of ZnO decorated VOC sensors can be further enhanced by optimizing both the composition of the sensitive material and the operating parameters of the sensor. By altering the properties of ZnO nanostructures via noble metallic doping, highly selective VOC sensors can be fabricated, showing great potential for the development of robust, and multifunctional VOC sensors in the future.

5.2.2 Nickel oxide

Nickel oxide (NiO), an intrinsic semiconductor, popularly known for its thermal stability, wide band gap, and strong chemical resistance, has been generally utilized for VOC sensing. The performance of NiO sensors depends on the porosity, dimension, and surface morphology.

(Patil, 2021) investigated the properties of NiO nanosheets for VOC sensing developed by a hydrothermal chemical method by mixing sodium hydroxide and nickel nitrate solution in the de-ionized water. It is reported that the ultra-thin nanosheets of NiO may provide efficient active sites and a high surface-to-volume ratio, which is important for the high absorption

FIGURE 5.2 Material characterization of SnO_2/ZnS cubes. (A) SEM and (B) TEM images of hollow SnO_2/ZnS cubes, (C) SEM, (D) TEM, (E) HRTEM images, and (F) SAED pattern of hollow SnO_2/ZnO cubes. *From Dong, Z. M., Xia, Q., Ren, H., Shang, X., Lu, X., Joo, S. W., & Huang, J. (2023). Preparation of hollow SnO_2/ZnO cubes for the high-performance detection of VOCs.* Ceramics International, *49(3), 4650–4658. https://doi.org/10.1016/j.ceramint.2022.09.352.*

of gases and hence results in improved sensing efficiency (Mokoena et al., 2019). Nakate et al. (2020) report an excellent gas response of 191% along with 150 seconds of average response time under the exposure of 150 ppm hydrogen concentration recorded at 250°C. Li et al. (2022) synthesized porous and ultrathin NiO nanosheets utilizing a facile solvothermal approach for acetone monitoring. They have developed different types of NiO thin nanosheets coated over interdigitated electrodes. The nanosheets were synthesized by controlling the ethanol to deionized water ratios such as 0:8, 1:7, 1:3, 1:1, 3:1, and 8:0, which were referred to as NiO-0, NiO-12.5, NiO-25, NiO-50, NiO-75, NiO-100, respectively . Among all the synthesized ratios, the NiO-75 (75% ethanol solution-based NiO gas sensor) depicted extremely high selectivity for acetone at 225°C along with a low LOD (0.8 ppb), good stability, and a rapid response. The excellent performance of the device was observed due to its large surface area, porosity, and crystal facet compared to other NiO nanosheets (Li et al., 2022). Its low-cost production and controllable synthesis approach make this NiO-75 material a potential candidate for a commercial acetone gas sensor. In 2019, Yin and Zhu reported the synthesis of mesoporous NiO via a sol-gel method to identify various concentrations of ammonia (Yin & Zhu, 2019). The reported mesoporous NiO structure exhibited a uniform rod structure comprising a pore size of 4 nm. Their investigation of the sensor stability showed a long-term steady response up to 21 days for ammonia concentration ranging from 0.2 to 0.4 ppm. Yu et al. (2022) developed an innovative mesoporous oxide semiconductor gas sensor which utilizes bio-metallic MOFs-derived core-shell mesoporous Sn-doped NiO sensitivity materials, synthesized through hydrothermal and ion exchange methods. The fabricated sensor demonstrated high sensitivity toward a 100 ppm concentration of xylene monitored at 250°C, along with a low LOD of 63 ppb because of their high pore interconnectivity, leading to improved gas diffusion through the surface.

Ayyala and Covington (2021) have reported NiO based chemoresistive sensors to identify seven different types of most commonly utilized VOC biomarkers, including isobutylene, ethanol, acetone, toluene, methanol, hexane, and 2-propanol vapors ranging from 5 to 25 ppm at 350°C. In the reported work, the authors utilize the spin coating technique to deposit NiO films of two different thicknesses (5 and 10 μm) on the sensor substrate. The surface morphology of the coated NiO film exhibited a uniform microporous structure throughout the deposited area. With the combination of these two thicknesses, the fabricated NiO microporous sensor demonstrated a rapid response time (80 seconds) and recovery time (120 seconds) when exposed to all the tested VOCs. The highest response of 1.5 was achieved for both sensors, calculated as the ratio of the electrical resistance in air to that in the tested VOC (5 ppm ethanol) at 350°C. More recently, John et al., 2022 synthesized manganese-doped NiO nanostructures using the co-precipitation technique for the detection of several VOC biomarkers such as toluene, xylene, n-butyl, 2-methoxyethanol, ammonia, ethanol, methanol, acetone, and formaldehyde. The reported work included extensive material characterization, i.e., structural analysis, elemental analysis, morphological study, and optical properties of the fabricated gas sensor. Their results showed that the manganese-doped NiO coated sensor demonstrates high sensitivity, excellent response time (5 seconds), low LOD (1 ppm), and excellent selectivity for 100 ppm concentration of formaldehyde monitored at RT. Their approach offered a cost-effective and facile method to produce NiO and manganese-doped NiO macroporous structures.

In summary, NiO has been shown to enhance the sensitivity, selectivity, and stability of VOC sensors and can be fabricated through various methods such as sol-gel, solvothermal, hydrothermal, spin coating, and doping. The pore size of NiO is primarily adjusted for VOC detection based on the size of the target VOC molecules. Highly responsive sensors typically have a mesoporous structure, which offers excellent volatile gas adsorption properties.

5.2.3 Molybdenum trioxide

Molybdenum trioxide (MoO_3) is an interesting n-type wide-bandgap semiconductor MO with three distinct polymorphous crystalline forms, i.e., monoclinic β-MoO_3 (low temperature metastable phase), orthorhombic α-MoO_3 (thermodynamically stable phase), and hexagonal h-MoO_3 phases (Malik et al., 2021). MoO_3 is considered to be a highly sensitive gas sensing material which can effectively detect various gases, e.g., hydrogen Zhang et al., 2017a, carbon monoxide (Yang et al., 2020), methanol (Comini et al., 2005), etc. The sensing performance of MoO_3 based devices can be determined based on the variation of the resistance with respect to the adsorption and desorption of target gaseous molecules on the sensing layer. The adsorption phenomenon of the gaseous molecule relies on the surface morphology of the MoO_3 film which can be categorized in three types – (1) one dimensional (1D) nanostructures such as nanoribbons, nanorods and nanofibers which possess high aspect ratio but limited surface area and their adsorption sites are generally exposed to ambient environments, (2) two-dimensional (2D) nanostructure which includes nanosheets and nanoplates and provides a larger surface area to improve the adsorption and device sensitivity, and (3) three-dimensional (3D) nanostructures, e.g., nanoflowers, nanoarrays, and nanospheres which are generally assembled from low dimensional structures i.e., 0D and 1D thereby providing the largest surface area and possessing more reaction sites for gas adsorption. The larger dimensional (3D) structure has more gas diffusion channels, enabling target gases to penetrate and reach to the reaction sites, resulting in improved gas absorption and sensing performance for gaseous molecules.

Despite their several advantages, MoO_3-based sensors suffer from poor selectivity. This issue could potentially be addressed by modifying the surface of MoO_3 to improve target selectivity. The property of MoO_3 can be changed by doping the MoO_3 with various types of transition metals, e.g., chromium (Cr), tungsten (W), silver (Ag), gold (Au), iron (Fe), zinc (Zn), nickel (Ni), or other types of materials to lower the activation energy of chemisorption reaction. Moreover, the hybrid structure, which is synthesized by doping Au in MoO_3 (Angiola et al., 2015), α-MnO_2, and h-MoO_3 (Shafi et al., 2017), significantly affects the sensing property of the device by improving its grain boundary barrier, carrier concentration, energy band, and depletion layer (Wang et al., 2020a). Yang et al., (2016) synthesized MoO_3 based nanobelts using a hydrothermal method to identify trimethylamine. The sensing response of the device was calculated by determining the ratio between the electrical resistance of the fabricated sensor obtained for air and target gas. The sensor exhibited response values of 6, 36, and 582 for 1, 5, and 50 ppm of trimethylamine at a measured temperature of 240°C, respectively, along with a rapid response time of 15 seconds (Yang et al., 2016). Additionally, the reported sensor demonstrated a higher selectivity to trimethylamine when tested with various other gases, i.e., trimethylamine, methanol, ethanol, ammonia, acetone, and toluene. In other investigations, it has been reported that both the surface morphology of MoO_3 and doping with noble materials play significant roles in improving the sensing performance (Liu et al., 2022a). Shen et al. (2017) proposed the doping of Ni to α-MoO_3 to synthesize nano-lamella using a solvothermal method to detect formaldehyde concentrations. In the reported work, α-MoO_3 was doped with Ni at various doping concentrations of 2.5, 5, and 10 mol%. The surface morphology of α-MoO_3 doped with various Ni concentrations is shown in Fig. 5.3. They all have the same lamellar shape with a lateral length of 220–500 nm. With 100 ppm concentration of formaldehyde at 255°C, the 5 mol% doping concentration showed the best response with a response time of 37 seconds and a recovery time of 2 seconds (Shen et al., 2017).

Later, in 2018, Wang et al. synthesized a multi-component Au doped α- MoO_3 structure and prepared Au nanoparticle decorated α- MoO_3 nanobelts using a hydrothermal method to monitor ethanol concentrations (Wang et al., 2018b). The surface morphology of rod structures was observed to be uniform and homogeneous, comprising an average rod length of 5–10 μm along with a 4.3 nm diameter of Au nanoparticle (Wang et al., 2018b). In the presence of 200 ppm ethanol concentration, the Au nanoparticle-decorated α-MoO_3 sensor exhibited a quick response time of 4.3 seconds at 200°C, compared to traditional α-MoO_3 sensors. The synthesized Au nanoparticle decorated α-MoO_3 was tested over various VOCs, including acetone, ethanol, xylene, ammonia, toluene, and formaldehyde, and reported excellent selectivity for ethanol among other VOCs. Shen et al. (2019) developed

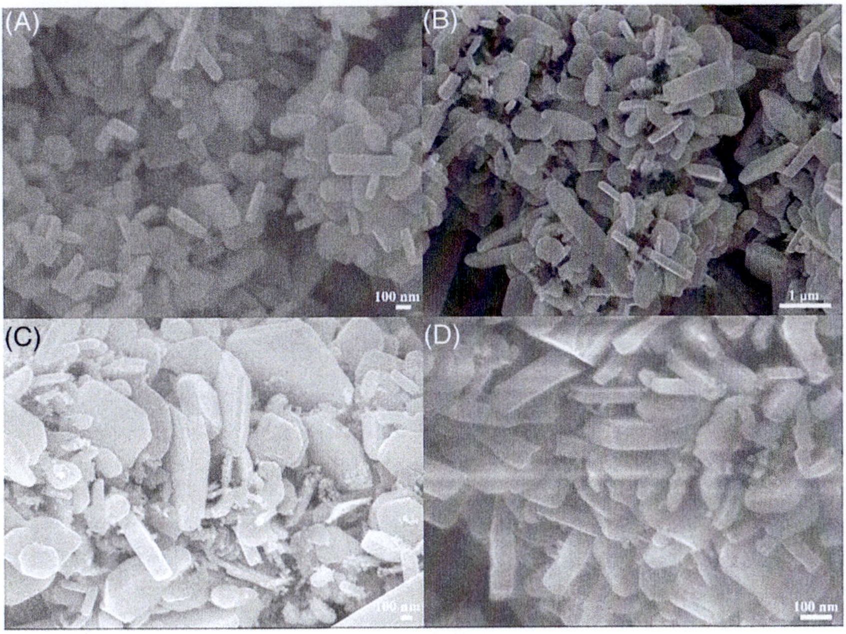

FIGURE 5.3 Material characterization of Ni doped α- MoO_3. SEM image of Ni doped α-MoO_3 (A) 0 mol%, (B) 2.5 mol%, (C) 5 mol%, and (D) 10 mol%. *From Shen, J., Guo, S., Chen, C., Sun, L., Wen, S., Chen, Y., & Ruan, S. (2017). Synthesis of Ni-doped A-MoO_3 nanolamella and their improved gas sensing properties.* Sensors and Actuators, B: Chemical, 252, *757–763. https://doi.org/10.1016/j.snb.2017.06.040.*

porous α-MoO_3 nanosheets to monitor the concentration of trimethylamine (TMA), demonstrating rapid response and recovery times of 12 seconds and 200 seconds, respectively. The fabricated sensor exhibited rapid response to TMA when investigated in the presence of other VOCs such as formaldehyde, acetone, ammonia, ethanol, dimethylamine, and TMA. The sensor demonstrated an excellent performance for TMA concentrations ranging from 198 to 50 ppm and achieved a LOD of 20 ppb. Ji et al., 2019 reported a micro flower-like MoO_3 structure assembled from nanosheets for the detection of ethanol. The authors characterized three different types of microflowers synthesized via a hydrothermal method: (1) rose-like, (2) sphere-like, and (3) plate-flower like structures, as shown in Fig. 5.4. The sensing response of all the synthesized nanostructures was examined at a controlled ethanol concentration of 300 ppm, characterized at temperatures ranging from 50°C to 350°C. The obtained results showed that among the various nanostructures the rose-like microflower achieved maximum response of 37.1 as compared to the plate and sphere flower nanostructures, showing the highest response of 30.9 and 27.3, respectively.

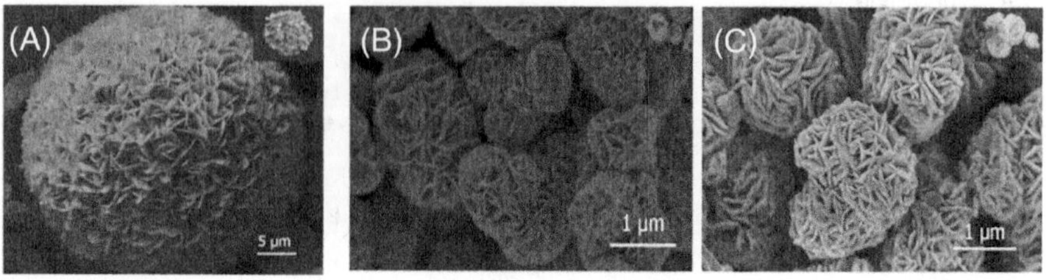

FIGURE 5.4 MoO_3 based various nanostructures. SEM image of MoO_3 samples (A) sphere-like, (B) rose-like, and (C) plate flower. *From Ji, H., Zeng, W., & Li, Y. (2019). Assembly of 2D nanosheets into flower-like MoO_3: New insight into the petal thickness affect on gas-sensing properties.* Materials Research Bulletin, 118. *https://doi.org/10.1016/j.materresbull.2019.05.001.*

The above overview clearly demonstrates the potential of MoO_3 as a highly selective VOC gas sensor based on its surface morphologies and doping material. This material can be utilized to significantly enhance the sensitivity of the device by lowering the activation energy of the chemisorption reaction, allowing more interaction between the target gas and the reaction sites. Furthermore, doping MoO_3 with noble metals in order to form hybrid structures is also reported to improve both the sensing performance and response time of the sensing device.

5.2.4 Titanium dioxide

Titanium dioxide (TiO_2) is an n-type high resistance organic compound with a band gap around 3.1–3.2 eV. The material attracts great attention for a broad range of applications such as photocatalysis (He et al., 2021), chemical sensing (Pathak et al., 2017), solar cells (Li et al., 2020a), and gas sensing (Chen et al., 2017). The unique material characteristics of TiO_2, such as catalytic properties, chemical stability, and ability to tune its structural and optical properties, make it a promising candidate in gas sensing applications (Pozos et al., 2018). In general, TiO_2 occurs in three crystal structures namely (1) brookite, (2) anatase, and (3) rutile. Among these, the rutile crystal structure is widely known as the most stable, whereas the anatase and brookite structures are present in metastable form and can be irreversibly converted to the stable rutile structure by annealing it at a higher temperature of 600°C–800°C (Liao et al., 2012). Despite its metastable characteristics, anatase is broadly utilized in gas sensing due to the presence of high oxygen vacancies, leading to increased gas reaction capacity (Tian et al., 2021). Nonetheless, the TiO_2-based gas sensors still suffer several limitations such as high operating temperatures, poor selectivity, stability, and repeatability in the presence of oxidizing and toxic gases. In the past decade, various types of TiO_2 nanomaterials have been synthesized and reported with new composites and surface structures utilized for various hazardous gases and VOC detection (Tian et al., 2021). The surface morphology of the TiO_2 nanostructures can be categorized into 0D, 1D, 2D, and even 3D nanostructures (Reghunath et al., 2021). This section provides an overview of different types of TiO_2 structures and their role in improving the performance of VOC sensing.

0D-TiO_2 nanostructures can be synthesized via widely known techniques such as hydrothermal (Navale et al., 2018), sol gel (Dubey, 2018), laser ablation in liquid (Lim & Park, 2017), and metal organic decomposition (Sugahara et al., 2020) and can be obtained in the form of quantum dots, nanospheres, and nanoparticles. These 0D-TiO_2 nanostructures are well exploited in the development of ultrasensitive, highly selective, and portable gas sensors because of their large surface area, offering numerous active sites for gas adsorption (Pathak & Viphavakit, 2021). Li et al. (2015) synthesized hollow TiO_2 microspheres covered over chemiresistive gas sensors to monitor various concentrations of formaldehyde. The reported device demonstrated good sensitivity, superior selectivity, and rapid response time (40 seconds) toward formaldehyde monitored at RT even in sub-ppm levels. The responsiveness of the TiO_2-based sensing devices was further improved by utilizing other forms of nanostructures (Tian et al., 2021). A rapid responsive (10 seconds) and low LOD (500 ppb) chemiresistive sensor coated with anatase TiO_2 nanoparticles for acetone detection was reported by Navale et al. (2018), occurring at 270°C. The fabricated device showed excellent selectivity for acetone among other types of VOC vapors, including methanol, ethanol, nitrogen dioxide, ammonia, and cyclohexanol. The high selectivity of the device toward acetone is attributed to the greater number of adsorption sites available on the TiO_2 surfaces. Later in 2020, Sugahara et al. synthesized anatase TiO_2 spherical nanostructures by utilizing the metal organic decomposition method, which involves coating and photosintering for VOC detection (Sugahara et al., 2020). In this work, the authors demonstrated the successful growth of a thin TiO_2 nanocrystalline film over a polyimide substrate. The reported sensor exhibited the highest response to 1-propanol, followed by ethanol and methanol, at concentrations ranging from 50 to 350 ppm. The sensor demonstrated high sensitivity of 40%–60% along with extremely rapid response time (1 second) at 350°C. The unique property of the material, in terms of its highly crystalline structure and large surface area, is attributed to the improved sensing performance.

The selectivity of TiO_2 for formaldehyde was later improved by Gakhar and Hazra, 2021 by encapsulating TiO_2 nanoparticles in fullerene-C_{60}. The material was synthesized using a chemical method combining hydrated fullerene C_{60} with undoped TiO_2 nanoparticles obtained from the well-known sol-gel method. From the reported method, the maximum particle size of C_{60}-encapsulated TiO_2 nanoparticles was determined to be 150 nm. The fabricated device demonstrated an improved sensing performance along with high target selectivity and rapid response (12 seconds) toward formaldehyde ranging from 1 to 1000 ppm. The improved sensing performance was attributed to the electronic exchange ability between formaldehyde and the TiO_2 nanoparticles via C_{60} layers, and the highly reactive surface created by the oxygen-functionalized C_{60}. The 1D-TiO_2 nanostructure generally involves nanotubes, nanorods, nanowires, nanofibers, and nanobelts. The 1D-TiO_2 has been found to exhibit better performance compared to 0D-TiO_2 because of their natural material properties, including large specific surface area, strong adsorption, excellent electron transportation rate, as well as presence of porous sites offering rapid diffusion to gaseous molecules, therefore, it is considered an ideal nanostructure for VOC detection (Bindra et al., 2020; Lee et al., 2014). To obtain these 1D nanostructures, various approaches have been reported, including electrochemical anodization (Rho et al., 2015), hydrothermal (Wang et al., 2017b), electrospinning (Landau & Rothschild, 2015), matrix-assisted pulsed laser evaporation (Caricato et al., 2011), and template-assisted synthesis (Xu & Gao, 2012). Among all the reported methods, the TiO_2 nanotubes synthesized via electrochemical anodization provide greater active porous sites for gas diffusion. The pure form of 1D-TiO_2 has been widely exploited to detect various VOC gases, e.g., isopropanol (Alev, Şennik, Kılınç, & Öztürk, 2015; Sennik, Kilinc, & Ozturk, 2014), acetone(Kumar et al., 2020), formaldehyde (Lin et al., 2011), ethanol (Gakhar & Hazra, 2020), etc. In 2015, Dutta et al., 2015 synthesized TiO_2 nanotubes using electrochemical anodization in order to perform real-time monitoring of benzene, toluene, and xylene (BTX), providing improved sensing performance at a low temperature (50°C). Among all the tested VOCs, the sensor exhibited rapid response toward benzene, followed by toluene and xylene, ranging from 20 to 400 ppm.

In 2019, Bindra and Hazra introduced a sandwich-structured VOC sensor comprising Au/TiO_2 nanotubes/Ti for the selective identification of methanol among other VOCs including acetone, methanol, ethanol, and 2-propanol investigated at RT (Bindra & Hazra, 2019b). In the reported work, the growth of TiO_2 nanotubes was performed using the electrochemical anodization method and demonstrated maximum sensing response for methanol compared to the other tested VOCs. Although other tested gases also show significant performance, the sensor fails to recover its response. Using a voltage pulse-assisted anodization method, several researchers have synthesized multiple-layered TiO_2 nanotube arrays to improve the performance by increasing the porosity volume (Bindra & Hazra, 2019a). In 2020, Zhao et al. reported selective detection of ammonia by using double layered TiO_2 nanorods array decorated over fluorine-doped tin dioxide (Zhao et al., 2020). The growth time of TiO_2 nanorods was obtained to be 6 hours in the article and exhibited 102% response under 100 ppm concentration of ammonia. Recently, Cai and Park reported a novel synthesis method to obtain cobalt (II,III) oxide (Co_3O_4)-decorated porous TiO_2 nanofibers by utilizing the electrospinning method to identify acetone VOC concentration at 250°C (Cai & Park, 2022). Co_3O_4 nanoparticles with a particle size of 30–50 nm were decorated to TiO_2 nanofibers through a solvothermal method. The fabricated device demonstrated good sensing response for acetone (71.88) compared to other VOCs, including ethanol (7.58), toluene (2.55), benzene (2.01), and p-xylene (3.01). In addition to excellent performance, the fabricated sensor also showed an average response time of 122 seconds under the exposure of 100 ppm acetone.

After the successful demonstration of 0D and 1D TiO_2, several researchers also utilized 2D-TiO_2 VOC sensors due to their versatile characteristics, including flexibility, large surface area, quantum Hall effect, and excellent mechanical strength (Kumari et al., 2020). The 2D-TiO_2 nanostructure can be obtained in the form of nanosheets, nanoplates, and thin films. These 2D-TiO_2 nanostructures can be synthesized by using well-known methods explored by various researchers, including hydrothermal methods (Hayashi et al., 2020), one-step annealing process (Ge et al., 2020), chemical decomposition methods (Sonker et al., 2020), and reactive magnetron sputtering methods (Horprathum et al., 2012). Among all of the reported techniques, the hydrothermal method is the most appealing technique used in the synthesis of 2D-TiO_2 nanostructures as it provides controlled morphology and high purity due to the closed-vessel reaction. The pristine 2D-TiO_2 nanostructure based VOC sensors have been employed for a broad range of volatile gases, e.g., methanol (Bhowmik and Bhattacharyya, 2017), isopropanol (Pathak & Viphavakit, 2021), hydrogen sulfide (Chinh et al., 2019), acetone (Wang et al., 2018a), and ammonia (Wilson et al., 2018). Wang et al. (2019b) reported MEMS-based acetone sensor coated with the TiO_2 nanoplates. The synthesis of TiO_2 nanoplates was performed by carefully varying the ratio of hydrofluoric and hydrochloric acid during the synthesis. In the reported work, the authors compared the performance of two TiO_2 nanoplate syntheses using complete (001) and defective crystal facets, among which the nanoplates synthesized via complete (001) crystal facets exhibited a 70% higher response compared to the defective TiO_2 nanoplates due to greater adsorption sites. The TiO_2 nanoplates with a volume ratio of hydrofluoric acid to hydrochloric acid of 2/50 exhibited good sensitivity of 21.8, 8.46, 4.32, 4.08, 6.64, and 2.73, for acetone, ethanol, toluene, methanol, isopropanol, and benzene, respectively at 400°C (Wang et al., 2019b).

Ge et al., 2020 reported sensors based on porous TiO_2 nanosheets for acetone detection. The TiO_2 nanosheets were synthesized from layered $TiSe_2$ nanosheet templates via a one-step annealing method. The fabricated device demonstrated superior selectivity and rapid response (0.75 seconds) toward a 200 ppm concentration of acetone tested at 400 °C. The rapid transport of electrons between TiO_2 and facet one was achieved by the unique porous hierarchical structure of the TiO_2 nanosheets, along with crystal facet engineering and robust interface coupling. Recently, Tian et al. (2023) reported a MEMS sensor for real-time monitoring of ethanol by using multi-layer TiO_2/SnO_2 nanosheets. In this work, SnO_2 was utilized along with TiO_2 due to their small particle size and smooth surface, which facilitates rapid adsorption of target VOCs. Although the fabricated TiO_2/SnO_2-based MEMS sensor showed a relatively long response time (33 seconds), it successfully identified low ethanol concentrations (50 ppm). The reported sensor also exhibited a response of 7.54, which was estimated from the ratio between the electrical resistance of the sensor recorded for ambient air and ethanol.

3D-TiO_2-based nanostructures have gained significant attention in VOC sensing because of their unique structural properties and associated surface chemistry (Zhao et al., 2024). The 3D-TiO_2 mainly occurs in the form of flower-like structures, facilitating the rapid diffusion and strong adsorption of gaseous molecules through their surface and allowing penetratation into these nanoflowers (Lee, 2009). Additionally, some studies also suggest that TiO_2 nanoflowers exhibit superior electronic properties in comparison to other low-dimensional TiO_2 nanostructures. This enhancement in performance is attributed to structural defects that serve as active sites between the gas and the TiO_2 material (Prabakar et al., 2009). The 3D-TiO_2 nanostructures can be obtained via a facile hydrothermal method using titanium (III) chloride (Gao et al., 2017), Ti powders (Wang et al., 2020b), titanium tetrachloride (Akhtar et al., 2019), Titanium (IV) isopropoxide (Wategaonkar et al., 2020), titanium (IV) oxysulfate (Liu et al., 2010), or Ti foil (Siva Prasad et al., 2020) during the synthesis. In 2016, Bhowmik et al. reported the synthesis of 3D-TiO_2 flowers made up of 2D nanosheets using a well known hydrothermal method (Bhowmik et al., 2016). The synthesized material exhibited rapid response and high selectivity toward acetone when tested at 60°C. In this work, the authors attributed the effect of the bond dissociative activation energy, vapor pressure of the individual VOCs, acetone selectivity, and the diffusivity of the gas species at the TiO_2 nanoflowers surface in order to describe its performance at low operating temperature. They investigated the dual effect of the low bond dissociative activation energy and the higher VOC vapor pressure of the acetone molecules compared to the other VOCs (toluene, propanol, methanol, and butanone). The fabricated device exhibited maximum response under exposure to acetone. Later Wang et al. (2020b) reported an ethanol sensor based on 3D hierarchical TiO_2 nanoflowers performing under various VOC concentrations ranging from 10 to 500 ppm at RT, as illustrated in Fig. 5.5. The reported TiO_2-nanoflower structure was synthesized via a hydrothermal method. The fabricated sensor showed excellent response in terms of

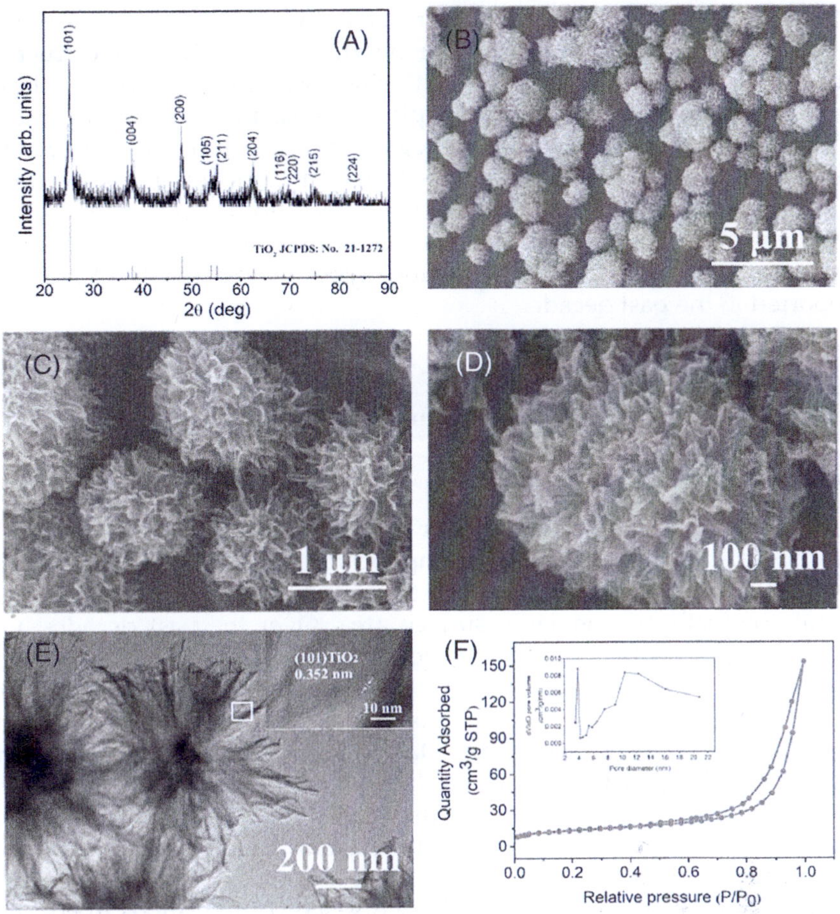

FIGURE 5.5 Material characterization of TiO_2 microstructures. Images shows (A) XRD, (B)–(D) FESEM, (E) TEM and HRTEM (inset), and (F) BET results for TiO_2 microstructures. *From Wang, M., Zhu, Y., Meng, D., Wang, K., & Wang, C. (2020). A novel room temperature ethanol gas sensor based on 3D hierarchical flower-like* TiO_2 *microstructures*. Materials Letters, 277. *https://doi.org/10.1016/j.matlet.2020.128372*.

stability, target selectivity, rapid response, and acceptable reproducibility in the presence of ethanol vapor. The nanoflower structures and their spacing lead to an increased number of adsorption sites for oxygen species and VOC gases, which significantly enhances the material's sensing capability.

To summarize, we note that TiO_2 nanostructures can be divided into 0D, 1D, 2D, and 3D based on their surface morphology. The synthesized nanostructures exhibit unique characteristics, allowing them to achieve improved sensing performance and selectivity toward the target VOCs.

We also learn that the higher-dimensional TiO_2 nanostructure exhibited a rapid response time. Although the low-dimensional TiO_2 nanostructure demonstrated poor response due to the increased charge carrier combination, it remains valuable for its simple synthesis procedure and compatibility with several existing technologies. It has been observed that TiO_2 itself can potentially detect various VOCs, it lacks selectivity in mixed gas environments. In order to improve selectivity for target VOCs, several researchers have utilized noble metal loading and/or composition techniques. Table 5.1 represents various types of MO based VOC sensors reported in the past decade.

5.3 Carbon and composites

5.3.1 Graphene oxide

Graphene oxide (GO) is another known potential material which is widely utilized in various sensing applications due to its novel material properties and desirable characteristics, i.e., thermal stability, high electrical conductivity, and large surface area. Over the past decades, GO has been well exploited in the development of various VOC sensors. The large surface area of GO facilitates easy adsorption and adhesion of VOC vapor particles, resulting in a variation in the conductivity of the GO (Ravi Kumar et al., 2022; Pathak et al., 2023; Tung et al., 2020). The change in the conductivity of GO can be utilized to identify VOC concentrations with a high sensitivity because of the excess electron mobility and large amounts of charge carriers. Given the current available technology, it is expensive and challenging to obtain large quantities of single-layer pure graphene for industrial use, and therefore many applications that utilize graphene-based structures typically rely on multi-layer GO sheets, which can be synthesized in large quantities via chemical exfoliation. Despite their several advantages, GO suffers from some limitations, including degradation of the conductivity because of oxygenated groups entrapped between the GO sheets lowering their sensing performance for VOCs. Additionally, these make the sensors more vulnerable to humidity in the environment (Wu et al., 2018). These can be mitigated by using reduced graphene oxide (rGO) instead of pure GO. rGO comprises lower levels of oxygenated groups inside the graphene matrix, allowing them to work more like pure graphene. However, the application of rGO alone is not considered to be a very efficient way to implement them in a mixed gaseous environment as other VOC molecules also adhere to the surface leading to poor selectivity. Therefore, the wisest approach reported in past decades is to utilize rGO along with other noble materials making the rGO surface more selective toward target VOCs.

TABLE 5.1 Summary of all metal oxide-based VOC sensors.

Material	Morphology	VOC	Sensing range	Response time	Operating temperature	References
ZnO based VOC sensors						
ZnO	Flower like microstructure	Ethanol	50 ppm	12 s	RT	Wang et al. (2019a)
$ZnFe_2O_4$/ZnO	Flower like microstructures	Acetone	50 ppm	2 s	250°C	Liu et al. (2017)
ZnO/CuO on graphene substrate	Nanoflower	Ammonia	5 ppm	4.1 s	RT	Jagannathan et al. (2022)
ZnO QDs	Nanoparticles	Isoprene	1 ppm	8 s	150°C	Park et al. (2019)
ZnO@CuO	Porous hollow sphere nanoparticle	H_2S	10 ppm	33 s	RT	Wang et al. (2021)
ZnO/Zn_2SnO_4	Microflowers	CH_4	400 ppm	10 s	250°C	Li et al. (2020b)
SnO_2/ZnO	Nanosphere	Formaldehyde	20 ppm	27 s	200°C	Lou et al. (2021)
ZnO/Au	Porous nanosheets	NO_2	1 ppm	1 s	RT	Mun et al. (2013)
ZnO	Thin film	NO_2	100 ppm	3 s	200°C	Patil et al. (2018)

(continued on next page)

TABLE 5.1 Summary of all metal oxide-based VOC sensors—cont'd

Material	Morphology	VOC	Sensing range	Response time	Operating temperature	References
NiO based VOC sensors						
NiO:Al	Thin films	Hydrogen Methane	100 ppm 100 ppm	388 s 1737 s	RT	Gagaoudakis et al. (2020)
NiO/Sn	Mesoporous	Xylene	100 ppm		250°C	Yu et al. (2022)
NiO-In_2O_3	nanosphere	N-propanol Formic acid Ethanol Methanol Isopropanol Acetone	100 ppm 100 ppm 100 ppm 100 ppm 100 ppm 100 ppm	541 s 215 s 263 s 38 s 139 s 141 s	210°C 200°C 240°C 300°C 250°C 250°C	Wu et al. (2022)
NiO	Mesoporous	Ammonia	0.4 ppm	54 s	RT	Yin and Zhu (2019)

(continued on next page)

MoO_3 based VOC sensor						
Pt-MoO_3	Nanobelts	Formaldehyde	200 ppm	8.8 s	27°C	Fu et al. (2019)
α-MoO_3	Nanorod	Methanol	500 pm	32 s	300°C	Cong et al. (2016)
α-MoO_3	Nanobelts	Ethanol	500 ppm	14 s	300°C	Mo et al. (2020)
Zn-MoO_3	Nanobelts	Ethanol	250 ppm	121 s	240°C	Yang et al. (2017)
Fe_2O_3/MoO_3	Nanobelts	Xylene	100 ppm	4 s	233.5°C	Qu et al. (2019)
Ni-MoO_3	Nano-pompon	Xylene	100 ppm	1 s	250°C	Jiang et al. (2019)

(continued on next page)

TABLE 5.1 Summary of all metal oxide-based VOC sensors—cont'd

Material	Morphology	VOC	Sensing range	Response time	Operating temperature	References
TiO_2 based VOC sensors						
TiO_2 (rutile)	Thin film	NH_3	12.5–100 ppm	2 min	210°C	Gardon et al. (2013)
α–Fe_2O_3–TiO_2	Heterostructure Thin film	Ethanol	100 ppm	46 s	325°C	Mei et al. (2020)
TiO_2–WO_3	Thin film (Nanoparticles)	Xylene	10 ppm	410 s	160°C	Chen et al. (2016)
Ag/TiO_2	Nanoparticles	Acetone	100 ppm	11 s	275°C	Wang et al. (2020c)
Fe/TiO_2	nanotubes	H_2S	50 ppm	22 s	100°C	Tong et al. (2020)
TiO_2-ZnO core shell	nanorods	N-butanol	100 ppm	17 s	200°C	Xu et al. (2020)
F/TiO_2	nanosheets	Acetone	400 ppm	162 s	RT	Feng et al. (2020)
TiO_2–SnO_2	nanofibers	Formaldehyde	100 ppm	13 s	RT	Chen et al. (2020)

From Pathak, A., Swargiary, K., Kongsawang, N., Jitpratak, P., Ajchareeyasoontorn, N., Udomkittivorakul, J., & Viphavakit, C. (2023). Recent advances in sensing materials targeting clinical volatile organic compound (VOC) biomarkers: A review. Biosensors, 13(1). *https://doi.org/10.3390/bios13010114.*

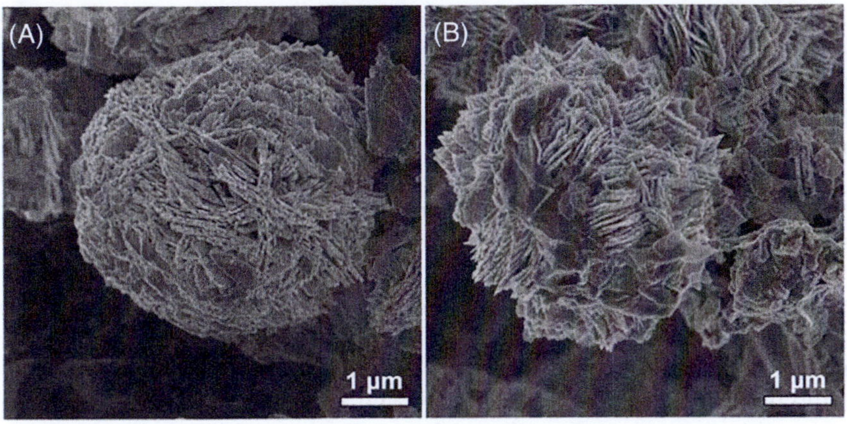

FIGURE 5.6 CuO hierarchical structure. SEM image of (A) CuO hierarchical structure and (B) 2% rGO and the CuO hierarchical structure. *From Modenes-Junior, M. A., Zito, C. A., Perfecto, T. M., & Volanti, D. P. (2019). Ethanol detection using composite based on reduced graphene oxide and CuO hierarchical structure under wet atmosphere.* Materials Science and Engineering: B, 248. *https://doi.org/10.1016/j.mseb.2019.114385.*

Zito et al. (2018) reported such a composite by combining rGO sheets along with NiO to synthesize a microflower composite nanostructure to identify methanol concentrations. The integration of NiO and rGO creates a heterojunction interface due to their mismatched work functions, i.e., 5 eV for NiO and 4.6 eV for RGO. This difference allows the transfer of the charge at the interface in order to equilibrate the Fermi energies (electrons transferred from NiO to rGO, whereas holes in the opposite direction). This electron transfer increases the hole accumulation layer (HAL) near the NiO surface. As a result, the RGO/FL-NiO composite exhibits an enhanced resistance change in the presence of methanol. Hence, the interfacial charge flow plays a crucial role in improving the sensing performance of RGO/FL-NiO composites compared to pure FL-NiO. Oxygen undergoes oxidation in the presence of methanol, enabling electrons to recombine with the holes. This process increases the resistance, which is utilized to monitor methanol concentration, resulting in the enhancement of the performance of the sensor. In 2019, Modenes et al. synthesized a hierarchical nanostructure by combining rGO with copper (II) oxide (CuO) to detect ethanol concentrations, as shown in Fig. 5.6 (Modenes-Junior et al., 2019). Fig. 5.6A illustrates a SEM image of the synthesized CuO nanostructure, whereas Fig. 5.6B depicts the rGO-CuO composite nanostructures obtained from 2% weight of the CuO in the entire solution. Although CuO has been utilized previously to determine VOC concentrations, it demonstrates lower sensitivity when applied alone. By combining rGO along with the CuO, the VOC sensors can exploit the properties of both materials to effectively detect VOC vapors. In the presence of ethanol, the ionized oxygen attached to the surface of rGO-CuO will be oxidized, leading to electron-hole pair recombination. This decreases the overall conductivity of CuO, which can be potentially used to identify ethanol concentrations. When oxygen atoms adhere to the surface of CuO, they turn to negatively charged ions and create holes in the CuO matrix, as it is a p-type semiconductor. These holes gather on the surface of CuO, improving its conductivity and hence making it more responsive toward ethanol.

Wu et al. (2020) have synthesized rGO and tin sulfide (SnS_2) nanocomposites to detect nitrogen (NO_2) vapor. The 3D porous rGO nanostructure improves the surface area and binding sites on the surface to facilitate the adhesion of NO_2 gaseous molecules on the surface. In this work, SnS_2 is combined with rGO in order to improve target specificity, as it is an n-type semiconductor that possesses high affinity, enabling NO_2 adsorption due to the high electronegativity of the semiconductor. Nonetheless, SnS_2 alone as a material for VOC sensing is generally not suitable at RT because of its high electrical resistance. Therefore, in the reported work, the authors combined the selective property of SnS_2 along with the rGO to realize an NO_2 sensor, where rGO benefits from the presence of SnS_2 to improve the selectivity, and SnS_2 takes advantage of the enhanced conductivity of rGO, allowing it to overcome its natural resistance and hence operate effectively even at RT.

There are several other approaches that have also been utilized to improve the sensitivity of rGO by integrating them within other nanostructures. Zhao et al. (2021) synthesized rGO/SnO_2 nanorods to fabricate a highly selective sensor to detect formaldehyde concentration. The sensing performance of the rGO/SnO_2 nanorods is based on the surface morphology. In the reported work, the fabricated device exhibited 150 times better performance toward formaldehyde due to the availability of a larger number of available binding sites on nanorods. Jia et al., 2022 synthesized silver nanoparticle modified Fe_3O_4/rGO via a hydrothermal method to determine various concentrations of acetone. The synthesized Ag/Fe_3O_4/rGO nanocomposite exhibited an extremely high sensing response of 35.81 at 50 ppm acetone, which was nearly 2.5 times higher than that of the Fe_3O_4/rGO nanocomposite. This high sensing response is achieved due to the electronic excitation effects of silver nanoparticles, which produce more oxygen species and binding sites for achieving improved sensing performance. Additionally, the incorporation of noble metals (i.e., Ag, Au, Cu, etc.) along with the rGO allows for rapid response, excellent sensitivity, high selectivity, and good responsiveness of the device for real-world applications.

In general, rGO based VOC sensors are fabricated to monitor only one VOC vapor. Nonetheless, considering the commercial need, it is highly crucial to fabricate sensors that can accurately monitor multiple VOC vapors at the same time and explore their potential for real-world applications. To address the issue, (Wang et al., 2019c) synthesized a ternary nanocomposite comprising rGO, ZnO and SnO_2 to detect NO_2 concentrations. In this work, the authors observed that the formation of heterostructures between ZnO and SnO_2 can significantly improve the NO_2 adsorption and therefore its sensing properties. This finding indicates that the electronic properties of the hybrids and the band structure are altered, enhancing effective charge transfer at the interfaces of heterostructures, which in turn improves gas sensing performance. Considering the potential of ternary structures, Sen and Kundu (2021) reported a similar ternary p-n-n heterojunction prepared by using rGO along with ZnO-SnO_2 nanocomposite. Later on, several other works have also emerged in this field, claiming performance improvements by utilizing ternary nanocomposites (Ahmed & Soltani, 2024; Gao et al., 2024). ZnO is known as a key material in VOC sensors due to its strong responsiveness to VOC molecules and high stability for a variety of chemical reactions. However, its high level of resistance leads to poor performance, requiring integration with rGO, which not only suppresses its resistivity but also improves sensitivity. Additionally, the p-n-n heterojunction fabricated by integrating ZnO within SnO_2/rGO provides a larger surface area for greater VOC absorption. This approach can significantly enhance the sensitivity as well as the response of the sensor, which is not only limited to one type of VOC but can also be implemented for multi-VOC detection.

5.3.2 Multi-wall carbon nanotube

Carbon possesses the ability to form various crystal structures and is considered one of the most promising materials for the fabrication of nano-electronic devices. Because of its four valence electrons, it can form various hybridizations, such as sp^2 and sp^3, which affect its electrical conductivity and electrochemical properties (Liang et al., 2022). Additionally, these carbon structures can be modified into different types of nanostructures including fullerene 0-D nanostructure, carbon nanotubes 1-D nanostructures, graphene 2-D nanostructures, and fullerite 3-D structures. Specifically, carbon nanotubes (CNT) hold great potential in VOC sensing.

CNT is a 2-D cylindrical structure in the form of a tube with an average wall thickness equal to one graphene layer arranged in the form of a hexagonal lattice, in which each atom creates a covalent bond with sp^2 hybridization which is stronger compared to a diamond lattice formed in sp^3 hybridization (Yang et al., 2018). CNTs have been widely explored in various applications due to their superior mechanical, electrical, and thermal properties compared to other leading engineering materials (Ibrahim, 2013). Because of this, CNTs have attracted significant attention, including for biological applications, surface modification, material enhancement, as well as electronic applications (De Volder et al., 2013). Various researchers have observed that the CNTs can transport electrons without any scattering due to their structure, which have diameters in the angstrom (Å) range. This enables electrons to be mainly confined on the circumference primarily moving only in the axial direction, with a resistivity which is a thousand times better than that of copper. As a result, CNTs exhibit an extremely low level resistivity (i.e., in the range of microohm/cm) (Lekawa-Raus et al., 2014). Their superior mechanical properties can be seen from their 63 GPa tensile strength and 1 TPa Young's modulus, which is an order of magnitude greater than industrial grade carbon fibers (Ruoff et al., 2003). Under ideal conditions, CNTs have an extremely high thermal conductivity of 3000 W/m.K, which is nearly 8 times higher than that of copper, and approximately twice that of diamond (Gulotty et al., 2013). Deng et al., 2014 reported thermal coefficient of CNTs to be as low as $1.9 \times 10^{-5}\ K^{-1}$ recorded at RT which makes them attractive for applications where thermal deformation needs to be low.

CNTs can be categorized into two types based on the number of layers, i.e., (1) single-walled carbon nanotubes (SWCNTs) and (2) multi-walled carbon nanotubes (MWCNTs). CNTs can also be categorized based on their chirality to be chiral, zigzag, and armchair, which affects their electronic properties to be metallic-like or semiconductor-like (Pathak et al., 2023). The MWCNTs have a large surface area, and exhibit high conductivity and excellent chemical stability compared to the SWCNTs (Tao et al., 2021). Hence, it is a popular material for use in VOC sensing applications. Bahoumina et al. (2017) fabricated a microwave VOC gas sensor by integrating poly (3,4-ethylenedioxythiophene) polystyrene sulfonate – MWCNTs (PEDOT:PSS-MWCNTs). The fabricated sensor is reported to be appropriate for wireless sensing, and is made of two stub resonators fabricated over a flexible Kapton substrate to achieve differential detection. The sensing performance of the device was evaluated after 4 and 10 minutes of ethanol exposure at concentrations of 0, 500, 1000, and 2000 ppm. The sensor achieves a maximum sensitivity of –642.9 Hz/ppm and 648.1 Hz/ppm at 4 and 10 minutes of ethanol exposure, respectively. Later, Liu et al. reported a MWCNT/poly-ethylene glycol (PEG) coated chemiresistive VOC sensor tested for acetone, isoprene, isopropanol, and ethanol (Liu, 2018). The performance of the sensor was estimated by monitoring the change in the relative resistance in the presence of various VOC concentrations ranging from 10 to 100 ppm. The reported sensors achieved maximum sensitivity for acetone, isopropanol, ethanol, and isoprene at RT, with an average response time of 110 ± 5 seconds and a LOD of 9 ppm.

Later, Janfaza et al., 2019 fabricated a highly responsive chemiresistive sensor coated with molecularly imprinted polymer (MIP)/MWCNTs composite to selectively detect hexanal. The obtained result showed an

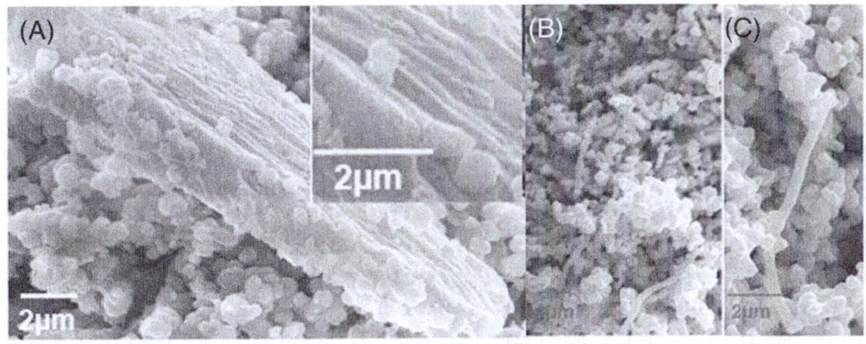

FIGURE 5.7 Material characterization of MoS_2 and MWCNT. SEM images (A) MoS_2 only and (B)–(C) MoS_2 and multi-walled CNT. *From Singh, S., Sharma, S., Singh, R. C., & Sharma, S. (2020). Hydrothermally synthesized MoS_2-multi-walled carbon nanotube composite as a novel room-temperature ammonia sensing platform.* Applied Surface Science, 532. *https://doi.org/10.1016/j.apsusc.2020.147373.*

increase in the electrical resistance under the exposure of various concentrations of hexanal. The MIP is a type of synthetic polymer engineered to have target specific binding sites, allowing it to be highly selective even in a mixed gaseous environment (Pathak et al., 2022). In the reported work, the MIP layer comprises numerous binding sites/porous structures throughout the surface of MIP which selectively entraps hexanal molecules, resulting in the swelling of the MIP layer. The process allows the decoration of MWCNTs to be separated from each other and hence vary the conducting path produced by the growth of MWCNTs, which leads to the variation in the electrical resistance of the sensor. The fabricated sensor exhibited linear sensing response along with a low LOD of 10 ppm measured in the presence of low hexanal concentration ranging from 10 to 200 ppm. In 2020, Singh et al. reported an ammonia sensor working in the range of 12–325 ppm by depositing molybdenum disulfide (MoS_2)/MWCNT composite over a quartz substrate with pre-deposited silver (Singh et al., 2020). The authors performed a comparative investigation over two different types of sensing configurations made of Ag/MoS_2/Ag and Ag/MoS_2/MWCNT/Ag and tested for ammonia as low as 12 ppm. Fig. 5.7 shows an SEM image of MoS_2 and MoS_2/MWCNT. The sensor exhibited a slow response time of 400 seconds for the MoS_2 coated sensor, which was later improved to 65 seconds by integrating MoS_2/MWCNT. The obtained results indicate that these improvements arise from the varying adsorption energies of different VOC molecules on the surfaces of MoS_2 and the MoS_2/MWCNT composite.

The performance of the MWCNTs for VOC sensing was later improved by Yan et al. (2022) by considering thermoplastic polyurethane (TPU) integrated with MWCNT-/polypyrrole (PPy). TPU was synthesized by utilizing TPU as a matrix and MWCNTs/PPy as a conductive filler. The fabricated sensors coated with this nanocomposite exhibited selective detection toward ammonia compared to ethanol, formaldehyde, ether, and benzene. The sensor demonstrated improved response for ammonia tested over the range of 90–100%, along with an average response time of 40–50 seconds.

From the above discussion, it is clear that the MWCNT is one of the promising candidates which can be used to selectively detect various VOCs because of their exotic mechanical, electrical, and thermal properties. Several researchers reported that the integration or doping of other materials along with these MWCNTs can significantly enhance their VOC sensing properties. Despite their excellent performance, these MWCNTs still suffer from major challenges related to their structural deterioration, which arises due to prolong VOC exposure, resulting in inaccurate measurements. However, these drawbacks also appear to be resolved by utilizing composite materials, such as PEG, PPy, MoS_2, and MIP, which not only address the deterioration of MWCNTs but also improve their sensing performance, including sensitivity, selectivity, and response time.

5.3.3 Activated carbon

Due to the rapid development of urbanization and industrialization, the presence of VOCs and their related health issues, arising from their toxicity, has become a concern not only for the environment but also for human health (Gałęzowska et al., 2016). As VOC emissions continue to rise, several groups within the research community are exploring and implementing numerous methods for their identification and removal. One such method is add-on control, which involves the absorption of VOCs by sensing materials that interact with the exhaust VOC molecules by physical and/or chemical means (Pina et al., 2018). This approach enables an efficient and cost-effective measurement (Li et al., 2020b, Li et al., 2020c).

ACs are widely utilized as a highly adsorptive material due to their unique physical and chemical properties. The ACs exhibit excellent chemical stability, a large surface area to volume ratio, high mechanical strength, and a rich variety of functional groups, and can function in both alkali and acids. Although ACs possess several advantages, they also suffer from some limitations because of their hydrophobic and thermal stability. ACs can spontaneously collapse into a porous structure or ignite at high temperatures because of their high surface area and carbon composition. Additionally, their nonpolar feature limits their ability to adsorb hydrophilic VOCs (Rakić et al., 2015; Twumasi et al., 2012; Zhang et al., 2017a). ACs are typically generated from the carbonization and activation process by carbon rich materials e.g., wood, coconut shell, lignin, sawdust, palm shell, coal, and petroleum pitch (Ma et al., 2020). Fig. 5.8 shows micrograph

FIGURE 5.8 Palm shell ACs. SEM image of the palm shell-based ACs. *From Arami-Niya, A., Wan Daud, W. M. A., Mjalli, F. S., Abnisa, F., & Shafeeyan, M. S. (2012). Production of microporous palm shell based activated carbon for methane adsorption: Modeling and optimization using response surface methodology.* Chemical Engineering Research and Design, 90*(6), 776–784. https://doi.org/10.1016/j.cherd.2011.10.001.*

images of a raw palm shell AC prepared under the optimized conditions as reported by Arami et al. (2012).

Many researchers have reported on the utilization of ACs for VOCs sensing, e.g., alcohols, alkanes, aldehydes, esters, ethers, ketones, aromatics, etc. Li et al. (2012) studied the effect of AC porosity and its influence on the adsorption of VOCs. They conducted fixed-bed thermostatic adsorption experiments under specific conditions to monitor the adsorption capacity of ACs for toluene, acetone, and 1,2-dichloroethane. The fabricated device exhibited highly selective adsorption of organic gases and VOCs based on the different pore diameters of ACs, with ranges of 1.27–1.49 nm for toluene, 0.67–0.84 nm for acetone, and 1.39–1.75 nm for 1,2-dichloroethane. In 2019, Shen et al. synthesized ACs from rice husks and studied the influence of the adsorption of phenol and toluene (Shen & Zhang, 2019). In the reported work, the authors observed enhanced adsorption capacity for toluene (i.e., 264 mg/g) compared to phenol (6.53 mg/g). The enhanced performance was observed due to the volatile nature of toluene, whereas phenol is a semi-volatile organic compound which exhibits weaker adsorption in gaseous form in comparison to its liquid phase.

The authors also observed that ACs can also be reused via thermal desorption. Later Kang et al., 2021 reported a VOC analyzer designed to monitor propionic acid in the exhaled breath of individuals by analyzing the adsorption and desorption performance using a commercial photoionization detector and a colorimetric tube detector. Its adsorption and desorption rates were approximately 90% and 70%, respectively, making it an excellent gas concentrator which is beneficial for low-concentration VOC detection (Pathak et al., 2023). Ma et al. (2020) also investigated the adsorption of chlorobenzene and toluene via commercially available ACs synthesized using medium high temperature (MHT) techniques. The utilized ACs were a synthesized form of coconut shell (CSAC) and wood (WAC) under MHT of 90°C–150°C. The results indicated that the adsorption characteristics of toluene and chlorobenzene under MHT conditions differed from those observed at lower temperatures. Specifically, the adsorption capacities for toluene on CSAC and WAC under MHT conditions ranged from 20.6 to 49.9 mg/g and 13.9 to 46.9 mg/g, respectively. For chlorobenzene, the capacities were 30.5 to 80.3 mg/g for CSAC and 26.0 to 63.9 mg/g for WAC.

From the above reports, it is clear that the AC-based VOC sensors are mainly reported based on the adsorption capacities, adsorption rates, and their synthesis method (i.e., MHT). Nonetheless, the response time of the AC based VOC sensors is yet to be fully explored. Furthermore, only a few studies are available targeting the selective adsorption capacity of ACs for VOC gases. It is observed that varying the physical and chemical properties of ACs by doping with other materials, can potentially improve VOC adsorption and enable VOC detection in the sub-ppb/ppm ranges. ACs have so far shown their potential in VOC sensing due to their variable surface chemical functionalization, large surface area, and tunable porosity (i.e., size and volume), and hence they can be considered as an excellent candidate in order to achieve high adsorption capacity and target selectivity. Table 5.2 provides a summary of various types of carbon based VOC sensors along with their sensing performance.

5.4 Polymers

Polymers are widely used organic materials made of large molecules, or macromolecules, formed by linking up repeating carbon chains. Over the past decades, several types of polymers have attracted significant attention for creating artificial sensing materials because of their unique mechanical and chemical properties, which allow them to expand and contract in the presence of gas molecules (Adhikari & Majumdar, 2004). These polymers are reported to be highly responsive, providing good sensitivity due to their permeable features, and also offer high selectivity. Several

TABLE 5.2 Summary of all carbon-based VOC sensors.

Material	Morphology	VOC	Sensing range	Response time	Operating temperature	References
GO based VOC sensors						
ZnO/GO	Nanorods/Sheet	Acetone, Benzene, Ethanol, Methanol	10–500 ppm	8.79–12.43 s	200–500°C	Vessalli et al. (2017)
rGo/NiO	Sheet/Hierarchical Flower-like	Methanol	61.51–100 ppm	16–30 s (Graph Estimate)	180°C	Zito et al. (2018)
PEDOT: PSS/GO	Nanowire	Acetone,p-Xylene, Ethanol, n-Hexane	5–25 ppm	8–12 s	RT	Tang et al. (2018)
CuO/rGO	Sheet/Hierarchical	Ethanol	5–200 ppm	23 s	250°C	Modenes-Junior et al. (2019)
SnS_2/rGO	3D Porous Structure	Nitrogen Dioxide	0.5–8 ppm	153 s	RT	Wu et al. (2020)
SnO_2/GO	Nanocomposite Heterojunction	Ethanol, Acetone, Ethylbenzene	1 ppm	120 s	RT (With UV light) 300°C	Pargoletti et al. (2020)
SnO_2/rGO	Nanorods Array/ 3D Hierarchical Structure	Formaldehyde	10–50 ppm	52 s	50–150°C	Zhao et al. (2021)
ZnO-SnO_2/rGO	Ternary Hybrid Nanocomposite with p-n-n heterojunction	Acetone, Ethanol	1–10 ppm	10 s	150°C	Sen and Kundu (2021)

(continued on next page)

TABLE 5.2 Summary of all carbon-based VOC sensors—cont'd

Material	Morphology	VOC	Sensing range	Response time	Operating temperature	References
MWCNT based VOC sensor						
WO_3/MWCNT	Nanobrick thin film	Ammonia	60 ppm	512 s	RT	Duong et al. (2021)
PDDA/MWCNT	Thin film	Carbon monoxide	20 ppm	18 s	RT	Roy et al. (2020)
ZIF-8/MWCNT	Sharp facets crystals thin film	Formaldehyde	5 ppm	12 s	RT	Jafari and Zeinali (2020)
Al_2O_3/MWCNT	Porous film	Carbon dioxide	450 ppm	53.7 s	RT	Ahmad et al. (2020)
In_2O_3/MWCNT	Nanobar SAED pattern	Acetone	250 ppm	9 s	300°C	Kohli et al. (2020)
PPy/N-MWCNT	Loose granular structure	Nitrogen dioxide	5 ppm	65 s	RT	Liu et al. (2019)
PANI/MWCNTs/MoS_2	Porous film	Ammonia	0.25 ppm	32 s	RT	Zhang et al. (2018)
PVA/MWCNT	Entangled porous, thin film	Ethanol	100 ppm	24 ± 3 s	RT	Maity et al. (2018)
PANI/MWCNTs	Entangled porous, thin film	Ammonia	2 ppm	6 s	RT	Abdulla et al. (2015)
NiO/MWCNT	Aggregated flake-like structure	Ethanol	100 ppm	27 s	180	Chen et al. (2015)

(continued on next page)

Activated carbon (AC) based VOC sensors						
AC	Non uniform sheet-like structure	Isopropanol	1–10 µL	100 s	RT	Kumar and Tyagi (2018)
Ag-AC-PEDOT: PSS	Nanoparticle composite conductor	Ethanol	(a) 250 mTorr (b) 500 mTorr (c) 1000 mTorr	(a) 470 s (b) 610 s (c) 609 s	RT	Lee et al. (2018)
AC	Pistachio shell base	Benzene	700 and 1000 ppm	NA	30–45°C	Mirzaie et al. (2021)
Granular AC	NA	Methanol, Butanone, Benzene, Ethanol, n-Propanol, o-xylene, Toluene	6000 ppmv	NA	RT	Oh et al. (2010)
TiO_2/activated carbon fiber felt	Porous composite structure	Toluene	<1150 ppm	NA	RT	Li et al. (2017)

From Pathak, A., Swargiary, K., Kongsawang, N., Jitpratak, P., Ajchareeyasoontorn, N., Udomkittivorakul, J., & Viphavakit, C. (2023). Recent advances in sensing materials targeting clinical volatile organic compound (VOC) biomarkers: A review. Biosensors, 13*(1). https://doi.org/10.3390/bios13010114.* Room temperature = RT* studies have demonstrated the replacement of classical sensing materials (e.g., MOs and carbons) with polymers involving nanotechnology, thereby exploiting the extrinsic and intrinsic properties of these materials. These polymers play an important role in identifying various chemical and biological targets due to their swelling and shrinking phenomenon (Pathak & Singh, 2017; Swargiary et al., 2022; Viphavakit et al., 2015).

5.4.1 Poly (dimethylsiloxane)

Among the several known polymers, poly (dimethylsiloxane) (PDMS) is one of the most widely used silicon-based elastomeric polymers, synthesized by the cross-linking of the repeating units of -OSi $(CH_3)_2$-. PDMS has been recognized as a material well suited for the fabrication of microfluidic, flexible electronics, and integrated circuit devices (Raj M & Chakraborty, 2020; Victor et al., 2019). Over the past several decades, PDMS has been well exploited in VOC sensing due to its ability to swell under VOC exposure (Sidek & Bin Afzal, 2011).

Reddy et al. (2012) utilized PDMS to fabricate a Fabry–Pérot interferometer (FPI) based VOC vapor sensor array to integrate with micro-gas chromatography. The sensor was fabricated by coating a uniform and thin layer of PDMS on a silicon wafer. The Fabry–Pérot was formed between the PDMS–silicon and air-PDMS interfaces, resulting in a resonant cavity. The resonance wavelengths are influenced by changes in the refractive index caused by the variations in VOC vapors, enabling rapid detection and quantification of VOCs. In a proof-of-concept study, two different polymers, PDMS and SU-8, were individually utilized and selectively placed within a microfluidic channel. The fabricated sensor exhibited different sensitivities for different VOC vapors, showing strong performance toward toluene and acetone for the concentrations varying from 0 to 20 ng and 0 to 150 ng, respectively. Later, in 2016, Ning et al. fabricated a highly sensitive VOC vapor sensor by integrating an FPI with a Sagnac interferometer coated with a thin layer of PDMS to identify multiple VOC targets simultaneously (Ning et al., 2016). The reported device successfully demonstrated selective identification of ethanol and 2-propanol by utilizing a second-order inverse matrix method. The sensing probe was formed by depositing a 15 μm thick PDMS film on the tip of a single-mode hollow core optical fiber, creating an FPI cavity between the fiber and the PDMS layer. The fabricated device demonstrated good sensitivity toward 2-propanol and ethanol to be 1.61×10^{-3} nm/ppm and 1.17×10^{-3} nm/ppm, respectively. Kacik and Martincek, 2017 also considered a similar approach to identify toluene using an FPI fabricated by forming a micro air-cavity inside the PDMS coated on the tip of an optical fiber. The FPI was formed between the fiber-air and air-PDMS interfaces, enabling two reflecting surfaces to create an interference pattern. The fabricated sensor demonstrated an average sensitivity of 0.15 to 1.4 $nm/g.m^{-3}$ for the toluene concentration ranging from 0.833 to 100 $g.m^{-3}$.

Zhao et al. (2019) also reported PDMS coated optical fiber VOC sensors based on the FPI. The sensing head comprises a small section of hollow core fiber (HCF) stubbed between single-mode fiber (SMF) and large mode area fiber (LMAF), making a total length of 1 mm. Here, three interfaces, SMF-HCF, HCF-LMAF, and LMAF-Air, work as reflecting surfaces allowing the production of interference. A thin layer of PDMS was coated, covering the HCF-LMAF. The sensitivity of the fabricated sensor was evaluated by analyzing the shift in the interference spectra. The sensor achieved a maximum sensitivity of 20 pm/ppm for isopropanol, observed over a concentration range from 0 to 500 ppm.

Most recently, in 2021, Sappati et al. fabricated a 3D printed acoustic flexural plate wave (FPW) VOC sensor printed on thin and flexible lead zirconate titanate-poly dimethyl siloxane (PZT-PDMS) using a composite film with silver ink (Sappati & Bhadra, 2021). The reported prototype FPW resonator possesses a resonant frequency of 22.65 MHz along with −1.552 dBm attenuation. The gravimetric mass sensitivity of the device was monitored by introducing PDMS layers between the input and the output interdigital transducers. The sensor demonstrated high sensitivity, achieving 0.66 and 160.63 kHz/ppm, along with a low LOD of 10.9 and 0.03 ppm for acetic acid and toluene, respectively.

5.4.2 Conducting polymers

After PDMS, conducting polymers (CPs) are the second most widely used polymers reported for VOC monitoring. The CPs are organic materials recognized for their novel electrical and optical properties similar to inorganic semiconductors and metals. They are easy to synthesize and process (Snook et al., 2011). The CPs are synthesized using various versatile, low-cost, and simple methods, making them an ideal candidate for VOC sensing applications. CPs are made of functional groups which exhibit pseudocapacitance and therefore contribute to the elecrical conductivity of the material e.g., polypyrrole, polyacetylene, polythiophene, polyaniline, poly(para-phenylene), poly(phenylenevinylene), and polyfuran. This section will provide a brief overview of two major CPs, polypyrrole and polythiophene, and their roles in VOC sensing.

5.4.2.1 *Polypyrrole*

Polypyrrole (PPy) has recently gained attention in VOC sensing because of their high sensitivity and target selectivity for rapid identification of organic and inorganic gases (Miah et al., 2022). PPy is utilized for a

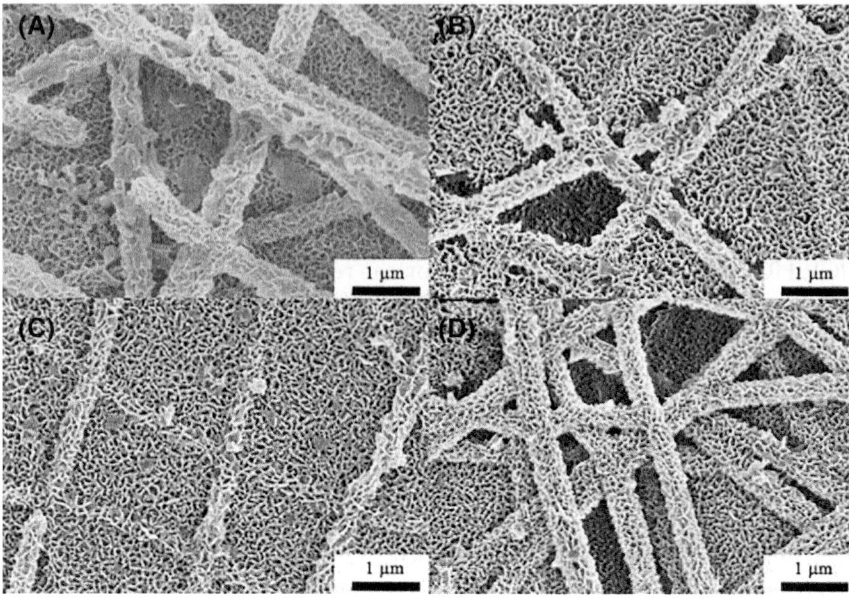

FIGURE 5.9 Material characterization of SnO_2/PPy nanocomposite. SEM images of SnO_2/PPy nanocomposite synthesized in presence of various doping acids: (A) PSSA, (B) HCl, (C) TSA, and (D) CSA. *From Li, Y., Ban, H., & Yang, M. (2016). Highly sensitive NH_3 gas sensors based on novel polypyrrole-coated SnO_2 nanosheet nanocomposites.* Sensors and Actuators, B: Chemical, 224, *449–457. https://doi.org/10.1016/j.snb.2015.10.078.*

broad range of gas sensing applications by depositing over electronic, electrochromic, and optical sensors and transducers. It is a promising alternative material in gas sensing technology as it possesses high chemical stability, ease of deposition, good conductivity, and provides excellent adhesion on a large scale of substrates.

A PPy-PVA composite material has been demonstrated with excellent sensitivity and rapid responsivity toward ethanol as compared to other VOCs, including chloroform, ammonia, acetone, and toluene (Das & Sarkar, 2018). Out of various CPs, PPy is well exploited to identify VOCs at RT (Miah et al., 2022). Li et al. (2016) fabricated ammonia sensors by deposition of SnO_2 over PPy, forming a composite material operating at RT. Fig. 5.9 illustrates an SEM image of the synthesized SnO_2/PPy composition. SnO_2 is a well-known inorganic semiconductor material widely utilized for VOC sensing. Although it exhibits poor conductivity and minimal response to ammonia at RT which can be further improved via hybridization, doping, or surface modification. One study highlighted a significant impact of the electrical properties of PPy on its ammonia-sensing performance (Li et al., 2016). In order to improve the sensor efficiency, the authors utilized SnO_2 over PPy, exploiting the advantageous features of both materials to maximize the overall performance. The synthesized SnO_2/PPy nanocomposite coated sensors showed an extremely high sensitivity of ~6.2%/ppm achieved for ammonia, varying from 1 to 10.7 ppm along with a low LOD ~257 ppb which is significantly higher than that of a sensor coated with PPy alone.

Wu et al. (2019) coated a polarization-maintaining optical fiber with a PPy film and utilized it to detect isopropanol at RT. The reported sensor comprises a combination of the Vernier effect in a single fiber Sagnac loop, along with the PPy as a gas sensing material. The basic mechanism relies on monitoring the strain variation which arises due to the swelling of PPy under the exposure to isopropanol. The sensor achieved a maximum sensitivity of 239 pm/ppm across a range of 0 to 42 ppm compared to a conventional Sagnac interferometer for the same range. Šetka et al. (2020) detected acetone concentration via cadmium telluride (CdTe)/PPy nanocomposites coated novel surface acoustic wave sensor. The sensing performance of the device was obtained by monitoring the frequency shift in the presence of acetone. The reported sensor demonstrated a high sensitivity to acetone compared to toluene and ethanol, along with a low LOD of 5 ppb. Shoeb et al. (2021) reported rapid responsivity and high selectivity of ammonia sensors by utilizing Gr/Ag–Ag_2O/PPy nanocomposites. The sensing performance of the device was estimated by measuring the variation in the electrical conductivity in the presence of various VOCs. In the reported work, Gr/Ag–Ag_2O/PPy composite demonstrated 40 times greater conductivity compared to pristine PPy when the sensor was exposed to 1000 ppm of ammonia. Furthermore, the sensor also exhibited high selectivity for ammonia over other VOCs, including acetaldehyde, ethanol, benzene, formaldehyde, m-xylene, and toluene.

Recently, Byeon et al. (2022) reported the integration of SWCNTs with PPy and phenyllactic acid, forming an excellent nanocomposite, as shown in Fig. 5.10, for the selective detection of acetone. The synthesized SWCNTs/PPy/phenyllactic acid nanocomposite featured an identical surface morphology as the heptadecafluorooctanesulfonic acid-doped PPy layer encasing SWCNT, with the phenyllactic acid layer surrounding the PPy functioning as a selective sensing layer for acetone. In this work, the PLA chemically bonded to the PPy, creating a cooperative sensing effect that significantly enhanced the device's sensitivity for acetone. The heptadecafluorooctanesulfonic acid doped PPy/PLA@SWCNT acetone sensor reliably identified low acetone concentrations down to 50 ppb at RT. Additionally, the sensor demonstrated clear signal detection across humidity levels from 0% to 80% at low temperatures, an essential feature for medical diagnostics in breath analysis applications.

From the above discussion, it is clear that PPy has gained significant attention in VOC sensing because of its rapid response time, high sensitivity, excellent stability, and target selectivity. The sensing performance of the

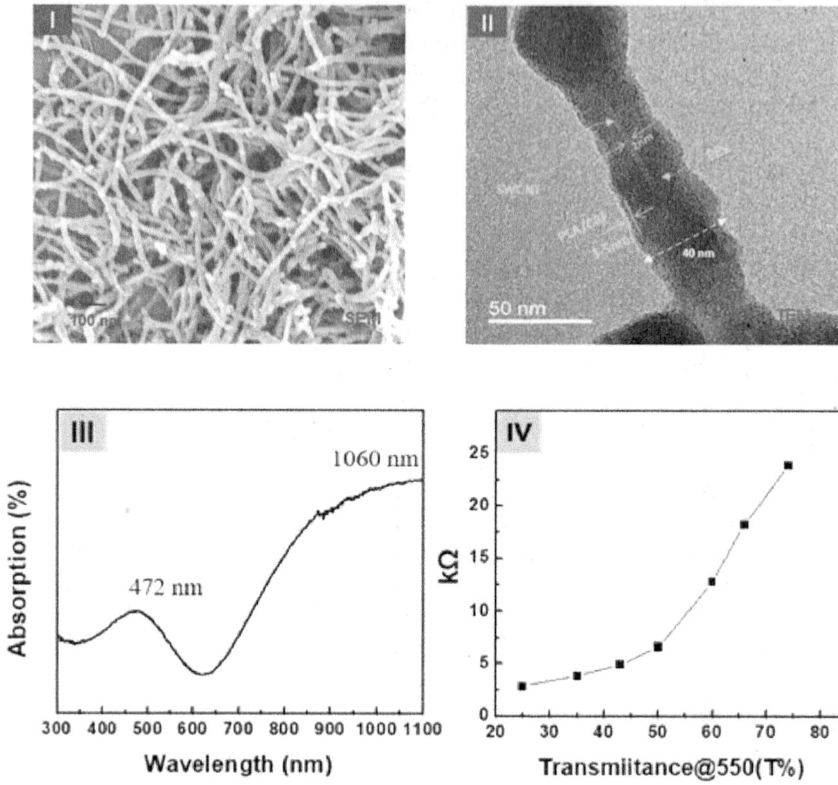

FIGURE 5.10 SEM micrograph of SWCNTs/PPy/ phenyllactic acid nanocomposite. The (A) SEM and (B) TEM images of core–shell-shaped nanorods. *From Byeon, J. H., Kim, J. S., Kang, H. K., Kang, S., & Kim, J. Y. (2022). Acetone gas sensor based on SWCNT/polypyrrole/ phenyllactic acid nanocomposite with high sensitivity and humidity stability.* Biosensors, 12(5). *https://doi.org/10.3390/bios12050354.*

material was also improved by integrating or doping other materials such as CdTe, SWCNT, SnO_2, etc. A major limitation of PPy lies in its inherent properties, making it challenging to integrate with other materials. This requires a deep understanding of chemistry and expertise in synthesis.

5.4.2.2 *Polythiophene*

The application of polythiophene (PTh) nanocomposites in VOC sensing technology has grown rapidly over the past few decades. Tripathi et al. (2015) reported the first PTh/aluminium oxide nanocomposite to investigate its potential in ammonia sensing. The PTh/Al_2O_3 nanocomposite was produced in pellet form via a chemical oxidation method. The sensing performance of PTh/Al_2O_3 coated device was evaluated by monitoring changes in the resistance of the synthesized pellets under the exposure of various ammonia concentrations ranging from 25 to 650 ppm. The obtained result indicated that the performance of the device can be improved by increasing the doping concentration of Al_2O_3. In 2016, Bai et al. reported rGO/PTh nanocomposite to identify NO_2 concentration (Bai et al., 2016). Due to the unique features of GO discussed earlier, the synthesized rGO/PTh nanocomposite with 5 wt% incorporation of rGO demonstrated four times higher sensitivity under 10 ppm NO_2 exposure, along with excellent selectivity as compared to the pure PTh coated sensor. Husain et al. synthesized a series of PTh nanocomposites for multiple organic vapor sensing (Husain et al., 2019, 2020a, 2020b, 2020c). In 2019, they synthesized PTh/ZrO_2 nanocomposite to identify various concentrations of ethane (Husain et al., 2019). The sensing property of the synthesized materials was estimated by analyzing the DC electrical conductivity retention under cyclic aging and isothermal conditions. In this study, the high surface area and effective conductive network enabled quantitative adsorption and desorption of ethane. When exposed to ethane gas, the PTh/ZrO_2 nanocomposite sensor demonstrated electrical conductivity 19 times greater than that of a pure PTh sensor, indicating faster adsorption and desorption of ethane due to the increased surface area. In 2020, they also reported the use of a PTh/SnO_2 nanocomposite in pellet form for alcohol detection, e.g., butan-2-ol, butan-1-ol, and 2-methyl propanol at RT (A. Husain et al., 2020). The PTh/SnO_2 nanocomposite pellet showed a high sensing response and good reproducibility for butan-1-ol. The sensing performance of the material was evaluated by monitoring changes in electrical conductivity. Additionally, they also synthesized a PTh/GO nanocomposite to detect ethanol concentration at RT (Husain et al, 2020b). The performance of the material was evaluated under exposure to ethanol concentrations ranging from 400 to 2000 ppm, and a significant enhancement in sensitivity was observed for the PTh/GO nanocomposite compared to pure PTh, along with a low LOD of 400 ppm. From the above discussion, we can clearly understand that a combination of PTh and an appropriate amount of composite may improve the sensing performance of the device, specifically in terms of selectivity.

5.4.3 Molecularly imprinted polymer

In the past decade, rapid growth in MIP-based VOC sensors has occurred, which has already been summarized in other reviews (Fang et al., 2021; Hasseb et al., 2022; Ramanavicius et al., 2021; Uzun & Turner, 2016; Vasapollo et al., 2011). MIPs are polymeric materials comprising specific nanocavities engineered to have pores with the identical size and shape of the target molecules (known as templates). Therefore, target specificity is an inherent engineered property of MIPs, making them a vital component of a sensing system. The first article on the successful imprinting of an amino acid in a polymer was reported by Andersson et al. (1984), and in 1985, G. Wulff used the term "imprinted polymer" for the first time in history (Wulff et al., 1985), although he had been publishing articles in a sequence entitled "Enzyme-Analog Built Polymers" since 1973 (Wulff et al., 1973). Andersson et al. (1984) reported a noncovalent approach for binding between the host and target; however, Wulff proposed a covalent binding technique to create the imprinted polymer. The chemistry required to remove the template molecules was the major difference between the two ideas. Specifically, the covalent technique is expected to produce an excellent and uniform set of rebinding cavities of specific shape and size, and hence provide a huge surface area to accommodate a large number of target molecules. Due to the emerging idea of molecular recognition based on the concept of molecular interactions, even fundamental biological systems could be mimicked by synthetic strategies. In 1984, a modern idea of molecular interactions was reported by Emil Fischer using the term "lock-and-key" (Fischer, 1894). It is a phenomenon of the interaction between substrate and enzyme where the substrate works like a key and the active site of the enzyme works as a lock. Similarly, other researchers have developed another promising strategy of mimicking the "lock-and-key" mechanism, popularly known as MIP (Viveiros et al., 2017). In general, MIP synthesis followed a similar approach, which involves (1) mixing monomers with target molecules, i.e., template, (2) degassing and polymerization under oven, hot air, or UV exposure, and (3) removing the template from the polymerized surface, leaving cavities of specific shapes behind. These MIPs can be formed for many types of target molecules, including proteins, enzymes, viruses, explosives, and VOC gases. With the ability to selectively bind target molecules, MIPs play a crucial role in developing customized sensors and have gained significant attention in VOC detection (Hua et al., 2022; Zhang et al., 2017c). The MIP-based VOC sensors demonstrate a broad range of properties, such as high sensitivity, target selectivity, reproducibility, cost-effectiveness, and physical and chemical stability (Carballido et al., 2022).

Mustafaa and Lieberzeit synthesized MIP/Ag_2S nanocomposite deposited over a quartz crystal microbalance (QCM) to detect aliphatic alcohol (Mustafa & Lieberzeit, 2014). Integrating MIPs with Ag_2S nanoparticles produces a hybrid nanocomposite material that exploits the properties of both materials collectively. The synthesized MIP/Ag_2S nanocomposite achieved a two time increment in normalized response, reaching −70 Hz under 400 ppm of 1-butanol, compared to the sensor coated with pure Ag_2S of similar thickness. The combined effect of the strong affinity of Ag_2S nanoparticles and the selective detection ability of MIPs for 1-butanol vapor allows the fabricated sensors to achieve a relatively low LOD of 2 ppm. Hussain et al., 2016 fabricated a MIP-coated QCM sensor to identify different concentrations of formaldehyde vapors in air streams. The sensor

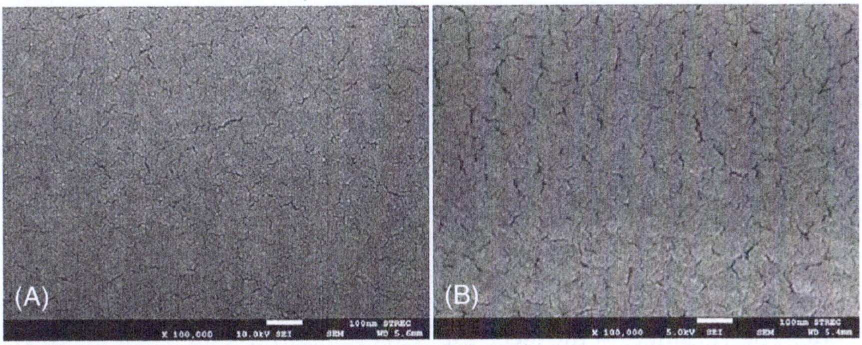

FIGURE 5.11 SEM micrographs of MIP and NIP. SEM images show (A) NIP and (B) IPA-MIP surface. *From Pathak, A. K., Limprapassorn, P., Kongruttanachok, N., & Viphavakit, C. (2022). Molecularly imprinted polymer-based optical sensor for isopropanol vapor.* Journal of Sensor and Actuator Networks, 11*(3). https://doi.org/10.3390/jsan11030046.*

demonstrated a low LOD of 500 ppb and exhibited excellent selectivity for formaldehyde over acetaldehyde, methanol, dichloromethane, and formic acid. Tang et al. (2017) reported another type of formaldehyde sensor for formaldehyde detection using MIP/TiO_2 nanotube array composite. The MIP/TiO_2 nanotube possesses a high surface-to-volume ratio, enabling improved sensitivity and response time of the material. With humidity effects neglected, the sensor can selectively detect formaldehyde in the ppm range at RT, therefore offering long-term stability.

Abdelghani et al. (2019) fabricated electrochemical sensors coated with SnO_2 integrated MIP to simultaneously identify acetone and ammonia. In the reported work, the authors deposited four SnO_2 thin films using a spin coater under different conditions of introducing ammonium hydroxide and acetone as target templates during the synthesis process. They achieved 500 nm thick nanosheets from SnO_2 nanoparticles, with the particle sizes ranging from 20 to 50 nm, depending on the synthesis conditions. The sensor's response was estimated in the presence of various gases, including ammonium hydroxide, liquid petroleum gas (LPG), oxygen, benzene, and acetone. The obtained results exhibited an improved response to ammonia and acetone as the concentration increased during synthesis. The results demonstrated SnO_2/MIP as a promising material to selectively detect ammonia with a high sensitivity of 89% at 260°C, while the same material achieved a maximum sensitivity of 77% for acetone at 230°C. Recently, Pathak et al., 2022 fabricated portable isopropanol optical sensors by utilizing a thin layer of MIP on a glass slide. Fig. 5.11 illustrates the surface morphology of non-imprinted polymer (NIP) and MIP. The regular texture observed in the NIP appeared due to the absence of the target template during synthesis, hence leading to a lack of binding sites for IPA molecules. However, cracks and porosity appeared on the surface of the MIP due to the successful removal of IPA templates. The increased wrinkling and surface roughness in the imprinted polymer are attributed to a larger surface area created by accommodating the IPA template. The fabricated sensor was tested under different concentrations of IPA solutions ranging from 50 to 100%. The sensor demonstrated a maximum sensitivity of 0.63 nm/%IPA vapor at 120 minutes exposure, along with good selectivity among a similar class of VOCs (Pathak et al., 2022).

Mousazadeh et al. (2022) fabricated a chemiresistive sensor in the same year using an Au/MIP nanocomposite to identify hexanal concentration. The synthesized nanocomposite was deposited over an interdigitated electrode. The variation in the electrical response and resistance of the device was analyzed by Fourier transform cyclic voltammetry in the presence of various concentrations of hexanal gas. The sensor achieved a low LOD of around 1.1 ppm for a linear range of hexanal concentration ranging from 2.5 to 300 ppm. Furthermore, the device also demonstrated a threefold higher sensitivity and superior selectivity toward hexanal in comparison to other VOCs with a similar carbon atom number. Recently, Pathak et al. (2025) also utilized a similar combination of sensing material coated over a no-core optical fiber to realize an isopropanol sensor for industrial applications. The reported article demonstrated the numerical and experimental optimization of the sensor and summarizes the advantages of employing a gold layer in sensitivity enhancement. The performance of the fabricated device was investigated under the exposure of various VOCs, including methanol, acetone, isopropanol, and ethanol, to evaluate the sensitivity and selectivity of the sensor. The sensors exhibited an excellent selectivity toward isopropanol measured at ppm levels showing a seven times higher sensitivity of 0.014 nm/ppm compared to the bare fiber coated only with the MIP layer (0.002 nm/ppm).

Considering the properties of MIP, it has been widely exploited in various sensing applications, specifically to obtain target selectivity. MIPs can play a crucial role in a mixed gaseous environment. Despite their ability to be designed for specific target selectivity, MIPs suffer from poor response time due to the difficulties associated with molecule diffusion (Neto et al., 2011). The sensitivity and efficiency of MIP-based sensors are directly related to the number of imprinted cavities available on the sensor surface. Table 5.3 summarizes various kinds of polymer based VOCs sensors and their sensing performances.

5.5 Other sensing materials

Besides these highly sensitive VOC sensing materials, several authors have also explored other materials beyond these categories, which not only possess superior adsorption capability for VOC molecules but also

TABLE 5.3 Summary of various polymer-based VOC sensors.

Material	Morphology	VOC	Sensing range	Response time	Operating temperature	References
PDMS based VOC sensors						
PDMS	Film	Toluene	0.833–100 $g.m^{-3}$	5 s	RT	Kacik and Martincek (2017)
PZT-PDMS	Composite film	Acetic acid Toluene	45–100 ppm 0.2–0.6 ppm	50 s	RT	Sappati and Bhadra (2021)
PDMS	Film	Isopropanol	0–500 ppm	50 s	RT	Zhao et al. (2019)
PDMS and PDMS-PDPS	Film	m-xylene and cyclohexane	0–1000 ppm 0–3500 ppm	150 s	RT	St-Gelais et al. (2013)
PDMS and SU-8	Film	Toluene acetone	0–20 ng 0–150 ng	0.9 s 0.5 s	RT	Reddy et al. (2011)
PDMS	Film	Alcohol, ethers, alkanes	0–24000 ppm	240 s	RT	Martínez-Hipatl et al. (2010)

(continued on next page)

TABLE 5.3 Summary of various polymer-based VOC sensors—cont'd

Material	Morphology	VOC	Sensing range	Response time	Operating temperature	References
Polypyrrole based VOC sensors						
PPy loaded $Sn_{1-x}Sb_xO_2$	Nanocubes	Ammonia	0–20 ppm	~ 4 s	RT	Chakraborty and Mondal (2022)
PPy	Single nanowire	Heptanal, Acetophenone, Isopropyl Myristate and 2-Propanol	0.3–5%	˂ 200 s	NA	Xu et al. (2013)
PPy/MoO_3	Film	Formaldehyde and acetaldehyde	100 ppm	140 s	RT	Hosono et al. (2005)
PPy doped do-decylbenzene sulfonic acid	Film	Methane	20 ppm	NA	RT	Campos et al. (2007)
Au/PPy	Nanorods	Acetic acid, Benzene, toluene	10–100 ppm	20 s	NA	Lee et al. (2015)
PPy	Nanoparticles	Methanol, acetic acid, Acetonitrile	NA	˂1 s	RT	Liu et al. (2022b)

(continued on next page)

Polythiophene based VOC sensor						
PTh/MoO_3	Sheet/Rectangular particle	Ammonia, methanol, acetone	1 M concentration	NA	RT	Husain (2021)
PTh/Al_2O_3	Nanocomposite	Ammonia	25–650 ppm	25–72 s	RT	Tripathi et al. (2015)
rGO-PTh	Hybrid Film	Nitrogen oxide	10 ppm	26.36 s	RT	Bai et al. (2016)
PTh/SnO_2	Nanocomposite	Nitrogen oxide	10–200 ppm	Few minutes	RT	Kong et al. (2008)
PTh/ZrO_2	Nanocomposite	Ethene	NA	NA	RT	Husain et al. (2019)
PTh/WO_3	Hybrid film	Nitrogen oxide	100 ppm	NA	70°C	Huang et al. (2011b)
PTh/ZnO	Nanocomposite	Liquefied petroleum	600–2400 ppm	NA	RT	Husain et al. (2020c)
PTh/GO	Nanocomposite	Ethanol	400–2000 ppm	Very short	RT	Husain et al. (2020b)

(continued on next page)

TABLE 5.3 Summary of various polymer-based VOC sensors—cont'd

Material	Morphology	VOC	Sensing range	Response time	Operating temperature	References
Molecularly Imprinted polymer-based VOC sensors						
MIP-Ag_2S	Nanoparticles	1-butanol	100–400 ppm	NA	NA	Mustafa and Lieberzeit (2014)
Au NPs/MIPs	Nanoparticles	Acetone	50–300 ppm	3.7 ± 1 s	180°C	Jahangiri-Manesh et al. (2022)
Ag–$LaFeO_3$	Nanoparticles	Methanol	1–5 ppm	40–47 s	175°C	Rong et al. (2017)
Polyaniline-fluoral-bilayer MIP	Film	Ammonia, formaldehyde	0–1 ppm	NA	NA	Antwi-Boampong et al. (2014)
MIP	Bulk	Hydroquinone, phenol, toluene, benzene, heptane	50–400 ppm	5–12 s	NA	Fu and Finklea (2003)
MIP	Bulk	Hexanal, hexanoic acid, octanoic acid	5 + 5 + 5 μL	5 s	NA	Jha and Hayashi (2015)
MIP	Bulk	Acetaldehyde	NA	˂30 s	NA	Hirayama et al. (2002)

From Pathak, A., Swargiary, K., Kongsawang, N., Jitpratak, P., Ajchareeyasoontorn, N., Udomkittivorakul, J., & Viphavakit, C. (2023). Recent advances in sensing materials targeting clinical volatile organic compound (VOC) biomarkers: A review. Biosensors, 13(1). *https://doi.org/10.3390/bios13010114.* Room temperature = RT* allow easy integration over various types of sensing platforms. Among several emerging novel materials, this section will cover some with the most potential as sensing materials.

5.5.1 Metal organic frameworks

Metal organic frameworks (MOFs) have gained significant attention among researchers due to their extreme porosity, crystalline nature, and high surface area, making them an exceptional material for VOC sensing (Linnane et al., 2022). These unique material characteristics of MOFs allow them to detect both inorganic and organic compounds. The MOFs primarily utilized for VOC adsorption belong to the Material of Institute Lavoisier (MIL) series and the isoreticular metal-organic frameworks (IR-MOF) series, which are summarized in the following subsections. Over the past decades, MOFs have gained attention in gas sensing; however, their application in VOC detection remains underexplored. Currently, more than 20,000 MOF materials are available, many of which are yet to be studied for sensing applications (Furukawa et al., 2015; Tayfuroglu et al., 2022; Thomas, 2009).

5.5.1.1 *Material of institute lavoisier series*

MIL-n series is a class of MOFs which was first introduced by Ferey's research group (Férey et al., 2005). The MIL-101 contains a mesoporous structure comprising a pore diameter of 2.9–3.4 nm along with a surface area of 6000m^2/g (Férey et al., 2005). Because of their porous structure, the MIL-101 has gained significant attention in gas sensing (Yin et al., 2022b). Considering the potential of MIL-101, Huang et al., 2011a reported the first MIL-101 coated quartz microbalance for VOC detection (Huang et al., 2011a). They reported that the device exhibited several affinities to detect multiple VOC molecules due to the energy heterogeneity of MIL-101. The adsorption isotherms of VOCs on MIL-101 were observed to follow the Dubinin–Astakhov (DA) equation, with the characteristic energy values ranging from 5.70 to 9.13 kJ/mol, for methanol and n-butylamine, respectively. The DA equation is a model utilized to explain the gas adsorption phenomenon on any porous sensing materials, specifically in relation to the microporous adsorbents (Sun et al., 2020). In this work, the Astakhov exponent varied from 0.50 to 3.03 for n-butylamine and n-hexane, respectively, while the limiting adsorption capacity spanned from 0.08 mmol/g to 12.8 mmol/g for n-hexane and n-butylamine, respectively. MIL-101 demonstrated a lower affinity for n-hexane and the strongest affinity for n-butylamine.

The adsorption in MOFs is mainly attributed to the metal sites available on MIL-101, resulting in a greater affinity and higher VOC adsorption capacity than AC. Recently, Mousavi and Zeinali proposed that forming a nano-composition of MIL-101 (Cr)/CNT can significantly improve the sensing performance of resistive gas nanosensors (Mousavi & Zeinali, 2022). The sensing performance of the fabricated device was tested in the presence of multiple VOCs such as isopropanol, methanol, ethanol, formaldehyde, acetonitrile, tetrahydrofuran, acetone, dichloromethane, and n-hexane. The sensor achieved an extremely low LOD of 0.07 ppm, along with a high sensitivity of 325.21% and excellent selectivity for formaldehyde. Although the reported sensor demonstrated a remarkable performance, the sensor exhibited a rather slow average response time of 6 minutes for wide VOC concentrations ranging from 10 to 1000 ppm. It has been observed that the chemical structures, molecular sizes, as well as polarity of the sensing materials play a crucial role in the sensing performance of the device.

Despite the unique characteristics of the MIL-n, the material has various limitations, e.g., poor stability, cost, and poor storage, which can significantly limit the potential of these materials. However, key factors such as π-π interactions, metal sites, specific surface area, pore structure, and modifiable groups play a crucial role in affecting the separation and adsorption of several VOCs. Currently, several research groups are focused on designing and improving the surface morphology of MOFs to produce new variants of MOFs that can offer specific functionalities as well as address the existing challenges of MOFs (Fang et al., 2025; Ma et al., 2025; Soudagar et al., 2025).

5.5.1.2 *Isoreticular metal-organic frameworks series*

IRMOFs are another class of MOFs that comprise an identical network topology. By substituting a small amount of organic linkers in IRMOF-1 (also known as MOF-5), various other types of IRMOFs can be synthesized, which may exhibit favorable characteristics, including large surface area and enhanced chemical stability, as monitored by the Brunauer-Emmett-Teller (BET) method (Yin et al., 2022a). Microporous MOFs were synthesized and reported by Yaghi et al. (1995) to selectively detect VOC molecules. The building block comprises a symmetric organic molecule that allows the selective binding of metal ions to produce a layer of metal-organic compound. This layer was alternatively formed with the layers whose composition was determined by functionalization of the starting molecules, creating channels which allow the selective binding of guest molecules (Pathak et al., 2023). The crystal lattice of IRMOF layers remained stable up to 350°C, even after the removal of guest aromatic molecules, and demonstrated selective adsorption of gaseous molecules. The rigidity and stability of IRMOF materials facilitate the size- and shape-selective absorption of organic molecules and ions. In 2015, Furukawa et al. reported an extremely absorptive MOF-5 (also known as IRMOF-1) by utilizing benzene and octahedral Zn-O-C clusters links

(Furukawa et al., 2015). In the reported work, the authors have demonstrated that the synthesized MOF-5 can be further functionalized via other organic groups, e.g., Br, NH_2, OC_3H_7, OC_5H_{11}, C_2H_4, and C_4H_4. Additionally, the pore size of MOF-5 can be expanded by utilizing a long molecular structure such as pyrene, terphenyl, tetrahydropyrene, and biphenyl. To enhance the adsorption performance and crystal and pore structure of MOF-5, Saha et al. (2009) replaced the traditional solvent with diethylformamide during the synthesis process. The obtained MOF-5 showed a larger surface area (3917 m^2/g) along with a pore size of 8.6 Å and a pore volume of 1.39 cm^3/g, respectively. Yang et al. synthesized MOF-177 at 85°C via a solvothermal method, demonstrating its exceptional adsorption capacity for organic molecules (Yang et al., 2013). In the reported work, the authors achieved an improved surface area (4170 m^2/g) for MOF-177, along with a pore size of 0.94 nm and a pore volume of 1.11 cm^3/g. The synthesized MOF-177 exhibited enhanced adsorption when tested under exposure to various VOCs, such as toluene, benzene, ethylbenzene, acetone, o-xylene, m-xylene, and p-xylene in the air. From these reports, it is evident that MOFs exhibit a superior surface area along with an appropriate solvent, thereby possessing excellent adsorption capacity, making them a promising material for future VOC sensing applications.

5.5.2 Zeolites

Zeolites are crystalline microporous aluminosilicates with a highly ordered porous structure. Over the past decade, zeolites have gained significant attention among researchers due to their tunable pore size, high hydrophobicity, and ease of functionalization. Due to these desirable characteristics, zeolites have been widely utilized in a broad range of applications, such as biosensing (Ozansoy Kasap et al., 2017), heavy metal detection (Elboughdiri, 2020), VOC gas sensing (Yin et al., 2022b), etc. Previous studies have reported that the average specific surface area of zeolites is approximately 800 m^2/g, with the potential to be enhanced up to 1769 m^2/g (2020c).

Bhatia et al., (2009) reported adsorption of butyl acetate using silver exchanged yttrium (AgY (IE)) and silver loaded zeolites (AgZSM-5 (IE)) at 28°C. The Ag was loaded in the zeolites via ion exchange and impregnation techniques. The adsorption capacity of AgY (IE) against butyl acetate was reduced by 42% due to the increase in the relative humidity of the VOCs. However, the adsorption capacity of AgZSM-5 (IE) for butyl acetate was negligibly affected by the humidity, and the adsorption capacity was only reduced by 7%. In 2016, Ning et al. utilized the optical fiber as a sensing platform for the detection of VOC traces (Ning et al., 2016). In this work, the authors utilized the feature of nanoporous zeolites for the selective adsorption of isopropanol. The device comprised of the spherical shaped fiber coated with a thin zeolite film at the edge. The zeolite layer and spherical shaped fiber formed an arc-shaped inline Fabry–Perot cavity, which improved the sensing performance. The proposed sensor exhibited enhanced sensitivity of 0.91 nm/ppm for IPA concentrations ranging from 0 to 70 ppm.

Sun et al. (2021) synthesized SnO_2/ZSM-5 zeolite composite coated over an electrochemical material for the selective detection of formaldehyde. The high selectivity and improved sensing performance were attributed to a large number of oxygen vacancies available throughout the surface of the composite, enabling the rapid adsorption of formaldehyde. The SnO_2/ZSM-5 composite demonstrated superior sensing performance compared to pure SnO_2.

Zeolites are commonly utilized in VOC recovery and commercial separation processes because of their unique material characteristics, which allow them to selectively absorb some specific gaseous molecules while excluding others (Baimpos et al., 2012). The selectivity and adsorption capacity are highly influenced by the size and structure of both the micropores and the molecular adsorbate. Sorbed VOC molecules can interact with (1) the zeolite framework, (2) extra-framework cations, and (3) other adsorbed molecules (Pathak et al., 2023). The polarity, dimensions, and partial pressure of the probe molecule determine the nature and strength of their interaction with a given adsorbent. Additionally, for a specific adsorbate, factors such as the number and location of free cations, the topology and size of the channels, and the potential pre-adsorption of other molecules (e.g., humidity) also impact the adsorbate-adsorbent interaction. The SBA-15 and MCM-41 are the two most popular types of zeolites, which are widely utilized in VOC detection (Hu et al., 2009; Serrano et al., 2004).

5.6 Prospects

Breath analysis is a multidisciplinary research area that integrates analytical techniques, materials chemistry, medical science, data processing, and engineering. Conventional methods for health monitoring and disease diagnosis generally rely on invasive blood tests, which are often painful. Over the past decades, exhaled breath analysis has gained significant interest as a non-invasive alternative for early disease detection, offering a more user-friendly approach. Despite its advantages, this method has several limitations. For instance, direct exhaled breath cannot be utilized for every breath analyzer; thus, collection and storage pose key challenges. In general, collected breath samples often degrade quickly, leading to changes in their composition. Researchers are still investigating which types of breath are most suitable for breath analyzers, such as exhaled air from the mouth or nose. Currently, it is advised not to use dead space air for VOC monitoring; instead, end-tidal breath is recommended for breath analysis. The limitation of breath collection can easily be resolved by utilizing on-site monitoring via an appropriate sensing tool.

Considering the growth in the market, significant advances in handheld VOC detection tools can be expected, specifically with rapid growth in low-cost, compact, and printable materials and devices. The most successful sensing configuration generally involves well-structured surface morphologies, typically hierarchical nanostructures with intentionally introduced "impurities" to form nanocomposites (Tomić et al., 2021). The addition of external elements to MOs to form nanocomposites plays a crucial role as either host or guest elements and has exhibited enhanced sensing performance in VOC monitoring. This improvement primarily includes rapid response times, lower operating temperatures, and improved stability. However, target selectivity remains unclear. For example, ZnO nanomaterials have exhibited excellent selectivity toward acetone (Chen et al., 2022), but they have also been observed to be selective toward ethanol (Liu et al., 2024) and several other VOCs (Peng et al., 2024; Zhang et al., 2024) simply by tuning the composition or surface morphology. Additionally, the selectivity assessments are often based on comparing responses to individual analytes, frequently tested at varying, and incomparable concentrations. From a sensing materials perspective, research is increasingly focused on developing novel hybrid sensing materials by combining nanoscale interfaces. The next generation of researchers must be focused on controlling the surface morphology and reproducing the MOs nanostructure at the nanoscale to obtain the repetitive response.

A wide range of emerging sensing materials, although less associated with gas sensing, have shown significant response to be used independently or along with any well-explored gas-sensitive materials. Nowadays, researchers are exploring (1) transition metal dichalcogenides (TMDs) (Joshi, 2018) and transition metal carbides/nitrides (MXenes) (Shahzad, 2020), (2) perovskite and spinel oxides with multication variations (Santos, 2021), and (3) nitrides, along with novel materials, e.g., phosphorene, germanene, and silicene (Gablech, 2018). For both conventional and emerging gas-sensitive materials, further research is required in several key areas: (1) regulating the composition and understanding their effects on VOC sensing, (2) improve materials selectivity in mix gaseous enviornment, and (3) developing scalable approaches to synthesize and integrate these materials along with miniaturized platforms for real-world applications. The solution to these challenges will create a path to producing inexpensive, target-selective, easy-to-use, highly sensitive detection systems in the future.

5.7 Conclusion

The exhaled human breath contains more than 1000 VOCs, among which certain VOCs play a crucial role in early disease diagnosis. The selective detection of VOCs is not only essential but also challenging due to the complex gaseous environment. For effective disease diagnosis, a VOC sensor must accurately detect the target VOCs, which can only be achieved by employing suitable sensing materials. This chapter provides in-depth information on various types of sensing materials (i.e., MOs, polymers, and carbon-based materials) along with strategies to enhance sensitivity and selectivity. Additionally, a summary table is also included at the end of each section, enabling readers to compare the performance of different materials and select the most appropriate one for their device and target application.

References

Abdelghani, R., Shokry Hassan, H., Morsi, I., & Kashyout, A. B. (2019). Nano-architecture of highly sensitive SnO_2–based gas sensors for acetone and ammonia using molecular imprinting technique. *Sensors and Actuators B: Chemical, 297,* 126668. doi:10.1016/j.snb.2019.126668.

Abdulla, S., Mathew, T. L., & Pullithadathil, B. (2015). Highly sensitive, room temperature gas sensor based on polyaniline-multiwalled carbon nanotubes (PANI/MWCNTs) nanocomposite for trace-level ammonia detection. *Sensors and Actuators, B: Chemical, 221,* 1523–1534. http://www.journals.elsevier.com/sensors-and-actuators-b-chemical/. doi:10.1016/j.snb.2015.08.002.

Adhikari, B., & Majumdar, S. (2004). Polymers in sensor applications. *Progress in Polymer Science (Oxford), 29*(7), 699–766. http://www.sciencedirect.com/science/journal/00796700. doi:10.1016/j.progpolymsci.2004.03.002.

Ahmad, Z., Naseem, M. S., Talib, M., Islam, S. S., & Mishra, P. (2020). Self-standing MWCNTs based gas sensor for detection of environmental limit of CO_2. *Materials Science and Engineering: B, 255,* 114528. doi:10.1016/j.mseb.2020.114528.

Ahmed, M., & Soltani, T. (2024). Biosynthesized composite for gas sensor application: Structural and sensing performance evaluation. *Nanotechnology Perceptions., 20.* doi:10.62441/nano-ntp.v20iS4.52.

Akhtar, M. S., Umar, A., Sood, S., Jung, I. N. S., Hegazy, H. H., & Algarni, H. (2019). Rapid growth of TiO_2 nanoflowers via low-temperature solution process: Photovoltaic and sensing applications. *Materials, 12*(4), 566. doi:10.3390/ma12040566.

Alev, O., Şennik, E., Kılınç, N., & Öztürk, Z. Z. (2015). Gas sensor application of hydrothermally growth TiO_2 nanorods. *Procedia Engineering, 120,* 1162–1165. doi:10.1016/j.proeng.2015.08.747.

Al-Hardan, N. H., Abdullah, M. J., Abdul Aziz, A., Ahmad, H., & Low, L. Y. (2010). ZnO thin films for VOC sensing applications. *Vacuum, 85*(1), 101–106. doi:10.1016/j.vacuum.2010.04.009.

Almeida, O. A. C., de Araujo, N. O., Dias, B. H. S., de Sant'Anna Freitas, C., Coerini, L. F., Ryu, C. M., & de Castro Oliveira, J. V. (2023). The power of the smallest: The inhibitory activity of microbial volatile organic compounds against phytopathogens. *Frontiers in Microbiology, 13.* https://www.frontiersin.org/journals/microbiology#. doi:10.3389/fmicb.2022.951130.

Andersson, L., Sellergren, B., & Mosbach, K. (1984). Imprinting of amino acid derivatives in macroporous polymers. *Tetrahedron Letters, 25*(45), 5211–5214. doi:10.1016/S0040-4039(01)81566-5.

Angiola, M., Alsaif, M. M. Y. A., Kalantar-zadeh, K., Wisitsoraat, A., Wlodarski, W., & Martucci, A. (2015). Optical hydrogen sensing based on hybrid 2D MoO_3/Au nanoparticles. *Procedia Engineering, 120*, 1141–1144. doi:10.1016/j.proeng.2015.08.830.

Antwi-Boampong, S., Peng, J. S., Carlan, J., & Belbruno, J. J. (2014). A molecularly imprinted fluoral-p/polyaniline double layer sensor system for selective sensing of formaldehyde. *IEEE Sensors Journal, 14*(5), 1490–1498. doi:10.1109/JSEN.2014.2298872.

Arami-Niya, A., Wan Daud, W. M. A., Mjalli, F. S., Abnisa, F., & Shafeeyan, M. S. (2012). Production of microporous palm shell based activated carbon for methane adsorption: Modeling and optimization using response surface methodology. *Chemical Engineering Research and Design, 90*(6), 776–784. doi:10.1016/j.cherd.2011.10.001.

Ayyala, S. K., & Covington, J. A. (2021). Nickel-oxide based thick-film gas sensors for volatile organic compound detection. *Chemosensors, 9*(9). https://www.mdpi.com/2227-9040/9/9/247/pdf. doi:10.3390/chemosensors9090247.

Bahoumina, P., Hallil, H., Lachaud, J. L., Abdelghani, A., Frigui, K., Bila, S., Baillargeat, D., Ravichandran, A., Coquet, P., Paragua, C., Pichonat, E., Happy, H., Rebière, D., & Dejous, C. (2017). Microwave flexible gas sensor based on polymer multi wall carbon nanotubes sensitive layer. *Sensors and Actuators B: Chemical, 249*, 708–714. doi:10.1016/j.snb.2017.04.127.

Bai, S., Guo, J., Sun, J., Tang, P., Chen, A., Luo, R., & Li, D. (2016). Enhancement of NO_2-sensing performance at room temperature by graphene-modified polythiophene. *Industrial & Engineering Chemistry Research, 55*(19), 5788–5794. doi:10.1021/acs.iecr.6b00418.

Baimpos, T., Gora, L., Nikolakis, V., & Kouzoudis, D. (2012). Greece Selective detection of hazardous VOCs using zeolite/Metglas composite sensors. *Sensors and Actuators, A: Physical, 186*, 21–31. doi:10.1016/j.sna.2011.12.030.

Bhatia, S., Abdullah, A. Z., & Wong, C. T. (2009). Adsorption of butyl acetate in air over silver-loaded Y and ZSM-5 zeolites: Experimental and modelling studies. *Journal of Hazardous Materials, 163*(1), 73–81. doi:10.1016/j.jhazmat.2008.06.055.

Bhowmik, B., & Bhattacharyya, P. (2017). Efficient gas sensor devices based on surface engineered oxygen vacancy controlled TiO_2 nanosheets. *IEEE Transactions on Electron Devices, 64*(5), 2357–2363. doi:10.1109/TED.2017.2686425.

Bhowmik, B., Manjuladevi, V., Gupta, R. K., & Bhattacharyya, P. (2016). Highly selective low-temperature acetone sensor based on hierarchical 3-D TiO_2 nanoflowers. *IEEE Sensors Journal, 16*(10), 3488–3495. doi:10.1109/JSEN.2016.2530827.

Bindra, P., Gangopadhyay, S., & Hazra, A. (2020). 1-D TiO_2 nanorods array-based parallel electrode sensor for selective and stable detection of organic vapors. *IEEE Sensors Journal, 20*(2), 664–671. http://ieeexplore.ieee.org/xpl/RecentIssue.jsp?punumber=7361. doi:10.1109/JSEN.2019.2944408.

Bindra, P., & Hazra, A. (2019a). Multi-layered TiO_2 nanotubes array-based highly sensitive room-temperature vapor sensors. *IEEE Transactions on Nanotechnology, 18*, 13–20. http://ieeexplore.ieee.org/xpl/RecentIssue.jsp?puNumber=7729. doi:10.1109/TNANO.2018.2876555.

Bindra, P., & Hazra, A. (2019b). Selective detection of organic vapors using TiO_2 nanotubes based single sensor at room temperature. *Sensors and Actuators, B: Chemical, 290*, 684–690. https://www.journals.elsevier.com/sensors-and-actuators-b-chemical. doi:10.1016/j.snb.2019.03.115.

Bora, T., Fallah, H., Chaudhari, M., Apiwattanadej, T., Harun, S. W., Mohammed, W. S., & Dutta, J. (2014). Controlled side coupling of light to cladding mode of ZnO nanorod coated optical fibers and its implications for chemical vapor sensing. *Sensors and Actuators, B: Chemical, 202*, 543–550. https://www.journals.elsevier.com/sensors-and-actuators-b-chemical. doi:10.1016/j.snb.2014.05.097.

Byeon, J. H., Kim, J. S., Kang, H. K., Kang, S., & Kim, J. Y. (2022). Acetone gas sensor based on SWCNT/polypyrrole/phenyllactic acid nanocomposite with high sensitivity and humidity stability. *Biosensors, 12*(5). https://www.mdpi.com/2079-6374/12/5/354/pdf?version=1652962756. doi:10.3390/bios12050354.

Cai, Z., & Park, S. (2022). Highly selective acetone sensor based on Co_3O_4-decorated porous TiO_2 nanofibers. *Journal of Alloys and Compounds, 919*, 165875. doi:10.1016/j.jallcom.2022.165875.

Campos, M., Simões, F. R., & Pereira, E. C. (2007). Influence of methane in the electrical properties of polypyrrole films doped with dodecylbenzene sulfonic acid. *Sensors and Actuators B: Chemical, 125*(1), 158–166. doi:10.1016/j.snb.2007.02.001.

Carballido, L., Karbowiak, T., Cayot, P., Gerometta, M., Sok, N., & Bou-Maroun, E. (2022). Applications of molecularly imprinted polymers and perspectives for their use as food quality trackers. *Chem, 8*(9), 2330–2341. http://www.cell.com/chem/home. doi:10.1016/j.chempr.2022.05.021.

Caricato, A. P., Buonsanti, R., Catalano, M., Cesaria, M., Cozzoli, P. D., Luches, A., Manera, M. G., Martino, M., Taurino, A., & Rella, R. (2011). Films of brookite TiO_2 nanorods/nanoparticles deposited by matrix-assisted pulsed laser evaporation as NO_2 gas-sensing layers. *Applied Physics A, 104*(3), 963–968. doi:10.1007/s00339-011-6462-5.

Chakraborty, N., & Mondal, S. (2022). Chemiresistive NH_3 detection at sub-zero temperatures by polypyrrole- loaded $Sn_1{-}xSb_xO_2$ nanocubes. *Materials Horizons, 9*(6), 1750–1762. http://pubs.rsc.org/en/journals/journal/mh. doi:10.1039/d2mh00236a.

Chen, K., Pi, M., & Zhang, D. (2020a). Ultraviolet irradiation enhanced formaldehyde-sensing performance based on SnO_2@TiO_2 nanofiber heteroarchitectures. *Journal of Physics D: Applied Physics, 53*(12), 125301. doi:10.1088/1361-6463/ab5f40.

Chen, N., Deng, D., Li, Y., Xing, X., Liu, X., Xiao, X., & Wang, Y. (2016). The xylene sensing performance of WO_3 decorated anatase TiO_2 nanoparticles as a sensing material for a gas sensor at a low operating temperature. *Royal Society of Chemistry, China RSC Advances, 6*(55), 49692–49701. http://pubs.rsc.org/en/journals/journalissues. doi:10.1039/c6ra09195d.

Chen, N., Li, Q., Li, Y., Deng, D., Xiao, X., & Wang, Y. (2015). Facile synthesis and gas sensing performances based on nickel oxide nanoparticles/multi-wall carbon nanotube composite. *Journal of Materials Science: Materials in Electronics, 26*(10), 8240–8248. doi:10.1007/s10854-015-3487-0.

Chen, N., Li, Y., Deng, D., Liu, X., Xing, X., Xiao, X., & Wang, Y. (2017). Acetone sensing performances based on nanoporous TiO_2 synthesized by a facile hydrothermal method. *Sensors and Actuators, B: Chemical, 238*, 491–500. doi:10.1016/j.snb.2016.07.094.

Chen, T., Xue, L., Zheng, P., Zhang, Y., Liu, Y., Sun, J., Han, G., Li, H., Zhang, X., Li, Y., Li, H., Dong, C., Xu, F., Zhang, Q., & Wang, W. (2020b). Volatile organic compounds and ozone air pollution in an oil production region in northern China. *Atmospheric Chemistry and Physics, 20*(11), 7069–7086. http://www.atmos-chem-phys.net/volumes_and_issues.html. doi:10.5194/acp-20-7069-2020.

Chen, Y., Li, H., Huang, D., Wang, X., Wang, Y., Wang, W., Yi, M., Cheng, Q., Song, Y., & Han, G. (2022). Highly sensitive and selective acetone gas sensors based on modified ZnO nanomaterials. *Materials Science in Semiconductor Processing, 148*, 106807. doi:10.1016/j.mssp.2022.106807.

Chinh, N. D., Kim, C., & Kim, D. (2019). UV-light-activated H_2S gas sensing by a TiO_2 nanoparticulate thin film at room temperature. *Journal of Alloys and Compounds, 778*, 247–255. https://www.journals.elsevier.com/journal-of-alloys-and-compounds. doi:10.1016/j.jallcom.2018.11.153.

Comini, E., Yubao, L., Brando, Y., & Sberveglieri, G. (2005). Gas sensing properties of MoO_3 nanorods to CO and CH_3OH. *Chemical Physics Letters, 407*(4–6), 368–371. doi:10.1016/j.cplett.2005.03.116.

Cong, S., Sugahara, T., Wei, T., Jiu, J., Hirose, Y., Nagao, S., & Suganuma, K. (2016). Diverse adsorption/desorption abilities originating from the nanostructural morphology of VOC gas sensing devices based on molybdenum trioxide nanorod arrays. *Advanced Materials Interfaces, 3*(14). http://onlinelibrary.wiley.com/journal/10.1002/(ISSN)2196-7350. doi:10.1002/admi.201600252.

Dai, C., Liu, Y., & Wei, D. (2022). Two-dimensional field-effect transistor sensors: The road toward commercialization. *Chemical Reviews, 122*(11), 10319–10392. http://pubs.acs.org/journal/chreay. doi:10.1021/acs.chemrev.1c00924.

Das, M., & Sarkar, D. (2018). Development of room temperature ethanol sensor from polypyrrole (PPy) embedded in polyvinyl alcohol (PVA) matrix. *Polymer Bulletin, 75*(7), 3109–3125. doi:10.1007/s00289-017-2192-y.

Deng, L., Young, R. J., Kinloch, I. A., Sun, R., Zhang, G., Noé, L., & Monthioux, M. (2014). Coefficient of thermal expansion of carbon nanotubes measured by Raman spectroscopy. *Applied Physics Letters, 104*(5). http://scitation.aip.org/content/aip/journal/apl. doi:10.1063/1.4864056.

De Volder, M. F. L., Tawfick, S. H., Baughman, R. H., & Hart, A. J. (2013). Carbon nanotubes: Present and future commercial applications. *Science, 339*(6119), 535–539. http://www.sciencemag.org/content/339/6119/535.full.pdf. doi:10.1126/science.1222453.

Dong, Z. M., Xia, Q., Ren, H., Shang, X., Lu, X., Joo, S. W., & Huang, J. (2023). Preparation of hollow SnO_2/ZnO cubes for the high-performance detection of VOCs. *Ceramics International, 49*(3), 4650–4658. https://www.journals.elsevier.com/ceramics-international. doi:10.1016/j.ceramint.2022.09.352.

Dubey, R. S. (2018). Temperature-dependent phase transformation of TiO_2 nanoparticles synthesized by sol-gel method. *Materials Letters, 215*, 312–317. doi:10.1016/j.matlet.2017.12.120.

Duong, V. T., Nguyen, C. T., Luong, H. B., Nguyen, D. C., & Nguyen, H. L. (2021). Ultralow-detection limit ammonia gas sensors at room temperature based on MWCNT/WO_3 nanocomposite and effect of humidity. *Solid State Sciences, 113*. http://www.journals.elsevier.com/solid-state-sciences/. doi:10.1016/j.solidstatesciences.2021.106534.

Dutta, K., Chattopadhyay, P. P., Lu, C.-W., Ho, M.-S., & Bhattacharyya, P. (2015). A highly sensitive BTX sensor based on electrochemically derived wall-connected TiO_2 nanotubes. *Applied Surface Science, 354*, 353–361. doi:10.1016/j.apsusc.2015.05.077.

Elboughdiri, N. (2020). The use of natural zeolite to remove heavy metals Cu (II), Pb (II) and Cd (II), from industrial wastewater. *Cogent Engineering, 7*(1). https://www.cogentoa.com/journal/engineering. doi:10.1080/23311916.2020.1782623.

Fang, L., Jia, M., Zhao, H., Kang, L., Shi, L., Zhou, L., & Kong, W. (2021). Molecularly imprinted polymer-based optical sensors for pesticides in foods: Recent advances and future trends. *Trends in Food Science & Technology, 116*, 387–404. doi:10.1016/j.tifs.2021.07.039.

Fang, X., Choi, J. Y., Lu, C., Reichert, E., Pham, H. T. B., & Park, J. (2025). From 0D to 2D: Microwave-assisted synthesis of electrically conductive metal-organic frameworks with controlled morphologies. *Chemical Science, 16*(7), 3168–3172. http://pubs.rsc.org/en/journals/journal/sc. doi:10.1039/d4sc07025a.

Feng, X., Qiu, F., Zheng, L., Zhang, Y., Wang, Y., Wang, M., Xia, H., Tang, B., Yan, C., & Liang, R. (2024). Exposure to volatile organic compounds and mortality in US adults: A population-based prospective cohort study. *Science of The Total Environment, 928*, 172512. doi:10.1016/j.scitotenv.2024.172512.

Feng, Z., Zhang, L. U., Chen, W., Peng, Z., & Li, Y. (2020). A strategy for supportless sensors: Fluorine doped TiO_2 nanosheets directly grown onto Ti foam enabling highly sensitive detection toward acetone. *Sensors and Actuators B: Chemical, 322*, 128633. doi:10.1016/j.snb.2020.128633.

Férey, G., Mellot-Draznieks, C., Serre, C., Millange, F., Dutour, J., Surblé, S., & Margiolaki, I. (2005). A chromium terephthalate-based solid with unusually large pore volumes and surface area. *Science, 309*(5743), 2040–2042. doi:10.1126/science.1116275.

Fischer, E. (1894). Einfluss der Configuration auf die Wirkung der Enzyme. *Berichte der deutschen chemischen Gesellschaft.*, *27*(3), 2985–2993. doi:10.1002/cber.18940270364.

Fu, X., Yang, P., Xiao, X., Zhou, D., Huang, R., Zhang, X., Cao, F., Xiong, J., Hu, Y., Tu, Y., Zou, Y., Wang, Z., & Gu, H. (2019). Ultra-fast and highly selective room-temperature formaldehyde gas sensing of Pt-decorated MoO_3 nanobelts. *Journal of Alloys and Compounds, 797*, 666–675. https://www.journals.elsevier.com/journal-of-alloys-and-compounds. doi:10.1016/j.jallcom.2019.05.145.

Fu, Y., & Finklea, H. O. (2003). Quartz crystal microbalance sensor for organic vapor detection based on molecularly imprinted polymers. *Analytical Chemistry, 75*(20), 5387–5393. doi:10.1021/ac034523b.

Furukawa, H., Müller, U., & Yaghi, O. M. (2015). Heterogeneity within order" in metal-organic frameworks. *Angewandte Chemie – International Edition, 54*(11), 3417–3430. http://onlinelibrary.wiley.com/journal/10.1002/(ISSN)1521-3773. doi:10.1002/anie.201410252.

Gablech, Imrich, et al. (2018). Monoelemental 2D materials-based field effect transistors for sensing and biosensing: Phosphorene, antimonene, arsenene, silicene, and germanene go beyond graphene. *TrAC Trends in Analytical Chemistry, 105*. doi:10.1016/j.trac.2018.05.008.

Gagaoudakis, E., Michail, G., Katerinopoulou, D., Moschovis, K., Iliopoulos, E., Kiriakidis, G., Binas, V., & Aperathitis, E. (2020). Transparent p-type NiO:Al thin films as room temperature hydrogen and methane gas sensors. *Materials Science in Semiconductor Processing, 109*, 104922. doi:10.1016/j.mssp.2020.104922.

Gakhar, T., Basu, S., & Hazra, A. (2022). Carbon-metal oxide nanocomposites for selective detection of toxic and hazardous volatile organic compounds (VOC) – A review. *Green Analytical Chemistry, 1*, 100005. doi:10.1016/j.greeac.2022.100005.

Gakhar, T., & Hazra, A. (2020). Oxygen vacancy modulation of titania nanotubes by cathodic polarization and chemical reduction routes for efficient detection of volatile organic compounds. *Nanoscale, 12*(16), 9082–9093. http://pubs.rsc.org/en/journals/journal/nr. doi:10.1039/c9nr10795a.

Gakhar, T., & Hazra, A. (2021). C_{60}-encapsulated TiO_2 nanoparticles for selective and ultrahigh sensitive detection of formaldehyde. *Nanotechnology, 32*(50), 505505. doi:10.1088/1361-6528/ac23f9.

Gałęzowska, G. Ż., Chraniuk, M., & Wolska, L. (2016). In vitro assays as a tool for determination of VOCs toxic effect on respiratory system: A critical review. *TrAC Trends in Analytical Chemistry, 77*, 14–22. doi:10.1016/j.trac.2015.10.012.

Gao, P., Wang, H., Li, X., & Meng, F. (2024). Flower-like SnO_2 nanorod aggregates modified with Co_3O_4 nanoparticles grown on reduced graphene oxide(rGO) sheets for xylene sensing. *Journal of Alloys and Compounds, 1004*, 175774. doi:10.1016/j.jallcom.2024.175774.

Gao, X., Li, Y., Zeng, W., Zhang, C., & Wei, Y. (2017). Hydrothermal synthesis of agglomerating TiO_2 nanoflowers and its gas sensing. *Journal of Materials Science: Materials in Electronics, 28*(24), 18781–18786. doi:10.1007/s10854-017-7827-0.

Gardon, M., Monereo, O., Dosta, S., Vescio, G., Cirera, A., & Guilemany, J. M. (2013). New procedures for building-up the active layer of gas sensors on flexible polymers. *Surface and Coatings Technology, 235*, 848–852. doi:10.1016/j.surfcoat.2013.09.011.

Ge, W., Jiao, S., Chang, Z., He, X., & Li, Y. (2020). Ultrafast response and high selectivity toward acetone vapor using hierarchical structured TiO_2 nanosheets. *ACS Applied Materials and Interfaces, 12*(11), 13200–13207. http://pubs.acs.org/journal/aamick. doi:10.1021/acsami.9b23181.

Gulotty, R., Castellino, M., Jagdale, P., Tagliaferro, A., & Balandin, A. A. (2013). Effects of functionalization on thermal properties of single-wall and multi-wall carbon nanotube-polymer nanocomposites. *ACS Nano, 7*(6), 5114–5121. doi:10.1021/nn400726g.

Han, B., Rupam, T. H., Chakraborty, A., & Saha, B. B. (2024). A comprehensive review on VOCs sensing using different functional materials: Mechanisms, modifications, challenges and opportunities. *Renewable and Sustainable Energy Reviews, 196*. https://www.sciencedirect.com/science/journal/13640321. doi:10.1016/j.rser.2024.114365.

Hasseb, A. A., Ghani, N. D. T. A., Shehab, O. R., & El Nashar, R. M. (2022). Application of molecularly imprinted polymers for electrochemical detection of some important biomedical markers and pathogens. *Current Opinion in Electrochemistry, 31*. www.journals.elsevier.com/current-opinion-in-electrochemistry. doi:10.1016/j.coelec.2021.100848.

Hayashi, K., Nozaki, K., Tan, Z., Fujita, K., Nemoto, R., Yamashita, K., Miura, H., Itaka, K., & Ohara, S. (2020). Enhanced antibacterial property of facet-engineered TiO_2 nanosheet in presence and absence of ultraviolet irradiation. *Materials, 13*(1), 78. doi:10.3390/ma13010078.

He, X., Wu, M., Ao, Z., Lai, B. o, Zhou, Y., An, T., & Wang, S. (2021). Metal–organic frameworks derived C/TiO_2 for visible light photocatalysis: Simple synthesis and contribution of carbon species. *Journal of Hazardous Materials, 403*, 124048. doi:10.1016/j.jhazmat.2020.124048.

Hirayama, K., Sakai, Y., Kameoka, K., Noda, K., & Naganawa, R. (2002). Preparation of a sensor device with specific recognition sites for acetaldehyde by molecular imprinting technique. *Sensors and Actuators, B: Chemical, 86*(1), 20–25. doi:10.1016/S0925-4005(02)00107-7.

Horprathum, M., Eiamchai, P., Chindaudom, P., Pokaipisit, A., & Limsuwan, P. (2012). Oxygen partial pressure dependence of the properties of TiO_2 thin films deposited by DC reactive magnetron sputtering. *Procedia Engineering, 32*, 676–682. doi:10.1016/j.proeng.2012.01.1326.

Hosono, K., Matsubara, I., Murayama, N., Woosuck, S., & Izu, N. (2005). Synthesis of polypyrrole/MoO_3 hybrid thin films and their volatile organic compound gas-sensing properties. *Chemistry of Materials, 17*(2), 349–354. doi:10.1021/cm0492641.

Hu, Q., Li, J. J., Hao, Z. P., Li, L. D., & Qiao, S. Z. (2009). Dynamic adsorption of volatile organic compounds on organofunctionalized SBA-15 materials. *Chemical Engineering Journal, 149*(1-3), 281–288. doi:10.1016/j.cej.2008.11.003.

Hua, Y., Ahmadi, Y., & Kim, K. I.-H. (2022). Molecularly imprinted polymers for sensing gaseous volatile organic compounds: Opportunities and challenges. *Environmental Pollution, 311*, 119931. doi:10.1016/j.envpol.2022.119931.

Huang, C. Y., Song, M., Gu, Z. Y., Wang, H. F., & Yan, X. P. (2011a). Probing the adsorption characteristic of metal-organic framework MIL-101 for volatile organic compounds by quartz crystal microbalance. *Environmental Science and Technology, 45*(10), 4490–4496. doi:10.1021/es200256q.

Huang, J., Kang, Y., Yang, T., Wang, Y., & Wang, S. (2011b). Preparation of polythiophene/WO_3 organic-inorganic hybrids and their gas sensing properties for NO_2 detection at low temperature. *Journal of Natural Gas Chemistry, 20*(4), 403–407. doi:10.1016/s1003-9953(10)60196-x.

Husain, A. (2021). Electrical conductivity based ammonia, methanol and acetone vapour sensing studies on newly synthesized polythiophene/molybdenum oxide nanocomposite. *Journal of Science: Advanced Materials and Devices, 6*(4), 528–537. www.journals.elsevier.com/journal-of-science-advanced-materials-and-devices. doi:10.1016/j.jsamd.2021.07.002.

Husain, A., Ahmad, S., & Mohammad, F. (2019). Thermally stable and highly sensitive ethene gas sensor based on polythiophene/zirconium oxide nanocomposites. *Materials Today Communications, 20*, 100574. doi:10.1016/j.mtcomm.2019.100574.

Husain, A., Ahmad, S., & Mohammad, F. (2020a). Electrical conductivity and alcohol sensing studies on polythiophene/tin oxide nanocomposites. *Journal of Science: Advanced Materials and Devices, 5*(1), 84–94. www.journals.elsevier.com/journal-of-science-advanced-materials-and-devices. doi:10.1016/j.jsamd.2020.01.002.

Husain, A., Ahmad, S., & Mohammad, F. (2020b). Synthesis, characterisation and ethanol sensing application of polythiophene/graphene nanocomposite. *Materials Chemistry and Physics, 239*, 122324. doi:10.1016/j.matchemphys.2019.122324.

Husain, A., Shariq, M. U., & Mohammad, F. (2020c). DC electrical conductivity and liquefied petroleum gas sensing application of polythiophene/zinc oxide nanocomposite. *Materialia, 9*. www.journals.elsevier.com/materialia. doi:10.1016/j.mtla.2020.100599.

Hussain, M., Kotova, K., & Lieberzeit, P. (2016). Molecularly imprinted polymer nanoparticles for formaldehyde sensing with QCM. *Sensors, 16*(7), 1011. doi:10.3390/s16071011.

Ibrahim, K. S (2013). Carbon nanotubes-properties and applications: A review. *Carbon letters, 14*(3), 131–144. doi:10.5714/cl.2013.14.3.131.

Jafari, N., & Zeinali, S. (2020). Highly rapid and sensitive formaldehyde detection at room temperature using a ZIF-8/MWCNT nanocomposite. *ACS Omega, 5*(9), 4395–4402. pubs.acs.org/journal/acsodf. doi:10.1021/acsomega.9b03124.

Jagannathan, M., Dhinasekaran, D., Rakkesh Rajendran, A., & Subramaniam, B. (2022). Selective room temperature ammonia gas sensor using nanostructured ZnO/CuO@graphene on paper substrate. *Sensors and Actuators B: Chemical, 350*, 130833. doi:10.1016/j.snb.2021.130833.

Jahangiri-Manesh, A., Mousazadeh, M., & Nikkhah, M. (2022). Fabrication of chemiresistive nanosensor using molecularly imprinted polymers for acetone detection in gaseous state. *Iranian Polymer Journal (English Edition), 31*(7), 883–891. http://www.springerlink.com/content/1026-1265. doi:10.1007/s13726-022-01044-w.

Janfaza, S., Banan Nojavani, M., Nikkhah, M., Alizadeh, T., Esfandiar, A., & Ganjali, M. R. (2019). A selective chemiresistive sensor for the cancer-related volatile organic compound hexanal by using molecularly imprinted polymers and multiwalled carbon nanotubes. *Microchimica Acta, 186*(3). http://www.springer/at/mca. doi:10.1007/s00604-019-3241-z.

Jha, S. K., & Hayashi, K. (2015). A quick responding quartz crystal microbalance sensor array based on molecular imprinted polyacrylic acids coating for selective identification of aldehydes in body odor. *Talanta, 134*, 105–119. https://www.journals.elsevier.com/talanta. doi:10.1016/j.talanta.2014.09.049.

Ji, H., Zeng, W., & Li, Y. (2019). Assembly of 2D nanosheets into flower-like MoO_3: New insight into the petal thickness affect on gas-sensing properties. *Materials Research Bulletin, 118*, 110476. doi:10.1016/j.materresbull.2019.05.001.

Jia, X., Yu, S., Cheng, C., Yang, J., Li, Y., Wang, S., & Song, H. (2022). Ag nanoparticles modified Fe_3O_4/reduced graphene oxide and their acetone sensing properties. *Materials Chemistry and Physics, 276*, 125378. doi:10.1016/j.matchemphys.2021.125378.

Jiang, W., Meng, L., Zhang, S., Chuai, X., Zhou, Z., Hu, C., Sun, P., Liu, F., Yan, X. U., & Lu, G. (2019). Design of highly sensitive and selective xylene gas sensor based on Ni-doped MoO_3 nano-pompon. *Sensors and Actuators B: Chemical, 299*, 126888. doi:10.1016/j.snb.2019.126888.

John, R. A. B., Shruthi, J., Ramana Reddy, M. V., & Ruban Kumar, A. (2022). Manganese doped nickel oxide as room temperature gas sensor for formaldehyde detection. *Ceramics International, 48*(12), 17654–17667. https://www.journals.elsevier.com/ceramics-international. doi:10.1016/j.ceramint.2022.03.036.

Joshi, Nirav, et al. (2018). A review on chemiresistive room temperature gas sensors based on metal oxide nanostructures, graphene and 2D transition metal dichalcogenides. *Microchimica Acta, 185*. doi:10.1007/s00604-018-2750-5.

Kacik, D., & Martincek, I. (2017). Toluene optical fibre sensor based on air microcavity in PDMS. *Optical Fiber Technology, 34*, 70–73. http://www.journals.elsevier.com/optical-fiber-technology/. doi:10.1016/j.yofte.2017.01.006.

Kang, H. L., Hong, D. K., Yoon, S., Song, S., Kim, Y. J., Kim, W. H., Cho, H., Kim, H. W., Seong, W. K., Shin, D. S., & Lee, K. N. (2021). Gas concentrator with an activated carbon adsorbent for breath analysis. *Sensors and Actuators, B: Chemical, 330*. https://www.journals.elsevier.com/sensors-and-actuators-b-chemical. doi:10.1016/j.snb.2020.129341.

Kohli, N., Hastir, A., Kumari, M., & Singh, R. C. (2020). Hydrothermally synthesized heterostructures of In_2O_3/MWCNT as acetone gas sensor. *Sensors and Actuators A: Physical, 314*, 112240. doi:10.1016/j.sna.2020.112240.

Kong, F., Wang, Y., Zhang, J., Xia, H., Zhu, B., Wang, Y., Wang, S., & Wu, S. (2008). The preparation and gas sensitivity study of polythiophene/SnO_2 composites. *Materials Science and Engineering: B, 150*(1), 6–11. https://www.journals.elsevier.com/materials-science-and-engineering-b. doi:10.1016/j.mseb.2008.01.003.

Kumar, K. G., Avinash, B. S., Rahimi-Gorji, M., & Majdoubi, J. (2020). Photocatalytic activity and smartness of TiO_2 nanotube arrays for room temperature acetone sensing. *Journal of Molecular Liquids, 300*, 112418. doi:10.1016/j.molliq.2019.112418.

Kumar, V., & Tyagi, P. K. (2018). Potential application of multi-walled carbon nanotubes/activated carbon/bamboo charcoal for efficient alcohol sensing. *Journal of Alloys and Compounds, 767*, 215–222. doi:10.1016/j.jallcom.2018.06.123.

Kumar, Y. R., Deshmukh, K., Kovářík, T., & Khadheer Pasha, S. K. (2022). A systematic review on 2D materials for volatile organic compound sensing. *Coordination Chemistry Reviews, 461*, 214502. doi:10.1016/j.ccr.2022.214502.

Kumar Pathak, A., Swargiary, K., & Viphavakit, C. (2025). SPR-enhanced detection of isopropanol vapor using Au/MIP-coated optical fiber sensor. *IEEE Sensors Journal, 25*(16), 30743–30750. doi:10.1109/JSEN.2025.3585846.

Kumari, A., Shanbogh, S. M., Udachyan, I., Kandaiah, S., Roy, A., Varade, V., & Ponnam, A. (2020). Interface-driven multifunctionality in two-dimensional TiO_2 nanosheet/poly(dimercaptothiadiazole-triazine) hybrid resistive random access memory device. *ACS Applied Materials and Interfaces, 12*(50), 56568–56578. http://pubs.acs.org/journal/aamick. doi:10.1021/acsami.0c16451.

Lai, C., Wang, Z., Qin, L., Fu, Y., Li, B., Zhang, M., Liu, S., Li, L., Yi, H., Liu, X., Zhou, X., An, N., An, Z., Shi, X., & Feng, C (2021). Metal-organic frameworks as burgeoning materials for the capture and sensing of indoor VOCs and radon gases. *Coordination Chemistry Reviews, 427*, 213565. doi:10.1016/j.ccr.2020.213565.

Landau, O., & Rothschild, A. (2015). Fibrous TiO_2 gas sensors produced by electrospinning. *Journal of Electroceramics, 35*(1–4), 148–159. www.kluweronline.com/issn/1385-3449/. doi:10.1007/s10832-015-0007-9.

Lee, J. H. (2009). Gas sensors using hierarchical and hollow oxide nanostructures: Overview. *Sensors and Actuators, B: Chemical, 140*(1), 319–336. doi:10.1016/j.snb.2009.04.026.

Lee, J. S., Yoon, N. R., Kang, B. H., Lee, S. W., Gopalan, S. A., Kim, S. W., Lee, S. H., Kwon, D. H., & Kang, S. W. (2015). Au-polypyrrole framework nanostructures for improved localized surface plasmon resonance volatile organic compounds gas sensing. *Journal of Nanoscience and Nanotechnology, 15*(10), 7738–7742. http://docserver.ingentaconnect.com/deliver/connect/asp/15334880/v15n10/s60.pdf?expires=1445298818&id=83131567&titleid=4286&accname=Elsevier+BV&checksum=BD7580F7B8D9CC6CCD9C8DE71D1663CD. doi:10.1166/jnn.2015.11194.

Lee, K., Kirchgeorg, R., & Schmuki, P. (2014). Role of transparent electrodes for high efficiency TiO_2 nanotube based dye-sensitized solar cells. *Journal of Physical Chemistry C, 118*(30), 16562–16566. http://pubs.acs.org/journal/jpcck. doi:10.1021/jp412351g.

Lee, M., Jang, Y., Oh, J., & Jung, J. (2018). A silver-nanoparticle-embedded activated-carbon–PEDOT:PSS composite conductor for enhancing electrical conductivity of a gas sensor module. *Digest Journal of Nanomaterials and Biostructures, 13*(2), 527–534. http://www.chalcogen.ro/527_LeeM.pdf.

Lekawa-Raus, A., Patmore, J., Kurzepa, L., Bulmer, J., & Koziol, K. (2014). Electrical properties of carbon nanotube based fibers and their future use in electrical wiring. *Advanced Functional Materials, 24*(24), 3661–3682. http://onlinelibrary.wiley.com/journal/10.1002/(ISSN)1616-3028. doi:10.1002/adfm.201303716.

Li, C., Choi, P. G, Kim, K., & Masuda, Y. (2022). High performance acetone gas sensor based on ultrathin porous NiO nanosheet. *Sensors and Actuators B: Chemical, 367*, 132143. doi:10.1016/j.snb.2022.132143.

Li, L. I., Song, J., Yao, X., Huang, G., Liu, Z., & Tang, L. (2012). Adsorption of volatile organic compounds on three activated carbon samples: Effect of pore structure. *Journal of Central South University, 19*(12), 3530–3539. doi:10.1007/s11771-012-1439-x.

Li, M., Lu, B., Ke, Q. F., Guo, Y. J., & Guo, Y. P. (2017). Synergetic effect between adsorption and photodegradation on nanostructured TiO_2/activated carbon fiber felt porous composites for toluene removal. *Journal of Hazardous Materials, 333*, 88–98. www.elsevier.com/locate/jhazmat. doi:10.1016/j.jhazmat.2017.03.019.

Li, T., Rui, Y., Zhang, X., Shi, J., Wang, X., Wang, Y., Yang, J., & Zhang, Q. (2020a). Anatase TiO_2 nanorod arrays as high-performance electron transport layers for perovskite solar cells. *Journal of Alloys and Compounds, 849*, 156629. doi:10.1016/j.jallcom.2020.156629.

Li, X., Li, Y., Sun, G., Zhang, B. O., Wang, Y., & Zhang, Z. (2020b). Enhanced CH_4 sensitivity of porous nanosheets-assembled ZnO microflower by decoration with Zn_2SnO_4. *Sensors and Actuators B: Chemical, 304*, 127374. doi:10.1016/j.snb.2019.127374.

Li, X., Li, X., Wang, J., & Lin, S. (2015). Highly sensitive and selective roomerature formaldehyde sensors using hollow TiO_2 microspheres. *Sensors and Actuators, B: Chemical, 219*, 158–163. https://www.journals.elsevier.com/sensors-and-actuators-b-chemical. doi:10.1016/j.snb.2015.05.031.

Li, X., Zhang, L. i, Yang, Z., Wang, P., Yan, Y., & Ran, J. (2020c). Adsorption materials for volatile organic compounds (VOCs) and the key factors for VOCs adsorption process: A review. *Separation and Purification Technology, 235*, 116213. doi:10.1016/j.seppur.2019.116213.

Li, Y., Ban, H., & Yang, M. (2016). Highly sensitive NH_3 gas sensors based on novel polypyrrole-coated SnO_2 nanosheet nanocomposites. *Sensors and Actuators, B: Chemical, 224*, 449–457. doi:10.1016/j.snb.2015.10.078.

Liang, S., Wang, H., & Tao, X. (2022). Effects of multiwalled carbon nanotubes and reduced graphene oxide of different proportions on the electrothermal properties of cationic cellulose nanofibril-based composites. *Journal of Materials Research and Technology, 17*, 2388–2399. doi:10.1016/j.jmrt.2022.02.004.

Liao, Y., Que, W., Jia, Q., He, Y., Zhang, J., & Zhong, P. (2012). Controllable synthesis of brookite/anatase/rutile TiO_2 nanocomposites and single-crystalline rutile nanorods array. *Journal of Materials Chemistry, 22*(16), 7937. doi:10.1039/c2jm16628c.

Lim, C. J., & Park, J. W. (2017). Luminescent oxygen-sensing films with improved sensitivity based on light scattering by TiO_2 particles. *Sensors and Actuators, B: Chemical, 253*, 934–941. doi:10.1016/j.snb.2017.07.045.

Lin, Q., Li, Y., & Yang, M. (2012). Tin oxide/graphene composite fabricated via a hydrothermal method for gas sensors working at room temperature. *Sensors and Actuators, B: Chemical, 173*, 139–147. doi:10.1016/j.snb.2012.06.055.

Lin, S., Li, D., Wu, J., Li, X., & Akbar, S. A. (2011). A selective room temperature formaldehyde gas sensor using TiO_2 nanotube arrays. *Sensors and Actuators, B: Chemical, 156*(2), 505–509. doi:10.1016/j.snb.2011.02.046.

Linnane, E., Haddad, S., Melle, F., Mei, Z., & Fairen-Jimenez, D. (2022). The uptake of metal-organic frameworks: A journey into the cell. *Chemical Society Reviews, 51*(14), 6065–6086. http://pubs.rsc.org/en/journals/journal/cs. doi:10.1039/d0cs01414a.

Liu, A., Liu, T., Fu, H., Yin, X., Liu, K., & Yu, J. (2022a). Enhanced performance of Zn and Co co-doped MoO_3 nanosheets as gas sensor for n-butylamine. *Ceramics International, 48*(22), 32986–32993. https://www.journals.elsevier.com/ceramics-international. doi:10.1016/j.ceramint.2022.07.230.

Liu, B., Liu, X., Yuan, Z., Jiang, Y., Su, Y., Ma, J., & Tai, H. (2019). A flexible NO_2 gas sensor based on polypyrrole/nitrogen-doped multiwall carbon nanotube operating at room temperature. *Sensors and Actuators, B: Chemical, 295*, 86–92. https://www.journals.elsevier.com/sensors-and-actuators-b-chemical. doi:10.1016/j.snb.2019.05.065.

Liu, C., Wang, B., Wang, T., Liu, J., Sun, P., Chuai, X., & Lu, G. (2017). Enhanced gas sensing characteristics of the flower-like $ZnFe_2O_4$/ZnO microstructures. *Sensors and Actuators, B: Chemical, 248*, 902–909. doi:10.1016/j.snb.2017.01.133.

Liu, M., Lu, W., Zhao, L., Zhou, C., Li, H., & Wang, W. (2010). Fabrication and photocatalytical properties of flower-like TiO_2 nanostructures. *Transactions of Nonferrous Metals Society of China, 20*(12), 2299–2302. doi:10.1016/s1003-6326(10)60644-9.

Liu, X., Zheng, W., Kumar, R., Kumar, M., & Zhang, J. (2022b). Conducting polymer-based nanostructures for gas sensors. *Coordination Chemistry Reviews, 462*, 214517. doi:10.1016/j.ccr.2022.214517.

Liu, Y. H., Young, S. J., Ji, L. W., Chu, Y. L., & Chang, S. J. (2024). High-sensitive ethanol gas sensor using Ag modified ZnO nanosheets. *Elsevier B.V., Taiwan Talanta Open, 10*. https://www.sciencedirect.com/science/journal/26668319. doi:10.1016/j.talo.2024.100386.

Liu, Z., et al. (2018). A Room Temperature VOCs Gas Sensor Based on a Layer by Layer Multi-Walled Carbon Nanotubes/Poly-ethylene Glycol Composite. *Sensors, 18*(9), 3113. doi:10.3390/s18093113.

Lou, C., Yang, C., Zheng, W., Liu, X., & Zhang, J. (2021). Atomic layer deposition of ZnO on SnO_2 nanospheres for enhanced formaldehyde detection. *Sensors and Actuators B: Chemical, 329*, 129218. doi:10.1016/j.snb.2020.129218.

Lu, G., Ocola, L. E., & Chen, J. (2009). Room-temperature gas sensing based on electron transfer between discrete tin oxide nanocrystals and multiwalled carbon nanotubes. *Advanced Materials, 21*(24), 2487–2491. http://www3.interscience.wiley.com/cgi-bin/fulltext/122267068/PDFSTART. doi:10.1002/adma.200803536. United States.

Ma, L., Ma, S. Y., Kang, H., Shen, X. F., Wang, T. T., Jiang, X. H., & Chen, Q. (2017). Preparation of Ag-doped ZnO-SnO_2 hollow nanofibers with an enhanced ethanol sensing performance by electrospinning. *Materials Letters, 209*, 188–192. doi:10.1016/j.matlet.2017.08.004.

Ma, X., Zhang, Z., Wu, H., Li, J., & Yang, L. (2020). Adsorption of volatile organic compounds at medium-high temperature conditions by activated carbons. *Energy & Fuels, 34*(3), 3679–3690. doi:10.1021/acs.energyfuels.9b03292.

Ma, Z., Zhang, Y. U., Yuan, T., Fan, Y. U., Wang, X., Xue, Z., Zhong, A., & Xu, J. (2025). Tailoring crystal facets of metal-organic frameworks to enhance sensing performance for aromatic vapors detection. *Journal of Hazardous Materials, 486*, 136859. doi:10.1016/j.jhazmat.2024.136859.

Maity, D., Rajavel, K., & Kumar, R. T. R. (2018). Polyvinyl alcohol wrapped multiwall carbon nanotube (MWCNTs) network on fabrics for wearable room temperature ethanol sensor. *Sensors and Actuators, B: Chemical, 261*, 297–306. doi:10.1016/j.snb.2018.01.152.

Malik, R., Joshi, N., & Tomer, V. K. (2021). Advances in the designs and mechanisms of MoO_3 nanostructures for gas sensors: A holistic review. *Materials Advances, 2*(13), 4190–4227. https://www.rsc.org/journals-books-databases/about-journals/materials-advances/. doi:10.1039/d1ma00374g.

Mandal, B., Aaryashree, R. S., & Mukherjee, S. (2018). Highly selective and sensitive methanol sensor using rose-like ZnO microcube and MoO_3 micrograss-based composite. *IEEE Sensors Journal, 18*(7), 2659–2666. http://ieeexplore.ieee.org/xpl/RecentIssue.jsp?punumber=7361. doi:10.1109/JSEN.2018.2803682.

Martínez-Hipatl, C., Muñoz-Aguirre, S., Beltrán-Pérez, G., Castillo-Mixcóatl, J., & Rivera-De la Rosa, J. (2010). Detection of volatile organic compounds by an interferometric sensor. *Sensors and Actuators, B: Chemical, 147*(1), 37–42. doi:10.1016/j.snb.2010.03.042.

Mei, H., Zhou, S., Lu, M., Zhao, Y., & Cheng, L. (2020). Construction of pine-branch-like α-Fe_2O_3/TiO_2 hierarchical heterostructure for gas sensing. *Ceramics International, 46*(11), 18675–18682. https://www.journals.elsevier.com/ceramics-international. doi:10.1016/j.ceramint.2020.04.181.

Miah, M. R., Yang, M., Khandaker, S., Bashar, M. M., Alsukaibi, A. K. D., Hassan, H. M. A., Znad, H., & Awual, M. R. (2022). Polypyrrole-based sensors for volatile organic compounds (VOCs) sensing and capturing: A comprehensive review. *Sensors and Actuators A: Physical, 347*. https://www.journals.elsevier.com/sensors-and-actuators-a-physical. doi:10.1016/j.sna.2022.113933.

Mirzaie, M., Talebizadeh, A. R., & Hashemipour, H. (2021). Mathematical modeling and experimental study of VOC adsorption by Pistachio shell–based activated carbon. *Environmental Science and Pollution Research, 28*(3), 3737–3747. https://link.springer.com/journal/11356. doi:10.1007/s11356-020-10634-1.

Mo, Y., Tan, Z., Sun, L., Lu, Y., & Liu, X. (2020). Ethanol-sensing properties of α-MoO_3 nanobelts synthesized by hydrothermal method. *Journal of Alloys and Compounds, 812*, 152166. doi:10.1016/j.jallcom.2019.152166.

Modenes-Junior, M. A., Zito, C. A., Perfecto, T. M., & Volanti, D. P. (2019). Ethanol detection using composite based on reduced graphene oxide and CuO hierarchical structure under wet atmosphere. *Materials Science and Engineering: B, 248*. https://www.journals.elsevier.com/materials-science-and-engineering-b. doi:10.1016/j.mseb.2019.114385.

Mokoena, T. P., Swart, H. C., & Motaung, D. E. (2019). A review on recent progress of p-type nickel oxide based gas sensors: Future perspectives. *Journal of Alloys and Compounds, 805*, 267–294. https://www.journals.elsevier.com/journal-of-alloys-and-compounds. doi:10.1016/j.jallcom.2019.06.329.

Mousavi, S., & Zeinali, S. (2022). VOCs detection using resistive gas nanosensor based on MIL-101(Cr) as a metal organic framework. *Sensors and Actuators A: Physical, 346*, 113810. doi:10.1016/j.sna.2022.113810.

Mousazadeh, M., Jahangiri-Manesh, A., Nikkhah, M., Abbasian, S., Moshaii, A., Javad Masroor, M., & Norouzi, P. (2022). Detection of hexanal gas as a volatile organic compound cancer biomarker using a nanocomposite of gold nanoparticles and selective polymers. *Journal of Electroanalytical Chemistry, 905*, 115962. doi:10.1016/j.jelechem.2021.115962.

Mun, Y., Park, S., An, S., Lee, C., & Kim, H. W. (2013). NO_2 gas sensing properties of Au-functionalized porous ZnO nanosheets enhanced by UV irradiation. *Ceramics International, 39*(8), 8615–8622. doi:10.1016/j.ceramint.2013.04.035.

Mustafa, G., & Lieberzeit, P. A. (2014). Molecularly imprinted polymer–Ag_2S nanoparticle composites for sensing volatile organics. *RSC Advances, 4*(25), 12723–12728. doi:10.1039/c3ra44208j.

Muthukrishnan, K., Vanaraja, M., Boomadevi, S., Karn, R. K., Singh, V., Singh, P. K., & Pandiyan, K. (2016). Studies on acetone sensing characteristics of ZnO thin film prepared by sol–gel dip coating. *Journal of Alloys and Compounds, 673*, 138–143. doi:10.1016/j.jallcom.2016.02.222.

Nakate, U. T., Ahmad, R., Patil, P., Yu, Y. T., & Hahn, Y. B. (2020). Ultra thin NiO nanosheets for high performance hydrogen gas sensor device. *Applied Surface Science, 506*. http://www.journals.elsevier.com/applied-surface-science/. doi:10.1016/j.apsusc.2019.144971.

Navale, S. T., Yang, Z. B., Liu, C., Cao, P. J., Patil, V. B., Ramgir, N. S., Mane, R. S., & Stadler, F. J. (2018). Enhanced acetone sensing properties of titanium dioxide nanoparticles with a sub-ppm detection limit. *Sensors and Actuators B: Chemical, 255*, 1701–1710. doi:10.1016/j.snb.2017.08.186.

Neto, J. D. R. M., Santos, W. D. J. R., Lima, P. R., Tanaka, S. M. C. N., Tanaka, A. A., & Kubota, L. T. (2011). A hemin-based molecularly imprinted polymer (MIP) grafted onto a glassy carbon electrode as a selective sensor for 4-aminophenol amperometric. *Sensors and Actuators, B: Chemical, 152*(2), 220–225. doi:10.1016/j.snb.2010.12.010.

Ning, X., Yang, J., Zhao, C. L., & Chan, C. C. (2016). PDMS-coated fiber volatile organic compounds sensors. *Applied Optics, 55*(13), 3543–3548. https://www.osapublishing.org/view_article.cfm?gotourl=https%3A%2F%2Fwww%2Eosapublishing%2Eorg%2FDirectPDFAccess%2F55AE38B5-A6D3-0469-F331178E592EB803_340243%2Fao-55-13-3543%2Epdf%3Fda%3D1%26id%3D340243%26seq%3D0%26mobile%3Dno&org=Elsevier%20Inc. doi:10.1364/AO.55.003543.

Ning, X., Zhao, C. L., Yang, J., & Chan, C. C. (2016). Zeolite thin film-coated spherical end-face fiber sensors for detection of trace organic vapors. *Optics Communications, 364*, 55–59. doi:10.1016/j.optcom.2015.11.023.

Oh, K.-J., Park, D.-W., Kim, S.-S., & Park, S.-W. (2010). Breakthrough data analysis of adsorption of volatile organic compounds on granular activated carbon. *Korean Journal of Chemical Engineering, 27*(2), 632–638. doi:10.1007/s11814-010-0079-9.

Ozansoy Kasap, B., Marchenko, S. V., Soldatkin, O. O., Dzyadevych, S. V., & Kurc, B. A. (2017). Biosensors based on nano-gold/zeolite-modified ion selective field-effect transistors for creatinine detection. *Nanoscale Research Letters, 12*(1). http://www.springer.com/materials/nanotechnology/journal/11671. doi:10.1186/s11671-017-1943-x.

Pargoletti, E., Hossain, U. H., Di Bernardo, I., Chen, H., Tran-Phu, T., Chiarello, G. L., Lipton-Duffin, J., Pifferi, V., Tricoli, A., & Cappelletti, G. (2020). Engineering of SnO_2-graphene oxide nanoheterojunctions for selective room-temperature chemical sensing and optoelectronic devices. *ACS Applied Materials and Interfaces, 12*(35), 39549–39560. http://pubs.acs.org/journal/aamick. doi:10.1021/acsami.0c09178.

Park, Y., Yoo, R., Park, S. r., Lee, J. H., Jung, H., Lee, H. S., & Lee, W. (2019). Highly sensitive and selective isoprene sensing performance of ZnO quantum dots for a breath analyzer. *Sensors and Actuators, B: Chemical, 290*, 258–266. https://www.journals.elsevier.com/sensors-and-actuators-b-chemical. doi:10.1016/j.snb.2019.03.118.

Pathak, A. K., Bhardwaj, V., Gangwar, R. K., De, M., & Singh, V. K. (2017). Fabrication and characterization of TiO_2 coated cone shaped nano-fiber pH sensor. *Optics Communications, 386*, 43–48. doi:10.1016/j.optcom.2016.11.021.

Pathak, A. K., Chaudhary, D. K., & Singh, V. K. (2018). Broad range and highly sensitive optical pH sensor based on Hierarchical ZnO microflowers over tapered silica fiber. *Sensors and Actuators, A: Physical, 280*, 399–405. doi:10.1016/j.sna.2018.08.013.

Pathak, A. K., Limprapassorn, P., Kongruttanachok, N., & Viphavakit, C. (2022). Molecularly imprinted polymer-based optical sensor for isopropanol vapor. *Journal of Sensor and Actuator Networks, 11*(3), 46. doi:10.3390/jsan11030046.

Pathak, A. K., & Singh, V. K. (2017). A wide range and highly sensitive optical fiber pH sensor using polyacrylamide hydrogel. *Optical Fiber Technology, 39*, 43–48. http://www.journals.elsevier.com/optical-fiber-technology/. doi:10.1016/j.yofte.2017.09.022.

Pathak, A. K., Swargiary, K., Kongsawang, N., Jitpratak, P., Ajchareeyasoontorn, N., Udomkittivorakul, J., & Viphavakit, C. (2023). Recent advances in sensing materials targeting clinical volatile organic compound (VOC) biomarkers: A review. *Biosensors, 13*(1), 114. doi:10.3390/bios13010114.

Pathak, A. K., & Viphavakit, C. (2021). VOC biomarker monitoring for diabetes through exhaled breath using Ag/P-TiO_2 composite plasmonic sensor. *IEEE Sensors Journal, 21*(20), 22631–22637. http://ieeexplore.ieee.org/xpl/RecentIssue.jsp?punumber=7361. doi:10.1109/JSEN.2021.3104766.

Pathak, A. K., & Viphavakit, C. (2022). A review on all-optical fiber-based VOC sensors: Heading towards the development of promising technology. *Sensors and Actuators A: Physical, 338*, 113455. doi:10.1016/j.sna.2022.113455.

Pathak, A. K., Viphavakit, C., Rahman, B. M. A., & Singh, V. (2021). A highly sensitive SPR refractive index sensor based on microfluidic channel assisted with graphene-Ag composite nanowire. *IEEE Photonics Journal, 13*(2), 1–8. doi:10.1109/jphot.2021.3069396.

Patil, N. B., Nimbalkar, A. R., & Patil, M. G. (2018). ZnO thin film prepared by a sol-gel spin coating technique for NO_2 detection. *Materials Science and Engineering: B, 227*, 53–60. https://www.journals.elsevier.com/materials-science-and-engineering-b. doi:10.1016/j.mseb.2017.10.011.

Patil, Pramila, et al. (2021). 2-D NiO nanostructured material for high response acetaldehyde sensing application. *Materials Letters, 293*, 129757. doi:10.1016/j.matlet.2021.129757.

Peng, C., Guo, J., Yang, W., Shi, C., Liu, M., Zheng, Y., Xu, J., Chen, P., Huang, T., & Yang, Y. (2016). Synthesis of three-dimensional flower-like hierarchical ZnO nanostructure and its enhanced acetone gas sensing properties. *Journal of Alloys and Compounds, 654*, 371–378. doi:10.1016/j.jallcom.2015.09.120.

Peng, H., Liu, Y., Shen, Y., Xu, L., Lu, J., Li, M., Lu, H. L., & Gao, L. (2024). Highly sensitive and selective toluene gas sensors based on ZnO nanoflowers decorated with bimetallic AuPt. *Molecules, 29*(7). http://www.mdpi.com/journal/molecules. doi:10.3390/molecules29071657.

Pina, C. D., De Gregorio, M. A., Clerici, L., Dellavedova, P., & Falletta, E. (2018). Polyaniline (PANI): An innovative support for sampling and removal of VOCs in air matrices. *Journal of Hazardous Materials, 344*, 308–315. www.elsevier.com/locate/jhazmat. doi:10.1016/j.jhazmat.2017.10.012.

Pozos, H. G., Krishna, K. T. V., de la Luz Olvera Amador, M., Kudriavtsev, Y., & Alvarez, A. M. (2018). TiO_2 thin film based gas sensors for CO-detection. *Journal of Materials Science: Materials in Electronics, 29*(18), 15829–15837. https://rd.springer.com/journal/10854. doi:10.1007/s10854-018-9477-2.

Prabakar, S., Bumby, C. W., & Tilley, R. D. (2009). Liquid-phase synthesis of flower-like and flake-like titanium disulfide nanostructures. *Chemistry of Materials, 21*(8), 1725–1730. http://pubs.acs.org/doi/pdfplus/10.1021/cm900110h. doi:10.1021/cm900110h. New Zealand.

Qu, F., Zhou, X., Zhang, B., Zhang, S., Jiang, C., Ruan, S., & Yang, M. (2019). Fe_2O_3 nanoparticles-decorated MoO_3 nanobelts for enhanced chemiresistive gas sensing. *Journal of Alloys and Compounds, 782*, 672–678. https://www.journals.elsevier.com/journal-of-alloys-and-compounds. doi:10.1016/j.jallcom.2018.12.258.

Raj, M. K., & Chakraborty, S. (2020). PDMS microfluidics: A mini review. *Journal of Applied Polymer Science, 137*(27). http://onlinelibrary.wiley.com/journal/10.1002/(ISSN)1097-4628. doi:10.1002/app.48958.

Rakić, V., Rac, V., Krmar, M., Otman, O., & Auroux, A. (2015). The adsorption of pharmaceutically active compounds from aqueous solutions onto activated carbons. *Journal of Hazardous Materials, 282*, 141–149. www.elsevier.com/locate/jhazmat. doi:10.1016/j.jhazmat.2014.04.062.

Ramanavicius, S., Jagminas, A., & Ramanavicius, A. (2021). Advances in molecularly imprinted polymers based affinity sensors (review). *Polymers., 13*(6), 974. doi:10.3390/polym13060974.

Reddy, K., Guo, Y., Liu, J., Lee, W., Khaing Oo, M. K., & Fan, X. (2011). On-chip Fabry-Pérot interferometric sensors for micro-gas chromatography detection. *Sensors and Actuators, B: Chemical, 159*(1), 60–65. doi:10.1016/j.snb.2011.06.041.

Reddy, K., Guo, Y., Liu, J., Lee, W., Khaing Oo, M. K., & Fan, X. (2012). Rapid, sensitive, and multiplexed on-chip optical sensors for micro-gas chromatography. *Lab on a Chip, 12*(5), 901–905. http://pubs.rsc.org/en/journals/journal/lc. doi:10.1039/c2lc20922e.

Reghunath, S., Pinheiro, D., & KR, S. D. (2021). A review of hierarchical nanostructures of TiO_2: Advances and applications. *Applied Surface Science Advances, 3*, 100063. doi:10.1016/j.apsadv.2021.100063.

Renganathan, B., Sastikumar, D., Gobi, G., Rajeswari Yogamalar, N., & Bose, A. C. (2011). Nanocrystalline ZnO coated fiber optic sensor for ammonia gas detection. *Optics & Laser Technology, 43*(8), 1398–1404. doi:10.1016/j.optlastec.2011.04.008.

Rho, W. Y., Jeon, H., Kim, H. S., Chung, W. J., Suh, J. S., & Jun, B. H. (2015). Recent progress in dye-sensitized solar cells for improving efficiency: TiO_2 nanotube arrays in active layer. *Journal of Nanomaterials, 2015*. http://www.hindawi.com/journals/jnm/. doi:10.1155/2015/247689.

Rong, Q., Zhang, Y., Wang, C., Zhu, Z., Zhang, J., & Liu, Q. (2017). A high selective methanol gas sensor based on molecular imprinted Ag-$LaFeO_3$ fibers. *Scientific Reports, 7*(1). www.nature.com/srep/index.html. doi:10.1038/s41598-017-12337-z.

Roy, N., Sinha, R., Daniel, T. T., Nemade, H. B., & Mandal, T. K. (2020). Highly sensitive room temperature CO gas sensor based on MWCNT-PDDA composite. *IEEE Sensors Journal, 20*(22), 13245–13252. http://ieeexplore.ieee.org/xpl/RecentIssue.jsp?punumber=7361. doi:10.1109/JSEN.2020.3004994.

Ruoff, R. S., Qian, D., & Liu, W. K. (2003). Mechanical properties of carbon nanotubes: Theoretical predictions and experimental measurements. *Comptes Rendus Physique, 4*(9), 993–1008. http://www.elsevier.com. doi:10.1016/j.crhy.2003.08.001.

Saeed, M., Marwani, H. M., Shahzad, U., Asiri, A. M., Rahman, M. M., et al. (2024). Recent advances, challenges, and future perspectives of ZnO nanostructure materials towards energy applications. *Chemical Record, 24*(1). http://onlinelibrary.wiley.com/journal/10.1002/(ISSN)1528-0691. doi:10.1002/tcr.202300106.

Saha, D., Wei, Z., & Deng, S. (2009). Hydrogen adsorption equilibrium and kinetics in metal-organic framework (MOF-5) synthesized with DEF approach. *Separation and Purification Technology, 64*(3), 280–287. doi:10.1016/j.seppur.2008.10.022.

Santos, Otavio Augusto L. dos, et al. (2021). Review-Perovskite/Spinel Based Graphene Derivatives Electrochemical and Biosensors. *Journal of The Electrochemical Society, 168*, 067506. doi:10.1149/1945-7111/ac0306.

Sappati, K. K., & Bhadra, S. (2021). Printed acoustic sensor for low concentration volatile organic compound monitoring. *IEEE Sensors Journal, 21*(8), 9808–9818. http://ieeexplore.ieee.org/xpl/RecentIssue.jsp?punumber=7361. doi:10.1109/JSEN.2021.3056002.

Sen, S., & Kundu, S. (2021). Reduced graphene oxide (rGO) decorated ZnO-SnO_2: A ternary nanocomposite towards improved low concentration VOC sensing performance. *Journal of Alloys and Compounds, 881*, 160406. doi:10.1016/j.jallcom.2021.160406.

Sennik, E., Kilinc, N., & Ozturk, Z. Z. (2014). Electrical and VOC sensing properties of anatase and rutile TiO_2 nanotubes. *Journal of Alloys and Compounds, 616*, 89–96. doi:10.1016/j.jallcom.2014.07.097.

Serrano, D. P., Calleja, G., Botas, J. A., & Gutierrez, F. J. (2004). Adsorption and hydrophobic properties of mesostructured MCM-41 and SBA-15 materials for volatile organic compound removal. *Industrial and Engineering Chemistry Research, 43*(22), 7010–7018. http://pubs.acs.org/journal/iecred. doi:10.1021/ie040108d.

Šetka, M., Bahos, F. A., Chmela, O., Matatagui, D., Gràcia, I., Drbohlavová, J., & Vallejos, S. (2020). Cadmium telluride/polypyrrole nanocomposite based love wave sensors highly sensitive to acetone at room temperature. *Sensors and Actuators B: Chemical, 321*, 128573. doi:10.1016/j.snb.2020.128573.

Shafi, P. M., Dhanabal, R., Chithambararaj, A., Velmathi, S., & Bose, A. C. (2017). α-MnO_2/h-MoO_3 hybrid material for high performance supercapacitor electrode and photocatalyst. *ACS Sustainable Chemistry and Engineering, 5*(6), 4757–4770. http://pubs.acs.org/journal/ascecg. doi:10.1021/acssuschemeng.7b00143.

Shahzad, Faisal, et al. (2020). 2D Transition Metal Carbides (MXenes): Applications as an Electrically Conducting Material. *Advanced Materials, 32*, 2002159. doi:10.1002/adma.202002159.

Shen, J., Guo, S., Chen, C., Sun, L., Wen, S., Chen, Y., & Ruan, S. (2017). Synthesis of Ni-doped A-MoO_3 nanolamella and their improved gas sensing properties. *Sensors and Actuators, B: Chemical, 252*, 757–763. doi:10.1016/j.snb.2017.06.040.

Shen, S., Zhang, X., Cheng, X., Xu, Y., Gao, S., Zhao, H., Zhou, X., & Huo, L. (2019). Oxygen-vacancy-enriched porous α-MoO_3 nanosheets for trimethylamine sensing. *ACS Applied Nano Materials, 2*(12), 8016–8026. https://pubs.acs.org/journal/aanmf6. doi:10.1021/acsanm.9b02072.

Shen, Y., & Zhang, N. (2019). Facile synthesis of porous carbons from silica-rich rice husk char for volatile organic compounds (VOCs) sorption. *Bioresource Technology, 282*, 294–300. www.elsevier.com/locate/biortech. doi:10.1016/j.biortech.2019.03.025.

Shoeb, M., Mobin, M., Ahmad, S., & Naqvi, A. H. (2021). Facile synthesis of polypyrrole coated graphene Gr/Ag–Ag_2O/PPy nanocomposites for a rapid and selective response towards ammonia sensing at room temperature. *Journal of Science: Advanced Materials and Devices, 6*(2), 223–233. www.journals.elsevier.com/journal-of-science-advanced-materials-and-devices. doi:10.1016/j.jsamd.2021.02.003.

Sidek, O., & Bin Afzal, M. H. (2011). A review paper on fiber-optic sensors and application of PDMS materials for enhanced performance. *2011 IEEE Symposium on Business, Engineering and Industrial Applications (ISBEIA)* (pp. 458–463). IEEE. doi:10.1109/ISBEIA.2011.6088858.

Singh, S., Sharma, S., Singh, R. C., & Sharma, S. (2020). Hydrothermally synthesized MoS_2-multi-walled carbon nanotube composite as a novel room-temperature ammonia sensing platform. *Applied Surface Science, 532*, 147373. doi:10.1016/j.apsusc.2020.147373.

Siva Prasad, M., Chen, R., Ni, H., & Kiran Kumar, K. (2020). Directly grown of 3D-nickel oxide nano flowers on TiO_2 nanowire arrays by hydrothermal route for electrochemical determination of naringenin flavonoid in vegetable samples. *Arabian Journal of Chemistry, 13*(1), 1520–1531. doi:10.1016/j.arabjc.2017.12.004.

Snook, G. A., Kao, P., & Best, A. S. (2011). Conducting-polymer-based supercapacitor devices and electrodes. *Journal of Power Sources, 196*(1), 1–12. doi:10.1016/j.jpowsour.2010.06.084.

Sonker, R. K., Hitkari, G., Sabhajeet, S. R., Sikarwar, S., & Rahul, S. S. (2020). Green synthesis of TiO_2 nanosheet by chemical method for the removal of Rhodamin B from industrial waste. *Materials Science and Engineering: B, 258*, 114577. doi:10.1016/j.mseb.2020.114577.

Soudagar, M. E. M., Marghade, D., Shelare, S., Karunanidhi, D., Prakash, C., Khan, T. M. Y., & Cao, W. (2025). Metal-organic framework nanocomposites: Engineering and morphological advances for photocatalytic CO_2 conversion into fuel. *Applied Energy, 379*. https://www.sciencedirect.com/science/journal/03062619. doi:10.1016/j.apenergy.2024.124977.

St-Gelais, R., MacKey, G., Saunders, J., Zhou, J., Leblanc-Hotte, A., Poulin, A., Barnes, J. A., Loock, H. P., Brown, R. S., & Peter, Y. A. (2013). Gas sensing using polymer-functionalized deformable Fabry-Perot interferometers. *Sensors and Actuators, B: Chemical, 182*, 45–52. doi:10.1016/j.snb.2013.02.016.

Sugahara, T., Alipour, L., Hirose, Y., Ekubaru, Y., Nakamura, J. I., Ono, H., Harada, N., & Suganuma, K. (2020). Formation of metal-organic decomposition derived nanocrystalline structure titanium dioxide by heat sintering and photosintering methods for advanced coating process, and its volatile organic compounds gas-sensing properties. *ACS Applied Electronic Materials, 2*(6), 1670–1678. pubs.acs.org/journal/aaembp. doi:10.1021/acsaelm.0c00237.

Sun, Y., Li, S., Sun, R., Yang, S., & Liu, X. (2020). Modified dubinin-astakhov model for the accurate estimation of supercritical methane sorption on shales. *ACS Omega, 5*(26), 16189–16199. pubs.acs.org/journal/acsodf. doi:10.1021/acsomega.0c01675.

Sun, Y., Wang, J., Du, H., Li, X., Wang, C., & Hou, T. (2021). Formaldehyde gas sensors based on SnO_2/ZSM-5 zeolite composite nanofibers. *Journal of Alloys and Compounds, 868*, 159140. doi:10.1016/j.jallcom.2021.159140.

Swargiary, K., Jolivot, R., & Mohammed, W. S. (2022). Demonstration of a polymer-based single step waveguide by 3D printing digital light processing technology for isopropanol alcohol-concentration sensor. *Photonic Sensors, 12*(1), 10–22. http://www.springer.com/physics/optics+%26+lasers/journal/13320. doi:10.1007/s13320-021-0626-5.

Swargiary, K., Metem, P., Kulatumyotin, C., Thaneerat, S., Ajchareeyasoontorn, N., Jitpratak, P., Bora, T., Mohammed, W. S., Dutta, J., & Viphavakit, C. (2022). ZnO nanorods coated single-mode–multimode–single-mode optical fiber sensor for VOC biomarker detection. *Sensors, 22*(16). http://www.mdpi.com/journal/sensors. doi:10.3390/s22166273.

Szulczyński, B., & Gębicki, J. (2017). Currently commercially available chemical sensors employed for detection of volatile organic compounds in outdoor and indoor air. *Environments, 4*(1), 21. doi:10.3390/environments4010021.

Tang, N., Jiang, Y., Qu, H., & Duan, X. (2018). Graphene oxide-doped conducting polymer nanowires fabricated by soft lithography for gas sensing applications. *IEEE Sensors Journal, 18*(19), 7765–7771. http://ieeexplore.ieee.org/xpl/RecentIssue.jsp?punumber=7361. doi:10.1109/JSEN.2018.2833146.

Tang, X., Raskin, J.-P., Lahem, D., Krumpmann, A., Decroly, A. É., & Debliquy, M. (2017). A formaldehyde sensor based on molecularly-imprinted polymer on a TiO_2 nanotube array. *Sensors, 17*(4), 675. doi:10.3390/s17040675.

Tao, F., Liu, Y., Ren, X., Jiang, A., Wei, H., Zhai, X., Wang, F., Stock, H.-R., & Wen, S. (2021). Carbon nanotube-based nanomaterials for high-performance sodium-ion batteries: Recent advances and perspectives. *Journal of Alloys and Compounds, 873*, 159742. doi:10.1016/j.jallcom.2021.159742.

Tayfuroglu, O., Kocak, A., & Zorlu, Y. (2022). A neural network potential for the IRMOF series and its application for thermal and mechanical behaviors. *Physical Chemistry Chemical Physics, 24*(19), 11882–11897. http://pubs.rsc.org/en/journals/journal/cp. doi:10.1039/d1cp05973d.

Teimoori, F., Khojier, K., & Dehnavi, N. Z. (2017). Investigation of sensitivity and selectivity of ZnO thin film to volatile organic compounds. *Journal of Theoretical and Applied Physics, 11*(2), 157–163. doi:10.1007/s40094-017-0253-0.

Thomas, K. M. (2009). Adsorption and desorption of hydrogen on metal–organic framework materials for storage applications: Comparison with other nanoporous materials. *Dalton Transactions*, (9), 1487. doi:10.1039/b815583f.

Tian, X., Cui, X., Lai, T., Ren, J., Yang, Z., Xiao, M., Wang, B., Xiao, X., & Wang, Y. (2021). Gas sensors based on TiO_2 nanostructured materials for the detection of hazardous gases: A review. *Nano Materials Science, 3*(4), 390–403. www.keaipublishing.com/en/journals/nano-materials-science/. doi:10.1016/j.nanoms.2021.05.011.

Tian, X., Hu, Z., Wang, T., Wang, H., Zhang, Q., & Wei, X. (2023). Influence of multi-layer TiO_2/SnO_2 heterojunctions on fast and sensitive ethanol detection. *Vacuum, 207*, 111620. doi:10.1016/j.vacuum.2022.111620.

Tomić, M., Šetka, M., Vojkůvka, L. Š., & Vallejos, S. (2021). VOCs sensing by metal oxides, conductive polymers, and carbon-based materials. *Nanomaterials, 11*(2), 552. doi:10.3390/nano11020552.

Tong, X., Shen, W., Zhang, X., Corriou, J.-P., & Xi, H. (2020). Synthesis and density functional theory study of free-standing Fe-doped TiO_2 nanotube array film for H_2S gas sensing properties at low temperature. *Journal of Alloys and Compounds, 832*, 155015. doi:10.1016/j.jallcom.2020.155015.

Tricoli, A., Righettoni, M., & Teleki, A. (2010). Semico nductor gas sensors: Dry synthesis andapplication. *Angewandte Chemie – International Edition, 49*(42), 7632–7659. doi:10.1002/anie.200903801.

Tripathi, A., Mishra, S. K., Bahadur, I., & Shukla, R. K. (2015). Optical properties of regiorandom polythiophene/Al_2O_3 nanocomposites and their application to ammonia gas sensing. *Journal of Materials Science: Materials in Electronics, 26*(10), 7421–7430. doi:10.1007/s10854-015-3373-9.

Tung, T. T., Tran, M. T., Feller, J. F., Castro, M., Van Ngo, T., Hassan, K., Nine, M. J., & Losic, D. (2020). Graphene and metal organic frameworks (MOFs) hybridization for tunable chemoresistive sensors for detection of volatile organic compounds (VOCs) biomarkers. *Carbon, 159*, 333–344. http://www.journals.elsevier.com/carbon/. doi:10.1016/j.carbon.2019.12.010.

Twumasi, E., Forslund, M., Norberg, P., & Sjöström, C. (2012). Carbon-silica composites prepared by the precipitation method. Effect of the synthesis parameters on textural characteristics and toluene dynamic adsorption. *Journal of Porous Materials, 19*(3), 333–343. doi:10.1007/s10934-011-9479-4.

Uzun, L., & Turner, A. P. F. (2016). Molecularly-imprinted polymer sensors: Realising their potential. *Biosensors and Bioelectronics, 76*, 131–144. www.elsevier.com/locate/bios. doi:10.1016/j.bios.2015.07.013.

Vasapollo, G., Sole, R. D., Mergola, L., Lazzoi, M. R., Scardino, A., Scorrano, S., & Mele, G. (2011). Molecularly imprinted polymers: Present and future prospective. *International Journal of Molecular Sciences, 12*(9), 5908–5945. http://www.mdpi.com/1422-0067/12/9/5908/pdf. doi:10.3390/ijms12095908Italy.

Vessalli, B. A., Zito, C. A., Perfecto, T. M., Volanti, D. P., & Mazon, T. (2017). ZnO nanorods/graphene oxide sheets prepared by chemical bath deposition for volatile organic compounds detection. *Journal of Alloys and Compounds, 696*, 996–1003. doi:10.1016/j.jallcom.2016.12.075.

Victor, A., Ribeiro, J., & Araújo, F. F. (2019). Study of PDMS characterization and its applications in biomedicine: A review. *Journal of Mechanical Engineering and Biomechanics, 4*(1), 1–9. doi:10.24243/jmeb/4.1.163.

Viphavakit, C., Komodromos, M., Themistos, C., Mohammed, W. S., Kalli, K., & Azizur Rahman, B. M. (2015). Optimization of a horizontal slot waveguide biosensor to detect DNA hybridization. *Applied Optics, 54*(15), 4881. doi:10.1364/ao.54.004881.

Viveiros, R., Lopes, M. I., Heggie, W., & Casimiro, T. (2017). Green approach on the development of lock-and-key polymers for API purification. *Chemical Engineering Journal, 308*, 229–239. www.elsevier.com/inca/publications/store/6/0/1/2/7/3/index.htt. doi:10.1016/j.cej.2016.09.040.

Wang, B., Deng, L., Sun, L., Lei, Y., Wu, N., & Wang, Y. (2018a). Growth of TiO_2 nanostructures exposed {001} and {110} facets on SiC ultrafine fibers for enhanced gas sensing performance. *Sensors and Actuators, B: Chemical, 276*, 57–64. https://www.journals.elsevier.com/sensors-and-actuators-b-chemical. doi:10.1016/j.snb.2018.08.080.

Wang, C., Wang, Z. G., Xi, R., Zhang, L., Zhang, S. H., Wang, L. J., & Pan, G. B. (2019a). In situ synthesis of flower-like ZnO on GaN using electrodeposition and its application as ethanol gas sensor at room temperature. *Sensors and Actuators, B: Chemical, 292*, 270–276. https://www.journals.elsevier.com/sensors-and-actuators-b-chemical. doi:10.1016/j.snb.2019.04.140.

Wang, H., Luo, Y., Liu, B. O., Gao, L., & Duan, G. (2021). CuO nanoparticle loaded ZnO hierarchical heterostructure to boost H_2S sensing with fast recovery. *Sensors and Actuators B: Chemical, 338*, 129806. doi:10.1016/j.snb.2021.129806.

Wang, J., Yang, J., Han, N., Zhou, X., Gong, S., Yang, J., Hu, P., & Chen, Y. (2017a). Highly sensitive and selective ethanol and acetone gas sensors based on modified ZnO nanomaterials. *Materials and Design, 121*, 69–76. doi:10.1016/j.matdes.2017.02.048.

Wang, J., Zhou, Q., Peng, S., Xu, L., & Zeng, W. (2020a). Volatile organic compounds gas sensors based on molybdenum oxides: A mini review. *Frontiers in Chemistry, 8*. http://journal.frontiersin.org/journal/chemistry. doi:10.3389/fchem.2020.00339.

Wang, L., Wang, S., Fu, H., Wang, Y., & Yu, K. (2018b). Synthesis of Au nanoparticles functionalized 1D α-MoO_3 nanobelts and their gas sensing properties. *Nano, 13*(10), 1850115. doi:10.1142/s1793292018501151.

Wang, M., Zhu, Y., Meng, D., Wang, K., & Wang, C. (2020b). A novel room temperature ethanol gas sensor based on 3D hierarchical flower-like TiO_2 microstructures. *Materials Letters, 277*, 128372. doi:10.1016/j.matlet.2020.128372.

Wang, W., Liu, F., Wang, B., & Wang, Y. (2019b). Effect of defects in TiO_2 nanoplates with exposed {001} facets on the gas sensing properties. *Chinese Chemical Letters, 30*(6), 1261–1265. http://www.elsevier.com/wps/find/journaldescription.cws_home/997/description#description. doi:10.1016/j.cclet.2018.12.030.

Wang, Y., Ge, S., Zhang, L., Yu, J., Yan, M., & Huang, J. (2017b). Visible photoelectrochemical sensing platform by in situ generated CdS quantum dots decorated branched-TiO_2 nanorods equipped with Prussian blue electrochromic display. *Biosensors and Bioelectronics, 89*, 859–865. www.elsevier.com/locate/bios. doi:10.1016/j.bios.2016.09.106.

Wang, Z., Ali Haidry, A., Xie, L., Zavabeti, A., Li, Z., Yin, W., Lontio Fomekong, R., & Saruhan, B. (2020c). Acetone sensing applications of Ag modified TiO_2 porous nanoparticles synthesized via facile hydrothermal method. *Applied Surface Science, 533*, 147383. doi:10.1016/j.apsusc.2020.147383.

Wang, Z., Gao, S., Fei, T., Liu, S., & Zhang, T. (2019c). Construction of ZnO/SnO_2 heterostructure on reduced graphene oxide for enhanced nitrogen dioxide sensitive performances at room temperature. *ACS Sensors, 4*(8), 2048–2057. http://pubs.acs.org/journal/ascefj. doi:10.1021/acssensors.9b00648.

Wategaonkar, S. B., Pawar, R. P., Parale, V. G., Pakhare, K. S., Sargar, B. M., & Mane, R. K. (2020). Synthesis and characterizations of 3D TiO_2 nanoflowers thin film: Hydrothermal method. *Macromolecular Symposia, 393*(1). http://onlinelibrary.wiley.com/journal/10.1002/(ISSN)1521-3900. doi:10.1002/masy.202000040.

Weisskopf, L., Schulz, S., & Garbeva, P. (2021). Microbial volatile organic compounds in intra-kingdom and inter-kingdom interactions. *Nature Reviews Microbiology, 19*(6), 391–404. doi:10.1038/s41579-020-00508-1.

Wilson, R., Simion, C., Blackman, C., Carmalt, C., Stanoiu, A., Di Maggio, F., & Covington, J. (2018). The effect of film thickness on the gas sensing properties of ultrathin TiO_2 films deposited by atomic layer deposition. *Sensors, 18*(3), 735. doi:10.3390/s18030735.

Wu, B., Chen, L., Zhao, C., & Chan, C. C. (2019). Polypyrrole-coated polarization maintaining fiber-based vernier effect for isopropanol measurement. *Journal of Lightwave Technology, 37*(11), 2768–2772. https://ieeexplore.ieee.org/xpl/mostRecentIssue.jsp?punumber=50. doi:10.1109/JLT.2019.2906648.

Wu, J., Li, Z., Xie, X., Tao, K., Liu, C., Khor, K. A., Miao, J., & Norford, L. K. (2018). 3D superhydrophobic reduced graphene oxide for activated NO_2 sensing with enhanced immunity to humidity. *Royal Society of Chemistry, China Journal of Materials Chemistry A, 6*(2), 478–488. http://pubs.rsc.org/en/journals/journal/ta. doi:10.1039/c7ta08775f.

Wu, J., Wu, Z., Ding, H., Wei, Y., Huang, W., Yang, X., Li, Z., Qiu, L., & Wang, X. (2020). Flexible, 3D SnS_2/Reduced graphene oxide heterostructured NO_2 sensor. *Sensors and Actuators B: Chemical, 305*, 127445. doi:10.1016/j.snb.2019.127445.

Wu, Z., Zhang, H., Ji, H., Yuan, Z., & Meng, F. (2022). Novel combined waveform temperature modulation method of NiO-In_2O_3 based gas sensor for measuring and identifying VOC gases. *Journal of Alloys and Compounds, 918*, 165510. doi:10.1016/j.jallcom.2022.165510.

Wulff, G., Oberkobusch, D., & Minárik, M. (1985). Enzyme-analogue built polymers, 18 chiral cavities in polymer layers coated on wide-pore silica. *Reactive Polymers, Ion Exchangers, Sorbents, 3*(4), 261–275. doi:10.1016/0167-6989(85)90017-0.

Wulff, G., Sarhan, A., & Zabrocki, K. (1973). Enzyme-analogue built polymers and their use for the resolution of racemates. *Tetrahedron Letters, 14*(44), 4329–4332. doi:10.1016/s0040-4039(01)87213-0.

Xu, C., & Gao, D. (2012). Two-stage hydrothermal growth of long ZnO nanowires for efficient TiO_2 nanotube-based dye-sensitized solar cells. *Journal of Physical Chemistry C, 116*(12), 7236–7241. doi:10.1021/jp300960r.

Xu, Y., Lee, H., Hu, Y., Huang, J., Kim, S., & Yun, M. (2013). Detection and identification of breast cancer volatile organic compounds biomarkers using highly-sensitive single nanowire array on a chip. *Journal of Biomedical Nanotechnology, 9*(7), 1164–1172. http://docserver.ingentaconnect.com/deliver/connect/asp/15507033/v9n7/s5.pdf?expires=1374287998&id=74921835&titleid=11665&accname=Elsevier+BV&checksum=C847FB6E9B0F8EF2B6CA20420C25D07C. doi:10.1166/jbn.2013.1651. United States.

Xu, Y., Zheng, L., Yang, C., Zheng, W., Liu, X., & Zhang, J. (2020). Chemiresistive sensors based on core-shell ZnO@TiO_2 nanorods designed by atomic layer deposition for n-butanol detection. *Sensors and Actuators B: Chemical, 310*, 127846. doi:10.1016/j.snb.2020.127846.

Yaghi, O. M., Li, G., & Li, H. (1995). Selective binding and removal of guests in a microporous metal–organic framework. *Nature, 378*(6558), 703–706. doi:10.1038/378703a0.

Yan, L., Xiong, T., Zhang, Z., Yang, H., Zhang, X., He, Y., Bian, J., Lin, H., & Chen, D. (2022). Facile preparation of TPU conductive nanocomposites containing polypyrrole-coated multi-walled carbon nanotubes for a rapid and selective response in volatile organic compounds applications. *Composites Part A: Applied Science and Manufacturing, 157*, 106913. doi:10.1016/j.compositesa.2022.106913.

Yang, G., Li, L., Lee, W. B., & Ng, M. C. (2018). Structure of graphene and its disorders: A review. *Science and Technology of Advanced Materials, 19*(1), 613–648. http://www.tandfonline.com/loi/tsta20#.V0KNDU1f3cs. doi:10.1080/14686996.2018.1494493.

Yang, K., Xue, F., Sun, Q., Yue, R., & Lin, D. (2013). Adsorption of volatile organic compounds by metal-organic frameworks MOF-177. *Journal of Environmental Chemical Engineering, 1*(4), 713–718. doi:10.1016/j.jece.2013.07.005.

Yang, S., Liu, Y., Chen, T., Jin, W., Yang, T., Cao, M., Liu, S., Zhou, J., Zakharova, G. S., & Chen, W. (2017). Zn doped MoO_3 nanobelts and the enhanced gas sensing properties to ethanol. *Applied Surface Science, 393*, 377–384. http://www.journals.elsevier.com/applied-surface-science/. doi:10.1016/j.apsusc.2016.10.021.

Yang, S., Liu, Y., Chen, W., Jin, W., Zhou, J., Zhang, H., & Zakharova, G. S. (2016). High sensitivity and good selectivity of ultralong MoO_3 nanobelts for trimethylamine gas. *Sensors and Actuators, B: Chemical, 226*, 478–485. doi:10.1016/j.snb.2015.12.005.

Yang, Z., Zhang, D., & Wang, D. (2020). Carbon monoxide gas sensing properties of metal-organic frameworks-derived tin dioxide nanoparticles/molybdenum diselenide nanoflowers. *Sensors and Actuators B: Chemical, 304*, 127369. doi:10.1016/j.snb.2019.127369.

Yin, D., Hu, X., Cai, M., Wang, K., Peng, H., Bai, J., Xv, Y., Fu, T., Dong, X., Ni, J., & Yin, X. (2022a). Preparation, characterization, and in vitro release of curcumin-loaded IRMOF-10 nanoparticles and investigation of their pro-apoptotic effects on human hepatoma $HepG_2$ Cells. *Molecules, 27*(12), 3940. doi:10.3390/molecules27123940.

Yin, F., Yue, W., Li, Y., Gao, S., Zhang, C., Kan, H., Niu, H., Wang, W., & Guo, Y. (2021). Carbon-based nanomaterials for the detection of volatile organic compounds: A review. *Carbon, 180*, 274–297. http://www.journals.elsevier.com/carbon/. doi:10.1016/j.carbon.2021.04.080.

Yin, M., & Zhu, Z. (2019). Mesoporous NiO as an ultra-highly sensitive and selective gas sensor for sensing of trace ammonia at room temperature. *Journal of Alloys and Compounds, 789*, 941–947. https://www.journals.elsevier.com/journal-of-alloys-and-compounds. doi:10.1016/j.jallcom.2019.03.143.

Yin, T., Meng, X., Wang, S., Yao, X., Liu, N., & Shi, L. I. (2022b). Study on the adsorption of low-concentration VOCs on zeolite composites based on chemisorption of metal-oxides under dry and wet conditions. *Separation and Purification Technology, 280*, 119634. doi:10.1016/j.seppur.2021.119634.

Yu, Q. I., Gong, X., Jiang, Y., Zhao, L., Wang, T., Liu, F., Yan, X. U., Liang, X., Liu, F., Sun, P., & Lu, G. (2022). Bimetallic MOFs-derived core-shell structured mesoporous Sn-doped NiO for conductometric ppb-level xylene gas sensors. *Sensors and Actuators B: Chemical, 372*, 132620. doi:10.1016/j.snb.2022.132620.

Zhang, D., Sun, Y., Jiang, C., & Zhang, Y. (2017a). Room temperature hydrogen gas sensor based on palladium decorated tin oxide/molybdenum disulfide ternary hybrid via hydrothermal route. *Sensors and Actuators, B: Chemical, 242*, 15–24. doi:10.1016/j.snb.2016.11.005.

Zhang, D., Wu, Z., Li, P., Zong, X., Dong, G., & Zhang, Y. (2018). Facile fabrication of polyaniline/multi-walled carbon nanotubes/molybdenum disulfide ternary nanocomposite and its high-performance ammonia-sensing at room temperature. *Sensors and Actuators, B: Chemical, 258*, 895–905. doi:10.1016/j.snb.2017.11.168.

Zhang, X., Gao, B., Creamer, A. E., Cao, C., & Li, Y. (2017a). Adsorption of VOCs onto engineered carbon materials: A review. *Journal of Hazardous Materials, 338*, 102–123. www.elsevier.com/locate/jhazmat. doi:10.1016/j.jhazmat.2017.05.013.

Zhang, Y., Wang, M., San, X., Zhang, L., Wang, N., Wang, G., Meng, D., & Shen, Y. (2024). Highly selective gas sensors for formaldehyde detection based on ZnO@ZIF-8 core-shell heterostructures. *Sensors and Actuators B: Chemical, 398*, 134689. doi:10.1016/j.snb.2023.134689.

Zhang, Y., Zhang, J., & Liu, Q. (2017c). Gas sensors based on molecular imprinting technology. *Sensors., 17*(7), 1567. doi:10.3390/s17071567.

Zhao, C., Han, F., Li, Y., Mao, B., Kang, J., Shen, C., & Dong, X. (2019). Volatile organic compound sensor based on PDMS coated fabry-perot interferometer with vernier effect. *IEEE Sensors Journal, 19*(12), 4443–4450. http://ieeexplore.ieee.org/xpl/RecentIssue.jsp?punumber=7361. doi:10.1109/JSEN.2019.2900080.

Zhao, G., Xuan, J., Gong, Q., Wang, L., Ren, J., Sun, M., Jia, F., Yin, G., & Liu, B. (2020). In situ growing double-layer TiO_2 nanorod arrays on new-type FTO electrodes for low-concentration NH_3 detection at room temperature. *ACS Applied Materials and Interfaces., 12*(7), 8573–8582. http://pubs.acs.org/journal/aamick. doi:10.1021/acsami.9b20337.

Zhao, J., Wang, H., Cai, Y., Zhao, J., Gao, Z., & Song, Y. Y. (2024). The challenges and opportunities for TiO_2 nanostructures in gas sensing. *ACS Sensors, 9*(4), 1644–1655. http://pubs.acs.org/journal/ascefj. doi:10.1021/acssensors.4c00137.

Zhao, Y., Li, H., Li, Y., Ma, Y., Yang, H., Liu, H., Ren, X., & Zhao, H. (2021). Layered SnO_2 nanorods arrays anchored on reduced graphene oxide for ultra-high and ppb level formaldehyde sensing. *Sensors and Actuators B: Chemical, 346*, 130452. doi:10.1016/j.snb.2021.130452.

Zhou, Xihe, et al. (2023). Environmental and human health impacts of volatile organic compounds: A perspective review. *Chemosphere, 313*, 137489. doi:10.1016/j.chemosphere.2022.137489.

Zhu, L., Zeng, W., Ye, H., & Li, Y. (2018). Volatile organic compound sensing based on coral rock-like ZnO. *Materials Research Bulletin, 100*, 259–264. doi:10.1016/j.materresbull.2017.12.043.

Zito, C. A., Perfecto, T. M., Fonseca, C. S., & Volanti, D. P. (2018). Effective reduced graphene oxide sheets/hierarchical flower-like NiO composites for methanol sensing under high humidity. *New Journal of Chemistry, 42*(11), 8638–8645. http://pubs.rsc.org/en/journals/journal/nj. doi:10.1039/c8nj01061g.

CHAPTER

6

Interferometric photonic sensors for VOC detection: Advancements and applications

6.1 Introduction

The real-time monitoring of volatile organic compounds (VOCs) at extremely low concentrations requires a selective, sensitive, and reliable sensing approach. Several VOC sensing methods have been published in the past decade to detect VOC molecules, for example, electrochemical gas sensors (Sekhar & Subramaniyam, 2013), photo-ionization detectors (Spinelle et al., 2017), micro gas chromatography(Tzeng et al., 2016), quartz crystal microbalances (Sofa et al., 2024), surface acoustic wave sensors (Nuhoglu et al., 2024), metal-oxide semiconductor gas sensors (Mirzaei et al., 2018), gravimetric-based gas sensors (Fanget et al., 2011), and optical sensors (Khan et al., 2020). Each detection technique has its advantages and limitations based on the type of VOC molecule and the surrounding environment. In general, metal oxide semiconductor VOC sensors operate at high temperatures and show cross-sensitivity toward external humidity and other gases such as CO_2. On the other hand, the electrochemical sensors suffer from aging and zero drift. A photo-ionization detector shows poor selectivity and reduced sensitivity to some specific VOCs, leading to inaccurate measurements. Surface acoustic-based VOC sensors contain large noise and are cross-sensitive to ambient temperature and humidity.

Photonic Sensors for Detection of VOC Biomarkers: Technological Principles and Medical Applications
DOI: https://doi.org/10.1016/B978-0-443-29166-1.00004-4

Copyright © 2026 Elsevier Inc. All rights are reserved, including those for text and data mining, AI training, and similar technologies.

With the invention of optical fiber technology, optical sensors primarily focused on utilizing optical fibers to enhance sensing capabilities and miniaturize conventional sensors. The sensing devices based on optical fibers appear as the most promising tool in gas monitoring due to their small size, remarkable sensing performance, chemical inertness, and immunity toward electromagnetic interference, ability of multiplexing, and notable durability under harsh environments (Pathak & Viphavakit, 2022). There have been numerous efforts to improve the performance of optical fiber sensors, among which sensors coated with a thin sensing layer are considered the most common approach in VOC monitoring. In such devices, the change in the optical property of the coated sensing layer is used to assess the absorption of the target molecule on the sensing layer and then to deduce the concentration of gases once it is calibrated. The interferometry technique for gas detection has gained significant attention in sensing layer coated optical sensors due to its unique characteristics, including high sensitivity, non-destructive operation and contactless nature. The real-time detection capabilities and versatility in target-specific detection make them suitable for accurate and precise identification of target gases in mixed gaseous environments, making them reliable for use in exhaled breath monitoring.

In general, interferometry-based optical fiber sensors launch light into two arms of the interferometer, one serving as the sensing arm and the other as the reference arm. The launched light interacts with the surrounding environment through the sensing arm (Miliou, 2021; Pathak & Singh, 2017). Interferometry is a well-known investigation technique widely used for various sensing applications, including physical and chemical parameters (Ghosh & and Rahman, 2017). The principle behind this technique comprises the splitting of the coherent beam into two or more transmitted coherent beams and then recombining them. As the light propagates through different sections of the fiber, it undergoes constructive or destructive interference when it recouples with itself. Various detection modulations, such as intensity, phase, wavelength, and polarization, can be utilized to detect the target parameters. For VOC sensing, various interferometer configurations, such as Michelson, Fabry–Perot, Mach–Zehnder, and Sagnac interferometers, have been utilized to improve the detection accuracy and sensing performance. The arrangement of each type of interferometry is shown in Fig. 6.1.

There are several interferometric arrangements that have been reported and employed for VOC monitoring. This chapter aims to establish a detailed discussion over various interferometry techniques utilized in VOC sensing, along with their designs, operating principles, sensing performance, and their advantages and limitations. This chapter also discusses potential improvements and further opportunities for the miniaturization of such devices.

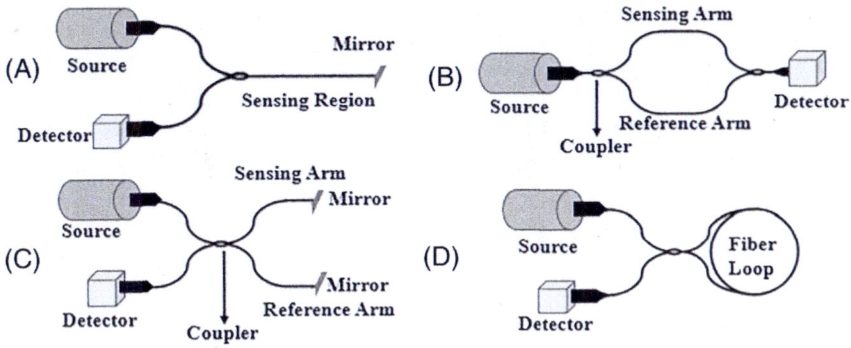

FIGURE 6.1 Various types of interferometers: The schematic diagram shows the basic configuration of (A) FPI, (B) MZI, (C) MI, and (D) SI. *From Pathak, A. K., & Viphavakit, C. (2022). A review on all-optical fiber-based VOC sensors: Heading towards the development of promising technology.* Sensors and Actuators A: Physical, 338. *https://doi.org/10.1016/j.sna.2022.113455.*

6.2 Principles of interferometry

The optical interferometry technique is a widely known investigative technique used for a broad range of applications for accurate detection. Recent developments in data acquisition, data interpretation, and computation have facilitated exceptionally reliable and sensitive measurements. The use of such digital computers makes the phase retrieval and fringe analysis process faster, and further improvement in the future is foreseen (Zuo et al., 2018). For example, interferometry has been utilized for the accurate pressure (Newport et al., 2008), temperature (Garvey et al., 2008), vibration (Aizen et al., 2018), gas concentration (Hübert et al., 2011; Pathak et al., 2023), bio-sensing (Makarona et al., 2016), etc.

In general, the interferometry technique utilizes a two-beam amplitude-division phenomenon in gas sensing applications (Khan et al., 2020). In the operation of two-beam amplitude-division interferometry, the basic working principle involves the recombination of two beams with identical frequency, a constant phase difference, and the same polarization, resulting in the formation of the interference pattern. The relationship can be expressed as follows:

$$I = I_1 + I_3 + 2\sqrt{I_1 I_3 \cos(\delta)} \tag{6.1}$$

$$\delta = 2yksin\theta - \Delta\Theta \tag{6.2}$$

where y denotes the fringe position, while k is the wavenumber of propagated light, δ is the phase difference, and Θ is the angle of interference between I_1 and I_3. Interferometry quantifies the phase shift due to the change

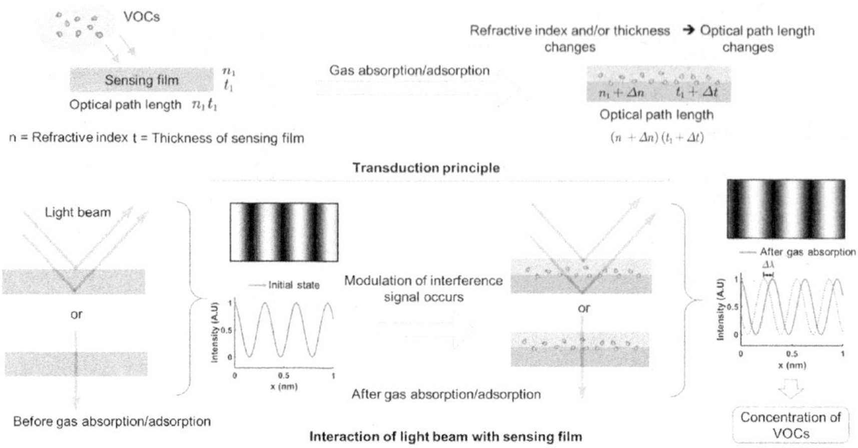

FIGURE 6.2 Schematic diagram of VOC detection: The schematic diagram shows the interferometry mechanism for gas detection. *From Khan, S., Le Calvé, S., & Newport, D. (2020). A review of optical interferometry techniques for VOC detection.* Sensors and Actuators A: Physical, 302. *https://doi.org/10.1016/j.sna.2019.111782.*

in the surrounding conditions, such as pressure, temperature, refractive index, and concentration.

Interferometry for VOC monitoring generally evaluates the changes in the sensing materials that take place under the exposure of VOC molecules. The sensing materials generally used in VOC sensing include micro-porous silicates, polymer, or hydrogel, which is selected based on the properties of the VOC molecule. When the VOC molecule interacts with the sensing material, it changes the refractive index or volume of the material due to the swelling and shrinkage phenomenon. Variations in the property of the material change the effective length, resulting in a quantifiable change in its phase, as shown in Fig. 6.2. The concentration level of VOCs can therefore be determined by monitoring and analyzing the induced phase shift.

Several interferometry-based optical configurations have been reported so far. The most common interferometry techniques involved in VOC detection are the FPI, MZI, SI, MI, and Pohl interferometers, as explained in the next sections.

6.3 Mach–Zehnder interferometer

Due to the flexible design, the Mach–Zehnder interferometer (MZI) based VOC sensors have gained huge interest (Lee et al., 2012). In the early stage, the MZIs require two independent arms, among which one works

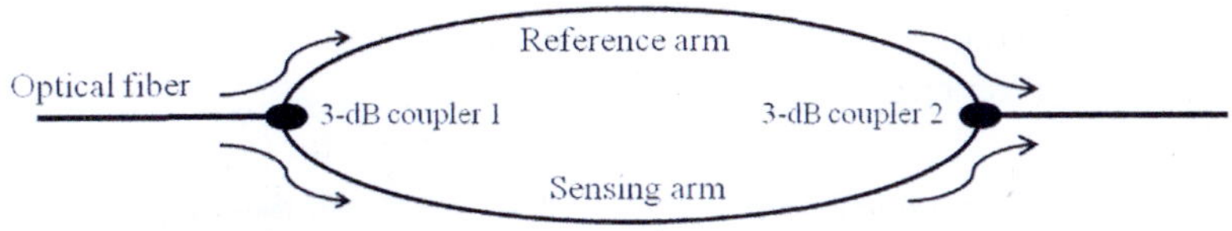

FIGURE 6.3 Mach–Zehnder interferometer: Schematic diagram showing the arrangement of MZI. *From Lee, B. H., Kim, Y. H., Park, K. S., Eom, J. B., Kim, M. J., Rho, B. S., & Choi, H. Y. (2012). Interferometric fiber optic sensors.* Sensors, 12(3), 2467–2486. *https://doi.org/10.3390/s120302467.*

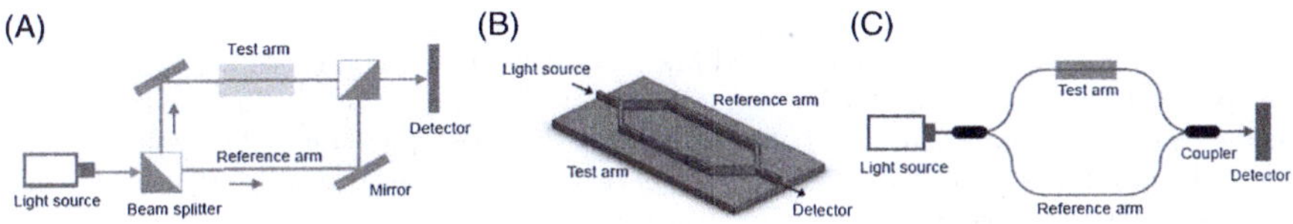

FIGURE 6.4 MZI actual sensing configurations: Schematic of Mach–Zehnder interferometer. (A) Open space interferometer, (B) chip-based MZ interferometer, and (C) fiber-based MZ interferometer. *From Khan, S., Le Calvé, S., & Newport, D. (2020). A review of optical interferometry techniques for VOC detection.* Sensors and Actuators A: Physical, 302. *https://doi.org/10.1016/j.sna.2019.111782.*

as the reference and the other as the sensing arm, as shown in Fig. 6.3. The input light is divided into two paths by a 3-dB fiber coupler and later recombined using another 3-dB coupler. The recoupled light contains the interference component due to the phase difference between the two arms. In general, the reference arm is kept separated from the measurand, while the sensing arm is allowed to be exposed to the target parameters. Then, the changes in the sensing arm due to other factors, i.e., strain, temperature, and RI, induce the changes in the phase difference of MZI, which can easily be detected by examining shifts in the interference pattern (Bhardwaj et al., 2020).

The traditional two-arm configuration in MZIs has been modified in several ways, such as adiabatically tapered fiber with reduced diameter, core-offset fiber, grating-based, non-adiabatic tapering, etc. (Bhardwaj et al., 2020). Three different types of MZIs have been reported so far, namely (1) open path, (2) planar waveguide, and (3) fiber-based, as shown in Fig. 6.4.

Drapp et al. (1999) reported a planar MZI that utilizes the phase transition and birefringence characteristics of the nematic liquid crystal (NLC) for VOC monitoring. This LC possesses a rod-shaped structure, leading to distinct dipole moments along and perpendicular to its molecular long axis. The aligned orientation of these molecules leads to the formation of a strongly birefringent LC structure. In reported work, the authors studied the variation in the ordered LC under the exposure to mixed VOC

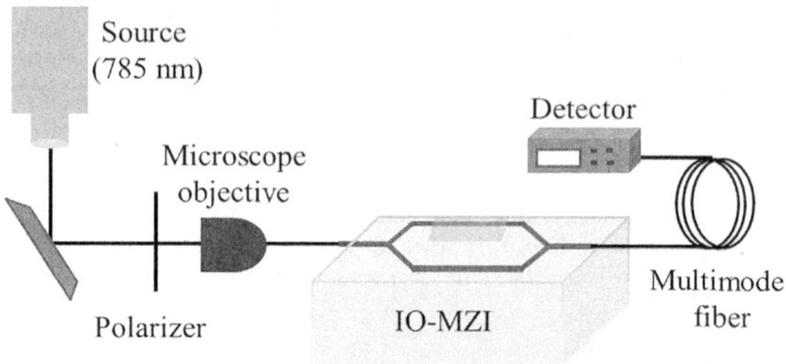

FIGURE 6.5 Liquid crystal embedded MZI: The schematic diagram of the experimental set-up used to characterize the MZI integrated chip with an embedded liquid crystalline layer. *From Drapp, B., Pauluth, D., Krause, J., & Gauglitz, G. (1999). The application of the phase transition in nematic liquid crystals for the optical detection of volatile organic compounds.* Fresenius' Journal of Analytical Chemistry, *364(1–2), 121–127. https://doi.org/10.1007/s002160051310.*

solvents. Here, two different transducer types were utilized in the device fabrication, namely, (1) an integrated optical MZI working as a miniature refractometric sensor, sensitive to the mixing of the LC with the VOC as well as changes in the ordering of the liquid crystal, and (2) an oriented LCs film between crossed polarizers. The NLC was also utilized as a sensing film over the integrated MZI. The schematic of the reported device is shown in Fig. 6.5. The calibration experiments performed with the NLC as a sensing layer over MZI exhibited a non-linear sensing performance to the various concentrations of *m*-xylene and *p*-xylene, varying from 400 to 2000 ppm.

Luo et al. (2003) reported an integrated photonic VOC sensor based on the single-mode MZI, as shown in Fig. 6.6. The sensor comprises a patterned waveguide configuration made of a lower silica (SiO_2) cladding, a phosphosilicate core, and an upper borophosphosilicate cladding. To operate the device, the input port of the planar waveguide was coupled with a white light source, while the output end of the waveguide was connected to an optical spectrum analyzer (OSA). Due to the presence of ethanol, the variation in the refractive index around the sensing arm showed good red-shifts in the output spectra along with a sensitivity of $\sim 4.0 \times 10^{-3}$, which is low as compared to the sensitivity recorded by other MZI-based VOC sensors. However, the first experimental demonstration of such a device with ethanol solutions solves other limitations of ambiguity and signal fading.

These advancements in miniaturizing technology offer a robust approach for developing optical waveguide VOC sensing configurations in the era of smart Internet of Things (IoT). Benefiting from the technology, Liao et al. (2019) reported a new type of fiber MZI fiber sensor for selective detection of ethanol concentration ranging from 0.3 to 0.7 v/v. The sensing

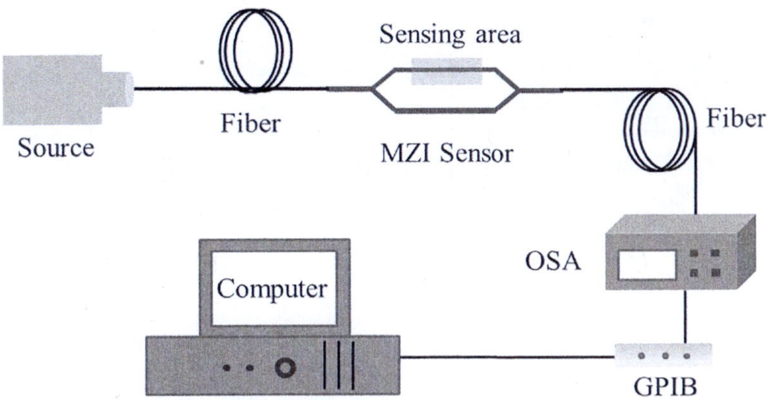

FIGURE 6.6 Patterned waveguide MZI: The experimental set-up comprises a patterned waveguide VOC sensor connected with a single-mode fiber. *From Luo, D. H., Levy, R. A., Hor, Y. F., Federici, J. F., & Pafchek, R. M. (2003). An integrated photonic sensor for in situ monitoring of hazardous organics.* Sensors and Actuators B: Chemical, 92(1–2), 121–126. *https://doi.org/10.1016/s0925-4005(03)00123-0.*

head comprises a tapered single-mode optical fiber along with a pair of air bubbles situated in the path of the launched light, as shown in Fig. 6.7. The device was fabricated in a three-step process, i.e., (1) initially, the center of the single-mode fiber (SMF) face was exposed to femtosecond laser pulses, using 1 mW of laser power, (2) the shutter of the laser was kept open for 1 second to destroy the end face of the optical fiber, hence appeared like a black dot, as depicted in Fig. 6.7A and (3) lastly, another ablated point appeared in the image under the end face due to the self-focusing effect. During the propagation, the laser pulse was visualized as comprising multiple thin power or intensity slices in time. If the peak power is higher compared to the critical power required for self-focusing, then according to Eq. 6.3, the central slice at the peak of the pulse will self-focus at a distance Zf from where it begins to propagate in the medium. The obtained result demonstrated a nonlinear response between the wavelength shift (1485 nm) and the ethanol concentrations. However, for the ethanol concentration ranging from 0.3 to 0.7, the sensor exhibited a linear sensitivity of 28 nm/vol.

$$Z_f = \frac{0.376Kr^2}{\left[\left(\sqrt{\frac{p}{p_c}} - 0.825\right)^2 - 0.0219\right]^{1/2}} \tag{6.3}$$

where, k, r, p, and p_c represent the wave number and radius of the beam profile at the 1/e level of intensity, peak power of the slice, and threshold power, respectively.

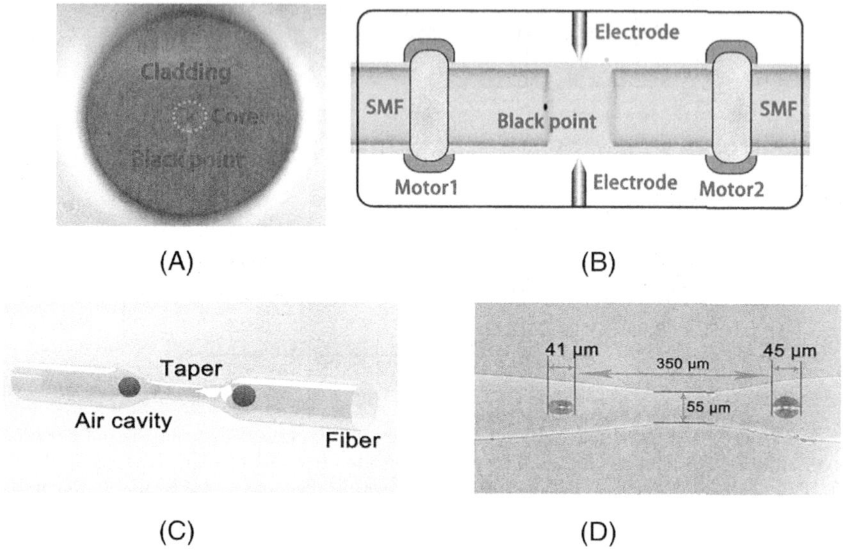

FIGURE 6.7 MZI fiber sensor embedded with air bubbles: (A) Fiber end face cleaved by utilizing femtosecond laser, (B) schematic of splicing of ablated SMF with a normal SMF, (C) diagram of fabricated Mach–Zehnder interferometer, and (D) optical microscope image of fabricated sensor. *From Liao, C., Zhu, F., Zhou, P., & Wang, Y. (2019). Fiber taper-based Mach–Zehnder interferometer for ethanol concentration measurement.* Micromachines, 10*(11). https://doi.org/10.3390/mi10110741.*

Ma et al. (2021) reported a 4 μm × 4 μm on-chip metal-organic framework (MOF) nanomaterial-coated planar waveguide for sensitive detection of ethanol. The self-assembled deposition of the ZIF-8 layer on top of the SIO_2 waveguide plays a crucial role in the reported work. The one self-assembly cycle required immersion of the waveguide in a mixed solution of imidazole and zinc nitrate for 30 minutes, followed by rinsing with methanol. Initially, the self-assembly process of ZIF-8 is performed on the SiO_2 substrate for 2, 4, and 8 cycles before proceeding to self-assembly on the waveguide. Based on surface characterization, it was observed that the 8 self-assembly cycles appear extremely dense compared to 4 cycles of ZIF-8. The fabricated device exhibited a low insertion loss of 13.3 dB and an ultra-high extinction ratio of 28.6 dB. The sensor also showed excellent response toward ethanol along with a high sensitivity (19 pm/ppm from 0 to 50 ppm or 41 pm/ppm from 600 to 1000 ppm), and low detection limit (1.6 ppb or 740 ppb) for concentration varying from 0 to 1000 ppm. Another self-assembled optical fiber MZI based on liquid crystals was reported by Li et al. (2020) for real-time detection of methanol, acetone, and ethanol. For the characterization of the sensor, a beam of light from a broadband source (BBS) was launched into the LC-based fiber MZI, and then the output beam was recorded on an OSA. The fabricated MZI comprises a sandwich arrangement of microtubules between two SMFs of core size 8.3 μm. After splicing, the ECG of 300 μm length, 50 μm width, and 87 μm depth was formed in a microtubule using a femtosecond laser (fs) with a central wavelength of 795 nm, which was later filled with LCs after a vertical polyimide coating. The fabricated device exhibited a good sensitivity of −83, −31, and −101.4 pm/ppm for methanol, acetone, and ethanol vapors, respectively. The sensor also exhibited good linear response and fast response times of 270, 180, and 390 seconds, for the mentioned gases, respectively.

Niu et al. (2022) reported an MZI-based integrated-photonic acetone sensor approaching the sub-ppm level detection limit. The sensor was fabricated on a silicon wafer with a size of 1 cm × 3 cm, following a 2 μm SiO_2 as buffer layer. The core of the waveguide was made of 250 nm thin Si_3N_4, while the cladding layer was considered to be 1 μm-thick SiO_2. A 50 μm wide and 18,000 μm open area was fabricated on the MZI sensing arm for exposure of VOC molecules, which was later filled with a composite film of amido-graphene oxide and polyethyleneimine. The fabricated device is illustrated in the schematic diagram shown in Fig. 6.8. As the light beam is launched, it splits into two arms, one of which propagates through the reference arm, while the other propagates through the sensing arm. The reference arm was covered with silicon dioxide; therefore, it was not exposed to the gas. The sensor arm was coated with a gas-sensitive layer. As the surrounding gaseous molecule interacts with the gas-sensitive film, it leads to a change in the refractive index of the material. Both beams from the reference arm and sensing arm were recoupled in the Y-coupler, resulting in an interference signal, which was recorded at the photo detector. The sensor exhibited a high sensitivity under 20 ppm of acetone, along with a low limit of detection (LOD) of 0.76 ppm. The performance of the device makes it feasible to be used for the clinical diagnosis of heart failure and diabetes.

Recently, Yang et al. reported an optical fiber MZI induced by multimodal interference to detect various concentrations of methane by utilizing ZIF-8/Polydimethylsiloxane (PDMS) as a sensing layer (Yang et al., 2025). The sensing configuration comprises splicing of a small length of ZIF-8/PDMS coated SMF (20–60 mm) between two small sections of multimode fiber (MMF), which was later sandwiched between two SMFs on both ends, as shown in Fig. 6.9. The variation in the methane concentration leads to a change in the cladding mode of the ZIF-8/PDMS deposited SMF, which induces a change in the interference spectrum, which was analyzed to detect the methane concentration in the surrounding environment. When the launched light propagates from SMF1 into the MMF1, the significant core mismatch between the two leads to the excitation of multiple propagation modes within the MMF1, each with a different propagation constant. Out of all the excited modes,

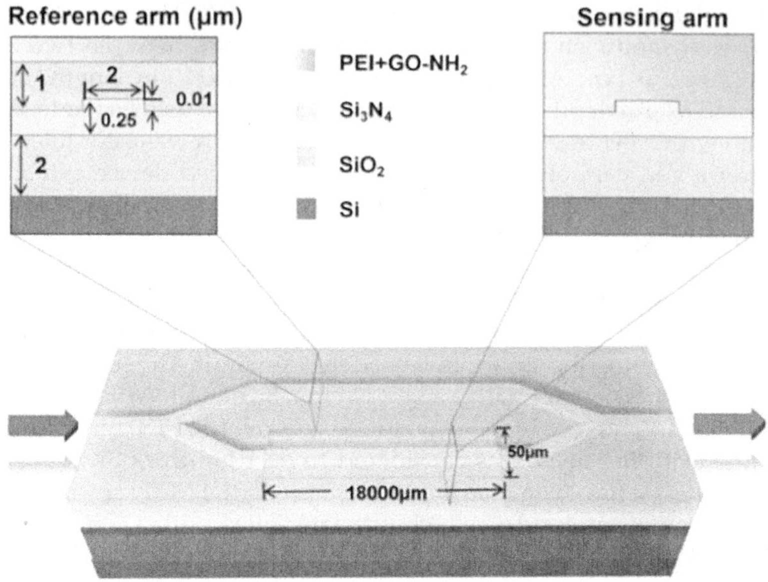

FIGURE 6.8 Integrated-photonic acetone sensor: The schematic 3D diagram of the MZI sensor along with the cross-sectional view of the sensing and the reference arm. *From Niu, H., Yu, P., Zhu, Y., Jing, Z., Li, P., Wang, B., Ma, C., Wang, J., Wu, J., Govorov, A. O., Neogi, A., & Wang, Z. M. (2022). Mach–Zehnder interferometer based integrated-photonic acetone sensor approaching the sub-ppm level detection limit.* Optics Express, *30(16), 29665–29679. https://doi.org/10.1364/OE.459805.*

mainly the higher-order modes are excited and show detectable phase differences. After MMF1, as the light propagates through the sensing head, these modes, along with the associated phase shifts, get coupled to the other MMF2, which was later detected on the OSA. The mode mismatch between fibers, along with the associated phase shift, give rise to a periodic interference pattern, the spacing of which is governed by the accumulated phase differences between the modes within the MS-SMF. The fabricated sensor was tested under different concentrations of methane ranging from 0% to 4%. The experiment exhibited an excellent sensitivity of 2.364 nm/% for the sensing length of 42 mm, along with a low LOD of 338 ppm. In addition to achieving good sensing performance, the sensor also demonstrates a recovery and response time of 45 and 30 seconds, respectively.

6.4 Fabry–Perot interferometer

The Fabry–Perot (FP) etalon or FP interferometer comprises two reflecting mirrors separated by a medium/cavity of length L. The FP

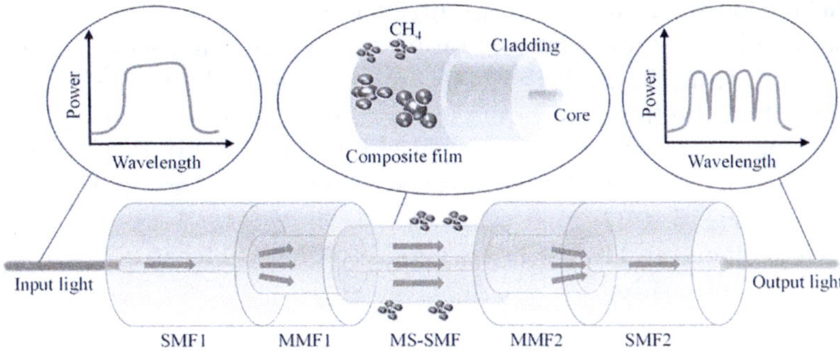

FIGURE 6.9 ZIF-8/PDMS coated MZI: Schematic diagram of methane sensor coated on ZIF-8/PDMS sensing film. *From Optical fiber methane sensor based on Mach–Zehnder interferometer induced by multimode interference. (2025).* Micromachines, 16(4), 406. *https://doi.org/10.3390/mi16040406.*

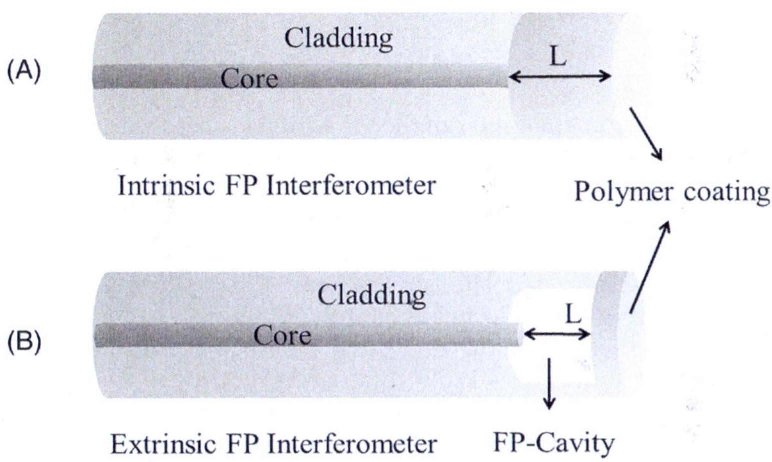

FIGURE 6.10 Schematic of FP interferometer: Basic mechanism of (A) intrinsic and (B) extrinsic-type optical fiber interferometers.

interferometer-based sensor works as a reflective sensing device using a two-beam interferometer (Khan et al., 2020). The FP based sensors are known to be highly sensitive to any perturbations that change the optical path length between the two mirrors. Due to its excellent sensitivity and fast response time, it has been utilized in various applications, including temperature, humidity, pressure, stress, and strain, bio-sensing, vibration, magnetic fields, gas detection, etc. (Huang et al., 2016). The FP interferometer can be classified into two types (1) intrinsic and (2) extrinsic, as shown in Fig. 6.10. In both types of FPI, a fiber is required to launch the light from the source to the interferometric device, and the reflected signal is collected on the detector. In an intrinsic type FP, the reflecting mirrors are isolated by the fiber, and the target measurand interacts with guided light propagating through the fiber, resulting in a change in the optical path length of the fiber, as shown in Fig. 6.10A. In the case of an extrinsic interferometer, both mirrors are isolated by some materials or an air gap, and the measurand introduces a change in the optical path in the medium, as shown in Fig. 6.10B.

The reflectance and transmittance for low-finesse FP can be given by the following relationship:

$$R_{FP} = \frac{R_1 + R_2 + 2\sqrt{R_1 R_2 \cos\Phi}}{1 + R_1 + R_2 + 2\sqrt{R_1 R_2 \cos\Phi}} \tag{6.4}$$

Where, R_{FP} represents the ratio of reflected power to the incident power produced by FPI, and ϕ represents the phase shift in the interferometer, which can be defined by the following relation:

$$\Phi = \frac{4\pi nL}{\lambda} \tag{6.5}$$

Here, L represents the cavity of the FP, while n is the refractive index of the cavity. The VOC concentrations can be detected by analyzing the shift in the reflective output spectrum.

Liu et al. (2009) reported a methanol sensor comprising a thin silver film and polyethylene glycol 400 and Norland optical adhesive 81 polymers as a vapor-sensitive layer deposited on the cleaved face of fiber. The interference produced from the reflected light at the silver-polymer and polymer-air interfaces varies when the deposited polymer is exposed to VOCs. The experiment shows drastically different responses to various VOCs, including methanol, hexanol, and acetone. The maximum sensitivity of the device for methanol was obtained at 3.5 and 0.1 pm/ppm for polyethylene glycol 400 and Norland optical adhesive 81, respectively, along with a low LOD of 1–10 ppm. St-Gelais et al. (2013) reported an FPI based on-chip VOCs sensor deposited with PDMS-polydiphenylsiloxane to detect cyclohexane and *m*-xylene, as shown in Fig. 6.11. The performance of the sensor was compared with PDMS and PDMS-polydiphenylsiloxane composite. The authors utilized silicon-air Bragg reflectors filled with PDMS/PDMS–polydiphenylsiloxane, serving both as the sensing layer and as the dielectric medium between the vertically aligned reflective surfaces of the SMFs. The device filled with PDMS–polydiphenylsiloxane composite exhibited 93% and 64% more sensitivity to cyclohexane and m-xylene, respectively, compared to the sensor containing PDMS only. The key limitation of the reported device lies in its sensing performance in mixed VOC environments due to the lack of selectivity. However, integrating a microfluidic system with the optical fiber enables the design of various interferometers

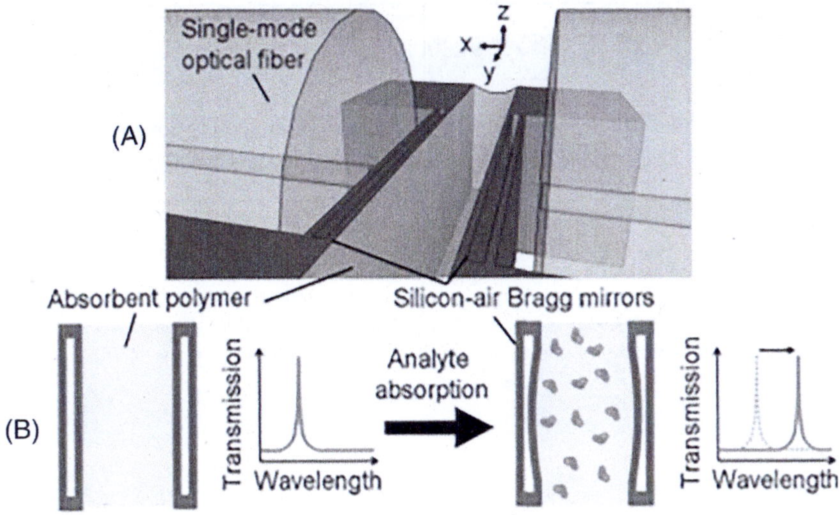

FIGURE 6.11 Deformable interferometric sensor: Schematic diagram of (A) deformable FPI configuration and (B) working mechanism of the sensor. *From St-Gelais, R., MacKey, G., Saunders, J., Zhou, J., Leblanc-Hotte, A., Poulin, A., Barnes, J. A., Loock, H. P., Brown, R. S., & Peter, Y. A. (2013). Gas sensing using polymer-functionalized deformable Fabry–Perot interferometers.* Sensors and Actuators, B: Chemical, 182, *45–52. https://doi.org/10.1016/j.snb.2013.02.016.*

using different types of sensing materials, polymers, and nanocomposites for target-specific detection.

Kanawade et al. (2019) reported a specialized negative axicon tip filled with PDMS to precisely detect multiple VOCs simultaneously. The sensing head was fabricated via a wet etching method, which generally requires concentrated hydrofluoric acid (HF) to form a negative axicon cavity at the tip of the SMF comprising refractive-reflective surfaces inside the cladding, as shown in Fig. 6.12A and C. The authors utilized a standard SMF comprising a germanium-doped core, while the cladding is made of fused silica. The configuration of optical fiber accelerated the etching rate of the core material in the presence of HF acid solution compared to silica cladding. The length of the negative axicon cavity is dependent on the etching time. The working mechanism of the device is based on two reflection steps, which occur from (1) fiber-PDMS interface (R_1) and (2) PDMS-air interface (R_2), as shown in Fig. 6.12B and D. The interference between R_1 and R_2 takes place due to the change in path difference between the beams reflecting from both surfaces. The fabricated sensor exhibited a maximum sensitivity of 3.2, 8.8, 17.8, 9.1, 21.6, and 23.7 pm/ppm in the presence of methanol, isopropanol, hexane, acetone, toluene, and chloroform gases, respectively, varying from 5 to 200 ppm. The major advantage

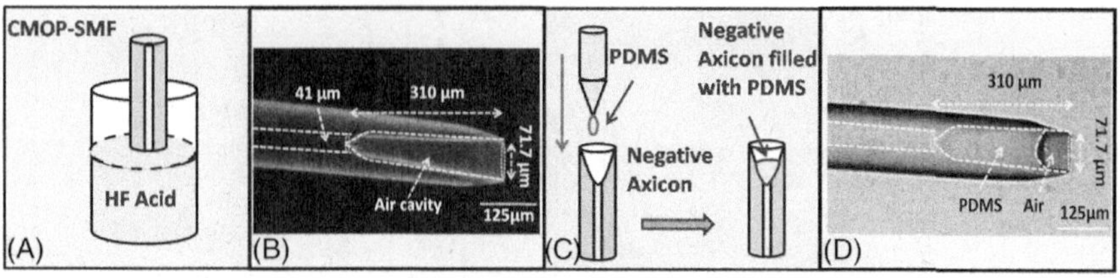

FIGURE 6.12 Negative axicon tip based FPI sensor: Schematic diagram shows the formation of negative axicon-tip (A) etching process, (B) SEM image of the fabricated axicon cavity, (C) PDMS filled cavity, and (D) sensing platform. *From Kanawade, R., Kumar, A., Pawar, D., Vairagi, K., Late, D., Sarkar, S., Sinha, R. K., & Mondal, S. (2019). Negative axicon tip-based fiber optic interferometer cavity sensor for volatile gas sensing.* Optics Express, *27(5), 7277–7290. https://doi.org/10.1364/OE.27.007277.*

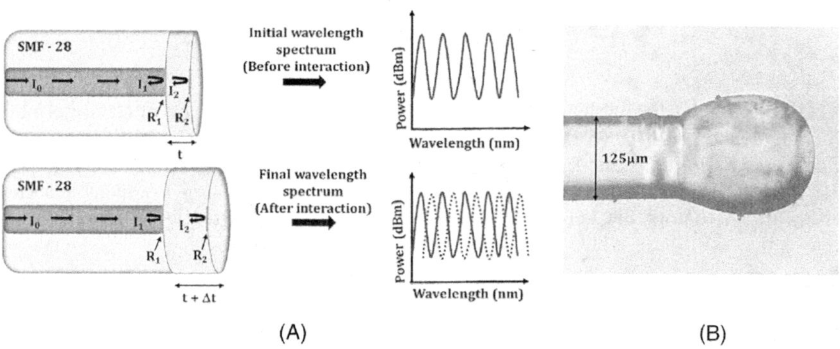

FIGURE 6.13 Polystyrene coated FP cavity: (A) the sensing mechanism of fabricated FPI and (B) the microscopic image of fabricated polystyrene polymer FP cavity. *From Kaushal, K., & Das, B. (2023). Polymer transduction cavity based optical fiber Fabry–Perot interferometer for VOC sensing.* Results in Optics, 12. *https://doi.org/10.1016/j.rio.2023.100479.*

of the axicon-tip sensor lies in its robustness and small size, allowing only the tip to be exposed to the ambient environment for the VOC detection.

Kaushal and Das (2023) reported a polystyrene-coated optical fiber FPI sensor to detect acetone. The sensing configuration was formed by using a thin layer of polystyrene polymer coated over the cleaved face of the SMF, as shown in Fig. 6.13. The deposition of polystyrene onto the fiber involves mixing of a separate solution of 25 wt% polystyrene with tetrahydrofuran and heating the mixture at 50°C for 1 minute until it turns into a viscous solution. The fibers were immersed in the solution, and multiple coatings were avoided to prevent a thicker layer. Subsequently, the fiber was placed in a desiccator for 24 hours to facilitate the polymer curing process at room temperature. The absorption of acetone on the polymer results in a change in the mechanical properties of the polymer in terms of its volume and

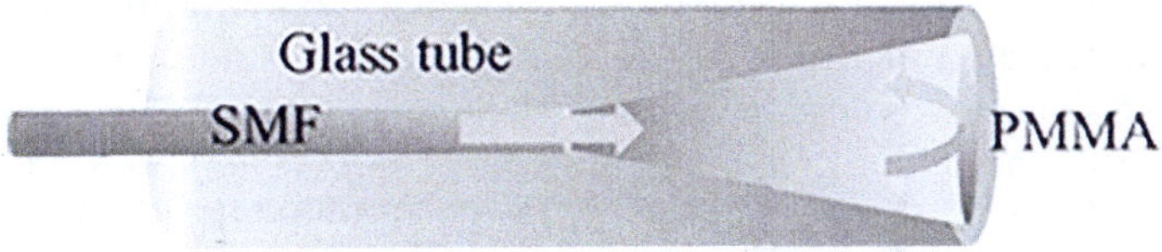

FIGURE 6.14 PMMA based FPI sensor: The schematic diagram of the FPI sensor containing glass tuber along with PMMA. *From Yu, C. B., Wu, Y., Li, C., Wu, F., Zhou, J. H., Gong, Y., Rao, Y. J., & Chen, Y. F. (2017). Highly sensitive and selective fiber-optic Fabry–Perot volatile organic compounds sensor based on a PMMA film.* Optical Materials Express, *7(6), 2111–2116. https://doi.org/10.1364/OME.7.002111.*

refractive indices, which significantly influences the shift in the interference pattern. The reported sensor exhibited an excellent sensitivity of 5.78 pm/ppm with good repeatability.

To further improve the sensitivity and target specificity of the sensor, Cai-bin et al. reported a novel configuration of SMF coated with polymethyl methacrylate (PMMA) for acetone and ethanol detection in 2017 (Yu et al., 2017). In general, PMMA reacts with several organic compounds, as most of them have a similar atomic structure of $-CH_2C-$. Therefore, the exposure of PMMA to such VOC gases leads to the swelling effect due to the penetration of gaseous molecules via the large molecular gap of the PMMA. The swelling of the PMMA significantly affects the length of the FP cavity, leading to a measurable wavelength shift in the reflected spectrum. The fabricated sensor exhibited a maximum sensitivity of 2.7 and 2.17 pm/ppm for ethanol and acetone, respectively, along with good selectivity. Fig. 6.14 illustrate the schematic representation of the sensing configuration. The reported FPI sensor can work as a key tool in VOC monitoring because of its unique characteristics in distinguishing among various VOCs and other harmful gases.

Wang et al. (2025) reported a 1040 nm thick ZIF-8 functionalized open-cavity FPI for VOC detection. The sensing performance of the device was investigated in the presence of multiple VOCs, including methanol, methylbenzene, and ethanol, and exhibited an excellent sensitivity of 0.177, 0.118, and 0.412 p.m./ppm, respectively. Sensitivity differences are analyzed and demonstrated at the molecular level. In the same year, Grantzioti et al. (2025) utilized a novel approach to fabricate an FPI configuration on the tip of SMF by forming a 32 μm cavity on the cleaved face of SMF by utilizing photopolymerizable resin. The sensing head was fabricated in two steps, in which the first step comprises the placement of UV photopolymerizable urethane-based, mercaptoester resin (NOA 65, Norland Products Inc.) between the gap of two aligned SMFs; and the second step requires irradiation of UV beam (370 nm, Intertronics, IUV101) through the fiber for 10 minutes to cure the resin. Following resin polymerization, one of the SMFs was carefully pulled and detached, allowing the resin cavity to remain attached at the end face of the SMF. The performance of the fabricated sensor was tested in the presence of a 200 ppb concentration of methanol, and a low LOD of 1 ppb was observed. Following the same approach, Leça et al. (2025) also utilized a UV light-induced microtip fabrication method to fabricate a microtip on the cleaved end of SMF to form an FPI configuration to detect various concentrations of 2-propanol. The fabrication of the microtip was performed by launching an ultraviolet light from one end of an optical fiber while the other end was inserted into a photosensitive polymer. When the light exits the fiber, the interaction of ultraviolet light with the polymer results in the formation of a microscopic tip structure at the end of the fiber. The optical fiber sensor with a 15 μm tip on the fiber demonstrated good sensitivity of -71.1 ± 2.1 pm/ppm and excellent selectivity toward 2-propanol. Additionally, the sensor also exhibited remarkable stability, with a maximum deviation of 0.15 nm and an estimated resolution of 3.18 ppm. Ahn et al. (2025) utilized a porous cholesteric liquid crystal film coating on an optical fiber ferrule for the selective and rapid detection of benzene, toluene, and acetone vapors. In this work, the authors used a broadband laser source along with an OSA to track the variations in the reflected signal of the cholesteric liquid crystal film under the exposure of various concentrations of toluene, benzene, and acetone vapors. From the obtained result, the fabricated sensor exhibited a maximum sensitivity of 6.11 pm/ppm for benzene, 4.48 pm/ppm for toluene, and 3.19 pm/ppm for acetone, with benzene demonstrating the highest selectivity and greatest sensitivity. Additionally, the sensor also exhibited strong repeatability, with only a 0.228 nm standard deviation in the reflection after five exposures to 25% of the lower explosive limit of benzene vapor. The authors reported a sixfold improvement in the device's sensing performance when employing a porous cholesteric liquid crystal film compared to the nonporous material.

6.5 Pohl interferometer

The Pohl interferometer functions as an amplitude-splitting interferometer. In this apparatus, light emitted from a point source undergoes reflection from surfaces characterized by distinct optical properties, specifically refractive index and reflectivity. This interaction results in the formation of an interference pattern (Barthwal & Vudayagiri, 2015). This technique has been utilized for phase measurement and parallelism measurement of transparent surfaces (Ewbank, 2010; Wasilik et al., 1971). In VOC sensing, the Pohl interferometric sensors are fabricated by depositing a VOC-sensitive layer on a substrate, as shown in Fig. 6.15. Pohl interferometry

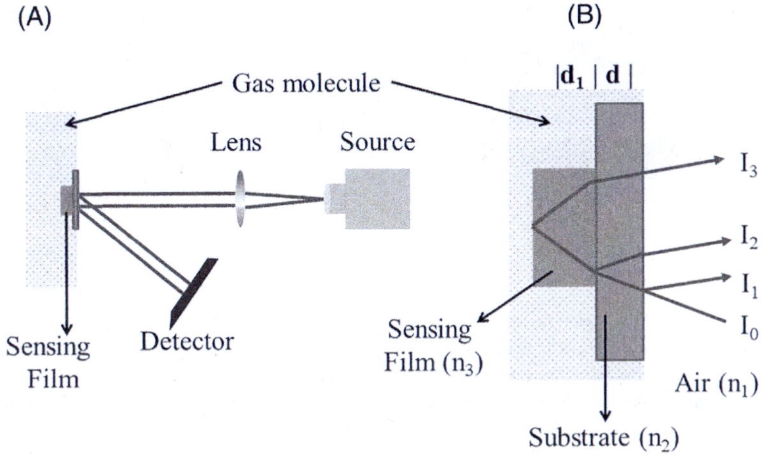

FIGURE 6.15 Pohl interferometer: Schematic of the (A) arrangement for Pohl interferometer and (B) reflection from three surfaces to form an interference pattern.

based on the reflection of light from the three interfaces results in the formation of an interference pattern, which can be written as follows:

$$I = I_1 + I_2 + 2\sqrt{I_1 I_1} \cos \delta \tag{6.6}$$

$$\delta = 2yk \sin(\Theta_i - \Delta\Phi) \tag{6.7}$$

where, k represents the wavenumber, y is the position of the interference fringe, θ_i is the angle between I_1 and I_3, and $\Delta\phi$ represents the phase difference.

Carlos et al. proposed an idea of monitoring the VOC gases based on Pohl interferometry in 2010 (Martínez-Hipatl et al., 2010). The fabricated sensor works on an intensity interrogation technique. The authors utilized PDMS as a sensing material deposited over a glass substrate. As discussed in Chapter 5, the unique characteristics of PDMS allow it to change its refractive index under exposure to VOC concentrations, leading to a shift in the fringe pattern within a Pohl interferometric. Fig. 6.16A demonstrated the schematic diagram of the Pohl interferometer used for the characterization, in which the light reflected from three interfaces, i.e., air–substrate, substrate–PDMS, and PDMS–air. The pattern of interference fringes evaluated in the function of the position on the screen is shown in Fig. 6.16B. The continuous line in Fig. 6.16B represents the shift observed for the PDMS layer over the glass substrate, whereas the dashed lines show the fringe pattern variations arising due to the swelling and refractive index changes in the PDMS.

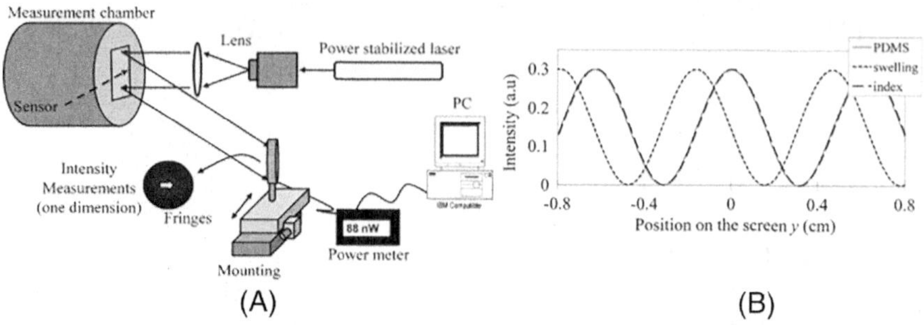

FIGURE 6.16 Pohl interferometric based VOC sensor: Schematic diagram of (A) Pohl interferometric configuration and (B) fringe shift due to PDMS. *From Martínez-Hipatl, C., Muñoz-Aguirre, S., Beltrán-Pérez, G., Castillo-Mixcóatl, J., & Rivera-De la Rosa, J. (2010). Detection of volatile organic compounds by an interferometric sensor.* Sensors and Actuators, B: Chemical, *147(1), 37–42. https://doi.org/10.1016/j.snb.2010.03.042.*

6.6 Sagnac interferometer

The Sagnac interferometer (SI) is mostly used in communications, filtering, and sensing applications because of its simple fabrication, and low sensitivity toward external noise (Fu et al., 2008). It has been utilized in several sensing applications, including pressure (Fu et al., 2008), humidity (Ascorbe et al., 2017), temperature (Shao et al., 2016), strain measurements (Gu et al., 2012), and gas detection (Pathak & Viphavakit, 2022). The SI generally involves a fiber loop mirror or a ring configuration for gas detection. Their working mechanism involves the splitting and recoupling of the counter-propagating beams. When the sensing region engages with its surrounding environment, a small polarization-maintaining fiber (PMF) is introduced into the loop to govern the interference arising from the interaction between the two counter-propagating beams and the measurand, as shown in Fig. 6.17. The transmitted intensity can be defined as:

$$I_{SI} = \frac{1}{2}\left(1 - \cos\frac{2\pi BL}{\lambda}\right) \tag{6.8}$$

where, L represents the length of the polarization maintaining fiber, while B is the refractive index difference between the slow and fast axes.

$$\frac{2\pi BL}{\lambda_{dip}} = 2k\pi \tag{6.9}$$

following the equation, a resonance dip can be achieved in the propagated spectrum (Xu et al., 2016), which can be defined:

$$\lambda_{dip} = BLk \tag{6.10}$$

here, k is an integer value. The values of B and L vary due to the interaction of the fabricated sensing film and the surrounding VOCs, which leads to a

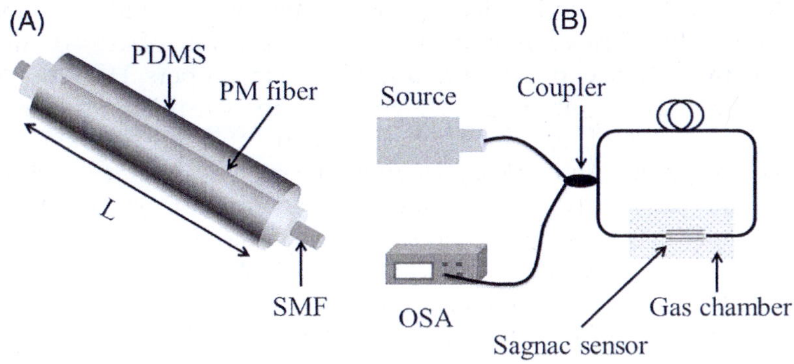

FIGURE 6.17 SI based VOC sensor: Schematic diagram of (A) SI fabricated sensor using polarization maintaining fiber and (B) schematic of the experimental setup utilized for VOC detection.

significant shift in the wavelength of polarization mode interference. The sensitivity of the device toward the target VOCs can be determined by analyzing the demodulated wavelength shift.

Ning et al. (2016a) reported a sensitive VOC sensor based on two interferometric configurations, namely FPI and SI, for the detection of multiple VOCs simultaneously. The schematic diagrams of PDMS coated SI sensing configuration and experimental setup is shown in Fig. 6.17A and B. In the reported work, an in-line FP configuration was fabricated using a 50 μm single-mode hollow optical fiber with a 15 μm-thick PDMS coating on the facet of the fiber. The FP cavity was fabricated between the fiber end face and the air-PDMS interface. The reported FPI configuration shows a good sensitivity of 1.61×10^{-3} nm/ppm for 2-propanol, while an average sensitivity of 1.17×10^{-3} nm/ppm for ethanol. In addition to FPI, the authors also fabricated SI-based VOC sensors to compare the performance of both designs. The SI was fabricated by splicing a 10 cm long PMF between the SMF and coating it with PDMS, similar to the FPI configuration. The result of the SI showed a non-linear trend for varying concentrations of VOCs. A maximum sensitivity of 2.71×10^{-3} nm/ppm was obtained for 2-propanol, while 9.02×10^{-4} nm/ppm was obtained for ethanol. The major limitation of SI based VOC sensor was the diffusion of 2-propanol from the PDMS, limiting its repeated use. PDMS was considered as a sensitive layer in both sensors, whose degree of swelling varies as the different VOC concentrations interact with it. Among both designs, the FPI-based sensor exhibited a rapid and reproducible response compared to SI, due to the thin layer of PDMS and miniature size.

In previous investigations, the response time for ethanol sensors was found to be extremely slow. Therefore, to address this issue, Li et al. (2023)

reported a NLC film embedded in optical fiber SI for rapid detection of ethanol. For the fabrication of the device, the NLC was infiltrated into the C-type ceramic tube connecting two single-mode jumpers. Here, the NLC film, possessing high birefringence, worked as the sensing material for the absorption of ethanol, as well as helped the generation of spectral interference as it penetrated into the Sagnac ring. The experimental result showed a maximum sensitivity of 2.22 $pm.ppm^{-1}$ for various concentrations of ethanol vapor ranging from 1210 to 10,000 ppm. The response and recovery times of the sensor were reported to be 15 and 20 seconds, respectively. Recently, Gu et al. (2025) reported a highly sensitive ethanol gas sensor utilizing a fiber ring laser along with the NLC integrated with the Sagnac loop. The sensing principle of the fabricated device relies on the wavelength modulation technique, where the shift in the output spectra is measured to determine the ethanol concentration. The basic principle involves splitting the launched light into two separate beams that propagate in the clockwise and counterclockwise directions via the fiber coupler, which are then recombined within the same coupler. Due to the birefringence effect of NLC after orientation, the two beams of light develop a phase difference, leading to SI. In general, variations in external environmental properties (such as refractive index, temperature, pressure, and the presence of gaseous molecules) influence the birefringence coefficient of the optical fiber, thereby changing the phase difference. From the fabricated sensor, the authors achieved a maximum sensitivity of 0.97 pm/ppm along with a low LOD of 247.42 ppm, a high resolution of 0.24 nm, and a high Q-factor of 40.69. In the same year, Zhang et al. (2025) utilized SI based on the vernier effect to fabricate a highly sensitive ethanol sensor. The fabricated sensor comprises PDMS coated PMF stubbed between two SMFs. Two PMFs with slightly different lengths which are utilized to form the sensing arm and reference arm in the two SIs, and thus, VE is excited. PDMS coating over fiber works as an absorbent for ethanol vapor, resulting in enhanced stress due to swelling-induced strain. To investigate the effect of PDMS coating length on sensor sensitivity, the authors performed three experiments considering PDMS coating lengths of 6, 8, and 10 cm on a 25 cm long PMF. In all experiments, the free spectral range of the interference spectrum remained constant at 13 nm. Experimental results exhibited that the measurement sensitivities based on cascaded and parallel SI were 11.27 and 11.34 pm/ppm, respectively, which is significantly improved by 13 times compared to the sensitivity obtained for a single SI. Meanwhile, the LOD for the cascaded and paralleled SIs were as low as 2.049 and 2.156 ppm, respectively.

Table 6.1 represents a summary of various interferometric VOC fiber sensors along with their sensing principles, target VOCs, sensitivity, detection limits, and operational ranges.

TABLE 6.1 Comparative table showing various interferometric based optical fiber VOC sensors.

References	Principle	Target VOCs	Sensing layer	Sensitivity	LOD	Operating range
Liu et al. (2009)	FPI	Methanol, hexanol, and acetone	Norland adhesive (NA) and Polyethylene Glycol (PEG)	3.53 pm/ppm for methanol using PEG and 0.1 pm/ppm using NA	1 ppm with PEG 10 ppm with NA	NA
Liu et al. (2010)	FPI	Toluene, methanol, decane, and dimethyl-methylphosphonate (DMMP)	PDMS and PEG	4.75 mV/ng for decane and 77 mV/ng for DMMP	50 pg (105.6 ppb)	NA
Ning et al. (2016a)	FPI/SI	Ethanol and 2-propanol	PDMS	9.02×10^{-4} to 5.14×10^{-4} nm/ppm for ethanol and 2.71×10^{-3} to 7.78×10^{-4} nm/ppm for 2-propanol	NA	NA
Ning et al. (2016b)	FPI	Isopropanol	Zeolite	0.91 nm/ppm in range of 0–70 ppm	NA	0–90 ppm
Yu et al. (2017)	FPI	Acetone and ethanol	PMMA	For acetone: 2.17 pm/ppm, For ethanol: 2.7 pm/ppm	NA	0–1800 ppm

(continued on next page)

TABLE 6.1 Comparative table showing various interferometric based optical fiber VOC sensors—cont'd

References	Principle	Target VOCs	Sensing layer	Sensitivity	LOD	Operating range
Wu et al. (2017)	SI	Isopropanol	PDMS	1.03 pm/ppm in range of 0–6000 ppm	19.3 ppm	0–6000 ppm
Kanawade et al. (2019)	FPI	Chloroform, isopropanol, hexane, acetone, methanol, and toluene	PDMS	23.7, 8.8, 17.8, 9.1, 3.2, 21.6 pm/ppm	0.84, 2.26, 1.11, 2.19, 6.10, 0.92 ppm	0–200 ppm
Martínez-Hipatl et al. (2010)	Pohl interferometer	Methanol (Me), ethanol (Et), propanol (Pp), octane (Oct), heptane (Hep), methyl acetate (MA), propyl acetate (PA)	PDMS	0.14 nW/ppm for PA 0.12 nW/ppm for Oct 0.04 nW/ppm for Hep 0–0.02 nW/ppm for rest	1500 ppm	0–24,000 ppm
Li et al. (2020)	MZI	Acetone, ethanol, methanol	Liquid crystals	–83 pm/ppm-methanol –31 pm/ppm-acetone –101.4 pm/ppm-ethanol	NA	0–2000 ppm
Ma et al. (2021)	MZI	Ethanol	ZIF-8	41 pm/ppm^{-1}	740 ppb	600–1000 ppm
Tabassum et al. (2021)	MZI	Ethylene	Copper (I)	60 pm/ppm	~4.7 ppm	0–140 ppm

From Pathak, A. K., & Viphavakit, C. (2022). A review on all-optical fiber-based VOC sensors: Heading towards the development of promising technology. Sensors and Actuators A: Physical, 338. *https://doi.org/10.1016/j.sna.2022.113455.*

6.7 Future prospects and challenges

In general, the interferometric-based VOC sensors comprise a flexible structure, excellent mechanical properties, and a simple fabrication approach to detect a wide range of VOCs in ambient environments. However, the major limitations of these interferometric sensors lie in the requirement for interference spectrum demodulation. For example, the MZI-based devices are prone to errors during the extraction of peaks from the spectrum. In the FPI, the limitations typically include high transmission losses, misalignment between fiber end-faces, and cavity size limitations due to coupling losses. For SI, the sensor is sensitive to other physical parameters, i.e., humidity, temperature, and vibration, which can significantly introduce significant errors in the measurement. The interferometry technique has considerable potential to achieve selective and sensitive VOC detection for domestic, industrial, and medical applications. In VOC sensing, it is crucial to consider appropriate techniques of interferometry, which will allow precise monitoring and improved performance of the specific VOCs. It has been observed from the above studies that the sensing performance of the fiber VOC sensors not only depends on the sensing approach but also depends on the utilized sensing elements, i.e., the types of waveguide, sensing materials, interaction lengths, etc. The interferometry-based optical VOC sensors demonstrated its potential in real-time and continuous monitoring of VOCs related to the exhaled breath. This capability of interferometry is highly useful in medical scenarios where dynamic variations in VOC concentrations may provide crucial information about a patient's health status. The emerging growth in the fields of electronics, materials science, optics, and data analysis has opened new research directions to broaden their horizons and realize an actual interferometric device in medical applications.

6.8 Conclusion

This chapter provides a detailed discussion of various interferometry technologies and their applications in VOC sensing. The chapter covers the basic definitions of various interferometries such as MZI, FPI, Pohl interferometry, and SI, along with their working principles, sensor designs, and different analytical performance parameters, including linearity, sensitivity, selectivity, and response time, are presented in detail. The chapter also discusses the key advantages and limitations of each method, along with a summary table to enable readers to understand and compare their performance, applicability, and suitability for VOC detection.

References

Ahn, S., Hwang, N. H., Kim, M. S., Kim, J. S. U., Choi, B. K., Cho, S. Y., Yoo, J., Bak, N.-H., Kim, M., Kim, J. H., Kim, M.-D., & Jeon, M. Y. (2025). Porous cholesteric liquid crystal film-integrated optical fiber sensor for real-time detection of high-concentration volatile organic compounds. *Sensors and Actuators B: Chemical, 444*, 138509. https://www.sciencedirect.com/science/article/pii/S0925400525012857. doi:10.1016/j.snb.2025.138509.

Aizen, A., Ney, M., Safrani, A., & Abdulhalim, I. (2018). A compact real-time high-speed high-resolution vibrometer, surface profiler and dynamic focus tracker using three wavelengths parallel phase-shift interferometry. *Optics and Lasers in Engineering, 107*, 304–314. doi:10.1016/j.optlaseng.2018.04.007.

Ascorbe, J., Corres, J., Arregui, F., & Matias, I. (2017). Recent developments in fiber optics humidity sensors. *Sensors, 17*(4), 893. doi:10.3390/s17040893.

Barthwal, S., & Vudayagiri, A. (2015). Construction of an inexpensive molecular iodine spectrometer using a self-developed Pohl wavemeter around 670 nm wavelength. *European Journal of Physics, 36*(5), 055014. doi:10.1088/0143-0807/36/5/055014.

Bhardwaj, V., Kishor, K., & Sharma, A. C. (2020). Tapered optical fiber geometries and sensing applications based on Mach–Zehnder interferometer: A review. *Optical Fiber Technology, 58*. http://www.journals.elsevier.com/optical-fiber-technology/. doi:10.1016/j.yofte.2020.102302.

Drapp, B., Pauluth, D., Krause, J., & Gauglitz, G. (1999). The application of the phase transition in nematic liquid crystals for the optical detection of volatile organic compounds. *Fresenius' Journal of Analytical Chemistry, 364*(1–2), 121–127. doi:10.1007/s002160051310.

Ewbank, D. E. (2010). *Single arm interferometer system for reflective micro-device phase measurement*. International Optical Design Conference and Optical Fabrication and Testing.

Fanget, S., Hentz, S., Puget, P., Arcamone, J., Matheron, M., Colinet, E., Andreucci, P., Duraffourg, L., Myers, E. D., & Roukes, M. L. (2011). Gas sensors based on gravimetric detection—A review. *Sensors and Actuators B: Chemical, 160*(1), 804–821. doi:10.1016/j.snb.2011.08.066.

Fu, H. Y., Tam, H. Y., Shao, L. Y., Dong, X., Wai, P. K. A., Lu, C., & Khijwania, S. K. (2008). Pressure sensor realized with polarization-maintaining photonic crystal fiber-based Sagnac interferometer. *Applied Optics, 47*(15), 2835–2839. http://ao.osa.org/Issue.cfm. doi:10.1364/AO.47.002835.

Garvey, J., Newport, D., Lakestani, F., Whelan, M., & Joseph, S. (2008). Full field measurement at the micro-scale using micro-interferometry. *Microfluidics and Nanofluidics, 5*(1), 77–87. doi:10.1007/s10404-007-0228-6.

Ghosh, S., & Rahman, B. M. A. (2017). A compact Mach–Zehnder interferometer using composite plasmonic waveguide for ethanol vapor sensing. *Journal of Lightwave Technology, 35*(14), 3003–3011. doi:10.1109/JLT.2017.2703827.

Grantzioti, E., Pissadakis, S., & Konstantaki, M. (2025). Optical fiber volatile organic compound vapor sensor with ppb detectivity for breath biomonitoring. *IEEE Sensors Journal, 25*(5), 8224–8229. doi:10.1109/JSEN.2025.3527765.

Gu, B., Yuan, W., He, S., & Bang, O. (2012). Temperature compensated strain sensor based on cascaded sagnac interferometers and all-solid birefringent hybrid photonic crystal fibers. *IEEE Sensors Journal, 12*(6), 1641–1646. doi:10.1109/JSEN.2011.2175725.

Gu, M., Chen, H., & Li, L. (2025). High performance ethanol gas sensor based on fiber ring laser with a nematic liquid crystal film embedded Sagnac interferometer. *Sensors and Actuators A: Physical, 387*. doi:10.1016/j.sna.2025.116481.

Huang, Y. I., Tao, J., & Huang, X. U. (2016). Research progress on F-P interference—based fiber-optic sensors. *Sensors, 16*(9), 1424. doi:10.3390/s16091424.

Hübert, T., Boon-Brett, L., Black, G., & Banach, U. (2011). Hydrogen sensors – A review. *Sensors and Actuators B: Chemical, 157*(2), 329–352. doi:10.1016/j.snb.2011.04.070.

Kanawade, R., Kumar, A., Pawar, D., Vairagi, K., Late, D., Sarkar, S., Sinha, R. K., & Mondal, S. (2019). Negative axicon tip-based fiber optic interferometer cavity sensor for volatile gas sensing. *Optics Express, 27*(5), 7277–7290. https://www.osapublishing.org/oe/abstract.cfm?uri=oe-27-5-7277. doi:10.1364/OE.27.007277.

Kaushal, K., & Das, B. (2023). Polymer transduction cavity based optical fiber Fabry–Perot interferometer for VOC sensing. *Results in Optics, 12*, 100479. doi:10.1016/j.rio.2023.100479.

Khan, S., Le Calvé, S. T., & Newport, D. (2020). A review of optical interferometry techniques for VOC detection. *Sensors and Actuators A: Physical, 302*, 111782. doi:10.1016/j.sna.2019.111782.

Leça, J. M., Antunes, P., Costa, F. M., Teixeira, A. J. S., & Ferreira, M. S. (2025). A novel fiber-optical Fabry–Perot microtip sensor for 2-propanol. *Sensors, 25*(7). doi:10.3390/s25072178.

Lee, B. H., Kim, Y. H., Park, K. S., Eom, J. B., Kim, M. J., Rho, B. S., & Choi, H. Y. (2012). Interferometric fiber optic sensors. *Sensors, 12*(3), 2467–2486. http://www.mdpi.com/1424-8220/12/3/2467/pdf. doi:10.3390/s120302467. South Korea.

Li, Y., Chen, H., Li, H., Fan, X., & Li, B. (2023). High sensitivity ethanol vapor sensor based on a nematic liquid crystal film embedded optical fiber Sagnac interferometer. *Journal of Physics D: Applied Physics, 56*(30), 305101. doi:10.1088/1361-6463/acd00d.

Li, Y., Chen, Y., Yi, D., Du, Y., Luo, W., Hong, X., Li, X., Geng, Y., & Luo, D. (2020). A self-assembled fiber Mach–Zehnder interferometer based on liquid crystals. *Journal of Materials Chemistry C, 8*(32), 11153–11159. http://pubs.rsc.org/en/journals/journal/tc. doi:10.1039/d0tc02482a.

Liao, C., Zhu, F., Zhou, P., & Wang, Y. (2019). Fiber taper-based Mach–Zehnder interferometer for ethanol concentration measurement. *Micromachines, 10*(11). doi:10.3390/mi10110741.

Liu, J., Sun, Y., & Fan, X. (2009). Highly versatile fiber-based optical Fabry–Pérot gas sensor. *Optics Express, 17*(4), 2731. doi:10.1364/oe.17.002731.

Liu, J., Sun, Y., Howard, D. J., Frye-Mason, G., Thompson, A. K., Ja, S.-J., Wang, S.-K., Bai, M., Taub, H., Almasri, M., & Fan, X. (2010). Fabry–Pérot cavity sensors for multipoint on-column micro gas chromatography detection. *Analytical Chemistry, 82*(11), 4370–4375. doi:10.1021/ac902956d.

Luo, D. H., Levy, R. A., Hor, Y. F., Federici, J. F., & Pafchek, R. M. (2003). An integrated photonic sensor for in situ monitoring of hazardous organics. *Sensors and Actuators B: Chemical, 92*(1–2), 121–126. doi:10.1016/s0925-4005(03)00123-0.

Ma, X., Wu, J., Jiang, L., Wang, M., Deng, G., Qu, S., & Chen, K. (2021). On-chip integration of a metal-organic framework nanomaterial on a SiO_2 waveguide for sensitive VOC sensing. *Lab on a Chip, 21*(17), 3298–3306. http://pubs.rsc.org/en/journals/journal/lc. doi:10.1039/d1lc00503k.

Makarona, E., Petrou, P., Kakabakos, S., Misiakos, K., & Raptis, I. (2016). Point-of-need bioanalytics based on planar optical interferometry. *Biotechnology Advances, 34*(3), 209–233. doi:10.1016/j.biotechadv.2016.02.005.

Martínez-Hipatl, C., Muñoz-Aguirre, S., Beltrán-Pérez, G., Castillo-Mixcóatl, J., & Rivera-De la Rosa, J. (2010). Detection of volatile organic compounds by an interferometric sensor. *Sensors and Actuators, B: Chemical, 147*(1), 37–42. doi:10.1016/j.snb.2010.03.042.

Miliou, A. (2021). In-fiber interferometric-based sensors: Overview and recent advances. *Photonics., 8*(7), 265. doi:10.3390/photonics8070265.

Mirzaei, A., Kim, J. H., Kim, H. W., & Kim, S. S. (2018). Resistive-based gas sensors for detection of benzene, toluene and xylene (BTX) gases: A review. *Journal of Materials Chemistry C, 6*(16), 4342–4370. http://pubs.rsc.org/en/journals/journal/tc. doi:10.1039/c8tc00245b.

Newport, D., Sobhan, C. B., & Garvey, J. (2008). Digital interferometry: Techniques and trends for fluid measurement. *Heat and Mass Transfer, 44*(5), 535–546. doi:10.1007/s00231-007-0267-2.

Ning, X., Yang, J., Zhao, C. L., & Chan, C. C. (2016a). PDMS-coated fiber volatile organic compounds sensors. *Applied Optics, 55*(13), 3543–3548. https://www.osapublishing.org/view_article.cfm?gotourl=https%3A%2F%2Fwww%2Eosapublishing%2Eorg%2FDirectPDFAccess%2F55AE38B5-A6D3-0469-F331178E592EB803_340243%2Fao-55-13-3543%2Epdf%3Fda%3D1%26id%3D340243%26seq%3D0%26mobile%3Dno&org=Elsevier%20Inc. doi:10.1364/AO.55.003543.

Ning, X., Zhao, C. L., Yang, J., & Chiu Chan, C. (2016b). Zeolite thin film-coated spherical end-face fiber sensors for detection of trace organic vapors. *Optics Communications, 364*, 55–59. doi:10.1016/j.optcom.2015.11.023.

Niu, H., Yu, P., Zhu, Y., Jing, Z., Li, P., Wang, B., Ma, C., Wang, J., Wu, J., Govorov, A. O., Neogi, A., & Wang, Z. M. (2022). Mach–Zehnder interferometer based integrated-photonic acetone sensor approaching the sub-ppm level detection limit. *Optics Express, 30*(16), 29665–29679. https://opg.optica.org/directpdfaccess/5bbc8953-8f94-46be-b4dc34d1c4047fad_484211/oe-30-16-29665.pdf?da=1&id=484211&seq=0&mobile=no. doi:10.1364/OE.459805.

Nuhoglu, D., Tasaltin, C., & Gurol, I. (2024). *QCM and SAW gas and VOC sensors based on metal oxide composites (principles, fabrication, sensing materials, and performances)* (pp. 223–252). Elsevier BV. doi:10.1016/b978-0-323-95385-6.00016-7.

Yang, F., Zong, S., Li, X., Hu, Y., Wang, Z., Qu, Y., … Li, Y. (2025). Optical fiber methane sensor based on Mach–Zehnder interferometer induced by multimode interference. *Micromachines, 16*(4), 406. https://www.mdpi.com/2072-666X/16/4/406. doi:10.3390/mi16040406.

Pathak, A. K., & Singh, V. K. (2017). A wide range and highly sensitive optical fiber pH sensor using polyacrylamide hydrogel. *Optical Fiber Technology, 39*, 43–48. http://www.journals.elsevier.com/optical-fiber-technology/. doi:10.1016/j.yofte.2017.09.022.

Pathak, A. K., Verma, S., Sakda, N., Viphavakit, C., Chitaree, R., & Rahman, B. M. A. (2023). Recent advances in optical hydrogen sensor including use of metal and metal alloys: A review. *Photonics, 10*(2). http://www.mdpi.com/journal/photonics. doi:10.3390/photonics10020122.

Pathak, A. K., & Viphavakit, C. (2022). A review on all-optical fiber-based VOC sensors: Heading towards the development of promising technology. *Sensors and Actuators A: Physical, 338*, 113455. doi:10.1016/j.sna.2022.113455.

Tzeng, T. H., Kuo, C. Y., Wang, S. Y., Huang, P. K., Huang, Y. M., Hsieh, W. C., Huang, Y. J., Kuo, P. H., Yu, S. A., Lee, S. C., Tseng, Y. J., Tian, W. C., & Lu, S. S. (2016). A portable micro gas chromatography system for lung cancer associated volatile organic compound detection. *IEEE Journal of Solid-State Circuits, 51*(1), 259–272. doi:10.1109/JSSC.2015.2489839.

Wu, B. Q., Zhao, C. L., & Ding, Z. M. (2017). Sagnac interferometer fiber volatile organic compounds sensor based on PDMS. *Acta Photonica Sinica, 46*(1). https://www.researching.cn/articles/OJ685348819891678a. doi:10.3788/gzxb20174601.0106003.

Sekhar, P. K., & Subramaniyam, K. (2013). Detection of harmful benzene, toluene, ethylbenzene, xylenes (BTEX) vapors using electrochemical gas sensors. *ECS Electrochemistry Letters, 3*(2), B1. doi:10.1149/2.005402eel.

Shao, L. I.-Y., Zhang, X., He, H., Zhang, Z., Zou, X., Luo, B., Pan, W., & Yan, L. (2016). Optical fiber temperature and torsion sensor based on Lyot-Sagnac interferometer. *Sensors, 16*(10), 1774. doi:10.3390/s16101774.

Sofa, S. A., Roto, R., Aflaha, R., Natsir, T. A., Humairah, N. A., Kusumaatmaja, A., Triyana, K., & Gupta, R. (2024). Formaldehyde gas sensors based on a quartz crystal microbalance modified with aniline-doped polyvinyl acetate nanofibers. *The Analyst., 149*(4), 1262–1270. doi:10.1039/d3an01856c.

Spinelle, L., Gerboles, M., Kok, G., Persijn, S., & Sauerwald, T. (2017). Review of portable and low-cost sensors for the ambient air monitoring of benzene and other volatile organic compounds. *Sensors., 17*(7), 1520. doi:10.3390/s17071520.

St-Gelais, R., MacKey, G., Saunders, J., Zhou, J., Leblanc-Hotte, A., Poulin, A., Barnes, J. A., Loock, H. P., Brown, R. S., & Peter, Y. A. (2013). Gas sensing using polymer-functionalized deformable Fabry–Perot interferometers. *Sensors and Actuators, B: Chemical, 182*, 45–52. doi:10.1016/j.snb.2013.02.016.

Tabassum, S., Kumar, D. P., & Kumar, R. (2021). Copper complex-coated nanopatterned fiber-tip guided mode resonance device for selective detection of ethylene. *IEEE Sensors Journal, 21*(16), 17420–17429. doi:10.1109/JSEN.2021.3057619.

Wang, Z., Li, Y., Gao, Y. A., Mu, Z., Zong, S., Han, X. F., Yang, Z., & Sun, X. (2025). Metal-organic framework modified open-cavity optical fiber Fabry–Pérot interferometer for volatile organic compound detection. *Talanta, 281*. doi:10.1016/j.talanta.2024.126901.

Wasilik, J. H., Blomquist, T. V., & Willett, C. S. (1971). Measurement of parallelism of the surfaces of a transparent sample using two-beam nonlocalized fringes produced by a laser. *Applied Optics, 10*(9), 2107. doi:10.1364/ao.10.002107.

Xu, B., Zhao, C. L., Yang, F., Gong, H., Wang, D. N., Dai, J. X., & Yang, M. (2016). Sagnac interferometer hydrogen sensor based on panda fiber with Pt-loaded WO_3/SiO_2 coating. *Optics Letters, 41*(7), 1594. doi:10.1364/OL.41.001594.

Yu, C. B., Wu, Y., Li, C., Wu, F., Zhou, J. H., Gong, Y., Rao, Y. J., & Chen, Y. F. (2017). Highly sensitive and selective fiber-optic Fabry–Perot volatile organic compounds sensor based on a PMMA film. *Optical Materials Express, 7*(6), 2111–2116. https://www.osapublishing.org/DirectPDFAccess/0AE28BDC-F7C9-11B6-FBF4DEB2FB2D80FD_367219/ome-7-6-2111.pdf?da=1&id=367219&seq=0&mobile=no. doi:10.1364/OME.7.002111.

Zhang, S. A., Chen, H., Li, Y., Gu, M., Ullah, S., Li, L., & Li, S. (2025). Sensitivity-enhanced ethanol vapor sensor based on vernier effect in Sagnac interferometers with a poly-dimethylsiloxane film-coated polarization maintaining fiber. *IEEE Sensors Journal*. doi:10.1109/JSEN.2025.3545754.

Zuo, C., Feng, S., Huang, L., Tao, T., Yin, W., & Chen, Q. (2018). Phase shifting algorithms for fringe projection profilometry: A review. *Optics and Lasers in Engineering, 109*, 23–59. doi:10.1016/j.optlaseng.2018.04.019.

CHAPTER

7

Fundamentals and applications of fiber Bragg gratings in VOC sensing

7.1 Introduction

In the broad field of photonic sensing, fiber Bragg gratings (FBGs) hold great importance where precision is required, casting their impact across a wide range of applications. These FBGs are made from standard single-mode optical fibers by creating a periodic change in the refractive index (RI) inside the core, which satisfies the Bragg condition for coupling between the forward- and backward-propagating fundamental core modes (Gangwar et al., 2023). The first FBG was fabricated by Hill et al. (1978) by introducing a periodic RI variation inside a germanium-doped core. The grating was produced using the light-induced refractive-index variation in the core of the waveguide that can be broadly attributed to the photosensitivity of the material. FBGs are currently being used as sensing devices across various sectors to accurately monitor a wide range of physical and chemical parameters (Liu et al., 2025; Xing et al., 2025; Zhang et al., 2025). FBGs have found extensive use in structural health monitoring (SHM) to assess the integrity and performance of structures(Venugopalan, Sun, & Grattan, 2008). These sensors are typically either surface-mounted or embedded within structures to detect variations in physical properties like vibration, temperature, and strain (Grattan & Meggitt, 2010). FBG sensors are highly sensitive, immune to electromagnetic interference, and capable of being multiplexed, making them ideal for real-time monitoring of infrastructure such as buildings, bridges, pipelines, and dams.

Nowadays, FBGs have also gained significant attention in medical applications (Song et al., 2025). Advancements in FBG technology have

Photonic Sensors for Detection of VOC Biomarkers: Technological Principles and Medical Applications
DOI: https://doi.org/10.1016/B978-0-443-29166-1.00007-X

Copyright © 2026 Elsevier Inc. All rights are reserved, including those for text and data mining, AI training, and similar technologies.

significantly contributed to the rapid growth of the sensing market, offering numerous medical applications. These developments have created new opportunities for enhancing in-vivo monitoring, improving health outcomes, and enhancing patient care and quality of life. FBGs provide high accuracy, a wide detection range, and the potential to overcome many challenges associated with conventional medical tools used for physiological monitoring and breath analysis (Yan, Feng, Xu, & Zhang, 2025). In addition to strain sensing, FBG-based sensors are also utilized in biosensing and gas sensing. To use FBGs as biosensing elements, researchers have reported coating target sensing materials onto FBGs, which selectively interact with the measurand, resulting in a variation in the properties of the transmitted light (Rohan et al., 2024). In the biomedical field, FBGs are mainly applied in areas such as cardiovascular diagnostics, volatile organic compound (VOC) detection, neurological diagnostics, urology, blood pressure monitoring, gastroenterology, and endoscopic procedures.

The aim of this chapter is to introduce the fundamentals of FBGs and their application in VOC sensing. It covers the fabrication process of FBGs, their classification, basic sensing principles, as well as their advantages and limitations. The structure of the chapter is organized as follows: Section 7.2 discusses the different types of gratings, while Section 7.3 focuses on the fabrication process. Section 7.4 covers the various grating types, and Section 7.5 explains their sensing mechanisms. Section 7.6, 7.7 and 7.8 provides a brief overview of their application in sensing of VOC, semi-VOCs, and other organic pollutants. Finally, Section 7.9 outlines future prospects and challenges, and concludes the chapter in Section 7.10.

7.2 Photosensitivity types of fiber Bragg gratings

Photosensitivity in optical fibers refers to the permanent modification of the core's RI when illuminated with light of a specific wavelength and intensity, a response that varies with the composition of the core material. At first, this effect was believed to occur only in fibers with high germanium concentrations in the core, excited by ultraviolet (UV) light in the 200–250 nm range. However, extensive research over the years has revealed that photosensitivity can be induced at various UV wavelengths in a wide range of fiber types, including those with little or no germanium doping. FBGs can be categorized into three classes, i.e., Type I, Type II, and Type IIA. The difference between these types depends on the formation mechanisms, which influence their physical behavior. The properties of each type of FBGs are generally identified through their growth rate and by examining their thermal stability. Generally, Type IA gratings exhibit the lowest thermal stability, while Type II gratings are known to be resistant to high temperatures. A brief discussion of various types is discussed in following subsections whereas in depth infromation on each types can be found in other books (Kashyap, 2013; Mihailov, 2012, 2023):

7.2.1 Type I grating

Type I gratings are generally known as standard gratings, which were first fabricated using germanosilicate fibers, via single-photon absorption of UV light, leading to small index variations. These fibers are mostly used in the industrial sector. This process excites oxygen deficiency centers, which have absorption bands around 244 nm. For larger index changes, the formation of defects is likely accompanied by a densification of the glass matrix. Type I gratings demonstrate a positive variation in the RI and exhibit a temperature-dependent decay that can be characterized using a power-law relationship (Erdogan et al., 1994). The decay takes place due to thermal depopulation of trapped excited states, which are formed during the fabrication of the gratings. At higher temperatures, carriers in the shallowest traps gain enough energy to escape and return to their ground state. The remaining carriers are associated with more stable states, which can also return to the ground state with further increase in temperature. Since the most unstable carriers decay at lower temperatures, Type I gratings are usually annealed at temperatures higher than their intended operating temperature to ensure long-term stability in reflectivity. Type I gratings are therefore not suitable for applications in a harsh environment (i.e., temperature above 450°C) as most of the RI variation is annealed out at higher temperatures.

7.2.2 Type II grating

Type II gratings can be fabricated using high-power pulsed UV laser sources, such as excimer lasers, enabling the inscription of precise gratings with high reflectivity (up to <99%) (Archambault et al., 1993). These FBGs are formed via a threshold-dependent multi-photon ionization technique, similar to the laser-induced damage as typically observed in bulk optics. Due to this reason, these gratings are known as "damage" gratings (Russell et al., 1993). These gratings possess excellent thermal stability at temperatures exceeding 1000°C, and have been widely utilized to obtain grating arrays during the fiber-drawing process. Single-shot exposure is often used to produce these grating structures, which may suffer substantial scattering losses due to the damaged region. Furthermore, this process also makes the fiber fragile, leading to poor strength and reliability of FBGs.

7.2.3 Type IIA grating

Type IIA gratings (also known as Type In) are fabricated in highly stressed Ge-doped cores using a UV laser. Under specific conditions, Type

I gratings have been observed to initially increase in optical reflectivity before decreasing as the grating approaches saturation. Under continuous exposure, the growth of a secondary grating is observed, which demonstrates a steady blue shift in the spectral response and, therefore, results in a negative RI variation (Xie et al., 1993). By optimizing the fabrication process, Type IIA grating structures can be formed which exhibit stability at moderate temperatures of up to 700°C (Groothoff & Canning, 2004).

7.3 Fabrication of fiber Bragg gratings

In past decades, various micromachining techniques have been reported in order to form different types of gratings. Among the several fabrication methods, this section provides a brief overview of the most popular approaches widely used in FBG fabrication, including the phase mask method, femtosecond laser writing, UV laser irradiation, and CO_2-induced inscription. Detailed explanations of the fabrication processes and operating principles of these fibers can be found in earlier publications (Arregui et al., 2001; Othonos et al., 2000). In general, three primary inscription methods are widely employed for FBG fabrication using fs-laser direct writing, commonly known as (1) point-by-point (P-b-P), (2) line-by-line (L-b-L), and (3) plane-by-plane (Pl-b-Pl). Among these techniques, the P-b-P method, consists of a series of periodic RI variation points, arranged parallel to the fiber axis. The P-b-P inscription method is considered one of the simplest and most straightforward approaches, which utilizes a single exposure of laser pulses with appropriate separation. In L-b-L technique, instead of a single pulse exposure, it consists of several lines made of multiple consecutive points along the length of the fiber axis. This method covers a broader cross-section of the core. The Pl-b-Pl technique primarily inscribes quasi-homogeneous 2-D RI modulation regions within the optical fiber cross-section. Due to the increased uniformity and broader modification area, this approach results in lower transmission loss, a narrower full-width half maximum (FWHM), and enhanced reflectivity.

7.3.1 Phase mask method

The phase mask method is considered as one of the most effective approaches of fabricating high-quality Bragg gratings in photosensitive optical fibers. This technique involves a diffractive optical element (also known as a phase mask) in order to spatially modulate the beam of a UV laser, thereby forming an interference pattern that is used to inscribe the grating structure in photosensitive optical fibers. Typically, these phase masks can be created via electron beam lithography (EBL) or holographically. EBL is considered more promising and advantageous over the holographic method due to its simple writing process, excellent reproducibility,

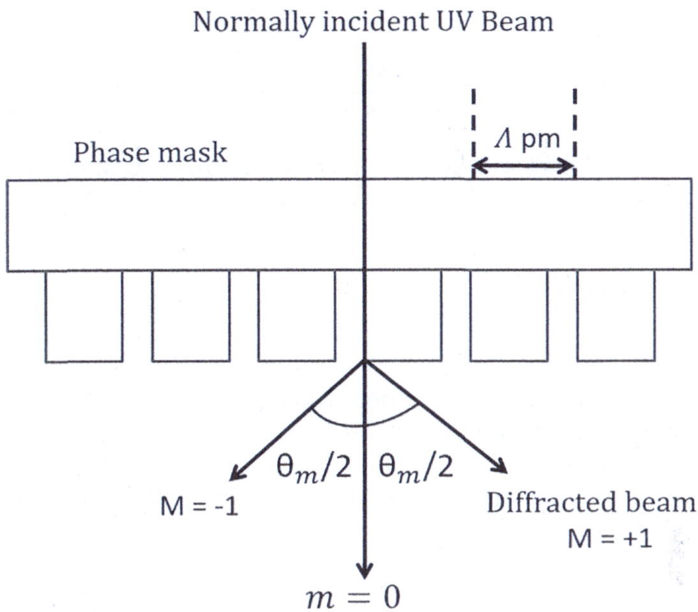

FIGURE 7.1 Phase mask: Schematic diagram of phase mask approach.

and ease of creating complex structures. Additionally, the phase mask reduces the strict requirement for the coherence of a UV light source. The light absorption is essential to induce the RI variation in the core. Atkins, (1992) investigated the absorption spectrum of a standard optical fiber comprising a germanium-doped core, extending to wavelengths as short as 200 nm. The results exhibited that UV light is the most effective way to induce RI variation in such types of optical fibers. Fig. 7.1 represents the schematic diagram of the fabrication of FBG using the phase mask method, where a phase mask is exposed to normally incident UV radiation and the diffracted radiation is split into m = 0 and ±1 orders. The interference pattern formed at the optical fiber by the combination of two beams of order ±1 exhibits a periodicity also known as the grating period Λ_g, which is associated with the diffraction angle $\theta_{m/2}$ through the following relationship:

$$\Lambda_g = \frac{\lambda_{uv}}{2\sin\left(\frac{\theta_m}{2}\right)} = \frac{\Lambda_{pm}}{2} \tag{7.1}$$

where Λ_{pm}, Λ_g, λ_{uv} represent the periodicity of the phase mask, period of the fringes, and the wavelength of the UV laser. The phase-mask grating period is defined based on the desired Bragg wavelength (λ_B) for the fiber grating, resulting in:

$$\Lambda_g = \frac{N\lambda_B}{2n_{eff}} = \frac{\Lambda_{pm}}{2} \tag{7.2}$$

where Λ_g, N, and n_{eff} represent the fiber grating period, Bragg order, and effective refractive index of the fiber core, respectively.

7.3.2 Femtosecond laser-induced gratings

The extreme peak power radiation produced by femtosecond (fs) laser systems has been extensively used to generate large index variations in bulk glasses to fabricate embedded waveguides (Mihailov, 2023). The mechanism responsible for this laser-induced index variation differs from the UV photosensitivity mechanism, which typically relies on color-center formation. The femtosecond infrared (fs-IR) laser technique is attributed to a nonlinear multiphoton absorption and ionization mechanism, which can induce either material densification or defect generation depending on the optical intensity. When the beam exposure surpasses the material-specific ionization threshold (I_{th}), multiphoton ionization can enable dielectric breakdown, resulting in the localized melting, densification, and formation of voids. This process introduces a permanent change in its RI, making it stable up to the glass transition temperature (T_g) of the host material. The resulting index change is closely identical to Type II gratings, which are generally produced via nanosecond laser pulses, with one key distinction: due to the ultrashort duration of femtosecond pulses, the breakdown is quenched almost instantaneously right after the exposure of the beam, preventing further damage outside the exposed area. Consequently, femtosecond-written gratings can exhibit superior spectral performance compared to those fabricated using UV sources.

Below I_{th}, another regime of induced index change has been observed that can be erased by annealing with temperatures below T_g (Mihailov, 2023). Below I_{th}, multiphoton absorption is expected to induce defect formation and material densification, identical to the Type I UV-induced RI changes observed in photosensitive germanosilicate glasses. The fs laser writing does not depend on the dopant material present in the formation of the core of the optical fiber, and it also does not require any hydrogen loading for the writing. Different strategies can be utilized to form gratings within the optical fibers, including interferometric exposure, line-by-line inscription, point-by-point writing, and plane-by-plane techniques. The first reported technique utilizes a bulk interferometer to induce periodic index changes in the bulk silica (Kawamura et al., 2001).

A specialty phase mask that was precision etched to maximize the coupling of the incident IR laser radiation into the ±1 order was utilized.

Gratings were written in standard telecom fiber and then later in "non-UV photosensitive" pure silica core fibers. Although the interference field produced by the phase mask is sinusoidal, the resulting refractive index modulation within the fibre can become non-sinusoidal due to nonlinear photo-induced index change processes during grating inscription. Using the phase mask approach, both Type I and Type II index changes can be induced in silica-based glass fibers.

The second approach to fs-IR laser-induced FBGs utilized a "point-by-point" writing technique where single pulses from the fs-IR laser were focused within the core region of the optical fiber using a high-power microscope objective. Highly localized changes to the RI were generated with each pulse to create each plane of the fiber grating. Subsequent planes were made in a step-and-repeat fashion by translating the beam using high-resolution mechanical translation stages (Martinez et al., 2004). As the index changes resulted from intensities greater than I_{th}, the grating structure consisted of material compaction surrounding void formation at each grating plane (Okamoto, 2021). The resulting structure had similar thermal stability to Type II UV gratings (Shu et al., 2002).

7.3.3 Carbon dioxide laser-induced grating

Besides phase mask and femtosecond laser-based writing, carbon dioxide (CO_2) laser-based inscription allows an alternative way for grating fabrication, offering a versatile and cost-effective method. Unlike other methods of grating fabrication, the CO_2 laser does not require any pretreatment of optical fiber, i.e., photosensitization or expensive phase masking. The CO_2 lasers are operated via a controlled translation stage, enabling periodic radiation along the axis of the sample fiber. Using a point-by-point technique, high-resolution grating structures can be achieved, identical to those fabricated via femtosecond lasers. The CO_2 lasers have been successfully utilized to write various gratings, including long-period gratings (LPG) in standard silica fiber (Xuan et al., 2009), as well as in the photonic crystal fibers (Lee & Chiang, 2009). However, the technique was found to be less effective for pure fused silica-based photonic crystal fibers due to their fragility, making them prone to structural damage.

7.3.4 Direct writing technique

The direct laser writing is a flexible method used to produce high-quality gratings inscribed directly within the optical fibers to form FBGs (Li, 2026). This method comprises two UV laser beams, focused and overlapped in order to form a micron-scale circular spot containing a one-dimensional interference pattern, identical to a conventional interferometric setup. The interference beam is focused onto the photosensitive fiber or waveguide, where the grating is formed by exposing, translating by one grating period, re-exposing, and moving to another location, systematically building the Bragg gratings. Unlike the traditional amplitude-splitting interferometer, direct writing employs continuous translation of the sample relative to a modulated laser beam that is synchronized with the motion, ensuring a smooth writing. The grating contrast can further be adjusted by controlling the exposure time. The whole process is computer-controlled, therefore allowing the precise fabrication of FBGs with arbitrary profiles, chirp, or apodization at any desired location within the waveguide.

7.4 Classification by grating structure

Optical fiber gratings are a well-established structure, with the conventional fabrication method involving the use of long-pulse lasers, such as UV or CO_2 lasers, to inscribe a one-dimensional periodic RI modification in the single mode fiber (Zhao et al., 2022). LPFGs and FBGs are the two most common types of fiber gratings available based on the grating period. It is important to highlight that a tilted FBG represents a novel type of grating, characterized by the coupling between the core and counter-propagating cladding mode (Sáez-Rodriguez, 2011).

7.4.1 Fiber Bragg grating

The standard FBG comprises periodic variations of the RI inscribed within the core of a single-mode optical fiber. This written grating allows for the reflection of some specific wavelengths of light, known as the Bragg wavelength, while transmitting the rest of the wavelengths. The strongest interaction takes place at the so-called Bragg wavelength (λ_B), given by:

$$\lambda_B = 2n_{core}(\lambda)\Lambda \tag{7.3}$$

where n_{core} (λ) and Λ represent the effective RI at a wavelength (λ) and the grating period, respectively (Zhao et al., 2022). In the fabrication of FBGs, it is important to note that designing the gratings of the lower Bragg wavelengths requires shorter grating periods. The FBGs are mainly written on photosensitive optical fibers by utilizing excimer or UV lasers. Martinez et al. (2004) first reported the fabrication of FBGs using the femtosecond laser direct writing method. Since then, several direct writing approaches have been utilized to produce high-resolution FBGs.

7.4.2 Long period grating

Long period gratings (LPGs), with a period much larger than the wavelength of light on the order of 10 to 100 of microns, do not support backward reflection and belong to the class of transmission band stop filters (Zhao et al., 2022). The basic principle of LPGs is based on the phase matching between the cladding and the fundamental modes of the optical fiber. The resonance wavelength (λ_L) can be defined as:

$$\lambda_L = \left[n_{core}(\lambda) - n^i_{cladding}(\lambda)\right]\Lambda \quad (7.4)$$

where, Λn_{core} (λ), and $n^i_{cladding}(\lambda)$, represent the grating period and effective RI of core and ith cladding mode at a specific wavelength λ.

In general, the core-cladding coupling occurs between cylindrically symmetric modes, which are commonly indicated with the $LP_{0,m}$, where LP represent "linearly polarized modes." For in-depth information, readers may refer to other sources such as Okamoto (2021). The key factor of LPG is that the effective RI of the cladding mode is significantly influenced by the RI of the ambient environment, such as temperature and mechanical deformation. As a result, the λ_L can be modulated by changing the external RI. This enables the use of LPG as a wavelength-modulated optical fiber sensor (OFS) (Shu et al., 2002). Among various methods of grating fabrication, femtosecond writing is considered to be one of the most popular ways to inscribe fine and high-resolution LPG grating through point exposure, line scanning, and path scanning.

7.4.3 Tilted fiber Bragg grating

Similar to standard FBGs, the tilted fiber Bragg gratings (TFBG) also comprise periodic RI variations along the length of the fiber axis. However, unlike FBGs, the TFBGs are inscribed to provide a certain tilt angle between the grating plane and the fiber cross-section, resulting in mode coupling to the cladding and an exterior environment, as shown in Fig. 7.2A and B.

Due to the tilt angle, the TFBGs exhibit not only the primary mode coupling as usually observed in conventional FBGs between forward and counter-moving fundamental modes, but also produce additional core-cladding coupling. This involves coupling between the fundamental mode propagating in the forward direction and the cladding mode propagating in the counter direction, as well as the interactions which occur between the forward-propagating core and radiation modes. In order to analyze the transmission characteristics of TFBGs, it is crucial to determine the expression of the cladding mode resonance wavelength. This resonance wavelength based on grating parameters can be obtained by solving the

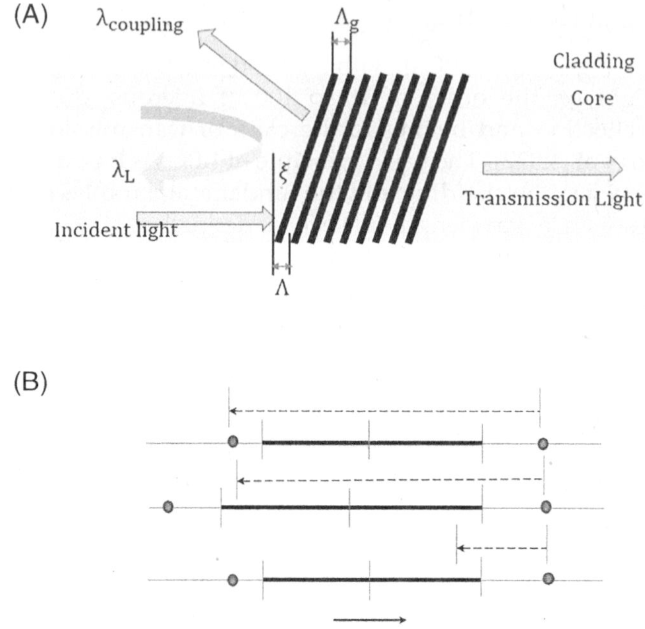

FIGURE 7.2 Tilted fiber Bragg grating: Schematic diagram of (A) TFBG working mechanism and (B) wave vectors illustrating mode coupling in TFBG.

coupled mode equation; however, this issue can also be dealt with conveniently by considering the analogy between the resonance conditions of FBGs and TFBGs. The Bragg resonance condition for TFBG could be defined as (Zhao et al., 2022):

$$\lambda_L = \left[n_{eff}^{core} + n_{eff}^{cladding}\right]\frac{\Lambda}{\cos\xi} \tag{7.5}$$

Due to the presence of the tilt angle, a part of the fundamental mode propagating in the forward direction will be coupled into the cladding mode propagating in the counter direction, and therefore the resonance wavelength (λ_L) of the cladding mode can be calculated by using the following relation:

$$\lambda_{Cl,i} = \left[n_{eff}^{core} + n_{eff,i}^{cladding}\right]\frac{\Lambda}{\cos\xi} \tag{7.6}$$

where, $n_{eff,i}^{cladding}$ represents the effective index of the ith cladding mode. The resemblance among the resonance conditions of conventional FBGs and TFBGs implies that the core mode in TFBGs possesses identical sensing performance towards temperature, strain, etc.

7.5 Sensing mechanism

The optical fibers are generally made from fused silica and germanium-doped glass to achieve a RI difference between core and cladding, enabling them to guide light through the process of the total internal reflection phenomenon between the higher-refractive index core and the lower-refractive-index cladding. With the rapid advancements of fiber technology, these fibers play a crucial role in telecommunication, which allows fast data transfer over longer distances compared to electronic communication. Interestingly, these optical fibers are capable of being transformed into various types of sensing tools, offering numerous advantages over conventional mechanical, electrical, or bulky optical sensing devices. For instance, changes in the physical parameters such as temperature or strain around the fiber lead to variations in the properties of the propagated optical signal in terms of phase, intensity, or spectral characteristics, which can be monitored through optoelectronic instruments (Zhu et al., 2017). Based on the distribution form of OFS, the sensing technology is divided into three categories such as (1) point sensing, (2) quasi-distributed sensing, and (3) fully-distributed sensing (Rovera, 2023). These OFSs can also be further categorized based on their working principle, i.e., reflection-based, intensity-based, and interferometry-based. FBG is a widely known quasi-distributed OFS based on the unique phenomenon of light reflection resulting from the Bragg grating inscribed in the fiber. The working principles of FBG were first defined by W. H. Bragg and his son W. L. Bragg, who were later awarded the Nobel prize in 1915 for their outstanding work (Zhu et al., 2017). Any variation in the n_{eff} or Λ produces a shift in the output spectra obtained from FBG, which are generally utilized for the accurate monitoring of several chemical and physical parameters, such as strain, voltage, current, temperature, index of refraction, biomolecules, etc. Fig. 7.3 illustrates the basic working principle of FBG-based sensors.

The Bragg wavelength is observed to be influenced by surrounding temperature and strain, which affect effective RI and the physical dimensions of the sensor. Strain results in the elongation of the FBG, whereas the change in temperature leads to thermal expansion and variation in the RI due to thermo-optic effects. In case of standard silica fibers, a change in strain ($\Delta\varepsilon$) or temperature (ΔT) leads to the shift in the Bragg wavelength (λ_B) as given by (Zhu et al., 2017; Othonos et al., 2000):

$$\Delta\lambda_B = 2n_{eff}\Lambda\left[\left(1-\frac{n_{eff}^2}{2}\right)[p_{12}-\upsilon(p_{11}-p_{12})]\Delta\varepsilon+\left(\alpha+\frac{\frac{dn_{eff}}{d\Delta T}}{n_{eff}}\right)\Delta T\right] \tag{7.7}$$

where, $\Delta\lambda_B$represents the change in the Bragg wavelength in presence of external strain and temperature, and p_{ij} (I, j = 1 or 2), ν, and α represent

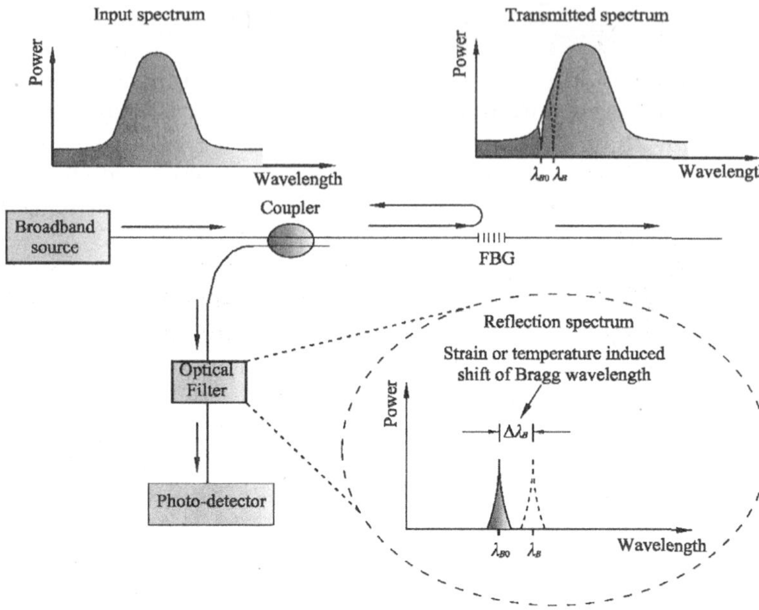

FIGURE 7.3 Working principle of FBG: Simple schematic diagram of the FBG sensor. *From Zhu, H.-H., Shi, B., & Zhang, C.-C. (2017). FBG-based monitoring of geohazards: Current status and trends. Sensors, 17(3). https://doi.org/10.3390/s17030452.*

Pockel's coefficients, Poisson's ratio, and the coefficient of thermal expansion of the FBG. Eq. (7.2) can also be written in a simpler form as given below:

$$\frac{\Delta\lambda_B}{\lambda_{B0}} = (1 - p_{eff})\Delta\varepsilon + (\alpha + \xi)\Delta T = c_\varepsilon \Delta\varepsilon + c_T \Delta T \tag{7.8}$$

where λ_{B0}, ξ, c_ε, c_T, and p_{eff}, represent the original Bragg wavelength without strain and temperature (0°C), thermo-optic coefficient, and calibration coefficients of strain and temperature, and the photo-elastic parameter, respectively. It can be described by the given relation:

$$p_{eff} = \frac{n_{eff}^2}{2}\left[p_{12} - \upsilon(p_{11} - p_{12})\right] \tag{7.9}$$

The FBG-based sensing devices are known for their extreme sensitivity due to the dependency of Bragg wavelength on the strain and temperature, which greatly influence the grating period. Since the invention of FBG, it has been broadly explored in several sensing applications, including SHM, biosensing, medical applications, gas sensing, etc. (Rohan et al., 2024; Theodosiou, 2024). Due to their special physical configurations, FBG can detect various physical and chemical parameters, allowing them to perform absolute measurements at low cost. Moreover, since the FBGs are passive sensors, they can be operated under wavelength or multiplexed, which enables quasi-distributed monitoring. Following these advantages, the FBG has gained significant attention in geohazard monitoring.

Based on its sensing principle, an FBG can be utilized either as a temperature, strain, or pressure sensor. To identify wavelength shifts resulting from the variation in the temperature or strain of the ambient environment, the FBG sensor requires an optical source that continuously measures the reflection spectrum, along with a detection system or interogator to track the shift in the peak reflectivity as a function of the wavelength. To monitor the shift in the Bragg wavelength, a range of interrogation systems has been designed, enabling accurate measurement of strain and temperature. Mainly, there are two methods that are widely utilized to precisely detect reflected FBG peaks. The first technique involves launching light in an optical fiber using a broadband source (BBS) and analyzing the shift in the output spectrum under various environmental conditions (e.g., varying temperature, strain, RI, etc.). This setup comprises an optical circulator allowing the propagation of outgoing beams along the length of the fiber while redirecting the reflected beam toward an optical spectrum analyzer (OSA). A Michelson interferometry-based OSA offers highly precise wavelength measurements. This technique has been well exploited in wavelength meters for accurate monitoring of the wavelength of laser sources. The second technique relies on a tunable narrowband laser source. A tunable laser source scans across the wavelength range comprising the FBG reflection peak. The reflected optical power is monitored using a power meter, and the wavelength corresponding to the maximum reflected intensity is identified as the Bragg wavelength.

7.6 Detection of volatile organic compound

This section reviews various VOC sensors based on FBG and LPG technologies reported in recent years. Most of the reported FBG sensors are mainly focused on the monitoring of physical parameters such as strain and temperature. Topliss et al. (2010) proposed a novel VOC sensor based on calixarene-coated LPG. The fabricated sensor employs a wavelength interrogation technique to identify the peak shift when the sensor is exposed to VOC gases. The calixarene coating on the LPG exhibits average sensitivity toward aromatic compounds (e.g., toluene and benzene) while remain insensitive to aliphatic hydrocarbons such as hexane. The nanoscale cavities within the calixarene molecules trap VOC molecules, inducing a change in their RI and, in turn, affecting the transmission spectrum of the LPG. Although the sensor shows an average sensitivity of 231 ppmv for toluene, it demonstrates an excellent reversible response

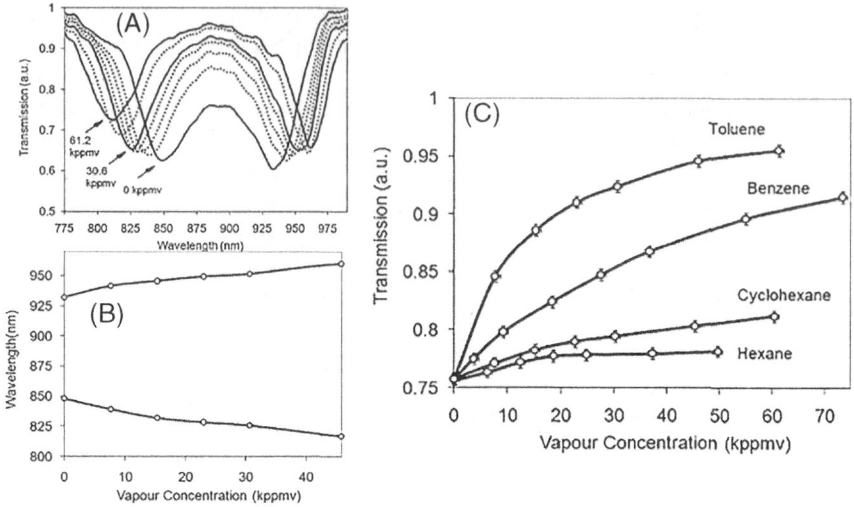

FIGURE 7.4 Sensing performance of the LPG sensor: (A) transmission spectrum in presence of different concentrations of toluene, (B) shift in the wavelength, and (C) sensitivity of the sensor towards toluene, benzene, hexane, and cyclohexane vapor. *From Topliss, S. M., James, S. W., Davis, F., Higson, S. P. J., & Tatam, R. P. (2010). Optical fibre long period grating based selective vapour sensing of volatile organic compounds.* Sensors and Actuators B: Chemical, 143*(2), 629–634. https://doi.org/10.1016/j.snb.2009.10.008.*

time of 15 seconds. Fig. 7.4A illustrates the device performance for various toluene concentrations ranging from 0 to 61.2 kppmv, while Fig. 7.4B shows the shifts in the spectra at longer and shorter wavelengths. The results indicate that the calixarene-coated LPG can selectively detect toluene among other VOCs, as shown in Fig. 7.4C. Notably, Fig. 7.4C clearly reveals that, up to a concentration of 10 kppmv, the device's sensitivity to hexane is only 7.5%, which is significantly lower compared to its sensitivity to toluene. For cyclohexane, the sensitivity is 14.8% lower than that observed for toluene. Moreover, the device shows excellent selectivity across a range of VOCs.

In the same year, Park et al. (2010) introduced a novel approach for the selective detection of multiple VOCs by integrating a thin layer of PDMS over FBG. The fabricated sensor was operated on a wavelength interrogation method, exploiting the swelling mechanism of PDMS in response to exposure to VOCs. This swelling of PDMS induces a tensile force on the FBG, which highly depends on the swelling ratio of the PDMS, resulting in accurate identification of the types of VOCs and their concentrations by analyzing the shift in the Bragg wavelength. The performance of the sensor was investigated in the presence of multiple VOCs, including acetone, 1-propanol, 2-propanol, and methanol, exhibiting a clear wavelength shift corresponding to the tested VOCs. In 2012 Konstantaki et al. reported the application of the ZnO nanorods grown over the LPG for the selective detection of ethanol (Konstantaki et al., 2012). Generally, ZnO in the nanostructured form exhibits a high surface-to-volume ratio, enabling it to selectively absorb small gaseous molecules, which enhances its sensitivity as compared to the ZnO film). When the fabricated sensor was placed in a VOC environment of varying concentration, its optical characteristics, including RI and absorption loss, changed accordingly. These variations in the optical properties were analyzed from the cladding modes of the LPG structure, which is manifested as strength and/or peak wavelength changes in its transmission spectrum. The fabricated sensor was based on a wavelength interrogation technique. To verify the growth of ZnO nanorods over LPG was characterized via scanning electron microscopy (SEM). The article reported that the synthesis conditions to obtain the ZnO nanorods play an important role in shaping the opto-geometrical feature of the random nanostructure overlayer, which influences the spectral response of the device. This feature was validated by a thorough investigation of ZnO growth under various growth conditions, deposited over LPG fiber, and tested in the presence of various concentrations of ethanol. The obtained spectral response was compared with the available simulation analysis. The fabricated device demonstrated the potential of ZnO in ethanol sensing. Moreover, the authors also observed a measurable variation of approximately 0.5 dB in grating strength, allowing the detection of low concentrations of ethanol. Despite several advantages and material characterization, several key factors, such as the detection limit, response time, and selectivity, were not considered during the device fabrication, posing a significant limitation to its practical application.

In 2018, Hromadka et al. reported LPG fiber coated with ZIF-8 for highly sensitive detection of VOCs. ZIF-8 is a porous sensing material which is a part of the zeolite imidazolate framework family (Hromadka et al., 2018). The working principle of the reported device was based on the wavelength interrogation method. The fabricated sensor was examined under exposure to different VOCs, including acetone, ethanol, and methanol, at high concentrations. In the reported work, the authors fabricated two different LPG sensors comprising grating periods of 109.0 and 109.5 μm, and denoted as Sensor A and Sensor B, respectively. Fig. 7.5 represents the transmission spectra obtained from sensor A (without ZIF-8 coating) and sensor B (coated with five layers of ZIF-8). The change in the U-shaped band of cladding mode (LP_{019}) into two distinct attenuation bands is clearly observed after depositing five layers of ZIF-8. In Fig. 7.5B, the attenuation bands corresponding to the LP_{019} mode are not visible before the deposition of ZIF-8, which concludes that these bands appear only after the deposition of ZIF-8. The highest sensitivity is obtained when the LPG operates exactly at the phase-matching turning point, where the bandwidth of the U-shaped attenuation band is evaluated. For ethanol, this

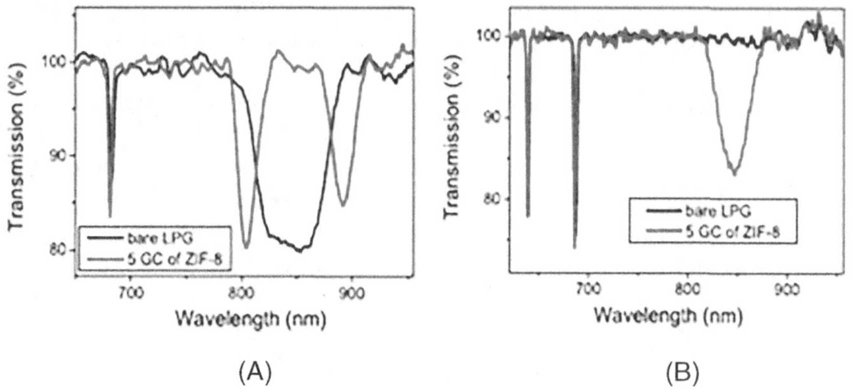

FIGURE 7.5 Transmission spectrum of a LPG with different period: Transmission spectrum of a LPG with period of (A) 109.0 μm–Sensor A and (B) 109.5 μm–Sensor B measured in air. *From Hromadka, J., Tokay, B., Correia, R., Morgan, S. P., & Korposh, S. (2018). Highly sensitive volatile organic compounds vapour measurements using a long period grating optical fibre sensor coated with metal organic framework ZIF-8.* Sensors and Actuators, B: Chemical, 260, *685–692. https://doi.org/10.1016/j.snb.2018.01.015.*

optimization reduces the calculated limit of detection (LOD) from 163 to 5.6 ppm. The optimized sensor demonstrates a high sensitivity of 0.018 ± 0.0015 and 0.015 ± 0.001 nm/ppm, with detection limits of 5.56 ppm for ethanol and 6.67 ppm for acetone, respectively.

In 2020, Corotti et al. introduced a novel material for detecting methanol vapor using an etched FBG sensor based on a wavelength interrogation technique (Corotti et al., 2020). The sensor was coated with diphenylalanine nanotubes (DNT) to enhance its capability to identify methanol in the methanol-ethanol vapor mixtures. The study was performed over two different phases of the DNT, i.e., hexagonal and orthorhombic. In a methanol-ethanol mixed environment, the hexagonal phase led to a redshift, whereas the orthorhombic phase induced a blueshift along with an excellent reversibility. When compared to the hexagonal phase, the orthorhombic phase demonstrated superior sensitivity, with a value of (−7.3 ± 0.8) pm/(%v/v) and a rapid response time. In the same year, Kumar et al. (2020) reported a graphene oxide coated etched FBG to detect adulteration of ethanol in petrol. In the reported study, the FBGs were fabricated in-house using a copper vapor laser operating at a repetition rate of approximately 5.5 kHz. The sensor was created by etching the FBG through a wet chemical etching method, followed by coating the etched region with a thin layer of graphene oxide. The performance of the graphene oxide-coated etched-FBG (EFBG) was compared with the non-coated EFBG, demonstrating a significant improvement in sensor performance due to the enhanced surface-to-volume ratio of the graphene oxide.

The fabricated graphene oxide-coated sensor was able to detect a minimum of 0.5% of ethanol in the petrol compared to the 5% using a non-coated EFBG. The sensor exhibited a maximum sensitivity of 0.18 dBm/%.

Konstantaki et al., 2020 fabricated silk fibroin coated LPG for the selective detection of methanol vapor. The fabricated sensor comprises a thin layer of silk fibroin coated over LPG via the drop casting method. The experimental configuration includes a light source, a gas chamber, and an OSA to examine the sensing performance of the device. The sensor operates on a wavelength interrogation technique. The setup was equipped with a sensitive manometer installed within the gas chamber to monitor the pressure variation caused by the evaporation of methanol, which was introduced as a liquid through the inlet. The sensing response of the device can be estimated by monitoring the shift in the output spectra, which takes place due to the interaction of methanol with Fibroin, leading to changes in its intermolecular hydrogen bonds. The fabricated sensor exhibited a notable sensitivity of 0.22 nm/mbar for methanol concentrations ranging between 80 and 100 mbar. This study marks the first exploration of the role of fibroin protein in sensing VOCs with a strong reversible response. In the same year, Wu et al. (2020) uniformly deposited a 300 nm thick layer of ZIF-8 on the LPG via in situ crystallization in order to selectively detect ethanol vapor. When the LPG deposited ZIF-8 was exposed to varying concentrations of ethanol, the fabricated sensor exhibited an excellent shift in resonant wavelength, varying from 1535.08 to 1535.04 nm for the concentration ranging from 0 to 9.8 ppm. The sensor demonstrated a linear response ($R^2 = 0.97$) for ethanol concentrations ranging from 9.8 to 540 ppm, along with the high selectivity for ethanol compared to the other VOCs, i.e., methanol, ethanol, acetone, ethyl ether, ethyl acetate, and dichloromethane. Yang et al. (2022) reported an ultrasensitive Ta_2C-MXene immobilized Au-coated TFBG sensor to detect high-risk molecules. Fig. 7.6 depicts the schematic diagram of the fabricated sensor along with the experimental setup used to monitor the sensor's performance. The integration of Ta_2C-MXene significantly improves sensitivity by 23%, demonstrating excellent linearity ($R^2 = 0.99$) and a high figure of merit (FOM = 2586.11 RIU^{-1}).

In 2023, Diego et al. reported the use of ZnO coating on FBGs for the detection of several common VOCs, including ethanol, methanol, acetone, and isopropanol at a low concentration of 25 ppm (Lopez-Torres et al., 2023). The working principle of the sensor was based on the chemostrictive effect. The chemostriction effect is generally the expansion or contraction of the sensing material due to the ambient environment or the chemical interaction, which can induce strain in optical fibers and shift the FBG resonance wavelength. When the fabricated sensor was exposed to target VOCs, it enabled the chemostrictive effect within the ZnO layer. Changes in VOC concentrations induce shifts in the output spectra, which were

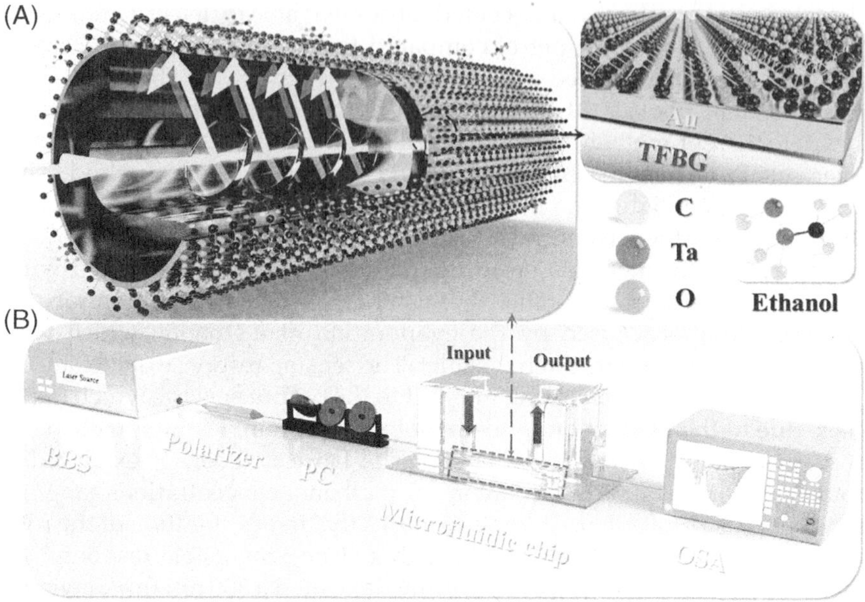

FIGURE 7.6 Ta_2C-M/Au coated TFBG sensor: (a) Schematic diagram of the Ta_2C-M/Au/TFBG sensor structure and optical detection path (b) experimental setup used to characaterize the sensor *From Yang, W., Cheng, Y., Jiang, M., Jiang, S., Liu, R., Lu, J., Du, L., Li, P., & Wang, C. (2022). Design and fabrication of an ultra-sensitive Ta_2C MXene/Au-coated tilted grating sensor.* Sensors and Actuators B: Chemical, 369. *https://doi.org/10.1016/j.snb.2022.132391.*

monitored to analyze the device's performance. The fabricated sensor demonstrated a maximum sensitivity of ~1 pm/ppm towards ethanol vapors, while other VOCs (i.e., isopropanol and methanol) exhibited lower sensitivities; also, the response to acetone vapors was obtained at ~0.5 pm/ppm. In the same year, Thirunavakkarasu et al. (2023) reported a carbon nanotube decorated over 45 nm-thick gold-coated FBG to detect ethanol in aqueous solution. The fabricated sensor demonstrated excellent sensitivity of 526 nm/RIU, providing a response more than 4 times higher than that of a sensor without carbon nanotubes. Later in 2024, the same group reported an improved performance by reducing the diameter of FBG (Thirunavakkarasu, 2024). The fabricated fiber comprises a 45 nm thick gold-coated over EFBG to detect ethanol based on the surface plasmon resonance (SPR) phenomenon. The thin layer of gold enhances sensitivity by improving the evanescent wave around the sensing region, leading to better interaction between the sensor and the analyte. The performance of the fabricated sensor was evaluated by analyzing the shift in the output spectra due to changes in ethanol concentrations in aqueous solutions ranging from 0% to 100%. The fabricated sensors exhibited a maximum

TABLE 7.1 Comparative table showing various grating-based optical fiber VOC sensors.

References	Target VOC	Type of grating	Sensing material	Sensitivity	LOD	Operating range
Topliss et al. (2010)	Toluene, cyclohexane, hexane, and benzene	LPG	Calixarene	—	—	0–80 kppmv
Park et al. (2010)	Methanol, 1-propanol, 2-propanol, acetone	FBG	PDMS	1.63 × 10–3 nm/% at $A_p = 2.25\ mm^2$ 2.9 × 10–3 nm/% at $A_p = 4\ mm^2$	–	0–1 relative volume ratio
Hromadka et al. (2018)	Ethanol, acetone, and methanol	LPG	Zeolite imidazolate framework (ZIF8)	0.015 ± 0.001 nm/ppm for acetone and 0.018 ± 0.0015 nm /ppm for ethanol	6.67 ppm for acetone 5.56 ppm for ethanol	987– 19,700 ppm for acetone 1240–24,800 ppm for ethanol
Corotti et al. (2020)	Methanol	FBG	Diphenilalanine nanotubes	–7:3 ± 0:8 pm/(%v/v)	–	
Konstantaki et al., 2020	Methanol	LPG	Fibroin	0.22 nm/mbar	–	
Barnes et al. (2008)	Xylene and cyclohexane	LPG	PDMS	19 nm/50% for xylene	300 ppm for Xylene	

(continued on next page)

TABLE 7.1 Comparative table showing various grating-based optical fiber VOC sensors—cont'd

References	Target VOC	Type of grating	Sensing material	Sensitivity	LOD	Operating range
Barnes et al. (2010)	Cyclohexane, m-xylene, trichloroethylene, and commercial gasoline	LPG	Copolymer PDMS /polymethyl-octylsiloxane (PMOS)		Xylene:134 ppm Trichloroethylene: 3300 ppm Cyclohexane: 1850 ppm Gasoline: 10,500 ppm	
Rodríguez-Garciapiña et al. (2021)	Acetone	LPG	PDMS	9.4×10^{-4} ppm^{-1}	170 ppm	
Partridge et al. (2014)	Toluene	LPG	Calix (Zhang et al., 2025) resorcin arene	–	~100 ppm	1–400 ppm
Konstantaki et al., 2012	Ethanol	LPG	ZnO nanorod	Measure in RI variation	–	100 min exposure time
González-Vila et al. (2017)	Ethanol, acetone, and toluene	TFBG	Molecularly imprinted polymer	0.44, 0.38, and 0.28 pm/ppm	Few ppm	0–17 ppm
Wales et al. (2016)	Hydrocarbons	4 FBG	Hydrophobic siloxane co-polymer	–	–	–
Latino et al. (2012)	Ethanol	FBG	PMMA	Linear response for 3% concentration	–	–
Lowder et al. (2007)	Dichloromethane and acetone	D shape FBG	PDMS	4000 and 6000 ppm	–	0–90,000 ppm

sensitivity of 20 pm/RIU. In addition to these FBG sensors, several other grating-based VOC sensors have been reported in the past decades, which are listed in Table 7.1.

7.7 Detection of semi-volatile organic compounds

Semi-VOCs (SVOCs) are considered more persistent and difficult to degrade as compared to the VOCs, and remain in the environment for an extended period of time. These SVOCs are generally found in soil, water, air, and other environmental media (see detailed discussion in chapter 10). Due to their poor degradation, these SVOCs pose a serious challenge to human health, ecological balance, and environmental safety (Feng et al., 2023).

Zhong et al. (2018) reported a photocatalytic LPG coated with Er^{3+}:$YAlO_3/SiO_2/TiO_2$ composite to detect phenol in aqueous solution. A polymer membrane with high phenol perm selectivity was synthesized by doping PEBA2533 with β-cyclodextrin and wrapped around the Er^{3+}:$YAlO_3/SiO_2/TiO_2$, creating a microchannel between the polymeric membrane and PLPG to facilitate the flow of target analytes. Furthermore, a Z-shaped microchannel was also fabricated within the PMMA plate, which is utilized to store the phenol analytes. In order to determine the sensing performance of the device, the sensitivity and selectivity of the sensors were tested for multiple parameters. The study includes the analysis of the thickness of sensing material, temperature variation, and various concentrations of phenolic analytes. The fabricated sensor exhibited high sensitivity, a rapid response time, and a low LOD of 0.0492 $nm_h(mg_hL^{-1})^{-1}$, 395 seconds, 7.5 μg/L for the phenol concentration ranging from 7.5 μg_hL^{-1} to 100 mg_hL^{-1}, for the pH values varying from 2.0 to 14.0, under moderate temperature of 10°C–48°C. Mishra and Chiang (2020) reported an LPG-based phenol sensors by immobilizing tyrosinase (polyphenol oxidase) in polyacrylamide. In the presence of a phenolic pollutant, the enzyme-functionalized LPG induces a change in the RI of the sensing material through an enzyme-mediated reaction. This results in a spectral shift in the resonance wavelength of the LPG, facilitating phenol detection. The obtained results exhibited a maximum sensitivity of 0.0088 nm/μM and LOD of 7.6 μM for the phenolic concentrations ranging from 0 to 1000 μM.

It has been reported that reducing the diameter of the optical fiber improves the sensing performance of the device by enhancing the evanescent field (Pathak et al., 2023b). Eisner et al. (2022) utilized the reduced fiber approach to detect 2,4-dinitrotoluene. The coating, doped with 3-aminopropyltrimethoxysilane and tetrabutylammonium hydroxide, forms a charge transfer complex with 2,4-dinitrotoluene. This interaction alters the surrounding RI when airborne 2,4-dinitrotoluene is adsorbed onto the silica-sol receptor coating (Zhong et al., 2024). The fabricated sensor exhibited a low LOD of 35.8 ± 0.9 ppb for 2,4-dinitrotoluene in the ambient environment. It has also been reported that the LOD can be improved by further reducing the diameter of FBG. Although reducing the diameter may improve the LOD, it can significantly weaken the strength of the fiber, making it fragile for robust performance. To overcome this challenge, several authors utilize the coating of plasmonic materials without compromising the fiber strength. Heidemann et al. (2018) reported FBG sensors functionalized with cysteamine and a thin layer of gold nanoparticles. The fabricated sensor was based on the reaction mechanism occuring between glyphosate and cysteamine, which alters the effective RI of the LPFG modes. When water containing glyphosate interacts with the sensor, it induces spectral shifts, allowing it to be interrogated via wavelength and/or intensity modulation techniques. The fabricated sensor achieves a low LOD of 0.02 μM. Ahad et al. (2018) fabricated polyaniline-coated FBG sensor to detect low concentrations of chloroform ranging from 10–100 ppm. In this work, the authors synthesized polyaniline utilizing aniline, sodium dioctyl sulfosuccinate, and ammonium persulfate as monomer, dopant, and oxidant, respectively. They utilize a wet chemical etching technique (i.e., 46% of HF) to reduce the fiber diameter, and the synthesized material was drop casted on the tapered region of the fiber. The interaction of chloroform with polyaniline induces a change in the RI of the material, which in turn affects the surrounding area of the FBG, resulting in a variation in the Bragg wavelength. This fabricated sensor exhibited excellent repeatability (up to 10 cycles) along with the rapid response and recovery times of 7 and 8 seconds, respectively. Additionally, the sensor also demonstrated a low LOD of 9.22 ppm. The sensing performance can also be improved (1.5 pm/ppm) by simply tuning the synthesis parameters of polyaniline Ahad et al. (2017).

7.8 Detection of other organic pollutants

In addition to VOCs and SVOCs, the environment is filled with other types of organic compounds, such as phthalates, bisphenol A, and other persistent organic pollutants, which are widely used in industrial and consumer products but can pose significant environmental and health risks. This section highlights the application of FBG sensors in the detection of additional organic pollutants.

Liu and team reported a plasmonic TFBG-based biosensor for the identification of estrogens in environmental water (Liu et al., 2021). Gold-coated TFBG was first immobilized by desthiobiotin–polyethylene glycol–thiol (DTB-PEG-SH). Later, TFBG was immersed in a bovine serum albumin mixture in order to perform the so-called blocking treatment. By utilizing molecular dynamics models, the 17β-estradiol–streptavidin complex was synthesized to work as a competitor for environmental estrogens. It was developed to release increased amounts of 17β-estradiol–streptavidin in the presence of large concentrations of estrogens in the measurand samples. The 17β-estradiol–streptavidin was selectively entrapped desthiobiotin functionalized over gold-coated TFBG, and the blocking treatment improved the specific streptavidin and desthiobiotin interaction. The interaction of the measurand can affect the surrounding RI of the TFBG, leading to a change in the plasmonic signal, which forms the basis of the device's sensing mechanism. The sensors exhibited a low LOD of 1.5×10^{-3} ng mL^{-1}.

Bisphenol A (BPA) is known to be an endocrine-disrupting compound that can result in serious health issues. It has been utilized in a broad range of applications, including polycarbonate plastics fabrication, thermal paper, and epoxy resins (Bousoumah et al., 2021). Considering health-related issues from BPA, it requires real-time and precise monitoring for health safety and environmental protection. Recently, Huang et al. (2023) reported LPFG functionalized with a DNA aptamer for real-time and label-free detection of BPA at extremely low concentration. In the reported work, a long DNA aptamer against Bisphenol molecules was selected and immobilized on the surface of the LPFG via chemical modifications. Upon exposure to BPA solutions, the long aptamer undergoes a significant conformational change, leading to a variation in the RI around the LPFG. This variation induces a detectable shift in the output spectra, resulting in improved device sensitivity and low LOD of 0.232 ng/mL towards BPA for the concentration ranging from 10^{-11} to 10^{-4} g/mL

7.9 Challenges and prospects

Over the past decades, several researchers have demonstrated the effectiveness, reliability, and accuracy of FBG-based sensing systems with their broad range of applications in physical, chemical, and industrial sectors. However, it is crucial to understand that the FBG-based sensors are not always applicable. One limitation associated with the FBG is that the sensors provide only quasi-distributed monitoring, with a limited number of sensors that can be connected in series. In contrast, fully distributed sensing techniques, e.g., BOTDR and Rayleigh backscattering optical frequency domain reflectometry, offer advantages in terms of the amount of data collection and the monitoring length.

Grating-based VOC gas sensors generally present numerous challenges that can limit device performance and real-world applications. In the absence of any sensing material, FBG-based VOC sensors exhibit low sensitivity to gas concentrations, making them less effective for detecting trace amounts of volatile gases, especially at low concentrations. To address this issue, several researchers have improved sensitivity by functionalizing the sensor surface with sensing materials. For example, applying metallic nanoparticle layers or polymers can significantly enhance the sensitivity of such sensors for target VOCs. In addition to sensitivity, FBG-based VOC sensors also suffer from poor selectivity when placed in mixed gaseous environments. Since they primarily detect variations in RI caused by gas absorption, it becomes challenging to distinguish between gases with similar optical characteristics in a mixed gaseous environment. This issue can be addressed by considering an appropriate selective material for the target VOC and by writing multiple gratings on the same fiber in which each grating can be selectively coated to produce different peaks corresponding to different target gases. In addition, several other factors play a crucial role in poor sensing performance, including cross-sensitivity to pressure, temperature, and humidity. A possible solution to this issue is the integration of temperature compensation mechanisms, for instance, writing an additional grating on the same fiber to facilitate temperature monitoring. Temperature monitoring is more challenging as the fluctuations in the surrounding temperature can influence shifts in the RI of the FBG, making it difficult to distinguish between wavelength shifts arising from gas absorption and those caused by temperature variations. The issue can be solved by utilizing temperature-insensitive optical fibers or by utilizing machine learning (ML) or other advanced algorithms, which can differentiate between the two sources of variation. Moreover, the grating-based VOC sensors have limited durability in toxic gaseous environments, where exposure to toxic gases and high humidity can degrade their sensing performance and, therefore, reduce the overall performance of the device. The issue can be solved by considering an additional protective layer, such as several types of polymer [e.g., teflon, polydimethylsiloxane, poly(methyl methacrylate), etc. (Coelho et al., 2015)] coatings that can help improve their durability and lifespan and, in some cases, improve the response time of the sensor. Although this approach can significantly improve the device performance, a thicker layer of protecting film may decrease the light-matter interaction and hence, affect the sensing performance (Pathak et al., 2023b). Therefore, the optimization of the protective layer is highly recommended to achieve the maximum output.

Every sensor design and approach offers tremendous advantages and certain limitations. In terms of achieving high sensitivity and remarkable performance, the optical fiber based devices can be ranked as SPR > evanescent field > FBG > microstructured optical fiber, based on their sensing mechanism and primarily influenced by the detection accuracy of the sensor configuration (Pathak et al., 2023b). In terms of the working mechanism, the surface plasmon based sensing device only requires a thin metallic layer (less than tens of nanometers) over the sensing head to produce the SPR effect, and exhibits high sensitivity and rapid response time. The evanescent field-based VOC sensors are similar to SPR sensors, utilize a thin layer of sensing material, and produce a rapid response and high sensitivity. When the light passes through the fiber (tapered, etched, side-polished fiber, etc.), it produces a higher evanescent field at the interface of the fiber and the sensing material. The absorption of evanescent fields varies according to the variation in the concentration of VOCs. The concentration of VOCs can be determined by monitoring the variations in the intensity of the transmitted light. Despite achieving a rapid response, the tapering of the fiber makes the sensor fragile, which may weaken the sensing system and affect the mechanical stability of the device. However, these limitations can be reduced significantly by utilizing TFBG and LPG-based VOC sensors.

Overall, integrating FBG into VOC sensing not only helps miniaturize the sensing technology but also incorporates multiple gratings to detect multiple parameters simultaneously, improving both sensing performance and device response time. It is well known that exhaled breath may contain more than 1000 VOCs along with respiratory droplets (Pathak et al., 2023a). Wang et al. (2023) reported that molecularly imprinted polymers (MIPs) can identify and distinguish multiple targets simultaneously, making them a potential material for selectively identifying multiple VOCs. However, the limited availability of binding sites for selective target binding may reduce sensor selectivity. In general, we note that the most effective approach for utilizing FBGs in VOC detection is to write multiple gratings on a single fiber, each corresponding to a specific VOC, which can significantly enhance device performance while maintaining high selectivity.

7.10 Conclusion

In summary, this chapter provides a detailed overview of the FBGs and their application in VOC sensing. Considering the advances in fiber sensing, FBGs appear to be one of the most promising tools for achieving high sensitivity and rapid response time when the surface is coated with appropriate sensing material. Compared to the traditional fiber, the light-matter interaction between the FBG and the sensing material can be improved in multiple ways, e.g., reducing the diameter of FBG, other types of gratings (i.e., TFBG and LPFG), etc. This chapter also provides in-depth information about the types, fabrication, and principles of optical FBGs, along with their applications in VOC and semi-VOC sensing.

References

Ahad, I. Z. M., Harun, S. W., Gan, S. N., & Phang, S. W. (2017). Application of fiber bragg grating sensor coated with polyaniline as an optical sensor for chloroform detection. *Polymers and Polymer Composites*, *25*(7), 555–562. http://www.polymerjournals.com/pdfdownload/1262914.pdf. doi:10.1177/096739111702500707.

Archambault, J.-L., Reekie, L., & Russell, P. J. (1993). High reflectivity and narrow bandwidth fibre gratings written by single excimer pulse. *Electronics Letters, 29*(1), 28–29. doi:10.1049/el:19930018.

Arregui, F. J., Claus, R. O., Cooper, K. L., Fernandez-Valdivielso, C., & Matias, I. R. (2001). Optical fiber gas sensor based on self-assembled gratings. *Journal of Lightwave Technology, 19*(12), 1932–1937. doi:10.1109/50.971687.

Atkins, R. M. (1992). Measurement of the ultraviolet absorption spectrum of optical fibers. *Optics Letters, 17*(7), 469. doi:10.1364/OL.17.000469.

Barnes, J., Dreher, M., Plett, K., Brown, R. S., Crudden, C. M., & Loock, H. P. (2008). Chemical sensor based on a long-period fibre grating modified by a functionalized polydimethylsiloxane coating. *Analyst, 133*(11), 1541–1549. http://pubs.rsc.org/en/journals/journal/an. doi:10.1039/b806129g.

Barnes, J. A., Brown, R. S., Cheung, A. H., Dreher, M. A., Mackey, G., & Loock, H.-P. (2010). Chemical sensing using a polymer coated long-period fiber grating interrogated by ring-down spectroscopy. *Sensors and Actuators B: Chemical, 148*(1), 221–226. doi:10.1016/j.snb.2010.04.007.

Bousoumah, R., Leso, V., Iavicoli, I., Huuskonen, P., Viegas, S., Porras, S. P., Santonen, T., Frery, N., Robert, A., & Ndaw, S. (2021). Biomonitoring of occupational exposure to bisphenol A, bisphenol S and bisphenol F: A systematic review. *Science of the Total Environment, 783*, 146905. doi:10.1016/j.scitotenv.2021.146905.

Coelho, L., de Almeida, J. M. M. M., Santos, J. L., & Viegas, D. (2015). Fiber optic hydrogen sensor based on an etched Bragg grating coated with palladium. *Applied Optics, 54*(35), 10342. doi:10.1364/ao.54.010342.

Corotti, R. D. P., Cunha, B. B., Barreto, R. C., Conceicao, A. L. C., & Kamikawachi, R. C. (2020). Diphenylalanine nanotube coated fiber Bragg grating for methanol vapor detection. *IEEE Sensors Journal, 20*(3), 1290–1296. http://ieeexplore.ieee.org/xpl/RecentIssue.jsp?punumber=7361. doi:10.1109/JSEN.2019.2946123.

Eisner, L., Flachenecker, G. Ü., & Schade, W. (2022). Doped silica sol layer coatings on evanescent field fiber Bragg gratings for optical detection of nitroaromate based explosives. *Sensors and Actuators A: Physical, 343*, 113687. doi:10.1016/j.sna.2022.113687.

Erdogan, T., Mizrahi, V., Lemaire, P. J., & Monroe, D. (1994). Decay of ultraviolet-induced fiber Bragg gratings. *Journal of Applied Physics, 76*(1), 73–80. doi:10.1063/1.357062.

Feng, C., Zhang, J., Bian, C., Li, L., Hu, R., Chang, H., Peng, F., Peng, X., & Zhong, N. (2023). Solid–liquid-core optical fiber biosensor for highly sensitive and selective detection of 4-chlorophenol in water. *Chinese Chemical Letters, 34*(12), 108457. doi:10.1016/j.cclet.2023.108457.

Gangwar, R. K., Kumari, S., Pathak, A. K., Gutlapalli, S. D., & Meena, M. C. (2023). Optical fiber based temperature sensors: A review. *Optics, 4*(1), 171–197. www.mdpi.com/journal/optics. doi:10.3390/opt4010013.

González-Vila, Á., Debliquy, M., Lahem, D., Zhang, C., Mégret, P., & Caucheteur, C. (2017). Molecularly imprinted electropolymerization on a metal-coated optical fiber for gas sensing applications. *Sensors and Actuators, B: Chemical, 244*, 1145–1151. doi:10.1016/j.snb.2017.01.084.

Grattan, K., & Meggitt, B. T. (2010). *Optical fiber sensor technology : advanced applications-bragg gratings and distributed sensors* (Softcover reprint of hardcover 1st ed. 2000, pp. 0–395). Springer.

Groothoff, N., & Canning, J. (2004). Enhanced type IIA gratings for high-temperature operation. *Optics Letters, 29*(20), 2360. doi:10.1364/OL.29.002360.

Heidemann, B. R., Chiamenti, I., Oliveira, M. M., Muller, M., & Fabris, J. L. (2018). Functionalized long period grating-plasmonic fiber sensor applied to the detection of glyphosate in water. *Journal of Lightwave Technology, 36*(4), 863–870. https://ieeexplore.ieee.org/xpl/mostRecentIssue.jsp?punumber=50. doi:10.1109/JLT.2017.2739243.

Hill, K. O., Fujii, Y., Johnson, D. C., & Kawasaki, B. S. (1978). Photosensitivity in optical fiber waveguides: Application to reflection filter fabrication. *Applied Physics Letters, 32*(10), 647–649. doi:10.1063/1.89881.

Hromadka, J., Tokay, B., Correia, R., Morgan, S. P., & Korposh, S. (2018). Highly sensitive volatile organic compounds vapour measurements using a long period grating optical fibre sensor coated with metal organic framework ZIF-8. *Sensors and Actuators, B: Chemical, 260*, 685–692. doi:10.1016/j.snb.2018.01.015.

Huang, C., Kou, F., Tu, T., Peng, K., Li, S., Guo, M., Wang, J., Zhou, Y., & Wang, J. (2023). Reusable aptamer sensors based on long period fiber gratings for detecting Bisphenol A at ultralow concentrations. *Optics & Laser Technology, 165*, 109572. doi:10.1016/j.optlastec.2023.109572.

Kashyap, R. (2013). *Fiber Bragg Gratings* (pp. 0–484). Academic Press.

Kawamura, K., Sarukura, N., Hirano, M., & Hosono, H. (2001). Holographic encoding of fine-pitched micrograting structures in amorphous SiO_2 thin films on silicon by a single femtosecond laser pulse. *Applied Physics Letters, 78*(8), 1038–1040. doi:10.1063/1.1347007.

Konstantaki, M., Klini, A., Anglos, D., & Pissadakis, S. (2012). An ethanol vapor detection probe based on a ZnO nanorod coated optical fiber long period grating. *Optics Express, 20*(8), 8472. doi:10.1364/OE.20.008472.

Konstantaki, M., Skiani, D., Vurro, D., Cucinotta, A., Selleri, S., Secchi, A., Iannotta, S., & Pissadakis, S. (2020). Silk fibroin enabled optical fiber methanol vapor sensor. *IEEE Photonics Technology Letters, 32*(9), 514–517. doi:10.1109/lpt.2020.2982451.

Kumar, P., Kumar, S., Kumar, J., Purbia, G. S., Prakash, O., & Dixit, S. K. (2020). Graphene-oxide-coated fiber Bragg grating sensor for ethanol detection in petrol. *Measurement Science and Technology, 31*(2), 025109. doi:10.1088/1361-6501/ab2d63.

Latino, M., Montanini, R., Donato, N., & Neri, G. (2012). Ethanol sensing properties of PMMA-coated fiber Bragg grating. *Procedia Engineering, 47*, 1263–1266. doi:10.1016/j.proeng.2012.09.383.

Lee, H. W., & Chiang, K. S. (2009). CO_2 laser writing of long-period fiber grating in photonic crystal fiber under tension. *Optics Express, 17*(6), 4533. doi:10.1364/OE.17.004533.

Li, T., et al. (2026). Review of Recent Advances in Femtosecond Laser Direct Writing Technology of Fiber Bragg Gratings. *Photonics, 13*, 215-215. doi:10.3390/photonics13030215.

Liu, H., Cai, J., Zhang, L., & Shu, X. (2025). Femtosecond laser inscribed small-period long-period fiber grating coated with a nano-assembled polyelectrolyte film for independent sensing of ammonia gas and temperature. *Sensors and Actuators B: Chemical, 423*, 136860. doi:10.1016/j.snb.2024.136860.

Liu, L., Zhang, X., Zhu, Q., Li, K., Lu, Y., Zhou, X., & Guo, T. (2021). Ultrasensitive detection of endocrine disruptors via superfine plasmonic spectral combs. *Light: Science and Applications, 10*(1). www.nature.com/lsa/. doi:10.1038/s41377-021-00618-2.

Lopez-Torres, D., Aguado, C. E., Pappas, G. A., Konstantaki, M., Klini, A., Lappas, A., Arregui, F. J., & Pissadakis, S. (2023). Piezotronic, ZnO overlaid Bragg grating organic vapor sensors. *IEEE Sensors Journal, 23*(12), 12536–12543. http://ieeexplore.ieee.org/xpl/RecentIssue.jsp?punumber=7361. doi:10.1109/JSEN.2023.3270966.

Lowder, T. L., Gordon, J. D., Schultz, S. M., & Selfridge, R. H. (2007). Volatile organic compound sensing using a surfacerelief D-shaped fiber Bragg grating and a polydimethylsiloxane layer. *Optics Letters, 32*(17), 2523–2525. doi:10.1364/OL.32.002523.

Martinez, A., Dubov, M., Khrushchev, I., & Bennion, I. (2004). Direct writing of fibre Bragg gratings by femtosecond laser. *Electronics Letters, 40*(19), 1170–1172. doi:10.1049/el:20046050.

Mihailov, S. J. (2012). Fiber bragg grating sensors for harsh environments. *Sensors, 12*(2), 1898–1918. http://www.mdpi.com/1424-8220/12/2/1898/pdf. doi:10.3390/s120201898Canada.

Mihailov, S. J. (2023). Femtosecond laser-induced Bragg gratings in silica-based fibers for harsh environment sensing. *APL Photonics, 8*(7). http://aip.scitation.org/journal/app. doi:10.1063/5.0142798.

Mishra, S. K., & Chiang, K. S. (2020). Phenolic-compounds sensor based on immobilization of tyrosinase in polyacrylamide gel on long-period fiber grating. *Optics and Laser Technology, 131*. https://www.journals.elsevier.com/optics-and-laser-technology. doi:10.1016/j.optlastec.2020.106464.

Mohamad Ahad, I. Z., Wadi Harun, S., Gan, S. N., & Phang, S. W. (2018). Polyaniline (PAni) optical sensor in chloroform detection. *Sensors and Actuators, B: Chemical, 261*, 97–105. doi:10.1016/j.snb.2018.01.082.

Okamoto, K. (2021). *Fundamentals of optical waveguides* (3rd ed.), pp. 1–713). Elsevier. https://www.sciencedirect.com/book/9780128156018. doi:10.1016/C2017-0-02432-1.

Othonos, A., Kalli, K., & Kohnke, G. E. (2000). Fiber Bragg gratings: Fundamentals and applications in telecommunications and sensing. *Physics Today., 53*(5), 61–62. doi:10.1063/1.883086.

Park, C. S., Han, Y., Joo, K. I., Lee, Y. W., Kang, S. W., & Kim, H. R. (2010). Optical detection of volatile organic compounds using selective tensile effects of a polymer-coated fiber Bragg grating. *Optics Express, 18*(24), 24753–24761. http://www.opticsinfobase.org/view_article.cfm?gotourl=http%3A%2F%2Fwww%2Eopticsinfobase%2Eorg%2FDirectPDFAccess%2F5B2288CE%2DF21B%2DE31D%2D34D7A151C1C9DD4F%5F207218%2Epdf%3Fda%3D1%26id%3D207218%26seq%3D0%26mobile%3Dno&org=. doi:10.1364/OE.18.024753.

Partridge, M., Wong, R., James, S. W., Davis, F., Higson, S. P. J., & Tatam, R. P. (2014). Long period grating based toluene sensor for use with water contamination. *Sensors and Actuators, B: Chemical, 203*, 621–625. doi:10.1016/j.snb.2014.06.121.

Pathak, A., Swargiary, K., Kongsawang, N., Jitpratak, P., Ajchareeyasoontorn, N., Udomkittivorakul, J., & Viphavakit, C. (2023a). Recent advances in sensing materials targeting clinical volatile organic compound (VOC) biomarkers: A review. *Biosensors, 13*(1), 114. doi:10.3390/bios13010114.

Pathak, A. K., Chaudhary, D. K., & Singh, V. K. (2018). Broad range and highly sensitive optical pH sensor based on Hierarchical ZnO microflowers over tapered silica fiber. *Sensors and Actuators, A: Physical, 280*, 399–405. doi:10.1016/j.sna.2018.08.013.

Pathak, A. K., Verma, S., Sakda, N., Viphavakit, C., Chitaree, R., & Rahman, B. M. A. (2023b). Recent advances in optical hydrogen sensor including use of metal and metal alloys: A review. *Photonics, 10*(2). http://www.mdpi.com/journal/photonics. doi:10.3390/photonics10020122.

Rodríguez-Garciapiña, J. L., Beltrán-Pérez, G., Castillo-Mixcóatl, J., & Muñoz-Aguirre, S. (2021). Application of the principal components analysis technique to optical fiber sensors for acetone detection. *Optics & Laser Technology, 143*, 107314. doi:10.1016/j.optlastec.2021.107314.

Rohan, R., Venkadeshwaran, K., & Ranjan, P. (2024). Recent advancements of fiber Bragg grating sensors in biomedical application: A review. *Journal of Optics, 53*(1), 282–293. doi:10.1007/s12596-023-01134-9.

Rovera, A., et al. (2023). Fiber Optic Sensors for Harsh and High Radiation Environments in Aerospace Applications. *Sensors, 23*(5), 2512. doi:10.3390/s23052512.

Russell, P. St. J., Archambault, J.-L., & Reekie, L. (1993). Fibre gratings. *Physics World, 6*(10), 41–48. doi:10.1088/2058-7058/6/10/25.

Sáez-Rodriguez, D., et al. (2011). Coupling between counterpropagating cladding modes in fiber Bragg gratings. *Optics Letters, 36*(8), 1518–1520. doi:10.1364/OL.36.001518.

Shu, X., Zhang, L., & Bennion, I. (2002). Sensitivity characteristics of long-period fiber gratings. *Journal of Lightwave Technology, 20*(2), 255–266. doi:10.1109/50.983240.

Song, X., Fan, Y., & Tang, X. (2025). FBG-based wearable sensors and devices in the healthcare field: A review. *Optics & Laser Technology, 181*, 111920. doi:10.1016/j.optlastec.2024.111920.

Theodosiou, A. (2024). Recent advances in fiber Bragg grating sensing. *Sensors, 24*(2), 532. doi:10.3390/s24020532.

Thirunavakkarasu, P., Khan, A. A., Shabaneh, A., Saidin, N. B., Sulaiman, M. S., & Noor, A. S. M. (2023). Carbon nanotubes enhancement of fiber Bragg grating surface plasmon resonance sensor. *2023 International Electronics Symposium: Unlocking the Potential of Immersive Technology to Live a Better Life* (pp. 155–158). Institute of Electrical and Electronics Engineers Inc. http://ieeexplore.ieee.org/xpl/mostRecentIssue.jsp?punumber=10242388. doi:10.1109/IES59143.2023.10242414.

Thirunavakkarasu, P. M., Khan, A. A., Noor, A. S. M., Saidin, N., & Waqas, N. (2024). Au coated etched FBG SPR sensor for the detection of ethanol in aqueous solution. *Optical Fiber Technology, 82*, 103584. doi:10.1016/j.yofte.2023.103584.

Thirunavakkarasu, P. M., et al. (2024). Au coated etched FBG SPR sensor for the detection of ethanol in aqueous solution. *Optical Fiber Technology, 82*, 103584. doi:10.1016/j.yofte.2023.103584.

Topliss, S. M., James, S. W., Davis, F., Higson, S. P. J., & Tatam, R. P. (2010). Optical fibre long period grating based selective vapour sensing of volatile organic compounds. *Sensors and Actuators B: Chemical, 143*(2), 629–634. doi:10.1016/j.snb.2009.10.008.

Venugopalan, T., Sun, T., & Grattan, K. T. V. (2008). Long period grating-based humidity sensor for potential structural health monitoring. *Sensors and Actuators A: Physical, 148*(1), 57–62. doi:10.1016/j.sna.2008.07.015.

Wales, D. J., Parker, R. M., Quainoo, P., Cooper, P. A., Gates, J. C., Grossel, M. C., & Smith, P. G. R. (2016). An integrated optical Bragg grating refractometer for volatile organic compound detection. *Sensors and Actuators, B: Chemical, 232*, 595–604. doi:10.1016/j.snb.2016.03.150.

Wang, S., Zhang, L. E., Zeng, J., Hu, X., Wang, X., Yu, L. I., Wang, D. U., Cheng, L., Ahmed, R., Romanovski, V., Li, P., & Zhang, Z. (2023). Multi-templates molecularly imprinted polymers for simultaneous recognition of multiple targets: From academy to application. *TrAC Trends in Analytical Chemistry, 166*, 117173. doi:10.1016/j.trac.2023.117173.

Wu, J., Zhang, W., Wang, Y., Li, B., Hao, T., Zheng, Y., Jiang, L., Chen, K., & Chiang, K. S. (2020). Nanoscale light-matter interactions in metal-organic frameworks cladding optical fibers. *Nanoscale, 12*(18), 9991–10000. doi:10.1039/c9nr09061d.

Xie, W. X., Niay, P., Bernage, P., Douay, M., Bayon, J. F., Georges, T., Monerie, M., & Poumellec, B. (1993). Experimental evidence of two types of photorefractive effects occuring during photoinscriptions of Bragg gratings within germanosilicate fibres. *Optics Communications, 104*(1–3), 185–195. doi:10.1016/0030-4018(93)90127-q.

Xing, R., Zhao, Z., He, C., Xu, P., Zhu, D., Li, Y., Li, Y., & Yang, Z. (2025). Optimizing FBG sensor layout of tunnel monitoring using improved multi-objective snow ablation optimizer based on radial basis function. *Measurement, 242*, 116289. doi:10.1016/j.measurement.2024.116289.

Xuan, H., Jin, W., & Zhang, M. (2009). CO_2 laser induced long period gratings in optical microfibers. *Optics Express, 17*(24), 21882. doi:10.1364/oe.17.021882.

Yan, X., Feng, Y., Xu, M., & Zhang, H. (2025). Monitoring Nasal Breathing Using an Adjustable FBG Sensing Unit. *Sensors, 25*(13), 4060. doi:10.3390/s25134060.

Yang, W., Cheng, Y., Jiang, M., Jiang, S., Liu, R., Lu, J., Du, L., Li, P., & Wang, C. (2022). Design and fabrication of an ultra-sensitive Ta_2C MXene/Au-coated tilted grating sensor. *Sensors and Actuators B: Chemical, 369*, 132391. doi:10.1016/j.snb.2022.132391.

Zhang, Z., Zhou, Z., & He, J. (2025). Development of FBG-based road ice thickness monitoring sensor and its application on the traffic road. *Optical Fiber Technology, 89*, 104070. doi:10.1016/j.yofte.2024.104070.

Zhao, J., Zhao, Y., Peng, Y., Lv, R. I., & Zhao, Q. (2022). Review of femtosecond laser direct writing fiber-optic structures based on refractive index modification and their applications. *Optics & Laser Technology, 146*, 107473. doi:10.1016/j.optlastec.2021.107473.

Zhong, N., Chen, M., Wang, Z., Xin, X., & Li, B. (2018). Photochemical device for selective detection of phenol in aqueous solutions. *Lab on a Chip, 18*(11), 1621–1632. http://pubs.rsc.org/en/journals/journal/lc. doi:10.1039/c8lc00317c.

Zhong, X., Xie, Q., Liu, Y., He, Y., Zhong, N., Zhang, Z., Karimi-Maleh, H., Peng, X., & Lichtfouse, E. (2024). Recent advances in optical fiber grating sensors for detection of organic substances. *Chemical Engineering Journal, 492*. https://www.sciencedirect.com/science/journal/13858947. doi:10.1016/j.cej.2024.152260.

Zhu, H.-H. U., Shi, B., & Zhang, C.-C. (2017). FBG-based monitoring of geohazards: Current status and trends. *Sensors., 17*(3), 452. doi:10.3390/s17030452.

CHAPTER

8

Microstructured optical fiber sensors for VOC detection

8.1 Introduction

Optical fiber sensors have a broad range of applications in the fields of environment, industry, and chemistry. Optical fiber sensors can be utilized to monitor various physical or chemical parameters such as refractive index, humidity, temperature, displacement, curvature, pressure, magnetic and electric field, and gases. The optical fiber sensor also provides several advantages, including immunity to electromagnetic interference, small size, cost-effectiveness, low transmission loss, and the ability to multiplex several sensors. Thanks to the extensive efforts of various research groups since the 1970s, the microstructured optical fibers (MOFs) were conceived, leading to significant advancements in sensing applications in terms of selectivity, sensitivity, and response time (Kaiser & Astle, 1974). The MOF is a class of specialty optical fiber consisting of several small air holes along the length of the optical fiber. In 1997, Russell et al. reported the term photonic crystal fiber (PCF) for periodically arranged air holes (Birks et al., 1997). The major difference between standard optical fiber (i.e., single-mode optical fiber and multimode optical fiber) and MOF is its structural design. The standard fibers generally comprise germanium (Ge) doped solid-core and solid-cladding made of fused silica to maintain the refractive index difference to allow total internal reflection (TIR). In the case of MOF, the core and cladding are made of only one material, but they also have periodically arranged air holes in the cladding region to lower the effective refractive index needed for TIR. The distribution of the air holes determines the specific propagation modes as well as the effective refractive index. Some of the modes are completely confined within the core of the MOF, and others extend into the cladding region, interacting with the air holes.

Photonic Sensors for Detection of VOC Biomarkers: Technological Principles and Medical Applications
DOI: https://doi.org/10.1016/B978-0-443-29166-1.00002-0

Copyright © 2026 Elsevier Inc. All rights are reserved, including those for text and data mining, AI training, and similar technologies.

Other than soft-glass and silica-based optical fibers, polymer-based optical fibers are also considered as one of the most interesting materials used for fiber manufacturing (Talataisong et al., 2019). Despite their relatively lower transmission efficiency compared to silica-based fibers, the plastic optical fiber maintains excellent flexibility even at a large core diameter. Polymers are generally favored for fabricating MOFs due to their low-temperature processing, cost-effective materials, mechanical ductility, and low manufacturing costs, making them suitable for a broad range of applications (Iezzi et al., 2023). Additionally, polymers are also considered advantageous for various sensing applications, more specifically, in biological and chemical contexts, as they are permeable to gases and can be easily functionalized directly. Therefore, the polymer-based optical fibers are widely used for local area network transmission and sensing applications (Monroy et al., 2003; Talataisong et al., 2019). The most commonly used polymer for the fabrication of plastic optical fiber is polymethylmethacrylate (PMMA), which exhibits a theoretical loss limit of 106 dB/km at $\lambda = 650$ nm (Zubia & Arrue, 2001). Other polymers which have been utilized for the fabrication of such fibers include polystyrene (Woyessa et al., 2017) and polycarbonate (Fasano et al., 2016); the latter is generally used in applications that require enhanced thermal stability. Both glass- and polymer-based MOFs are widely utilized to fulfill specific application needs. For example, the glass-based MOFs are generally employed in high-temperature or chemically stable environments, such as gas sensing and catalysis (Moeglen Paget et al., 2024), whereas the polymer-based MOFs are more considered in drug delivery, membrane separation, and wearable sensing devices (Argyros, 2010).

Based on the above discussed characteristics and versatility of MOFs, various researchers have demonstrated that the MOFs can be used as a good alternative platform to fabricate a broad range of sensing devices to measure strain (Shi et al., 2008), electric and magnetic field (Gangwar et al., 2016; Mathews et al., 2011), temperature (Gangwar et al., 2023), refractive index (Gangwar et al., 2023), gas sensors (Pathak & Viphavakit, 2022), etc. In particular, the MOF-based sensors are promising for the detection of various classes of gaseous molecules (Lopez-Torres et al., 2020). The presence of air holes in the MOF allows direct interaction between gaseous molecules and propagated light through the fiber, or with sensing material coated on the MOFs. In this chapter, the fabrication technique and light guiding principles of MOFs are first discussed, followed by applications to VOC sensing.

8.2 Fabrication of microstructured optical fibers

The fabrication of MOFs using a single material offers significant advantages compared to the standard core-clad based optical fibers

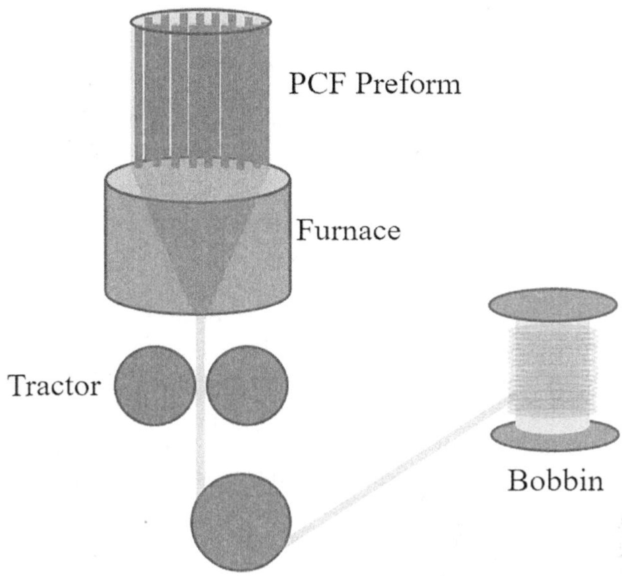

FIGURE 8.1 Fabrication of PCF: Schematic diagram illustrating stack-and-draw method used for the fabrication of MOF.

because issues related to the chemical and thermal compatibility between different materials are eliminated. Due to the unique feature of MOF's air hole arrangement, the manufacturing procedure of MOF is somewhat different than traditional optical fibers, which are fabricated using methods such as modified chemical vapor deposition (MCVD), vapor phase axial deposition (VAD), and outside vapor deposition (OVD), etc. (Amouzad Mahdiraji et al., 2014; Singh et al., 2023). The schematic diagram shown in Fig. 8.1 exhibits the general steps involved in the fabrication of MOFs. The fabrication of MOF includes several steps, which generally involve the fabrication of capillaries and rods with optimized diameter, cutting them into small pieces, and stacking them together to form a large-diameter glass tube called a preform (Amouzad Mahdiraji et al., 2014). The last step includes drawing the preform into a cane of about 2–3 mm diameter (for a longer length of optical fiber, this preform diameter may vary up to 6 mm), and finally pulling the cane into a fiber with the required diameter using a drawing tower (Amouzad Mahdiraji et al., 2014). During the pulling process, the preform is positioned at the top of a drawing tower, with its lower end inserted into a furnace. At high temperatures exceeding 1800°C (approximately 2100°C), the glass softens and is drawn by a tractor unit into the desired diameter and subsequently collected on a bobbin. More information about the preparation of the preform and the fabrication processes are briefly discussed in the next subsections.

8.2.1 Capillary and rod drawing

The formation of air holes along the length of MOFs involves the fabrication of capillary tubes with an optimized diameter. This step is considered to be the crucial step in the fabrication of MOF preforms due to their strong impact on the transmission characteristics of the final drawn optical fiber. These capillaries are generally fabricated from undoped, high-quality, low-OH-content silica tubes such as Suprasil F300, with inner and outer diameters of 19 and 25 mm, respectively (Amouzad Mahdiraji et al., 2014). A meter-long silica tube is drawn into capillaries with diameters of a few millimeters. Small capillaries possess handling challenges during the stacking process due to their flexible nature, ease of bending, and being prone to crossing each other while arranging them in a periodic order. To minimize structural imperfections during the fabrication process of the PCF, it is important to use capillaries of sufficient length to ensure stability during stacking and prevent collapse during the drawing process. However, the limited bore size of the furnace mouth restricts the diameter of the capillaries and the number of micro-structured cladding air-hole rings (Amouzad Mahdiraji et al., 2014). Additionally, having fewer rings potentially impacts the light confinement feature of MOFs. Therefore, in order to obtain an appropriate MOF, it is highly crucial to choose the preform size and capillaries. The sizes of capillaries are controlled by optimizing the preform feeding rate and the drawing speed of the capillary based on a simplified mass conservation law, as shown in the following relation

$$A_1 v_f = A_2 v_d$$

where, A_1, A_2, v_1, and v_2 represent the cross-sectional areas of preform, capillary, feeding rate, and drawing velocity, respectively. The ratios of outer and inner diameters of the capillaries are controlled by varying the surface tension of glass using controlled temperature and the pressure of the feeding gas. A glass rod matching the outer diameter of the capillary is required to serve as the core in a solid core PCF. Additionally, various rods of varying smaller diameters (fabricated in similar manners) are necessary for packing the capillary stack inside the large glass tube.

8.2.2 Sealing the capillary end

During the fabrication, the sealing of the ends of the capillaries is crucial in order to remove interstitial holes between them. However, for the fabrication of Kagome-type PCFs (i.e., hollow core fibers surrounded by a cladding with a lattice structure identical to a traditional Japanese Kagome basket pattern, the cladding generally consists of periodically arranged hexagonal air holes) this step is not required, as shown in Fig. 8.2A (Koehler et al., 2021). Generally, sealing of the ends of the capillary tube

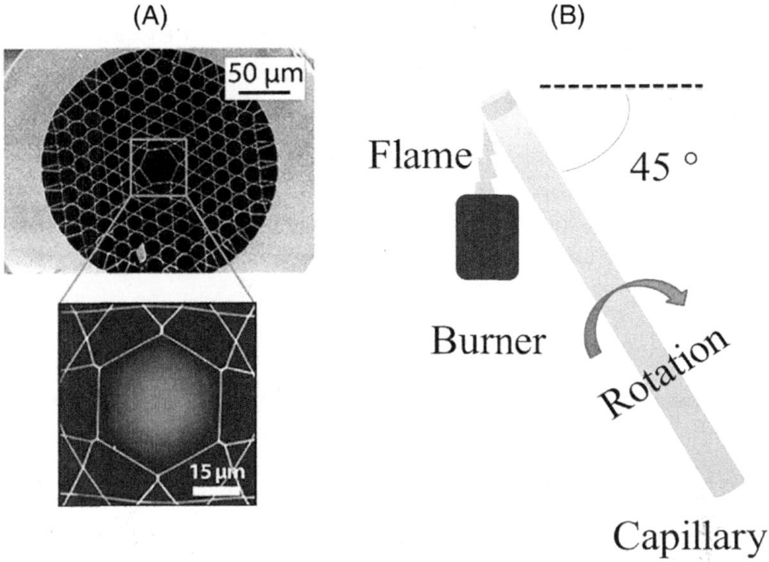

FIGURE 8.2 Kagome PCF and fabriaction: (A) Kagome PCF and confined modes and (B) sealing the end of the capillary tube using flame. *From Koehler, P., Lawson, T., Neises, J., Willkomm, J., Martindale, B. C. M., Hutton, G. A. M., Antón-García, D., Lage, A., Gentleman, A. S., Frosz, M. H., Russell, P. S. J., Reisner, E., & Euser, T. G. (2021). Optofluidic photonic crystal fiber microreactors for in situ studies of carbon nanodot-driven photoreduction.* Analytical Chemistry, *93(2), 895–901. https://doi.org/10.1021/acs.analchem.0c03546.*

is performed using a high-temperature Butanol-fuel hand torch (in the absence of a glass lathe) as shown in Fig. 8.2B. In order to obtain a uniform sealing, it is recommended to fix the torch on a stand. One of the primary challenges in sealing a capillary end is the slight increase in the diameter of the sealed head caused by the accumulation of molten glass. This issue can be minimized by slowly rotating the capillary tube during the sealing. Additionally, holding the capillary tube at an oblique angle relative to the flame helps to minimize the enlargement effect. Experimentally, it has been validated that the 45° angle is most effective and provides satisfactory sealing. However, longer sealing durations or processing at an extremely high temperature may lead to a larger sealed end.

8.2.3 Capillary cutting

To form a suspended core-PCF preform, all the capillaries are generally cut into 30 to 40 cm lengths. However, this approach cannot be considered ideal. Instead, cutting the capillaries to slightly varied lengths can make stacking them easier. This is because when utilizing the self-pressurization of capillaries (essential for closing the gaps between capillaries in the stack) and sealing them one by one (rather than sealing all the ends simultaneously with a glass lathe), the sealed ends have a slightly larger diameter than the capillaries themselves. If all the capillaries are cut to the same length, their sealed ends will sit directly on top of each other, making stacking difficult. However, by cutting the capillaries to slightly different lengths (e.g., ranging from 33 to 35 cm), the sealed ends of adjacent capillaries do not interfere with each other, preventing significant misalignment. When a glass lathe is used, all the capillaries are sealed simultaneously by locally heating the stack end due to surface tension. In this case, preform pressurization is applied from the side of the preform.

8.2.4 Cleaning

Before inserting the fabricated capillary tubes into a jacket tube, they require cleaning of their outer walls using acetone, followed by isopropanol with lint-free tissues (Amouzad Mahdiraji et al., 2014). Ensure that no residues, whether from external contaminants or the cleaning process, remain inside or on the surface of these capillaries. These residues may cause mechanical instability or burn at high temperature, producing trapped gases once fused into the glass during the fabrication process at high furnace temperature. The residue trapped inside may deform the cane structure during the pulling process.

8.2.5 Stacking

After cleaning, the capillaries are stacked together in a jacket tube in the form of a triangular lattice by utilizing a pair of hexagonal jigs. The hexagonal jigs are fabricated according to the internal diameter of the jacket tube, with the longest hypotenuse in the hexagon being equal to or slightly larger compared to the internal diameter of the tube. The hexagonal jigs are designed and fabricated in such a manner that one face of the hexagon remains parallel to the surface of the stacking table, so that the capillaries can be conveniently arranged layer by layer in the form of triangular lattice (Amouzad Mahdiraji et al., 2014). A 10–15 cm long tube is fitted between the fabricated jigs and is utilized during the stacking process, which allows the two sides of the capillaries to sit on the hexagonal jig. This tube has an internal diameter identical to the jacket tube, generally matching the portion of the tube from which the jacket is cut. The stacked capillaries will later be transferred to a longer tube. In order to facilitate the stacking process, parallel holders are utilized to keep the tube and the hexagonal holes concentrically aligned. The capillaries are arranged one by one, passing through the tube. Once all the capillaries are stacked, the central capillary will be replaced with a rod of the same diameter for an index-guiding PCF. In the case of a PBG fiber or hollow-core PCF with a core tube, a larger diameter capillary will be inserted, encompassing the central capillary and a few surrounding rings. The empty space between the hexagonal structure of the stacked capillaries and the circular aperture of the jacket tube is filled with packing rods. These rods secure and stabilize the hexagonal structure within the jacket tube. Without this, the structure could loosen and become misaligned during the transfer of the preform into the final jacket tube.

8.2.6 Transferring stacked capillaries into jacket tube

At this stage, the capillary stack needs to be transferred to the longer jacket tube from the shorter one fabricated above. In general, this longer jacket tube is slightly longer (~5–10 cm longer) or almost the same length as the stacked capillaries in order to maximize the utilization of the stacked preform. The stack is transferred by pushing the entire stacked preform from the shorter tube into the longer tube. The sealed ends of the capillaries are inserted into the longer tube first, while the pushing force is applied from the unsealed ends using a soft, wide pad. As previously mentioned, this transfer is easier if the heads of all capillaries and rods are aligned by their unsealed ends. Careful transfer of the stack is highly essential in order to avoid breaking or misalignment of the capillaries. During the transfer, it must be ensured that there is at least 5 cm of free space maintained between the end of the jacket tube and the sealed ends of the stacked capillaries. This spacing is important to prevent the capillary ends from collapsing together during handling. After the completion of the transfer, the stacked capillaries are tightened using thin rods inserted from both ends of the jacket tube. This tightening process continues until all capillaries and rods are securely fixed, ensuring no loose components remain in the preform.

8.2.7 Glass handle connection

Typically, a glass handle is attached to the preform using a glass working lathe with a high-temperature hydrogen-oxygen fuel burner (Amouzad Mahdiraji et al., 2014). To eliminate the need for a glass lathe, two alternative techniques are proposed. One technique involves using the furnace of the fiber drawing tower, where one end of the preform is connected to a 40 cm long glass tube of identical diameter, shown in Fig. 8.3A. The preform and tube are aligned concentrically in the hot zone area of the furnace while it is off. For instance, this can be done by holding the preform at the lower end of the furnace and fixing the tube at the top. After carefully aligning and ensuring that both parts are in contact without applying force, the furnace is turned on, and waiting untill the temperature reaches the glass melting point (i.e., 2000°C–2050°C) (Amouzad Mahdiraji et al., 2014). At this temperature, both parts fuse together within seconds. To ensure a strong, gap-free connection at the joint, the handle tube installed on top is gently pressed down by 1 mm. The furnace temperature is quickly reduced

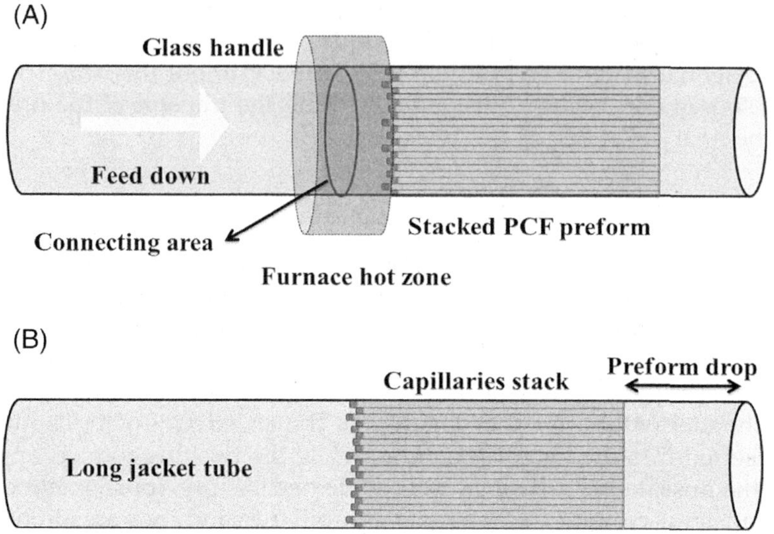

FIGURE 8.3 Handle connection: Schematic diagram of (A) handle connection along with the PCF preform using tower furnace and (B) handle connected with PCF preform.

to prevent excessive deformation of the preform. An alternative approach to the above method, which eliminates the need for the handle connection process altogether, involves using a longer jacket tube, approximately 1 m in length, instead of a shorter tube that requires a handle. In this case, the stack of capillaries is directly transferred from the initial short tube into the significantly longer one, as depicted in Fig. 8.3B. Care must be taken to prevent any breakage of the capillaries during the transfer into the longer jacket.

8.2.8 Photonic crystal fiber cane drawing and annealing

Before starting the drawing process, annealing is required in order to fix the stacked capillaries in the jacket as well as to provide heat polishing on the preform. This is performed by passing the preform vertically through the furnace multiple times while the furnace is set at the melting temperature ranging between 1500°C and 1700°C. Later, the preform is drawn down to a cane size of 2–3 mm in diameter. This cane is then jacketed with another glass tube to form PCF of varying lattice constants, core sizes, and outer diameters. Additionally, directly pulling a preform into a 125-μm-diameter fiber may result in trapping air, argon, or other processing gases inside or between the stacked capillaries. These trapped gases may expand excessively and possibly damage either the fabricated fiber or the preform itself, specifically near the end of the preform. During the cane pulling process, a high vacuum pressure is applied from the top of the preform to collapse the holes between the capillaries. The magnitude of the vacuum pressure mainly depends on the temperature of the furnace and the pulling speed. To obtain a high-quality cane, the vacuum pressure and temperature are precisely controlled by monitoring the fabricated cane with an optical microscope.

8.2.9 Photonic crystal fiber drawing

In this step, the fabricated canes are drawn to the desired size of PCF. The small output fiber size of 125 μm can easily cause the air holes in a PCF to collapse. In order to prevent any collapse or deformation of the air holes, the temperature and drawing speed are precisely monitored. There are two common techniques that have been reported to precisely control the air holes during the fabrication of PCFs. One method used the self-pressurizing approach on air holes by sealing the top end of the cane in order to trap any gas and air inside the holes during the pulling process. This leads to a gradual, uncontrollable increase in the size of the air holes in the fibers as the cane lengths shorten during the pulling process. While this technique is not considered appropriate for fabricating long-length fibers, it is a simple and effective method for producing large-hole small-core (LHSC) PCFs intended for shorter-length applications (tens of meters). The second technique is to maintain the cane end open and control the size of air-holes by applying positive pressure. In this case, the cane is either pulled directly into the fiber (without a jacket tube), or a complicated pressurizing system is required to close the gap between the cane and the jacket tube while positively pressurizing the air holes of the cane. The resulting fiber is later coated using a polymer, followed by UV curing during the pulling process to improve its mechanical strength.

8.2.10 Jacketing

The jacketing of PCF cane is done primarily to reduce the core of the optical fiber and to change the ratio of the desired radius of air holes (d), over the lattice spacing (ʃ) (i.e., d = ʃ). Generally, the jacketing is performed by fusing the cane and jacket together. The process takes place by carefully inserting the drawn cane into another type of capillary tube comprising slightly larger diameter compared to the cane. This procedure is typically done using two approaches: either by applying vacuum between the jacket tube and cane during the pulling, or by fusing the jacket with the cane in the furnace in a separate step before drawing the fiber. In the first method, the sealed top end of the cane is inserted into the larger diameter jacket. To prevent the cane from sliding through the jacket during the insertion, the cane is fixed to the jacket tube from the top without disturbing or blocking the pressurization path. The procedure is performed using a hand torch by deforming the cross-section of the jacket tube to an oval shape, close to the top end of the cane. The jacketed cane (fiber preform) is drawn to the desired fiber size while a low vacuum is applied from above to fuse the jacket and cane together. In this approach, the small amount of air trapped inside the cane enlarges the air-hole size, as discussed in the previous section. The second method is almost similar to the first, except this approach involves an annealing process at 1900°C before drawing the fiber (Amouzad Mahdiraji et al., 2014). During the annealing, a small vacuum pressure is applied to fuse the jacket with the cane. This process initiates from the lower end of the cane to avoid trapping of any gases or air between the jacket and cane. Later, the sealed top edge of the cane is removed to apply a positive pressure during the drawing process. In this manner, the air-hole size is controlled during the drawing process by adjusting the applied pressure.

8.3 Types of microstructured optical fibers

There are several ways to categorize MOFs besides the type of material, such as dimensions, geometries, shapes, and distributions of air holes. MOF structures exhibit unique properties that result in various propagation effects based on the fiber structure and arrangement of air holes. Small mode areas enable soliton propagation (Lopez-Torres et al., 2020), while large mode areas can handle high-power optical beams. Additionally, the birefringence and group velocity dispersion can also be controlled (Xuan et al., 2017; Zografopoulos et al., 2012).

8.3.1 Solid core and hollow core photonic crystal fiber

Fig. 8.4A and B exhibits the schematic of solid core and hollow core PCF, respectively. The core refractive index for the solid core PCF is stronger than the cladding refractive index, allowing the propagation of launched light due to the modified TIR, similar to the guiding mechanism of standard fibers. The solid-core PCFs are widely used due to the degrees of freedom offered by their manufacturing via manipulating the size of the air hole and their periodicity. This is the main reason that most of the PCF-based sensors are based on the solid core PCF. In contrast, the hollow core PCF comprises a hollow space in the middle of the fiber surrounded by several periodically arranged air holes, as shown in Fig. 8.4B. This configuration allows the great confinement of the launched light along the hollow cores due to the photonic band gap effect. Light is mostly confined in the hollow core region, which ensures a quasi-lossless propagation and allows the optical pulses to maintain their original form. In this design,

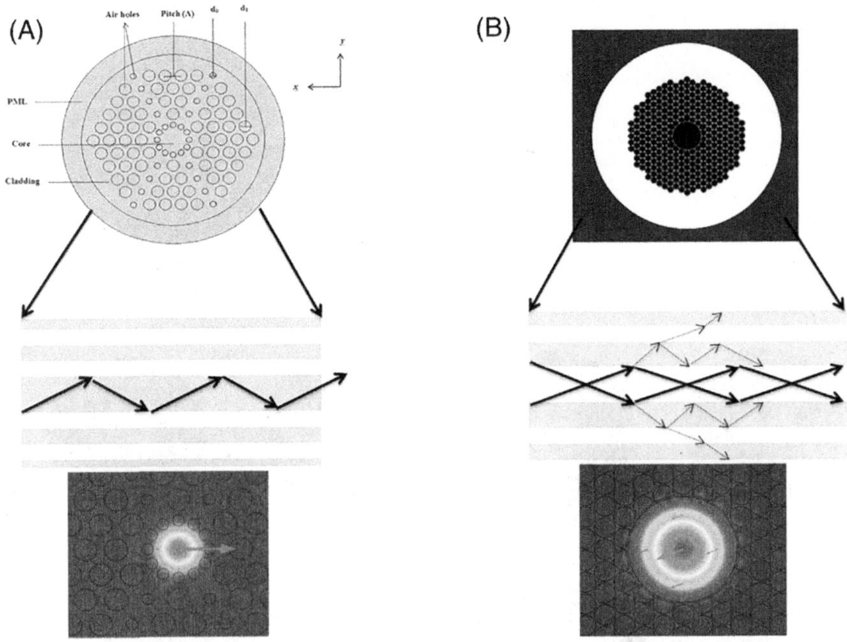

FIGURE 8.4 Solid and hollow core PCF: Schematic diagram and propagated modes for (A) solid core PCF and (B) hollow core PCF. *From Wan, B., et al (2021). Characteristic analysis and structural design of hollow-core photonic crystal fibers with band gap cladding structures.* Sensors *21(1), 284; N-yorbe, J. N., Akowuah, E. K., Danlard, I., Dede, A., & Haxha, S. (2023). Design and analysis of a hexagonal circular photonic crystal fibre with ultra-high negative dispersion and low confinement loss.* Results in Optics, 12. *https://doi.org/10.1016/j.rio.2023.100488.*

light cannot propagate unless its spectral components belong to some specific wavelength intervals determined by the spacing and the size of the air holes. The fact that the propagation medium is air prevents the hollow core fibers from being good candidates for nonlinear processes like second harmonic generation, since nonlinearities are weak in the air, and therefore, spectral broadening is difficult to produce.

8.3.2 Twin-core photonic crystal fiber

A twin-core PCF has been reported as an alternative to standard optical fibers and hollow core PCF (Liu & Tam, 2018). Fig. 8.5A shows the basic schematic diagram of twin-core PCF. The twin-core PCF comprises the insertion of two pure silica rods by replacing two capillary tubes during the stacking arrangement. The fabrication of twin-core PCF using a stacked process is illustrated in Fig. 8.5B. The air holes are fixed on a silica substrate. The pitch and cladding air holes diameter are defined by Λ and d,

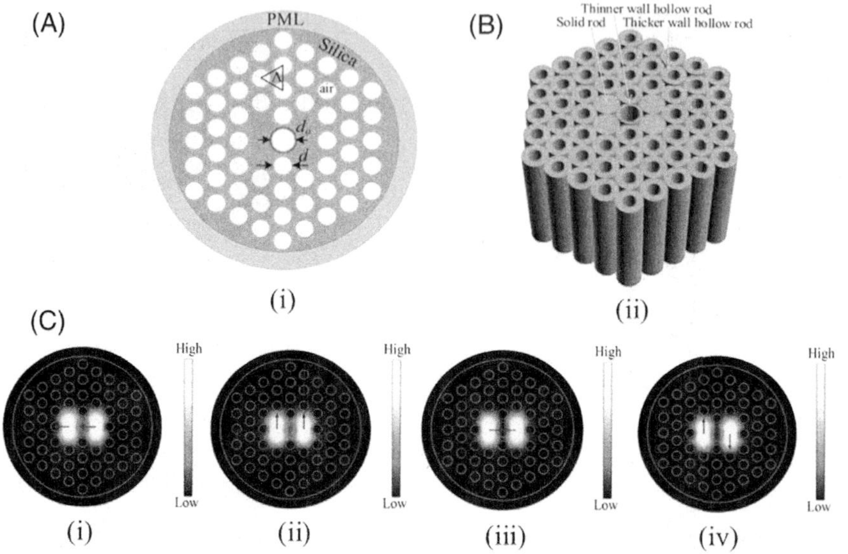

FIGURE 8.5 Twin core PCF along with the field distribution: (A) Cross sectional view of twin core PCF, (B) stacking of capillaries and solid rod for the fabrication of twin core PCF, and (C) field distribution for (i) x-even, (ii) y-even, (iii) x-odd, and (iv) y-odd mode at wavelength 1.8 μm. *From Mollah, M. A., Yousufali, M., Ankan, I. M., Rahman, M. M., Sarker, H., & Chakrabarti, K. (2020). Twin core photonic crystal fiber refractive index sensor for early detection of blood cancer.* Sensing and Bio-Sensing Research, 29. *https://doi.org/10.1016/j.sbsr.2020.100344.*

respectively. The big central air hole of the PCF, having a diameter of d_0 (marked in red circle) can be filled by the analytes for sensing purposes (Mollah et al., 2020). The large central air hole is positioned between the two cores to facilitate effective interaction of the electric field with the analyte during light propagation through the PCF. From the cross-section of twin-core PCF as shown in Fig. 8.5A, it is clearly visible that two separate waveguides are formed by the two fiber cores, which induce mode coupling within the PCF. According to mode coupling theory, the twin-core PCF exhibits coupling characterized by four supermodes (x-even, y-even, x-odd, and y-odd). Modes of the individual cores combine to form these supermodes, in which the electric field distribution is symmetric and asymmetric, respectively. Fig. 8.5C(i–iv) exhibits the electric field distribution of the x-even, y-even, x-odd, and y-odd modes at 1.8 μm, respectively. Therefore, it has been suggested that the twin-core PCF possesses a dual-core configuration, facilitating efficient mode coupling and hence significantly improving the light-matter interaction and enhancing the sensitivity and overall performance of the device (Abd-Elkader et al., 2025; Siddik et al., 2020).

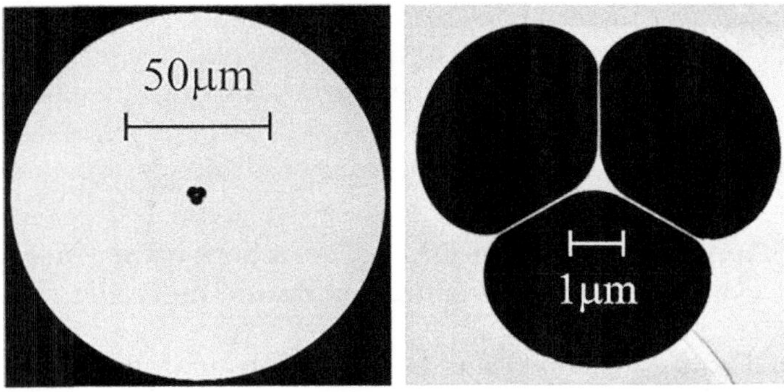

FIGURE 8.6 Suspended core PCF: SEM image illustrates cross section view of suspended-core PCF along with the central feature attached next to it. *From Monro, T. M., Warren-Smith, S., Schartner, E. P., Franois, A., Heng, S., Ebendorff-Heidepriem, H., & Afshar, S. (2010). Sensing with suspended-core optical fibers.* Optical Fiber Technology, *16(6), 343–356. https://doi.org/10.1016/j.yofte.2010.09.010.*

8.3.3 Suspended core photonic crystal fiber

The suspended-core PCF was first introduced by Monro et al. (2010). The suspended core PCFs are a form of MOF which utilizes a suspended core structure along the length of the fiber using thin bridges supported by the surrounding air holes, as shown in Fig. 8.6. Suspended core PCF offers enhanced light-matter interaction due to their small core size and air hole structure, enabling high sensitivity for sensing applications (Xuan et al., 2017; Yu et al., 2020). They provide tailored dispersion properties, making them ideal for nonlinear optical processes and advanced telecommunications. Additionally, their design allows for high power handling, suitable for high-power applications, i.e., high-power laser delivery, supercontinuum generation, and nonlinear optical signal processing, etc.

Talataisong et al. (2018) demonstrated the extrusion of polymer-based suspended core PCF using a 3D printer. In this work, the author fabricated a nozzle with an inverse structure profile, which was integrated with a 3D printer, for the extrusion of fiber. The structured nozzle design was separated into two pieces: the structured body and the cover. While the body forms the solid core of the extruded fiber and connects it to the outer cladding with three struts, the cover controls the thickness of the outer cladding. To fabricate suspended core PCF, the designed nozzle was attached to the 3D printer head while a standard 3D printer filament made of acrylonitrile butadiene styrene which was fed through a heated nozzle. The quality of the extruded polymer was assessed by extruding the fiber preform through the heated structured nozzle without any additional drawing. The fiber extrusion parameters, such as nozzle temperature and feeding speed, were investigated to explore their effect on the quality of the extruded fiber preform and achieve the optimal effective temperature experienced by the polymer optical fiber during drawing (Talataisong et al., 2019).

8.4 Application of photonic crystal fibers for monitoring various volatile organic compound molecules

The MOFs are especially useful in gas sensing applications due to their structural dimensions and design, which allow target gaseous molecules to diffuse through the structure, while the air holes act as vapor cells. One of the major differences between the detection of VOCs and other gases is that the absorption bands for VOC molecules are not found in the range of telecommunication windows (1300 and 1550 nm). Thus, the sensing mechanism of MOF-based VOC sensors is mostly due to the basic phenomenon of detecting changes in the effective refractive index of its surroundings. All the reported MOF-based VOC sensors are based on wavelength interrogation techniques. The use of PCF in gas sensing applications requires the utilization of additional sensing materials along with various designs of PCF to maximize the light-matter interaction. This section provides a brief discussion of PCF designs and their application in VOC sensing.

8.4.1 Novel designs and numerical findings of PCF-based VOC sensors

The finite element method (FEM) has been exploited to design novel PCF configurations, allowing for maximum light-matter interaction and improved sensitivity of the devices. Maidi et al. (2021) proposed a novel design of a hexagonal lattice PCF facilitated with a liquid-infiltrated core to detect ethanol and benzene. Fig. 8.7A shows a cross-sectional view of the design PCF, which comprises three different layers with the middle layer functioning as a liquid-infiltrated elliptical core surrounded by a cladding consisting of multiple air holes, and finally the top layer, which acts as a perfectly matched layer (PML), used to avoid reflection (see Chapter 4). The core of the PCF carries light along the length of the fiber, while the cladding surrounding the core works to confine the light within its region. The hexagonal PCF sensor is structured with cladding air holes having a uniform diameter of $d_a = 1.6$ μm. A single cladding air hole is removed from the center in order to insert three elliptical air holes, which work as the core of the proposed design. The air holes are periodically arranged with a pitch distance (P) = 1.633 μm. The elliptical hole has an

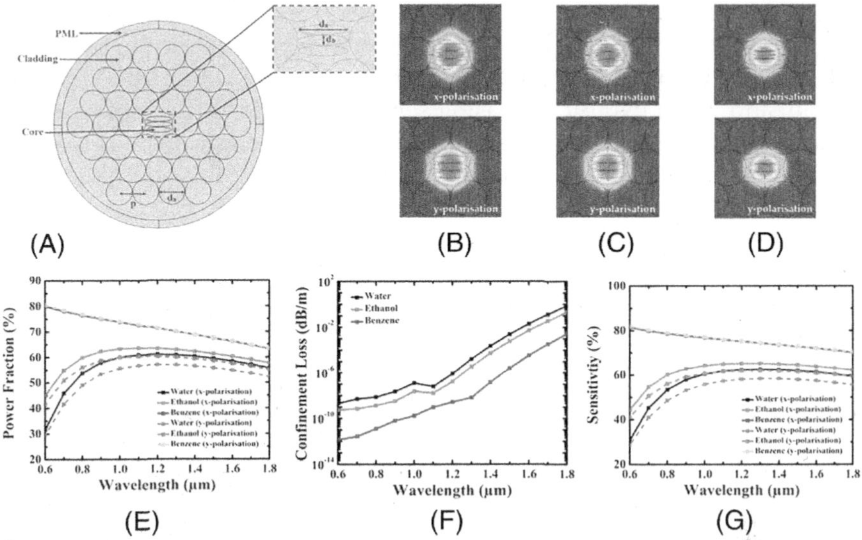

FIGURE 8.7 Hexagonal lattice PCF with liquid-infiltrated core for VOC sensing: (A) Schematic of the designed PCF, allowed mode profile for different polarization for, (B) water, (C) ethanol, (D) benzene, (E) power fraction of the PCF corresponding to wavelength for different analytes and different polarization, (F) Confinement loss observed for different analytes, and (G) relative sensitivity of the designed sensor in presence of various analytes. *From Maidi, A. M., Yakasai, I., Abas, P. E., Nauman, M. M., Apong, R. A., Kaijage, S., & Begum, F. (2021). Design and simulation of photonic crystal fiber for liquid sensing.* Photonics, *8(1), 1–14. https://doi.org/10.3390/photonics8010016.*

x-semiaxis (d_a) = 1.6 μm and a y-semiaxis (d_b) = 0.32 μm. The central elliptical hole is positioned directly at the center, surrounded by similar holes. These core holes facilitate the injection of various liquid analytes. The total fiber diameter is 13.06 μm, and the PML is optimized to be 10% of the cladding diameter to match the boundary condition (Maidi et al., 2021). Fig. 8.7B–D exhibits the mode profile for the various liquid analytes (i.e., water, ethanol, and benzene) analyzed at an operating wavelength of 1.3 μm for x- and y-polarization. Fig. 8.7E illustrates how power fraction varies with operating wavelength for water, ethanol, and benzene. Ethanol and water show increasing power fractions up to 1.0 μm, followed by stabilization and a slight decrease. In contrast, benzene exhibits a decreasing trend with increasing wavelength. Fig. 8.7F demonstrates the relationship between the confinement loss as a function of the wavelength for various analytes for the designed PCF. For all three analytes, the confinement losses generally increase with the increasing wavelength, leading to more light leaking out into the cladding. It can be seen that the confinement loss of benzene has a slightly lower value, as compared to ethanol and water. At an operating wavelength of 1.3 μm, confinement losses are 1.63×10^{-7}

dB/m for water, 3.55×10^{-8} dB/m for ethanol, and 6.76×10^{-11} dB/m for benzene. Fig. 8.7G illustrates how the relative sensitivity of the proposed PCF changes with operating wavelength for ethanol, water, and benzene. Ethanol and water exhibit increasing sensitivity up to 1.0 μm, followed by a plateau and a slight decrease with further wavelength increase. At 1.3 μm, the relative sensitivities are 60.90% for water, 63.06% for ethanol, and 70.07% for benzene.

Hossain et al. (2024) designed an octagonal shaped PCF based VOC sensor in order to detect ethanol and benzene in the terahertz (THz) spectrum. Fig. 8.8A exhibits the cross-section of the designed PCF. The design of PCF comprises five layers of air troughs and two layers of O-PCF arranged in an octagonal pattern and close to the center of the fiber. The distances between the air holes are defined by d_1 and P_1, as shown in Fig. 8.8A, whereas d_c and P_c define the dissimilarity in the core area, as shown in the inset of Fig. 8.8A. In this work, the author utilizes three VOC chemicals by tuning the refractive indices of the channel to match the refractive indices of ethanol, benzene, and water. The Zeonex in this work was considered as background material of PCF, to contribute to low confinement loss and high relative sensitivity. When the constraints ratio $d1/P1$ is increased, it can be considered as a filling ratios, which help reduce the production issues in the cover zone. Fig. 8.8B demonstrates that all the launched light is clearly propagating through the central regions for both types of polarization (i.e., x and y polarization). Both modes show excellent relative sensitivity and low confinement loss at 1 THz. It has been noted that the FEM strategy for the designed structure relies on the PML layers. From the simulation results, it has been observed that there are minimal variations in the confinement loss for various VOCs tested across the 0.80 to 3.0 THz frequency range. The designed sensor exhibited a high relative sensitivity of 91.25%, 93.80%, and 89.95% along with low confinement losses of 5.95×10^{-8} dB/m, 6.25×10^{-8} dB/m, and 5.94×10^{-8} dB/m, for ethanol ($n = 1.354$), benzene ($n = 1.366$), and water ($n = 1.330$), respectively.

8.4.2 Experimental validation of photonic crystal fiber based volatile organic compounds sensors

Villatoro et al. (2009) reported a highly sensitive VOC vapor sensor based on an in-reflection PCF interferometer. The major advantage of the device is that it selectively detects acetonitrile (CH_3CN) and tetrahydrofuran (THF) without requiring any permeable sensing materials. The sensor described in the work comprises PCF spliced to a standard SMF, as shown in Fig. 8.9. During the splicing process, the air holes at one end of the PCF collapse due to extreme heat, forming standard silica glass

(A)

(B)

(C)

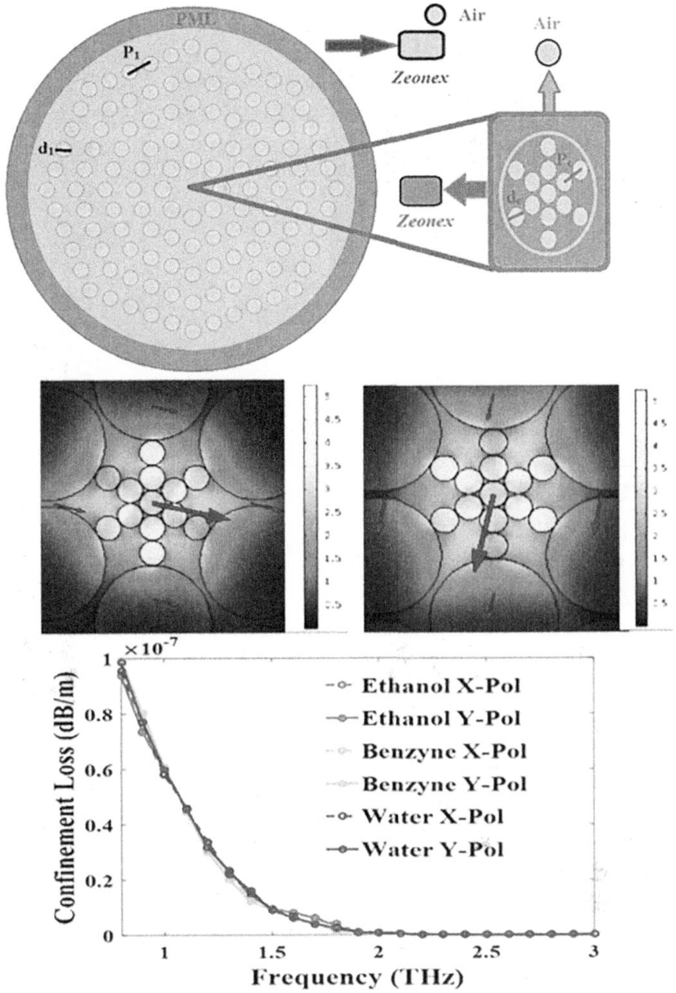

FIGURE 8.8 Octagonal PCF for VOC sensing: (A) Schematic diagram show the cross section of the designed octagonal PCF, (B) mode field circulation for x and y polarization at 1 THz, and (C) confinement loss corresponding to frequency of various VOCs. *From Hossain, M. S., Mollah, M. S. H., Rahman, M. H., & Sen, S. (2024). Design and performance analysis of PCF-based octagonal chemical sensor to detect benzene, ethanol, and water in terahertz spectrum.* Results in Optics, 14. *https://doi.org/10.1016/j.rio.2023.100599.*

joints that enable the splitting and recombination of launched light along the fiber's length. The other end of the PCF is cleaved perpendicularly, leaving the holes open so that propagated light can directly interact with VOC molecules through these openings. Fig. 8.9A illustrates the scanning

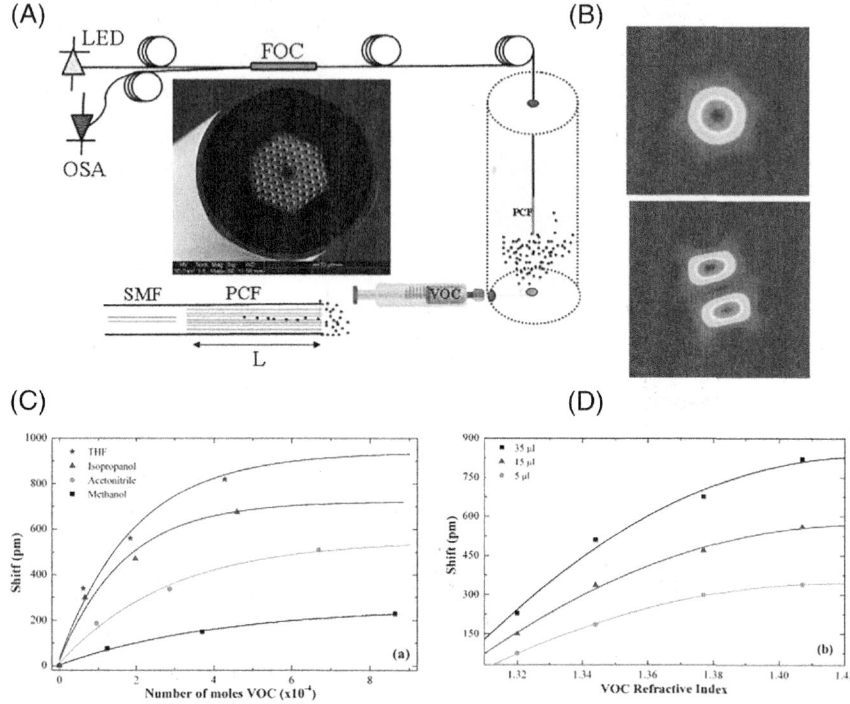

FIGURE 8.9 In-reflection PCF interferometer for VOC sensors: (A) Diagram of the experimental setup, a micrograph of the PCF used, and schematic drawing of the interferometer and (B) Transverse component of the electric field of the LP01- and LP11-like modes (C) Interference pattern shift versus VOC volume; (D) Refractive index dependence in liquid phase. *From Villatoro, J., Kreuzer, M. P., Jha, R., Minkovich, V. P., Finazzi, V., Badenes, G., & Pruneri, V. (2009). Photonic crystal fiber interferometer for chemical vapor detection with high sensitivity.* Optics Express, *17(3), 1447–1453. https://doi.org/10.1364/OE.17.001447.*

electron microscope (SEM) image of the reported PCF-based VOC sensor, the schematic of the interferometer configuration, and a line graph of the experimental setup used to characterize the device's performance. When light is launched from the LED, it propagates through the sensing configuration via a fiber optic coupler or circulator. The reflected light from the cleaved end of the PCF is directed into a high-resolution optical spectrum analyzer (OSA) integrated with a personal computer. This setup enables real-time monitoring of interference peaks on a millisecond timescale. In this work, the excitation and recombination of modes were performed using a single splice at the joint of SMF and PCF. The fundamental SMF mode begins to diffract upon entering the collapsed section of the PCF, leading to the excitation of two core modes within the PCF. Fig. 8.9B demonstrates the transverse component of the electric field of these modes, which plays a crucial role in interference. Mode analysis was conducted using FEM based on COMSOL Multiphysics. Fig. 8.9C and D depict the spectral shifts observed for varying numbers of moles and varying refractive indices of liquid VOCs, respectively. In Fig. 8.9C, it is evident that beyond a certain point, increasing the number of moles does not significantly impact the observed shifts. This plateau observed in the graph is attributed to the limited capacity of the PCF voids to accommodate gaseous molecules. Fig. 8.9D shows the high sensitivity to VOCs refractive indices in the liquid phase, indicating approximately a fourfold signal enhancement from methanol to THF for each volume evaporated. This underscores that the mechanism inducing wavelength shifts in the interference pattern is directly linked to variations in the refractive index due to the adsorption of molecules into the inner holes of the PCF structure.

Kim et al. (2014) reported a VOC sensor with hollow-core photonic crystal fiber (HC-PCF) based on reflection mode for the detection of acetone vapor without using any additional sensing material. In this method, MOF was fabricated using a preform comprising various thin silica capillaries. For the fabrication, the seven capillaries in the central part of the preform were replaced by seven solid rods in order to form a solid-core MOF. In this work, one end of the HC-PCF was spliced with SMF, and the other end was spliced to a solid-core PCF. The sensor operates on the basic principle of a Fabry–Perot interferometer, which arises from the interference of three different refractive indices that exist simultaneously. As the micro-cavity was filled with acetone vapors, the interference spectrum originating from the reflected light on the silica and air interfaces changed and showed a variation in the output wavelength recorded by an OSA. The final sensitivity of the designed sensor was analyzed using the fast Fourier transform technique and exhibited a shift of 1.3 rad/nm.

Later in the same year, Niu et al. (2014) reported an ethanol sensor based on the modified version of the above configuration. The reported sensor comprises a ~5.1 cm long HC-PCF placed between two SMFs (without splicing), as shown in the inset of Fig. 8.10 along with the associated experimental setup used to characterize the sensor. In this work, the HC-PCF was not spliced together; instead, it was left open at both ends of the HC-PCF to allow the VOC molecules to pass through these channels. The HC-PCF works on the transmission phenomenon based on the Rayleigh scattering effect in the HC-PCF instead of reflection. The whole arrangement was fixed in a sealed acrylic gas chamber, and various concentrations of ethanol in the chamber were controlled by a micropipette with a resolution of 0.05 mL. As various concentrations of

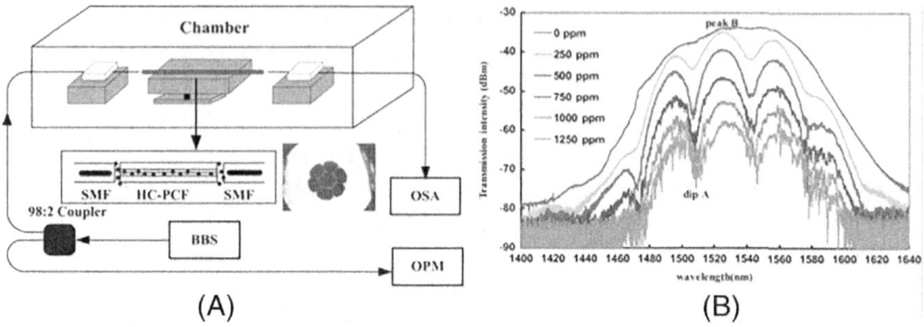

FIGURE 8.10 Rayleigh scattering based hollow core PCF: (A) illustrates the schematic of experimental set-up and (B) represent the variation of wavelength corresponding to various concentration of ethanol. *From Niu, L., Zhao, C. L., Kang, J., Jin, S., Guo, J., & Wei, H. (2014). A chemical vapor sensor based on Rayleigh scattering effect in simplified hollow-core photonic crystal fibers.* Optics Communications, *313, 243–247. https://doi.org/10.1016/j.optcom.2013.10.037.*

ethanol are released into the gas chamber, the ethanol molecules interact with the air holes of HC-PCF through the space between the SMF and the HC-PCF, leading to the excitation of Rayleigh scattering to produce an intensity loss of the transmitted light. When the propagated light reaches the end of the fibers, the modes recombine to form the fundamental mode. This sensing mechanism of the device is similar to a Mach–Zehnder interferometer. Fig. 8.10B illustrates the output transmission spectra of the fabricated sensor in the presence of various ethanol concentrations ranging from 0 to 1250 ppm in steps of 250 ppm. From the obtained spectra, it is clear that the transmission intensity is reduced with increasing concentration of ethanol. However, the transparent ranges become smaller as the concentration becomes larger due to the increased transmission intensity loss. The decline of the transmission intensity is mainly due to Rayleigh scattering. The fabricated sensor exhibited a maximum sensitivity of ~0.022 dB/ppm for ethanol concentrations ranging from 0 to 1250 ppm.

The major limitation of these MOF-based VOC sensors is their limited selectivity due to the lack of a functionalized layer. To solve this issue, Diego et al. developed an SCF-based VOC sensor utilizing a thin layer of ITO inside its micro-cavity in 2018 (Lopez-Aldaba et al., 2018; Lopez-Torres et al., 2018). In the proposed device, six Fabry–Perot sensing heads were multiplexed. Each one of them was fabricated by splicing different lengths of suspended core fiber (SCF) to the SMF in order to achieve specific interferometric patterns for every device. Among all the sensing heads, three of them were deposited with ITO thin film to enhance the sensitivity for ethanol, whereas two were deposited with SnO_2 to show a different response. The last sensing head was left without any metal deposition for temperature measurement. The sensing performance of the sensor was evaluated by analyzing the FFT phase variations in the interferometric pattern that occurred for each sensor. The device achieves a maximum phase shift of 0.31 π rad for methanol atmospheres. The study concluded that the final sensitivity was highly influenced by the geometry cavities, but mostly by one of the cores. The SCF with an air hole in the middle of its core exhibits five times higher sensitivity in comparison to the others (Lopez-Torres et al., 2018). Among them, MOF-1 exhibited the lowest sensitivity with a maximum variation of 0.685 radians, while MOF-3 gives the highest sensitivity with a maximum variation of 5 radians.

In addition to the discussed MOF-based VOC sensors, there are several other sensors that have been reported in past decades, which are summarized in Table 8.1.

8.5 Prospects and challenges

Throughout this chapter, it is evident that the MOF can be utilized in a broad range of VOC detection applications based on its structural design and sensing configurations. They are versatile and can be configured in numerous ways to create interferometers such as MZI, FPI, or Michelson interferometers, with desirable sensor properties including good kinetics, small size, and good time stability. The holes along the length of the MOFs play a crucial role, specifically in gas sensing, as they can host gases and VOC molecules, thereby improving the performance of the device. Additionally, the interaction between the guided light and measurand (liquid or gas) or deposited sensing materials (e.g., thin films) is improved, which in turn enhances the sensitivity and response time of the devices.

The role of MOF has not been fully explored yet for VOC monitoring in terms of their selectivity and sensitivity. Even though the non-functionalized surface sensor illustrates a good sensing response, there might be a selectivity issue when the device is placed in a mixed environment. In some designs, the splicing of MOF with SMF transmission fibers is required, causing the air hole to collapse during the splicing, which may distort the actual guided mode. Additionally, fabrication difficulties such as precisely controlling the length of the MOF or the length of the FP cavity can all lead to sensitivity loss. Another limitation of such devices might occur due to their air-hole size, which needs to be optimized before the fabrication for successful passing of VOC gases through the holes, as VOC molecules have different sizes.

TABLE 8.1 Comparative table showing various MOF based VOC sensors.

References	Target VOC	Type of Fiber	Sensing Material	Sensitivity
Villatoro et al. (2009)	THF and acetonitrile	PCF	–	–
Kim et al. (2014)	Allyl alcohol	PCF	–	A shift of 1.3 rad
Niu et al. (2014)	Ethanol	Hollow core PCF	–	0.022 dB/ppm
Niu et al. (2013)	Ethanol	Hollow core PCF	–	Shift 2.56 nm
Lopez-Aldaba et al. (2018)	Methanol	PCF	–	Shift 0.3π rad
Lopez-Torres et al. (2018)	Ethanol	Suspended core	–	Nearly 5π rad
Choudhary et al. (2022)	Ethanol	Hollow core PCF	–	Relative sensitivity 97.28%
Maidi et al. (2022)	Ethanol, water, benzene	Hexagonal PCF	–	94.47% for water, 96.32% for ethanol, 99.63% for benzene
Schwarm et al. (2022)	Ethyl chloride	Hollow core PCF	–	
Paixão et al. (2022)	Ethanol	Suspended core	–	3.9 pm/wt.%
Martan et al. (2014)	Hydrocarbons	Suspended core	–	1.20 ± 0.04 dB/mol%

8.6 Conclusion

In this chapter, in-depth information has been provided on MOFs, including the fabrication steps, types of MOFs, and their application in VOC sensing. This chapter discusses how numerical tools play a role in optimizing structural parameters and determining the mode of interest for VOC sensing. Additionally, it also explains how the designs are experimentally validated to meet the demands of commercialization. The sensing mechanisms of MOFs are largely based on variations in the optical properties of the photonic crystal caused by changes in the effective RI or lattice constant induced by the surrounding gas. Furthermore, the future of PCF-based VOC sensors will depend on the advancements in controllability, integration, and the exploration of new PCF designs and detection methods.

References

Abd-Elkader, A. E.-S., Hameed, M. F. O., Obayya, S. S. A., & Younis, B. M. (2025). Multifunctional dual-core liquid crystal photonic crystal fiber. *Optical Engineering, 64*(5). doi:10.1117/1.OE.64.5.057102.

Amouzad Mahdiraji, G., Chow, D. M., Sandoghchi, S. R., Amirkhan, F., Dermosesian, E., Yeo, K. S., Kakaei, Z., Ghomeishi, M., Poh, S. Y., Gang, S. Y., & Mahamd Adikan, F. R. (2014). Challenges and solutions in fabrication of silica-based photonic crystal fibers: An experimental study. *Fiber and Integrated Optics, 33*(1–2), 85–104. doi:10.1080/01468030.2013.879680.

Argyros, A. (2010). Applications of microstructured polymer optical fibers. *2010 Photonics Global Conference, PGC 2010*. IEEE. doi:10.1109/PGC.2010.5706084.

Birks, T. A., Knight, J. C., & Russell, P. S. J. (1997). Endlessly single-mode photonic crystal fiber. *Optics Letters, 22*(13), 961–963. http://ol.osa.org/Issue.cfm. doi:10.1364/OL.22.000961.

Choudhary K., Verma P., & Kumar A. (2022). Refractive index-based ethanol sensor using hollow core photonic crystal fiber in THz region. In Lecture Notes in Networks and Systems (pp. 669–677). Springer Science and Business Media Deutschland GmbH India. http://www.doi.org/10.1007/978-981-16-6246-1_57.

Fasano, A., Woyessa, G., Stajanca, P., Markos, C., Stefani, A., Nielsen, K., Rasmussen, H. K., Krebber, K., & Bang, O. (2016). Fabrication and characterization of polycarbonate microstructured polymer optical fibers for high-temperature-resistant fiber Bragg grating strain sensors. *Optical Materials Express, 6*(2), 649–659. https://www.osapublishing.org/DirectPDFAccess/EC419833-D929-72DF-2AF764C4061B690A_335842/ome-6-2-649.pdf?da=1&id=335842&seq=0&mobile=no. doi:10.1364/OME.6.000649.

Gangwar, R. K., Bhardwaj, V., & Singh, V. K. (2016). Magnetic field sensor based on selectively magnetic fluid infiltrated dual-core photonic crystal fiber. *Optical Engineering, 55*(2). http://www.spie.org/x867.xml. doi:10.1117/1.OE.55.2.026111.

Gangwar, R. K., Kumari, S., Pathak, A. K., Gutlapalli, S. D., & Meena, M. C. (2023). Optical fiber based temperature sensors: A review. *Optics, 4*(1), 171–197. www.mdpi.com/journal/optics. doi:10.3390/opt4010013.

Gangwar, R. K., Pathak, A. K., & Kumar, S. (2023). Recent progress in photonic crystal devices and their applications: A review. *Photonics, 10*(11). http://www.mdpi.com/journal/photonics. doi:10.3390/photonics10111199.

Gupta, P. S., & Faisal, M. (2017). Microstructure fiber with extremely high nonlinearity for supercontinuum generation. *2017 2nd IEEE International Conference on Telecommunications and Photonics, ICTP* (pp. 24–27). Institute of Electrical and Electronics Engineers Inc. doi:10.1109/ICTP.2017.8285920.

Hossain, M. S., Mollah, M. S. H., Rahman, M. H., & Sen, S. (2024). Design and performance analysis of PCF-based octagonal chemical sensor to detect benzene, ethanol, and water in terahertz spectrum. *Results in Optics, 14*. https://www.sciencedirect.com/science/journal/26669501. doi:10.1016/j.rio.2023.100599.

Iezzi, B., Coon, A., Cantley, L., Perkins, B., Doran, E., Wang, T., Rothschild, M., & Shtein, M. (2023). Polymeric photonic crystal fibers for textile tracing and sorting. *Advanced Materials Technologies, 8*(6). doi:10.1002/admt.202201099.

Kaiser, P., & Astle, H. W. (1974). Low-loss single-material fibers made from pure fused silica. *Bell System Technical Journal, 53*(6), 1021–1039. doi:10.1002/j.1538-7305.1974.tb02780.x.

Kim, B., Ahn, J. C., Chung, P. S., & Chung, Y. (2014). Microstructured optical fiber-based micro-cavity sensor for chemical detection. *Progress in Biomedical Optics and Imaging – Proceedings of SPIE*: Vol. 8938. SPIE. doi:10.1117/12.2041001.

Koehler, P., Lawson, T., Neises, J., Willkomm, J., Martindale, B. C. M., Hutton, G. A. M., Antón-García, D., Lage, A., Gentleman, A. S., Frosz, M. H., Russell, P. S. J., Reisner, E., & Euser, T. G. (2021). Optofluidic photonic crystal fiber microreactors for in situ studies of carbon nanodot-driven photoreduction. *Analytical Chemistry, 93*(2), 895–901. http://pubs.acs.org/journal/ancham. doi:10.1021/acs.analchem.0c03546.

Liu, Z., & Tam, H.-Y. (2018). *Fabrication and sensing applications of special microstructured optical fibers*. IntechOpen. doi:10.5772/intechopen.70755.

Lopez-Aldaba, A., Lopez-Torres, D., Aguado, C. E., Arregui, F. J., Auguste, J. L., Jamier, R., Roy, P., & Lopez-Amo, M. (2018). Real time measuring system of multiple chemical parameters using microstructured optical fibers based sensors. *IEEE Sensors Journal, 18*(13), 5343–5351. http://ieeexplore.ieee.org/xpl/RecentIssue.jsp?punumber=7361. doi:10.1109/JSEN.2018.2837159.

Lopez-Torres, D., Elosua, C., & Arregui, F. J. (2020). Optical fiber sensors based on microstructured optical fibers to detect gases and volatile organic compounds: A review. *Sensors., 20*(9), 2555. doi:10.3390/s20092555.

Lopez-Torres, D., Lopez-Aldaba, A., Elosua, C., Auguste, J. L., Jamier, R., Roy, P., Lopez-Amo, M., & Arregui, F. J. (2018). Comparison between different structures of suspended-core microstructured optical fibers for volatiles sensing. *Sensors, 18*(8), 2523. doi:10.3390/s18082523.

Maidi, A. M., Shamsuddin, N., Wong, W. R., Kaijage, S., & Begum, F. (2022). Characteristics of ultrasensitive hexagonal-cored photonic crystal fiber for hazardous chemical sensing. *Photonics, 9*(1). https://www.mdpi.com/2304-6732/9/1/38/pdf. doi:10.3390/photonics9010038.

Maidi, A. M., Yakasai, I., Abas, P. E., Nauman, M. M., Apong, R. A., Kaijage, S., & Begum, F. (2021). Design and simulation of photonic crystal fiber for liquid sensing. *Photonics, 8*(1), 16. https://www.mdpi.com/2304-6732/8/1/16. doi:10.3390/photonics8010016.

Martan, T., Aubrecht, J., Podrazký, O., Matějec, V. L., & Kašík, I. (2014). Detection of hydrocarbons using suspended core microstructured optical fiber. *Sensors and Actuators B: Chemical, 202*, 123–128. doi:10.1016/j.snb.2014.04.103.

Mathews, S., Farrell, G., & Semenova, Y. (2011). Liquid crystal infiltrated photonic crystal fibers for electric field intensity measurements. *Applied Optics, 50*(17), 2628–2635. http://www.opticsinfobase.org/view_article.cfm?gotourl=http%3A%2F%2Fwww%2Eopticsinfobase%2Eorg%2FDirectPDFAccess%2F59228D48%2DA19B%2DA0C7%2DD94F0E5EE42388FE%5F217272%2Epdf%3Fda%3D1%26id%3D217272%26seq%3D0%26mobile%3Dno&org=Elsevier%20Inc. doi:10.1364/AO.50.002628.

Moeglen Paget, B., Vinod Ram, K., Zhang, S., Perumal, J., Vedraine, S., Humbert, G., Olivo, M., & Dinish, U. S (2024). A review on photonic crystal fiber based fluorescence sensing for chemical and biomedical applications. *Sensors and Actuators B: Chemical, 400*. doi:10.1016/j.snb.2023.134828.

Mollah, M. A., Yousufali, M., Ankan, I. M., Rahman, M. M., Sarker, H., & Chakrabarti, K. (2020). Twin core photonic crystal fiber refractive index sensor for early detection of blood cancer. *Sensing and Bio-Sensing Research, 29*. https://www.journals.elsevier.com/sensing-and-bio-sensing-research/. doi:10.1016/j.sbsr.2020.100344.

Monro, T. M., Warren-Smith, S., Schartner, E. P., Franois, A., Heng, S., Ebendorff-Heidepriem, H., & Afshar, S. (2010). Sensing with suspended-core optical fibers. *Optical Fiber Technology, 16*(6), 343–356. doi:10.1016/j.yofte.2010.09.010.

Monroy, I. T., Boom, V. D., Koonen, K., Watanabe, K., & Ishigure, T (2003). Data transmission over polymer optical fibers. *Optical Fiber Technology, 9*(3), 159–171. doi:10.1016/S1068-5200(03)00006-3.

Niu, L., Zhao, C. L., Kang, J., Jin, S., Guo, J., & Wei, H. (2014). A chemical vapor sensor based on Rayleigh scattering effect in simplified hollow-core photonic crystal fibers. *Optics Communications, 313*, 243–247. doi:10.1016/j.optcom.2013.10.037.

Niu, L., Zhao, C. L., Kang, J., & Ye, M. P. (2013). Modal interferometer based on volatile organic compounds diffused in a simplified hollow-core photonic crystal fiber. *Proceedings of SPIE – The International Society for Optical Engineering*: Vol. 9044. SPIE http://spie.org/x1848.xml. doi:10.1117/12.2036968.

Paixão, T., Nunes, A. S., Bierlich, J., Kobelke, J., & Ferreira, M. S. (2022). Fabry-Perot interferometer based on suspended core fiber for detection of gaseous ethanol. *Applied Sciences (Switzerland), 12*(2). https://www.mdpi.com/2076-3417/12/2/726/pdf. doi:10.3390/app12020726.

Pathak, A. K., & Viphavakit, C. (2022). A review on all-optical fiber-based VOC sensors: Heading towards the development of promising technology. *Sensors and Actuators A: Physical, 338*, 113455. doi:10.1016/j.sna.2022.113455.

Schwarm, K. K., Nair, A. P., Wei, C., Spearrin, R. M., Ozen, E., Gonzalez, E., & Kriesel, J. (2022). Three-dimensional real-time mapping of CO and CO2 concentrations in active forest burns with a UAV spectrometer. *AIAA Science and Technology Forum and Exposition, AIAA SciTech Forum 2022*. American Institute of Aeronautics and Astronautics Inc. doi:10.2514/6.2022-2291.

Shi, Q., Lv, F., Wang, Z., Jin, L., Hu, J. J., Liu, Z., Kai, G., & Dong, X. (2008). Environmentally stable Fabry-Perot-type strain sensor based on hollow-core photonic bandgap fiber. *IEEE Photonics Technology Letters, 20*(4), 237–239. doi:10.1109/LPT.2007.913335.

Siddik, A. B., Hossain, S., Kumar Paul, A., Rahman, M., & Mollah, A. (2020). High sensitivity property of dual-core photonic crystal fiber temperature sensor based on surface plasmon resonance. *Sensing and Bio-Sensing Research, 29*. doi:10.1016/j.sbsr.2020.100350.

Singh, S., Chaudhary, B., Upadhyay, A., Sharma, D., Ayyanar, N., & Taya, S. A. (2023). A review on various sensing prospects of SPR based photonic crystal fibers. *Photonics and Nanostructures – Fundamentals and Applications, 54*, 101119. doi:10.1016/j.photonics.2023.101119.

Talataisong, W., Ismaeel, R., Beresna, M., & Brambilla, G. (2019). Suspended-core microstructured polymer optical fibers and potential applications in sensing. *Sensors, 19*(16), 3449. doi:10.3390/s19163449.

Talataisong, W., Ismaeel, R., Sandoghchi, S. R., Rutirawut, T., Topley, G., Beresna, M., & Brambilla, G. (2018). Novel method for manufacturing optical fiber: Extrusion and drawing of microstructured polymer optical fibers from a 3D printer. *Optics Express, 26*(24), 32007–32013. https://www.osapublishing.org/oe/abstract.cfm?uri=oe-26-24-32007. doi:10.1364/OE.26.032007.

Villatoro, J., Kreuzer, M. P., Jha, R., Minkovich, V. P., Finazzi, V., Badenes, G., & Pruneri, V. (2009). Photonic crystal fiber interferometer for chemical vapor detection with high sensitivity. *Optics Express., 17*(3), 1447–1453. http://www.opticsinfobase.org/DirectPDFAccess/4B6B1218-BDB9-137E-C461CB5C5073344B_176100.pdf?da=1&id=176100&seq=0&CFID=25114211&CFTOKEN=68780792. doi:10.1364/OE.17.001447.

Woyessa, G., Fasano, A., Markos, C., Stefani, A., Rasmussen, H. K., & Bang, O. (2017). Zeonex microstructured polymer optical fiber: Fabrication friendly fibers for high temperature, humidity insensitive Bragg grating sensing. *Optical Materials Express, 7*(1), 286–295. https://www.osapublishing.org/DirectPDFAccess/1BD1383F-F24D-B5B2-2CADA43935CF6855_356975/ome-7-1-286.pdf?da=1&id=356975&seq=0&mobile=no. doi:10.1364/OME.7.000286.

Xuan, K. D., Van, L. C., Long, V. C., Dinh, Q. H., Xuan, L. V., Trippenbach, M., & Buczynski, R. (2017). Dispersion characteristics of a suspended-core optical fiber infiltrated with water. *Applied Optics, 56*(4), 1012–1019. https://www.osapublishing.org/view_article.cfm?gotourl=https%3A%2F%2Fwww%2Eosapublishing%2Eorg%2FDirectPDFAccess%2FDB2E69A7-9003-90B8-7C0FB66E8FB35B50_357733%2Fao-56-4-1012%2Epdf%3Fda%3D1%26id%3D357733%26seq%3D0%26mobile%3Dno&org=Elsevier%20Inc. doi:10.1364/AO.56.001012.

Yu, R., Chen, Y., Shui, L., & Xiao, L. (2020). Hollow-core photonic crystal fiber gas sensing. *Sensors, 20*(10), 2996. doi:10.3390/s20102996.

Zografopoulos, D. C., Asquini, R., Kriezis, E. E., d'Alessandro, A., & Beccherelli, R. (2012). Guided-wave liquid-crystal photonics. *Lab on a Chip, 12*(19), 3598. doi:10.1039/c2lc40514h.

Zubia, J., & Arrue, J. (2001). Plastic optical fibers: An introduction to their technological processes and applications. *Optical Fiber Technology, 7*(2), 101–140. doi:10.1006/ofte.2000.0355.

CHAPTER

9

On-chip photonic sensors for volatile organic compound detection

9.1 Introduction

Integrated photonics is a branch of optics focused on the development of miniaturized optical technologies by combining numerous optical components onto a single substrate (Thylén & Wosinski, 2014). This field involves the fabrication of photonic devices such as modulators, light sources, optical waveguides, amplifiers, detectors, etc. (Butt, 2023). Integrated photonics enables precise manipulation and control of light at micro- and nanoscales, facilitating the production of highly efficient and miniaturized photonic circuits with a broad range of applications (Dai et al., 2015). By combining various optical functions onto a single substrate, integrated photonics offers key advantages, including small size, lower power requirements, and cost-effectiveness for large-scale production. This technology is important in advancing a wide range of applications, including data processing, telecommunication, imaging, biomedical imaging, quantum information processing, and next-generation computing architectures, allowing advancements in high-speed data exchange and target sensing (Butt et al., 2024; Kazanskiy et al., 2022).

Photonic and electronic sensors each have their unique advantages and limitations, making them more or less appropriate based on the required application. Integrated photonic sensors have gained significant attention for a wide range of applications due to their exceptional characteristics. The major advantage of integrated devices compared to electronic-based sensing technology is their inherent properties of immunity to electromagnetic interference. As the integrated photonic sensors solely rely on

Photonic Sensors for Detection of VOC Biomarkers: Technological Principles and Medical Applications
DOI: https://doi.org/10.1016/B978-0-443-29166-1.00003-2

Copyright © 2026 Elsevier Inc. All rights are reserved, including those for text and data mining, AI training, and similar technologies.

the light matter interactions rather than electrical signals, making them less susceptible to interference from external electromagnetic fields, e.g., power generation stations and industrial facilities, where the conventional electronic sensors may deliver inaccurate outputs. These sensors also offer excellent sensing response and improved detection accuracy for various applications requiring precise monitoring of the light intensity, wavelength, or the variation of refractive index. They are also highly useful for long-range monitoring because of the low loss of guided signals compared to the traditional electrical signals guided through the wires. Despite several advantages, the fabrication cost of photonic sensors is extremely high, along with precise integration challenges compared to electronic sensing devices. They may also require skilled personnel and specialized tools for proper calibration and periodic maintenance. Moreover, integrated photonic sensors are generally based on lasers or appropriate light sources, which can raise safety concerns and hence require additional training and careful handling. In contrast, the electronic sensing devices are considered to be straightforward, less expensive, and simpler to integrate with the electronic circuits, therefore allowing them to serve as potential tool for various sensing applications, due to device simplicity, low cost, and ease of integration. Ultimately, the selection between the electronic and photonic sensors depends on the application-specific requirements, such as environmental conditions, and available budget. In this chapter, we aim to explore the current landscape of integrated photonic sensing technologies along with the past literature, associated challenges, and possible solutions.

This chapter aims to provide in-depth information to its readers about integrated photonic devices and their application in volatile organic compounds (VOC) sensing. The chapter is organized as follows: Section 9.2 provides a basic overview of integrated photonic devices, along with the basic components of integrated photonic devices, including detailed explanations of waveguides, ring resonators, and grating couplers. Section 9.3 focuses on integrated photonic-based VOC sensors, and Section 9.4 describes various approaches used in the fabrication of such devices. Section 9.5 outlines future prospects and associated challenges, followed by the conclusion in Section 9.6.

9.2 Basic elements of integrated photonics

Fig. 9.1 represents a simple block diagram of chemical photonic sensors. A key component of chemical sensors is the receptor. This is generally achieved by depositing the sensor surface with a thin layer of target specific sensing materials. The main advantage of the receptor lies in its high selectivity, as it can precisely capture the target measurand from a complex mixture environment. Additionally, the optical waveguide serves

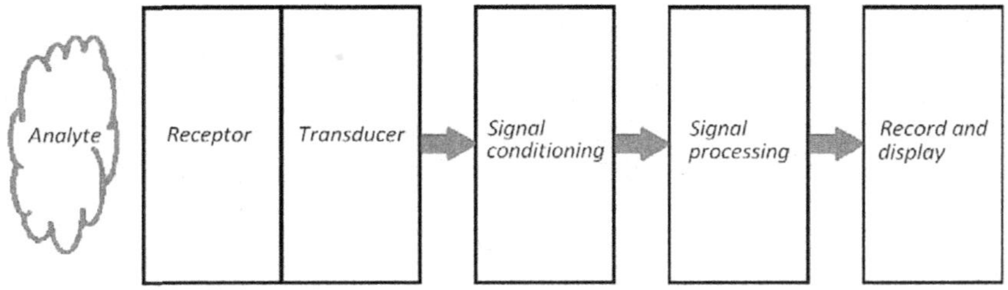

FIGURE 9.1 Basic working mechanism of chemical sensing based on integrated photonic devices. *From Passaro, V., Tullio, C., Troia, B., Notte, M., Giannoccaro, G., & Leonardis, F. (2012a). Recent advances in integrated photonic sensors.* Sensors, *12(11), 15558–15598. https://doi.org/10.3390/s121115558.*

as a transducer in the configuration. Any variation in the surrounding environment (either surface level or uniform) significantly changes the effective refractive index of the propagated optical modes, allowing change in the propagated signal. The main components of an integrated photonic sensor are discussed in detail in the subsequent subsections.

9.2.1 Waveguides

Optical waveguides for sensing applications are specifically designed to improve light-matter interaction and enhance the device sensitivity to allow accurate gas monitoring. The most useful sensor design is the planar waveguide, which allows the confinement of the light in a thin film on a substrate (Ho et al., 2013). In the planar waveguide, the evanescent field extends beyond the core of the waveguide and directly interacts with the ambient environment (Vitoria et al., 2023). To further improve the sensing performance, various materials like silicon nitride (SiN) or silicon (Si) are utilized due to their high refractive index contrast, which allows extreme optical confinement and low propagation loss at the cost of less interaction with the ambient environment. Furthermore, the selectivity of the waveguide can also be achieved by immobilizing the waveguide surface with the target sensing materials in order to selectively bind target molecules (Zheng et al., 2022). One simple example of a waveguide sensor is a planar waveguide, which was reported by Dominguez et al. (2021), utilizing a polydimethylsiloxane- and agarose-coated coverslip for the simultaneous detection of temperature and humidity, opening a new direction toward simultaneous detection of dual-parameter. In the reported work, the authors launched light onto the edge of a standard microscope glass slide, covered with a thin film of polydimethylsiloxane and agarose on both faces, which allows the excitation of two resonances corresponding to each polarization state (TE and TM) of the input light, which was utilized in the sensing of two different parameters (Butt & Piramidowicz, 2024). The reported

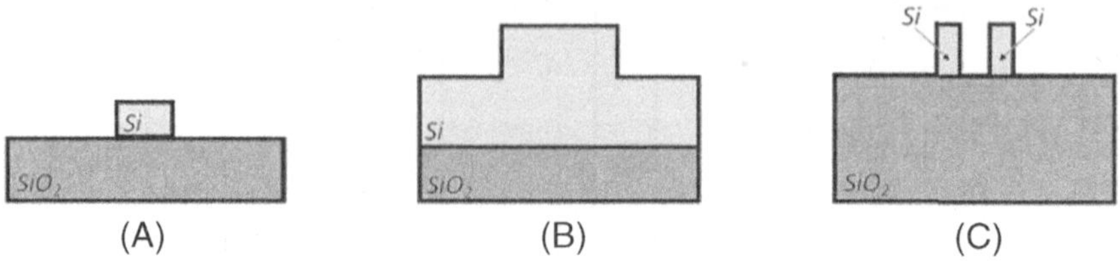

FIGURE 9.2 Types of optical waveguides: Architectures of various optical waveguide used in sensing applications (A) wire, (B) rib, and (C) slot. *From Passaro, V., Tullio, C., Troia, B., Notte, M., Giannoccaro, G., & Leonardis, F. (2012a). Recent advances in integrated photonic sensors.* Sensors, 12*(11), 15558–15598. https://doi.org/10.3390/s121115558.*

sensor was optimized to obtain two resonances in the visible region and two in the near-infrared (NIR), achieving a maximum sensitivity of 0.34 nm/°C and 0.23 nm/%RH for temperature and humidity, respectively, in the visible region, whereas in the NIR region the obtained sensitivities were 1.16 nm/°C and 0.34 nm/%RH, demonstrating the potential of reported sensor to operate in dual spectral ranges.

Integrated photonic-based sensors can be categorized into intrinsic and extrinsic sensors. In intrinsic sensors, the integrated waveguide possesses dual functionality, i.e., guiding light along the waveguide as well as the sensing element for the direct interaction of the measurand. In contrast, the extrinsic sensors require a separate sensing region from the actual waveguide, with no direct interaction between the target measurand and the waveguide. The extrinsic sensors are generally bulkier and costlier compared to the intrinsic sensors, therefore making them less suitable for mass-scale manufacturing. In the case of intrinsic photonic sensors, where the waveguide itself works as a transducer, a careful and optimized design is required in order to achieve high performance. On the silicon-on-insulator (SOI) wafer platform, several types of optical waveguides have been exploited to fabricate integrated chemical sensors (Butt et al., 2024). Fig. 9.2 illustrates various types of fabricated optical waveguides. Rib waveguides generally confine the launched light within the high-index core fabricated over SiO_2, resulting in minimal light-analyte interaction, hence exhibiting lower sensitivity to the variations of the analyte concentration. In contrast, wire or slab optical waveguides offer improved light-matter interaction by exposing a larger surface area to the surrounding environment. This sensing configuration can significantly enhance the transverse magnetic field at both the top and bottom interfaces, therby enhancing the sensing performance of the waveguide in response to external chemical variations.

In biosensing and chemical sensing, the slot waveguide is considered the most appropriate optical waveguide configuration to achieve high sensitivity and improve sensor performance. The slot waveguide possesses unique characteristics of confining a strong optical field within the low-index slot region, where the light-analyte interaction takes place (Almeida et al., 2004). Slot waveguide-based integrated photonic systems offer enhanced sensitivity and enable a low limit of detection (LOD) (i.e., the smallest detectable concentration of the measurand). In the case of quasi-E polarized modes, studies have demonstrated that the homogeneous sensitivity value exceeding 1 can be achieved, resulting in significant variations of the effective refractive index (n_{eff}) that can surpass the variation in the surrounding environment. Homogeneous sensitivity is a metric that is used to measure the change in the guided mode effective refractive index in response to slight variations in the refractive index of a medium surrounding the waveguide (Lukosz, 1995; Tyszkiewicz & Kielan, 2023). Furthermore, the structural parameter of the waveguide plays a crucial role in optimizing the sensing performance, as reported in past literature (Dell'Olio & Passaro, 2007; Gallinet et al., 2015).

Another key aspect in designing integrated optical sensors is the selection of the operating wavelength. For instance, various gases, including ethane (C_2H_6), carbon dioxide (CO_2), carbon monoxide (CO), and methane (CH_4), are some of the gases that exhibit distinct absorption spectra in the mid-IR region. Utilizing optical signals within this spectral range can significantly improve the sensor performance by simultaneously combining both homogeneous sensing and optical absorption. Homogeneous sensing is a method to measure the uniform variations in the refractive index of the surrounding medium, usually liquid or gas, which covers the sensing structure entirely. To achieve this goal, group IV photonics has recently been reported as an appropriate technological platform to realize an effective slot waveguide sensor for gas monitoring in mid-IR (Passaro et al., 2012b). The wavelengths selected for the sensor's operation are 2.883 and 3.39 μm. Group IV semiconductor materials and alloys such as GeSn, SiGeSn, SiGeC, and GeC demonstrate increased refractive indices (~4) in the mid-IR region. Consequently, extremely high sensing responses have been observed for homogeneous sensing and toxic gases. Furthermore, the simplified fabrication requirements and larger structural dimensions as compared to conventional SOI slot waveguides allow the propagation of second-order slot modes, offering promising performance in various sensing applications (Passaro et al., 2012).

9.2.2 Ring resonators

Combining photonic waveguides with resonant micro cavities allows an effective optical signal readout at the sensor output. Micro ring resonators (MRR) have been considered to be a potential sensing configuration across wide range applications including chemical, biological, and gas sensing (Mere et al., 2020; Sun & Fan, 2008; Zhang et al., 2016), where monitoring of the target measurand is determined by analyzing the shift in the resonance

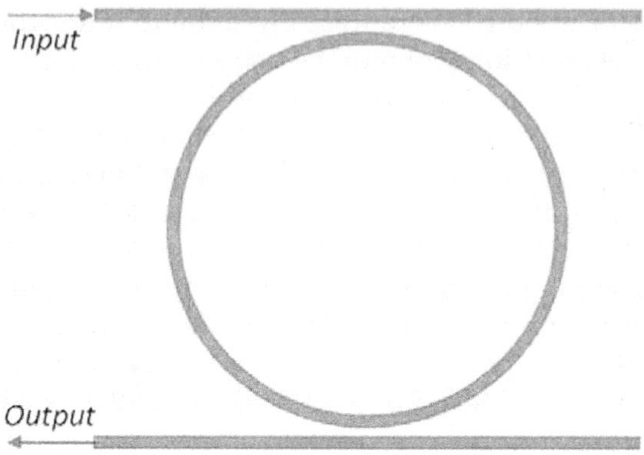

FIGURE 9.3 Ring resonator: Schematic diagram of the ring resonator along with two bus waveguides. *From Passaro, V., Tullio, C., Troia, B., Notte, M., Giannoccaro, G., & Leonardis, F. (2012a). Recent advances in integrated photonic sensors.* Sensors, *12(11), 15558–15598. https://doi.org/10.3390/s121115558.*

wavelength which occurs due to the variation in the refractive index of the waveguide or the coated sensing material in the presence of foreign substance or molecule. The resonant wavelength can be defined by the following relationship (Sun & Fan, 2011):

$$\lambda_0 = \frac{2\pi r n_{eff}}{m} \tag{9.1}$$

where r, n_{eff}, and m represent the ring radius, effective index, and an mode number (interger), respectively. Fig. 9.3 shows a schematic diagram of MRR where two straight lines represent the bus waveguide used for input and output, respectively, and a small gap between the ring and bus waveguides is introduced for evanescent coupling of light between both arms and the ring. Mason's rule and delay line signal processing methods can be utilized in order to analyze the transfer function of a single ring resonator coupled to two bus waveguides (Passaro et al., 2012b).

The sensitivity (S_{SR}) of the single ring resonators can be correlated with the sensitivity of the waveguide (S_W) and can be defined as:

$$S_{SR} = FSR_S \frac{S_w \lambda}{n_g} \tag{9.2}$$

where FSR is the free spectral range of the sensing cavity (i.e., the period of the resonant spectrum) and n_g is the group index. Integrated optical resonant microcavities have been well exploited for the development of highly sensitive bio and chemical sensors, achieving low LOD. Specifically, the resonant microcavities comprising low optical noises and high quality factors (Q-factor), provide unique advantages compared to the other sensing platforms.

Specifically, in resonant microcavities, the effective interaction length does not directly relate to the actual length of the device (Passaro et al., 2012a). This is because it propagates within the microcavity (i.e., racetrack resonator or ring), resulting in multiple trips in the waveguide. As a result, the overall path length of the device can be up to several orders of magnitude larger than the actual cavity size and proportional to the Q factor (which can be as high as $\sim 10^6$). The resonance wavelengths depend on the n_{eff} of the guided mode propagating into the cavity. When the device operates at one of these resonant wavelengths, any variation in the surrounding environment can lead to a shift in the resonance ($\Delta\lambda_R$). Since the sensitivity of the device is related to the cavity length, several researchers have reported spiral geometries in order to improve the performance and sensitivity of the sensor (Igarashi et al., 2023). Further improvements in the sensing performance were also reported by several authors utilizing multiple ring resonators to excite a well-known Fano resonance. To minimize the smallest detectable wavelength shift, this method is considered as a potential alternative to the conventional use of Lorentzian line shapes in single-cavity spectra. A low LOD of 10^{-8} RIU was observed for a 10 mm cavity length (Yi et al., 2010).

9.3 Integrated photonic-based volatile organic compounds sensors

Silicon-based photonic devices are considered a potential alternative for trace gas detection by utilizing evanescent field absorption sensing. These gas sensors are generally based on the presence of a specific absorption feature of the target gas at the operational wavelength (Kazanskiy et al., 2022). The optical attenuation at a target wavelength is directly proportional to the concentration of the target gas.

The power loss can be defined using the Beer–Lambert law, which depends on various key factors such as interaction length, target gas concentrations, and the evanescent field ratio (EFR) of the waveguide. As a result, the EFR works as a crucial parameter in analyzing the performance of gas sensors based on evanescent field absorption. The EFR can be defined as the ratio of the targeted region's intensity integration within the sensing region to the overall intensity integration and can be expressed as (Kazanskiy et al., 2022):

$$EFR = \frac{\int\int\int_{desired} |E(x,y,z)|^2 dxdydz}{\int\int\int_{total} |E(x,y,z)|^2 dxdydz}$$

Various optical waveguides-based gas sensors have been reported based on this phenomenon (Butt et al., 2020; Huang et al., 2014; Ranacher et al., 2018; Schmitt et al., 2008). To fabricate integrated photonic-based gas sensors, various types of waveguide systems integrated on different platforms have been reported over the past decades. Based on the fundamental concept of total internal reflection (TIR), these waveguides can direct the launched light in a high refractive index core. Waveguides with larger cross-sections offer strong optical mode confinement and generally exhibit low propagation losses. However, to enhance the sensing performance of the device in the evanescent field absorption-based gas sensors, a substantial overlap of the evanescent field with the surrounding medium is highly important. Reducing the geometry of the waveguide can significantly enhance this field interaction, but often leads to increased propagation losses. Moreover, the ability of these waveguides to confine light is fundamentally limited by diffraction.

Yebo et al. (2010) reported a SOI microring resonator (MRRs) coated with a thin layer of porous ZnO nanoparticles to selectively detect ethanol in the ambient environment. The MRRs, along with the grating couplers, were fabricated by utilizing standard complementary metal-oxide-semiconductor (CMOS) facilities. Typically, deep UV photolithography (193 nm) and dry etching methods were used to achieve high-Q resonators (> 25,000). To allow the localized growth of ZnO nanoparticles, the negative photoresist (AZ 2070) patterns were created onto the SOI chip. The patterning was achieved via a photolithographic process utilizing 365 nm ultraviolet (UV) light. After patterning, the sample was finally heated at 140°C for 2 minutes in order to improve the resistance of the photoresist against the ethanolic suspension. To create a thick layer of ZnO nanoparticles, the nanoparticle was coated via the drop-casting method on the fabricated structure and dried for 15 minutes at room temperature. Lastly, to complete the process, the photoresist was removed using a 1-methylpyrrolidine solution at 80°C, resulting in porous localized deposition on the MRRs. The performance of the fabricated device was tested under various concentrations of ethanol, varying from 120 to 1500 ppm. The fabricated sensor was able to detectconcentrations as low as 100 ppm, with a detection limit of 25 ppm. The sensor exhibited a response time of 15 seconds and a recovery time of 6 seconds when exposed to 300 ppm ethanol. In the same year, Ma et al. (2021) reported on-chip integration of a zeolitic imidazolate framework-8 (ZIF-8) coated SiO_2 waveguide for sensitive and selective detection of ethanol by exploiting evanescent wave sensing.

Butt et al. (2020) reported an ultrashort inverted tapered Si ridge-to-slot waveguide coupler for gas sensing optimized at 1.55 μm. The reported sensor comprises three different sections, including a ridge waveguide, an inverted taper segment, and a slot waveguide. The coupling region

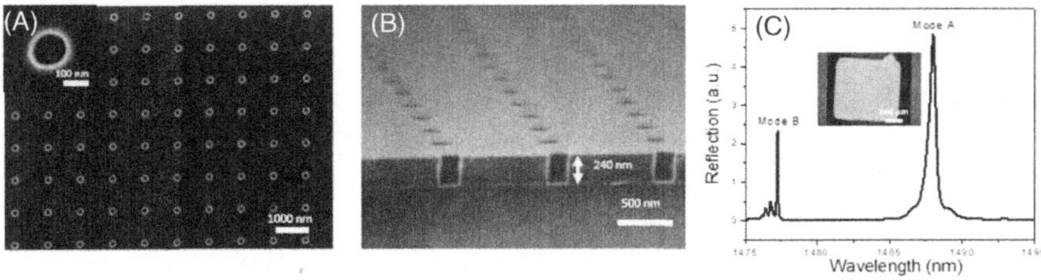

FIGURE 9.4 PCS sensor fabriacted on SOI and glass substrate: (A) SEM image of top view of the fabricated PCS sensor on SOI substrate, (B) cross-sectional SEM image of the PCS sensor, and (C) measured reflection spectrum of the PCS sensor on a glass substrate after transfer printing. *From Biswas, P., Zhang, C., Chen, Y., Liu, Z., Vaziri, S., Zhou, W., & Sun, Y. (2021). A portable micro-gas chromatography with integrated photonic crystal slab sensors on chip.* Biosensors, *11(9). https://doi.org/10.3390/bios11090326.*

was designed with a V-shaped geometry, which allows effective mode matching between the slot waveguide and the ridge. A tapered section of 100 nm in length was introduced between the ridge waveguide and slot waveguide, leading to a propagation loss of 1.69 dB/μm along with an excellent coupling efficiency of 91%. This overall loss accounts for contributions from all segments, i.e., taper transition, ridge waveguide, and slot waveguide. The reported device shows potential for integrated photonic circuits (IPC) where efficient light transfer into slot waveguides is highly crucial. The authors also demonstrated the potential of the fabricated sensor in CH_4 sensing via evanescent field absorption, taking place in the mid-infrared range of 3.392 μm. The strong optical confinement within the low-refractive-index slot waveguide region enhances light–matter interaction significantly. The optimized design exhibited an EFR of nearly 0.73, which allows a 3dB loss in the launched power at a CH_4 concentration of 60% in the surrounding medium.

Biswas et al. (2021) reported a portable micro-gas chromatography with an integrated photonic crystal slab (PCS) using transfer printing technology. The PCS was fabricated via e-beam lithography and the reactive-ion etching method onto a SOI substrate. The SEM images of the fabricated PCS are shown in Fig. 9.4A and B, whereas Fig. 9.4C shows the measured reflection spectrum of the PCS sensor after transfer printing. In the reported work, the authors transferred a 2D defect-free PCS printed on a borofloat glass, bonded it to a Si microfluidic gas cell or directly to a microfabricated GC column, and then deposited it with a gas-sensing polymer. Real-time shift in the Fano resonance was utilized to quantitatively identify 10 different VOCs, including pentane, hexane, toluene, trichloromethylene, benzene, hexanal, p-xylene, 2-heptanone, chlorobenzene, and ethylbenzene over a broad range of concentrations varying from 7 to 36 ng. The integrated micro gas chromatography-PCS was utilized

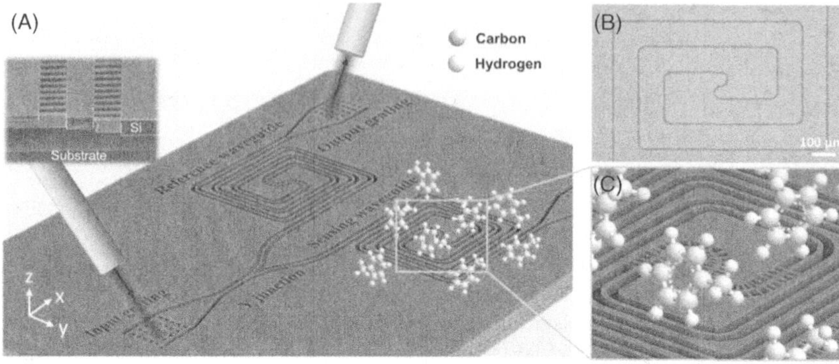

FIGURE 9.5 Suspended Si waveguide gas sensor: Schematic diagram of (A) suspended Si waveguide-based gas sensor (SEM shown in inset) and (B) microscope image of Si spiral waveguide. *From Liu, W., Ma, Y., Chang, Y., Dong, B., Wei, J., Ren, Z., & Lee, C. (2021). Suspended silicon waveguide platform with subwavelength grating metamaterial cladding for long-wave infrared sensing applications.* Nanophotonics, 10*(7), 1861–1870. https://doi.org/10.1515/nanoph-2021-0029.*

to demonstrate the separation and detection of a complex mixture of 10 chemicals, along with rapid separation and detection (4 minutes) and a low LOD (ng).

Long-wave infrared (6–14 μm) holds great potential in various sensing applications, including biochemical and gas, as it covers a broad range of molecular absorption footprints (Butt & Kazansky, 2020; Dinh et al., 2016;Kazanskiy et al., 2022). However, the research focused on photonic waveguides in this wavelength range is limited. Liu et al. (2021) reported a Si waveguide structure comprising subwavelength grating metamaterials cladding for long wave infrared (LWIR) sensing applications, as shown in Fig. 9.5. In reported work, the LWIR photonic platform was fabricated by utilizing suspended Si waveguides supported by subwavelength grating metamaterial cladding for rapid and ultra-sensitive on-chip gas detection. The fabricated platform allows an effective way to exploit the full transparency window of Si. The subwavelength grating structure enables precise control over the mode profile, as well as improving the light-matter interaction. Propagation and bending losses were investigated across a broad spectral range varying from 6.4 to 6.8 μm. In order to demonstrate the sensing response of the device, the fabricated platform was tested under the exposure of toluene vapor, achieving a LOD of 75 ppm. The system exhibited a rapid response and recovery times of 0.8 and 3.4 seconds, respectively, when exposed to 75 ppm toluene.

Chen et al. (2024) reported a hybrid plasmonic sensor made of large-area Au nanohole arrays integrated with ZIF-8 to detect ethanol vapor. The interaction between plasmonic modes of the nanoholes and the waveguide modes within the ZIF-8 layers results in a narrow-band hybrid mode that is highly sensitive to changes in ethanol concentration. The development of the gold nanohole arrays was performed using a cost-effective colloidal lithography method, allowing easy tuning of the LSPR modes by adjusting the diameter and periodicity of the nanoholes. Later, the cyclic synthesis method was utilized to deposit a thin layer of ZIF-8 on top of nanoholes. It was observed that the ZIF-8 layers not only allow the transportation of ethanol molecules but also support leaky modes that couple with the LSPR modes. The fabricated sensor exhibited an excellent sensitivity of 50.1 pm/ppm for the ethanol concentration ranging from 455 to 1543 ppm. The sensor also exhibited a low LOD of 11.8 ppm along with the lowest detected concentration of 47 ppm.

Recently, Xiahou et al. (2025) reported an on-chip array fluorescent sensor to detect multiple gases at the same time. Fluorescence based sensor arrays appear to be an efficient method to avoid the selectivity issue generally observed with single fluorescence materials via utilizing their high-dimensional response patterns and exceptional sensing performance. Traditional fluorescent sensors often suffer from bulky designs and complex configurations, leading to high costs and being unsuitable for combination within the compact sensing platforms. Integrated photonics, with its inherent advantage of high integration, presents a promising approach to fabricating a cost-effective and portable fluorescent-based gas sensor array. In the reported work, the authors developed an on-chip arrayed fluorescence sensing chip based on photonic waveguides, featuring a split optical path system. The schematic diagram shown in Fig. 9.6 illustrates the fabricated on-chip gas sensor along with the working mechanism. The integrated photonic device includes a 2.3 μm-thick silica layer grown on top of a Si substrate, serving as the bottom cladding for the SiN waveguide. Gratings, waveguides, and multimode interference (MMI) structures were then fabricated on a 180 nm-thick SiN layer using photolithography and dry etching. Finally, a 40 μm-thick layer of photoresist (SU-8) was deposited as the top cladding for the waveguide, with specific areas selectively etched to create windows for gas interaction. The performance of the device was tested in the presence of six different VOCs, including ammonia, cyclopentanone, ethyl acetate, ethyl alcohol, methylphenethylamine (MEPA), and toluene. Among the tested VOCs, the fabricated sensor exhibited an exceptional performance in accurately identifying all VOCs and achieving a remarkable LOD of 2.8 ppb for N-MEPA. Remarkably, the sensor demonstrateed accurate detection of MPEA even in a complex gaseous environment and high perfume concentrations.

Meziani et al. (2025) reported a mid-infrared integrated spectroscopic sensor. This liquid and gas VOC sensor was fabricated by utilizing two platforms, (1) porous silicon (PSi) and (2) chalcogenide glasses (ChGs). The ChGs layer was coated via sputtering machine, whereas the PSi layer

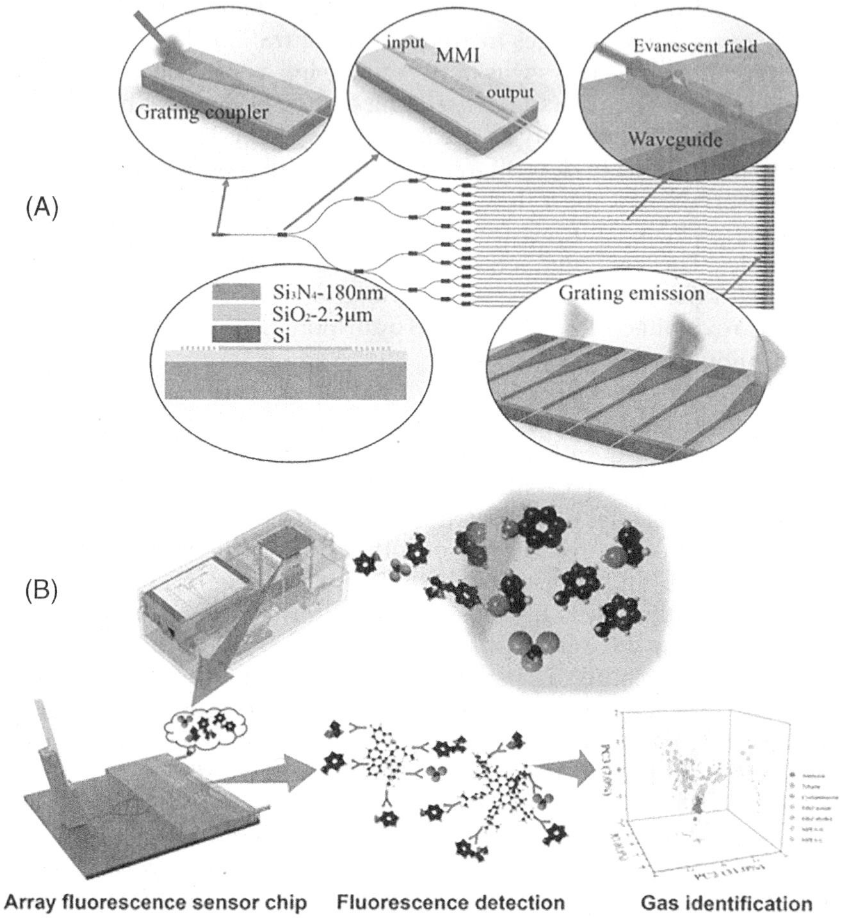

FIGURE 9.6 Fluorescence sensing chip: Schematic diagram of (A) design and (B) working principle of chip-based fluorescence sensor. *From Xiahou, Y., Wang, B., Li, H., Shen, Z., Jiang, Y., Li, H., Kerman, S., Wu, F., Fu, Y., Wang, T., Cheng, J., & Chen, C. (2025). On-chip array fluorescent sensor for high-sensitivity multi-gas detection.* ACS Sensors, *10(5), 3647–3657. https://doi.org/10.1021/acssensors.5c00460.*

was formed via electrochemical anodization. The ridge waveguide was fabricated utilizing i-line lithography method and developed via a reactive ion etching process. The ChGs waveguide supports a wide transmission spectrum ranging from $\lambda = 3.94$ to 8.95 μm, with a low propagation loss of 2.5 dB/cm at $\lambda = 7.58$ μm. In contrast, the PSi waveguide operates within a narrow spectral range of $\lambda = 3.94$ to 4.55 μm, along with a minimum propagation loss value of 9.1 dB/cm at $\lambda = 4.12$ μm. In order to assess the performance of the ChGs sensor, spectroscopic measurements were conducted to detect liquid analytes of acetonitrile and isopropanol. Due

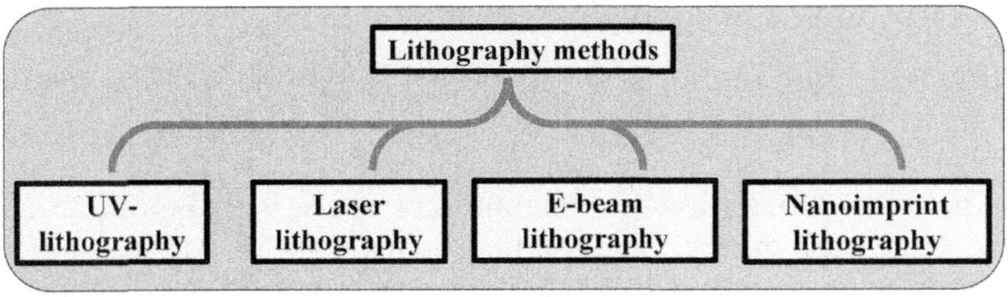

FIGURE 9.7 Lithography techniques: Different types of lithography. *From Butt, M. A. (2023). Integrated optics: Platforms and fabrication methods.* Encyclopedia, *3(3), 824–838. https://doi.org/10.3390/encyclopedia3030059.*

to strong evanescent field interactions, the obtained result exhibited a low LOD of 610 and 300 ppm for both acetonitrile and isopropanol, respectively. To test the potential of the device in the presence of gaseous samples, CO_2 was utilized as the target gas, and the low LOD of 17,000 ppm was obtained at $\lambda = 4.28$ μm using the ChGs platform. The sensing performance was improved by the PSi platform. Due to the available porous cavity, the launched light and target VOCs interact within the internal volume, unlike the ChGs platform, in which the evanescent field of the guided light allows the interaction with the entrapped gas. As a result, the PSi platform showed a significantly improved external confinement factor, exceeding that of the ChGs platform by more than 75 times, and allows CO_2 detection with a detection limit of 600 ppm at 4.26 μm. Quantification of the VOCs concentrations in the mixed solutions via the deconvolution technique of the measured spectra was also obtained with accurate concentration estimates, confirming its effectiveness in a complex environment by utilizing the ChGs platform.

9.4 Manufacturing strategies for integrated photonic systems

The development of integrated photonic sensors involves various advanced fabrication techniques that allow the miniaturization of the device and integration of multiple optical components onto a single platform (Butt, 2023). These processes exploit innovations in nanotechnology, materials science, and semiconductor techniques to produce such devices which can precisely control and manipulate the launched light within the device. Photolithography remains the key technology used in the fabrication of such miniaturized structures (Liaros & Fourkas, 2019). Several types of lithography techniques have been reported in the past decade to produce detailed structures, as shown in Fig. 9.7. A detailed discussion has been provided in the subsequent sections.

9.4.1 Ultraviolet lithography

Ultraviolet lithography (UVL) is a specialized photolithography technique which utilizes the exposure of UV beam to fabricate detailed miniaturized structures (Huang et al., 2015). It is one of the most widely utilized methods for the fabrication of semiconductor devices and other related industries to create a miniaturized pattern on a substrate. Although the UVL technology offers several advantages over the standard photolithography, it also offers unique characteristics. Compared with the visible light as utilized in standard lithography, the UV light comprises shorter wavelengths, hence offering the fabrication of high-resolution structures. These characteristics allow the patterning of smaller features and support increased integration density on a single substrate. Additionally, in comparison to the visible light the shorter wavelength generally provides a larger depth of focus. As a result, the lens can be placed farther from the substrate without compromising focus, allowing better control during the fabrication process (Butt, 2023). The UVL method is also compatible with a wider range of photoresists due to the strong interaction between photosensitive compounds and the UV radiation, therefore providing more material versatility for specific requirements (Benschop et al., 2008). Moreover, the higher photon energy associated with UV radiation allows faster exposure times, which provides improved manufacturing efficiency.

Implementing UVL typically requires specialized infrastructure, i.e., a UV light source and optical component, which can improve both the complexity as well as the cost of the setup. This technique is also prone to contamination, which means that even a small particle or any surface imperfection on the substrate or the photomask can significantly affect the pattern transfer, leading to poor device performance. Moreover, some materials, specifically certain polymers and some organic compounds, possess poor UV transparency, hence limiting their suitability for UVL applications. Additionally, the UV light utilized in lithography is considered to be harmful to human health and can damage skin and eyes under direct exposure. Therefore, the method requires special training and precautions before in order to ensure the safety of operators working with the UVL systems.

9.4.2 Laser lithography

Laser lithography (LL) is a fabrication method that utilizes a laser beam to expose and pattern photoresist materials. The LL is usually employed in micro- and nano-fabrication (Ocier et al., 2020). One can produce a high-resolution structure using LL, reaching submicron to nanometer-scale resolution, which is crucial for the development of intricate micro- and nanoscale integrated photonics devices. The key advantage of LL lies in its fabrication method, which allows the maskless and direct writing of patterns (Lü et al., 2001). This unique feature of LL allows rapid fabrication, quick prototyping, and the capability to fabricate customized and complex patterns. This method can be applied to a broad range of material types, including polymers, metals, biomaterials, and semiconductors making it highly versatile. Moreover, LL offers design flexibility, enabling the fabrication of complex structures and shapes which are usually challenging to obtain via other conventional lithography techniques.

Despite several benefits, the LL generally comprises lower production efficiency compared to the other lithography methods. The LL can be less effective for large-area applications due to its slow scanning process. The equipment used for LL typically includes a laser system and scanning stages, which are highly expensive with significant maintenance costs compared to the other standard lithography tools. These financial burdens can pose difficulties for smaller research groups and institutions with limited resources. Additionally, achieving accurate alignment in LL is also crucial for precise patterning. Precise alignment between the substrate, photoresist, and the laser beam is highly essential in order to obtain the desired structures. This alignment of LL appears to be highly demanding in the fabrication of complex, multilayer, or multistep lithography tasks, which are generally time-consuming and require technical expertise. Additionally, the patterning via LL can also be influenced by the substrate material, e.g., thermal conductivity and reflectivity. Substrates that either absorb or reflect a laser beam can significantly influence the accuracy and quality of structure formation. Thus, selecting and optimizing suitable substrate materials is crucial in order to produce desired structures. LL generally comprises a limited depth of focus, meaning that only a shorter range of distances from the objective lens can be precisely patterned in a single exposure. This limitation may constrain the maximum thickness of the sample that can be effectively patterned in a single exposure. Nonetheless, LL offers remarkable benefits, specifically in achieving high-resolution structuring and adaptable design flexibility.

9.4.3 Electron beam lithography

Electron beam lithography (EBL) is a known powerful nanostructuring tool utilized to produce nanometer-scale resolution and complex patterns (Broers, 1988; Broers et al., 1996). A key advantage of EBL is its ability to perform effectively with a variety of materials, such as metals, polymers, insulators, and semiconductors. This adaptability allows its application in numerous fields associated such as electronics, photonics, biology, and materials science (Geka et al., 2021). The major key feature of EBL involves maskless patterning on substrates, making them capable of the fabrication of rapid prototyping and highly customized designs (Tu et al., 2021). The

EBL technique is generally utilized for the fabrication of complex, high-resolution, and arbitrary geometries that are difficult to obtain via other traditional lithography methods, thereby providing greater controlled and enhanced flexibility for the research and development (Kumar Dey & Cui, 2013). Despite several advantages, EBL also comprises some limitations, including slow processing and taking longer times to pattern larger areas, which makes it less suitable for mass-scale production (Broers et al., 1996). In the EBL method, the electron beam scans across the sample point by point, hence taking an extended amount of time in production, which causes limited throughput (Bogdanov et al., 2012). For extensive patterning tasks or the fabrication of large device arrays, the duration required can become a significant bottleneck. EBL is considered an expensive tool as it requires specialized equipment, a high-resolution microscope, and a trained operator. The EBL setup is known for its expansiveness, along with high maintenance costs, making it less accessible for low-budget institutions. Furthermore, the parameter optimization, such as acceleration voltage, beam current, and dose of exposure, requires a good understanding and careful control.

9.4.4 Nanoimprint lithography

Nanoimprint lithography (NIL) is a unique nanofabrication method utilized to precisely produce complex structures at the nanoscale by transferring a design from a mold onto a substrate material (Butt, 2022). The NIL allows high-resolution patterning down to one nanometer, enabling it to be applicable in a broad range of applications, including photonics, electronics, and biotechnology. In general, the NIL utilizes three main steps, i.e., (1) fabrication of the key mold via traditional lithography or other nanofabrication techniques. The mold itself comprises the required structure to be transferred on the substrate (Grego et al., 2010), (2) the second step involves coating the substrate with a polymer film and a release agent to minimize adhesion with the template, (3) the final stage involves the imprinting process, where the mold is pressed against the polymer coated substrate. An optimized combination of pressure, temperature, and exposure to UV light or other curing approaches is utilized to soften the substrate, allowing it to flow within the patterned structures of the mold. Following solidification, the mold is carefully lifted, transferring the pattern onto the surface of the substrate, and (4) as a last and optional step, additional post-processing (via etching or annealing) may be required to further refine the structure (Prajzler et al., 2021).

NIL allows several advantages over conventional lithography methods, including large area patterning, high resolution, wide range of applications, and cost-effectiveness (Bruinink et al., 2008; Shneidman et al., 2018). Although NIL offers several advantages, it also comprises certain challenges, e.g., selection of the release, template fabrication, and material appropriateness.

9.5 Prospects and challenges

Integrated photonics-based sensors appear to be a promising tool in gas sensing, exploiting the unique characteristics of photonics in order to achieve remarkable sensing performance in terms of sensitivity, specificity, and miniaturization. These integrated photonic devices on a single chip can precisely detect and identify various gases along with their concentrations by utilizing light–matter interactions. Despite several advantages, the technology suffers from various limitations that need to be addressed in order to realize its full capabilities and enable broader implementation. The target selectivity of integrated photonic-based gas sensors poses several challenges in their real-world realization. One major concern lies in minimizing the interference from other toxic gaseous molecules present in the complex environment, which can lead to inaccurate detection or false readings. To achieve high target selectivity, sensor configuration must be modified to discriminate the desired gas from other similar compounds. Moreover, the fluctuations in the ambient conditions, i.e., humidity and temperature, can influence the behavior of the sensor, therefore complicating the accurate selectivity. Therefore, the combination of advanced signal processing via machine learning techniques and pattern recognition is crucial to improve the device selectivity and ensure consistent and reliable gas sensing in real-world applications.

Another major challenge associated with integrated photonics is the fabrication and integration complexity. Fabricating IPC capable of reliable and accurate gas detection requires advanced fabrication methods, which include precise lithographic patterning and controlled material deposition (Tolpygo et al., 2016). Maintaining uniform quality and ensuring reliable performance throughout the mass-scale production adds another layer of considerable challenge. Furthermore, the integration of various optical components, e.g., light sources, waveguides, detectors, and gas sensing materials, onto a single chip without compromising the performance of the device is a complex engineering task, typically requiring a trade-off between the performance and the miniaturization. This spectral overlap may lead to cross-sensitivity and potentially inaccurate readings. Therefore, developing materials and devices that can improve the target selectivity, possibly by utilizing functionalized surfaces or machine learning, is highly essential to overcome these challenges (Barik & Pradhan, 2022).

In addition to sensitivity and selectivity, the stability of the device is another key factor that plays a crucial role in achieving reliable performance and long-term operation. One major concern is the sensor drift, which may arise due to the material degradation, uneven deposition of the sensing material, or longer exposure to environmental factors. Such a drift in the sensor response can compromise both the device's sensitivity and accuracy, therefore affecting the overall performance of the sensor. Ensuring the stability generally requires precise calibration and periodic recalibration of the device in order to correct deviations in sensor output. Enhancing long-term stability may also require the utilization of robust sensing materials, or applying an additional protective layer on the sensor surface, and advanced encapsulation strategies to protect the sensor from the environmental factors (i.e., temperature and humidity) and ensure the functionality over extended use.

The LOD is a crucial factor that directly impacts the sensing performance of the photonics sensors, specifically when monitoring trace amounts of target gases. Achieving low LODs remains a significant challenge because of various limitations such as inherent system noise, interference from other surrounding gases, and the inherent sensitivity of the device's limit. Various strategies have been reported in past decades in order to address these limitations, including optimization of the sensor configuration, appropriate selection of the sensing material, utilizing advanced signal processing to improve the signal-to-noise ratio, and reducing the cross-sensitivity to background gases (Wang et al., 2023). Furthermore, the advancements in material science and nanotechnology have also opened new paths for further improvement of the sensing performance of the photonic gas sensors, potentially leading the LOD to reach even lower levels (Hänsel & Heck, 2020).

Despite these challenges, the future of integrated photonic VOC sensors appears to be promising. Progress in material science, specifically the introduction of various metamaterials and nanomaterials, is opening new possibilities for enhancing the performance of the integrated photonic sensors. The properties of these materials can be engineered to improve the light matter interactions, improve the gas absorption capabilities, as well as enhance the device sensitivity and target selectivity. Furthermore, the combination of such integrated photonic sensors along with the mature technologies, including microfluidic and microelectromechanical systems, can further help in the development of robust, compact, and multifunctional sensing platforms capable of real-time monitoring.

The integrated photonic sensor holds great potential for a wide range of applications, including industrial process control, environmental monitoring, and medical diagnostics (Butt & Piramidowicz, 2024). For example, these devices can provide on-site and real-time monitoring of the ambient environment and assess air quality with high accuracy, leading to better control over environmental and health conditions. In industries, these sensors can perform continuous monitoring of toxic gases and VOCs, maintaining workplace safety and operational efficiency. In healthcare applications, these devices allow non-invasive disease diagnosis via the detection of VOC biomarkers from exhaled breath, thereby offering a new pathway for early screening and disease detection.

9.6 Conclusion

Integrated photonic-based gas sensors play a crucial role in maintaining safety standards and preventing accidents by monitoring toxic gases in real time before they become hazardous. In industries, these gas sensors can detect leakage of various flammable and toxic gases, avoiding explosions and therefore protecting workers from serious poisoning. In residential and commercial places, these types of gas sensors play a crucial role in identifying various harmful VOCs and other gases, including CO, hydrogen, etc., providing real-time monitoring and alerts to save human lives. The importance of gas sensors also extends to medical applications. For example, these devices can be used to selectively monitor VOCs in exhaled breath, which not only helps healthcare professionals diagnose diseases at an early stage but also reduces hospital expenses for patients. The market for integrated photonic-based VOC sensors has shown substantial growth due to their compact size, high sensitivity, target selectivity, and robust platform for real-time and on-site VOC monitoring. These sensors exploit basic principles of photonic integration in order to improve sensing and reduce the cost and size of conventional VOC sensors. The future of integrated photonic VOC sensors remain promising, with the ongoing research mainly focused on improving the device performance in terms of selectivity, detection range, and robustness in various environmental conditions. Advances in nanofabrication, material science, and artificial intelligence (AI) are expected to play a vital role in the development of integrated photonic VOC sensors. Furthermore, integrating these sensors with the Internet of Things (IoT) could revolutionize industrial safety, environmental monitoring, and healthcare applications by providing real-time and continuous monitoring.

References

Almeida, V. R., Xu, Q., Barrios, C. A., & Lipson, M. (2004). Guiding and confining light in void nanostructure. *Optics Letters, 29*(11), 1209–1211. doi:10.1364/OL.29.001209.

Barik, P., & Pradhan, M. (2022). Selectivity in trace gas sensing: Recent developments, challenges, and future perspectives. *Analyst, 147*(6), 1024–1054. http://pubs.rsc.org/en/journals/journal/an. doi:10.1039/d1an02070f.

Benschop, J., Banine, V., Lok, S., & Loopstra, E. (2008). Extreme ultraviolet lithography: Status and prospects. *Journal of Vacuum Science & Technology B: Microelectronics and Nanometer Structures Processing, Measurement, and Phenomena, 26*(6), 2204–2207. doi:10.1116/1.3010737.

Biswas, P., Zhang, C., Chen, Y., Liu, Z., Vaziri, S., Zhou, W., & Sun, Y. (2021). A portable micro-gas chromatography with integrated photonic crystal slab sensors on chip. *Biosensors, 11*(9), 326. doi:10.3390/bios11090326.

Bogdanov, A. L., Lapointe, J., & Schmid, J. H. (2012). Electron-beam lithography for photonic waveguide fabrication: Measurement of the effect of field stitching errors on optical performance and evaluation of a new compensation method. *Journal of Vacuum Science & Technology B, Nanotechnology and Microelectronics: Materials, Processing, Measurement, and Phenomena, 30*(3). doi:10.1116/1.3700439.

Broers, A. N. (1988). Resolution limits for electron-beam lithography. *IBM Journal of Research and Development, 32*(4), 502–513. doi:10.1147/rd.324.0502.

Broers, A. N., Hoole, A. C. F., & Ryan, J. M. (1996). Electron beam lithography—Resolution limits. *Microelectronic Engineering, 32*(1–4), 131–142. doi:10.1016/0167-9317(95)00368-1.

Bruinink, C. M., Burresi, M., De Boer, M. J., Segerink, F. B., Jansen, H. V., Berenschot, E., Reinhoudt, D. N., Huskens, J., & Kuipers, L. (2008). Nanoimprint lithography for nanophotonics in silicon. *Nano Letters, 8*(9), 2872–2877. doi:10.1021/nl801615c.

Butt, M. A. (2023). Integrated optics: Platforms and fabrication methods. *Encyclopedia., 3*(3), 824–838. doi:10.3390/encyclopedia3030059.

Butt, M. A. (2022). Thin-film coating methods: A successful marriage of high-quality and cost-effectiveness—A brief exploration. *Coatings, 12*(8), 1115. doi:10.3390/coatings12081115.

Butt, M. A., & Kazansky, N. L. (2020). SOI suspended membrane waveguide at 3.39 μm for gas sensing application. *Photonics Letters of Poland, 12*(2), 67. doi:10.4302/plp.v12i2.1034.

Butt, M. A., Khonina, S. N., & Kazanskiy, N. L. (2020). Ultrashort inverted tapered silicon ridge-to-slot waveguide coupler at 1.55 μm and 3.392 μm wavelength. *Applied Optics, 59*(26), 7821. doi:10.1364/ao.398550.

Butt, M. A., Mateos, X., & Piramidowicz, R. (2024). Photonics sensors: A perspective on current advancements, emerging challenges, and potential solutions (invited). *Physics Letters A, 516*, 129633. doi:10.1016/j.physleta.2024.129633.

Butt, M. A., & Piramidowicz, R. (2024). Integrated photonic sensors for the detection of toxic gasses—A review. *Chemosensors, 12*(7). https://www.mdpi.com/journal/chemosensors. doi:10.3390/chemosensors12070143.

Chen, Z., Jao, C.-Y. U., Hu, K., Luo, Y., Ma, C., Jiang, R., Guo, T., & Chen, K. (2024). Plasmonic hybrid modes in a multifunctional ZIF-8 layer for high performance volatile organic compounds sensing. *Optics & Laser Technology, 176*, 110967. doi:10.1016/j.optlastec.2024.110967.

Dai, D., Liang, D., & Liu, L. (2015). Introduction for the integrated photonics: Challenges and perspectives feature. *Photonics Research., 3*(5), IP1. https://www.osapublishing.org/DirectPDFAccess/59A1ADC9-EE2A-BDE0-8D8E7999CAB7956B_327527/prj-3-5-IP1.pdf?da=1&id=327527&seq=0&mobile=no. doi:10.1364/PRJ.3.000IP1.

Dell'Olio, F., & Passaro, V. M. (2007). Optical sensing by optimized silicon slot waveguides. *Optics Express, 15*(8), 4977. doi:10.1364/OE.15.004977.

Dinh, T. V., Choi, I. Y., Son, Y. S., & Kim, J. C. (2016). A review on non-dispersive infrared gas sensors: Improvement of sensor detection limit and interference correction. *Sensors and Actuators, B: Chemical, 231*, 529–538. doi:10.1016/j.snb.2016.03.040.

Dominguez, I., Del Villar, I., Fuentes, O., Corres, J. M., & Matias, I. R. (2021). Dually nanocoated planar waveguides towards multi-parameter sensing. *Scientific Reports, 11*(1). www.nature.com/srep/index.html. doi:10.1038/s41598-021-83324-8.

Gallinet, B., Butet, J., & Martin, O. J. F. (2015). Numerical methods for nanophotonics: Standard problems and future challenges. *Laser and Photonics Reviews, 9*(6), 577–603. http://www3.interscience.wiley.com/journal/113511747/home. doi:10.1002/lpor.201500122.

Geka, G., Papageorgiou, G., Chatzichristidi, M., Karydas, A. G., Psycharis, V., & Makarona, E. (2021). Cuo/pmma polymer nanocomposites as novel resist materials for e-beam lithography. *Nanomaterials., 11*(3), 1–22. https://www.mdpi.com/2079-4991/11/3/762/pdf. doi:10.3390/nano11030762.

Grego, S., Huffman, A., Lueck, M., Stoner, B. R., & Lannon, J. (2010). Nanoimprint lithography fabrication of waveguide-integrated optical gratings with inexpensive stamps. *Microelectronic Engineering, 87*(10), 1846–1851. http://www.journals.elsevier.com/microelectronic-engineering/. doi:10.1016/j.mee.2009.11.003.

Hänsel, A., & Heck, M. J. R. (2020). Opportunities for photonic integrated circuits in optical gas sensors. *JPhys Photonics, 2*(1). https://iopscience.iop.org/article/10.1088/2515-7647/ab6742/pdf. doi:10.1088/2515-7647/ab6742.

Ho, W. F., Chan, H. P., & Yang, K. L. (2013). Planar optical waveguide platform for gas sensing using liquid crystal. *IEEE Sensors Journal, 13*(7), 2521–2522. doi:10.1109/JSEN.2013.2254596.

Huang, Y., Kalyoncu, S. K., Zhao, Q., Torun, R., & Boyraz, O. (2014). Silicon-on-sapphire waveguides design for mid-IR evanescent field absorption gas sensors. *Optics Communications, 313*, 186–194. https://www.journals.elsevier.com/optics-communications. doi:10.1016/j.optcom.2013.10.022.

Huang, Y., Zhao, Q., Kamyab, L., Rostami, A., Capolino, F., & Boyraz, O. (2015). Sub-micron silicon nitride waveguide fabrication using conventional optical lithography. *Optics Express, 23*(5), 6780. doi:10.1364/oe.23.006780.

Igarashi, A., Abe, M., Kuroiwa, S., Ohashi, K., & Yamada, H. (2023). Enhancement of refractive index sensitivity using small footprint S-shaped double-spiral resonators for biosensing. *Sensors, 23*(13), 6177. doi:10.3390/s23136177.

Kazanskiy, N. L., Khonina, S. N., & Butt, M. A. (2022). Advancement in silicon integrated photonics technologies for sensing applications in near-infrared and mid-infrared region: A review. *Photonics., 9*(5). https://www.mdpi.com/2304-6732/9/5/331/pdf?version=1652248249. doi:10.3390/photonics9050331.

Kumar Dey, R., & Cui, B. O. (2013). Stitching error reduction in electron beam lithography with in-situ feedback using self-developing resist. *Journal of Vacuum Science & Technology B, Nanotechnology and Microelectronics: Materials, Processing, Measurement, and Phenomena, 31*(6), 06F409. doi:10.1116/1.4831769.

Liaros, N., & Fourkas, J. T. (2019). Ten years of two-color photolithography [Invited]. *Optical Materials Express, 9*(7), 3006. doi:10.1364/ome.9.003006.

Liu, W., Ma, Y., Chang, Y., Dong, B., Wei, J., Ren, Z., & Lee, C. (2021). Suspended silicon waveguide platform with subwavelength grating metamaterial cladding for long-wave infrared sensing applications. *Nanophotonics, 10*(7), 1861–1870. http://www.degruyter.com/view/j/nanoph?rskey=PZEKdQ&result=1&q=Nanophotonics. doi:10.1515/nanoph-2021-0029.

Lü, L., Fuh, J. Y. H., & Wong, Y. S. (2001). Fundamentals of laser-lithography processes. *Laser-induced materials and processes for rapid prototyping* (pp. 9–38). Springer Nature. doi:10.1007/978-1-4615-1469-5_2.

Lukosz (1995). Integrated optical chemical and direct biochemical sensors. *Sensors and Actuators: B. Chemical., 29*(1–3), 37–50. doi:10.1016/0925-4005(95)01661-9.

Ma, X., Wu, J., Jiang, L., Wang, M., Deng, G., Qu, S., & Chen, K. (2021). On-chip integration of a metal-organic framework nanomaterial on a SiO_2waveguide for sensitive VOC sensing. *Lab on a Chip, 21*(17), 3298–3306. http://pubs.rsc.org/en/journals/journal/lc. doi:10.1039/d1lc00503k.

Mere, V., Muthuganesan, H., Kar, Y., Kruijsdijk, C. V., & Selvaraja, S. K. (2020). On-chip chemical sensing using slot-waveguide-based ring resonator. *IEEE Sensors Journal, 20*(11), 5970–5975. doi:10.1109/jsen.2020.2974502.

Meziani, S., Hammouti, A., Jaafar, A., Bodiou, L., Lorrain, N., Pirasteh, P., Chahal, R., Bénardais, A., Courson, R., Troles, J., Boussard-Pledel, C., Nazabal, V., Guendouz, M., & Charrier, J. (2025). Liquid and gas mid-infrared integrated spectroscopic sensor. *Optics Express, 33*(8), 17002. doi:10.1364/oe.553730.

Ocier, C. R., Richards, C. A., Bacon-Brown, D. A., Ding, Q., Kumar, R., Garcia, T. J., van de Groep, J., Song, J. H., Cyphersmith, A. J., Rhode, A., Perry, A. N., Littlefield, A. J., Zhu, J., Xie, D., Gao, H., Messinger, J. F., Brongersma, M. L., Toussaint, K. C., Goddard, L. L., & Braun, P. V. (2020). Direct laser writing of volumetric gradient index lenses and waveguides. *Light: Science and Applications, 9*(1). www.nature.com/lsa/. doi:10.1038/s41377-020-00431-3.

Passaro, V., Tullio, C., Troia, B., Notte, M., Giannoccaro, G., & Leonardis, F. (2012a). Recent advances in integrated photonic sensors. *Sensors, 12*(11), 15558–15598. doi:10.3390/s121115558.

Passaro, V. M. N., Troia, B., & De Leonardis, F. (2012b). A generalized approach for design of photonic gas sensors based on Vernier-effect in mid-IR. *Sensors and Actuators, B: Chemical, 168*, 402–420. doi:10.1016/j.snb.2012.04.044.

Prajzler, V., Chlupaty, V., Kulha, P., Neruda, M., Kopp, S., & Mühlberger, M. (2021). Optical polymer waveguides fabricated by roll-to-plate nanoimprinting technique. *Nanomaterials, 11*(3), 724. doi:10.3390/nano11030724.

Ranacher, C., Consani, C., Vollert, N., Tortschanoff, A., Bergmeister, M., Grille, T., & Jakoby, B. (2018). Characterization of evanescent field gas sensor structures based on silicon photonics. *IEEE Photonics Journal, 10*(5), 1–14. doi:10.1109/jphot.2018.2866628.

Schmitt, K., Oehse, K., Sulz, G., & Hoffmann, C. (2008). Evanescent field sensors based on tantalum pentoxide waveguides – A review. *Sensors, 8*(2), 711–738. http://www.mdpi.org/sensors/papers/s8020711.pdf. doi:10.3390/s8020711.

Shneidman, A. V., Becker, K. P., Lukas, M. A., Torgerson, N., Wang, C., Reshef, O., Burek, M. J., Paul, K., McLellan, J., & Lončar, M. (2018). All-polymer integrated optical resonators by roll-to-roll nanoimprint lithography. *ACS Photonics, 5*(5), 1839–1845. http://pubs.acs.org/journal/apchd5. doi:10.1021/acsphotonics.8b00022.

Sun, Y., & Fan, X. (2008). Analysis of ring resonators for chemical vapor sensor development. *Optics Express, 16*(14), 10254. doi:10.1364/oe.16.010254.

Sun, Y., & Fan, X. (2011). Optical ring resonators for biochemical and chemical sensing. *Analytical and Bioanalytical Chemistry, 399*(1), 205–211. doi:10.1007/s00216-010-4237-z.

Thylén, L., & Wosinski, L. (2014). Integrated photonics in the 21st century. *Photonics Research, 2*(2), 75–81. https://www.osapublishing.org/prj/home.cfm. doi:10.1364/PRJ.2.000075.

Tolpygo, S., Bolkhovsky, V., Weir, T., Wynn, A., Oates, D., Johnson, L., & Gouker, M. (2016). Advanced fabrication processes for superconducting very large scale integrated circuits. *IEEE Transactions on Applied Superconductivity, 26*(3), 1 1. doi:10.1109/tasc.2016.2519388.

Tu, M., Xia, B., Kravchenko, D. E., Tietze, M. L., Cruz, A. J., Stassen, I., Hauffman, T., Teyssandier, J., De Feyter, S., Wang, Z., Fischer, R. A., Marmiroli, B., Amenitsch, H., Torvisco, A., Velásquez-Hernández, M. D. J., Falcaro, P., & Ameloot, R. (2021). Direct X-ray and electron-beam lithography of halogenated zeolitic imidazolate frameworks. *Nature Materials, 20*(1), 93–99. http://www.nature.com/nmat/. doi:10.1038/s41563-020-00827-x.

Tyszkiewicz, C., & Kielan, P. Ł. (2023). Evanescent wave sensitivity of silica-titania rib waveguides in the single-mode propagation regime. *Photonics, 10*(9). doi:10.3390/photonics10091065.

Vitoria, I., Gallego, E. E., Melendi-Espina, S., Hernaez, M., Ruiz Zamarreño, C., & Matías, I. R. (2023). Gas sensor based on lossy mode resonances by means of thin graphene oxide films fabricated onto planar coverslips. *Sensors, 23*(3). http://www.mdpi.com/journal/sensors. doi:10.3390/s23031459.

Wang, H., Feng, Z., Zhang, Y., Han, D., Ma, J., Chai, X., & Sang, S. (2023). Highly sensitive and low detection limit NO_2 gas sensor based on In_2O_3 nanoparticles modified peach kernel-like GaN composites. *Sensors and Actuators B: Chemical, 382*, 133452. doi:10.1016/j.snb.2023.133452.

Xiahou, Y., Wang, B., Li, H., Shen, Z., Jiang, Y., Li, H., Kerman, S., Wu, F., Fu, Y., Wang, T., Cheng, J., & Chen, C. (2025). On-chip array fluorescent sensor for high-sensitivity multi-gas detection. *ACS Sensors, 10*(5), 3647–3657. http://pubs.acs.org/journal/ascefj. doi:10.1021/acssensors.5c00460.

Yebo, N. A., Lommens, P., Hens, Z., & Baets, R. (2010). An integrated optic ethanol vapor sensor based on a silicon-on-insulator microring resonator coated with a porous ZnO film. *Optics Express, 18*(11), 11859. doi:10.1364/oe.18.011859.

Yi, H., Citrin, D. S., & Zhou, Z. (2010). Highly sensitive silicon microring sensor with sharp asymmetrical resonance. *Optics Express, 18*(3), 2967. doi:10.1364/oe.18.002967.

Zhang, G., Feng, X. L., Liedberg, B., & Liu, A. Q. (2016). Gas sensor for volatile organic compounds detection using silicon photonic ring resonator. *Procedia Engineering, 168*, 1771–1774. doi:10.1016/j.proeng.2017.02.002.

Zheng, L., Keppler, N., Zhang, H., Behrens, P., & Roth, B. (2022). Planar polymer optical waveguide with metal-organic framework coating for carbon dioxide sensing. *Advanced Materials Technologies, 7*(12). doi:10.1002/admt.202200395.

CHAPTER

10

Advances in photonic sensors for semi-volatile organic compound detection

10.1 Introduction

A large part of human life is generally spent indoors; therefore, it is essential to maintain the quality of the indoor environment to prevent any health issues. The data related to contaminants, including pesticides, plasticizers, and flame retardants, are widely available for dust and indoor air; however, very little information is available on the release of semi-volatile organic compounds (SVOCs) from consumer products, including televisions, carpets, furniture, as well as computers. The dust particles present in the indoor environment are usually a mixture of complex elements which generally evolve from indoor aerosols, biological material, and soil particles.

House dust acts as a reservoir for SVOCs as well as for particle-bound contaminants (Butte & Heinzow, 2002). Therefore, the real-time monitoring of the chemical composition present in indoor dust not only provides information about the level of indoor pollution but also offers insights into potential human exposure levels. The contamination of house dust may take place via two routes, i.e., (1) adsorption of harmful chemicals present in the air and (2) direct interaction with various consumer products (Butte & Heinzow, 2002; Rauert & Harrad, 2015). Generally, the pollutants present in house dust, which sometimes include compounds that were banned a long time ago, can serve as a significant source of exposure to the general population, specifically for children. The SVOC pollutants related to house dust can enter the human body via several means, among which the most common pathway is the inhalation of airborne or resuspended particles,

Photonic Sensors for Detection of VOC Biomarkers: Technological Principles and Medical Applications
DOI: https://doi.org/10.1016/B978-0-443-29166-1.00008-1

Copyright © 2026 Elsevier Inc. All rights are reserved, including those for text and data mining, AI training, and similar technologies.

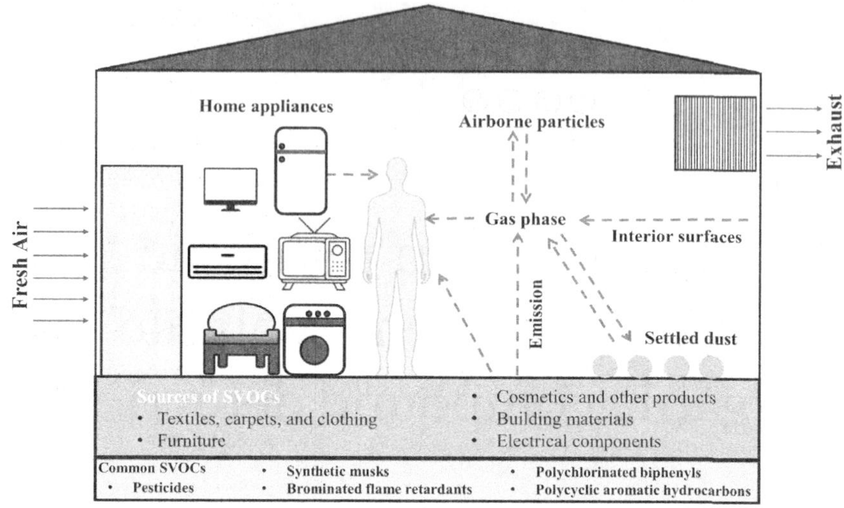

FIGURE 10.1 Sources of SVOCs in indoor environment. The schematic diagram summarizes the major sources of SVOCs, common SVOCs, and the routes of exposure to the human body.

while others include consuming food contaminated with dust particles or via dermal absorption. The amount of dust that settles in various regions of the respiratory system generally depends on the particle size and air flow properties. The schematic diagram shown in Fig. 10.1 represents the simplified version of the exposure of SVOCs to the human body, along with the major sources and most common SVOCs in indoor environments.

The potential risk of exposure to certain compounds present in house dust can be comparable to, or sometimes exceed, the exposure from food consumption, mainly in toddlers and infants (Hwang et al., 2008). The World Health Organization (WHO) recognizes SVOCs as key indoor pollutants having a boiling point varying from 240/260°C to 380/400°C (Lucattini et al., 2018). The SVOCs can be distinguishable from very volatile organic compounds (VVOCs) and volatile organic compounds (VOCs), which have boiling point ranges of <0 to 50/100°C and 50/100 to 240/260°C, respectively (*Text Version of the Indoor Air Quality House Tour*, 2025). In order to maintain indoor air quality and avoid excessive exposure, the US EPA recommends various guidelines (*Indoor Air Quality by Building Type*, n.d.; *Technical Overview of Volatile Organic Compounds*, n.d.).

One of the earliest efforts to link organic chemicals found in settled house dust with consumer products was first reviewed by Mercier et al. (2011). The reported study discusses the applications, uses, and sources of specific organic pollutants. As in previous chapters, detailed discussions are provided on VOCs and the technologies available to mitigate the issue and track the real-time monitoring of VOCs. This chapter will cover the basics of SVOCs along with their release sources, comparison with VOCs, types of SVOCs, and ways to precisely identify SVOCs in indoor and outdoor environments.

10.2 Semi-volatile organic compounds in consumer products

The SVOCs are mostly utilized as additives in electronics, furnishings, furniture, and building materials; hence, their concentration and presence are not disclosed publicly. Due to the lack of information regarding the concentration levels of SVOCs in consumer products, a general search is provided in this section. This section provides a concise summary of various consumer products on how materials and changes in chemicals have evolved over the past decades, which is relevant for the presence of phased-out or restricted chemicals in indoor dust and ambient air.

10.2.1 Textiles, carpets, and clothing

In the past few decades, extensive growth has occurred in the carpet and textile industries. During the 1950s, traditional woven carpets made of cotton and wool began to be replaced by tufted carpets, which were generally produced using synthetic fibers, such as rayon, nylon, and acrylic. In the late 1960s, the introduction of olefin-based carpets led to further utilization of polypropylene and polyester (Weschler, 2009). During the same period, various types of additives, such as fluorinated surfactants, became popular for their role as stain repellents and flame retardants, and were incorporated into carpets in the form of backing, padding, or adhesive (Weschler, 2009; Weil & Levchik, 2008). The use of flame retardants [mostly hexabromocyclododecane (HBCD)] in the carpet has already been reviewed by several researchers. A slow replacement of HBCD by advanced flame retardants, including organophosphate fire retardants (OPFRs), started to gain attention as early as in 2011 (Kajiwara et al., 2011). Various investigations carried out between 2008 and 2013 in Japan revealed the existence of fire retardants in textiles and curtains, mainly focusing on their photodegradation under sunlight. In this study, the investigations were conducted on 10 samples of polyester curtains (manufactured in Japan), which revealed excessive concentration levels of HBCD isomers in at least 9 samples, with concentrations ranging from 22×10^6 to 43×10^6 ng/g (Kajiwara et al., 2009).

The exact amounts of surfactant additives in consumer products, e.g., per- and polyfluoroalkyl substances (PFASs), are typically proprietary information and remain confidential. However, over the past decades, the compositions of PFASs in consumer products have been studied extensively, focusing on their release mechanisms and the used concentrations (Posner, 2012). In 2009, the US EPA tested 116 commercial products obtained from retail stores in the United States and divided them into 13 categories. Among these, textiles and carpets were identified as the primary sources for the release of perfluorocarboxylic acid (PFCA) contamination, with concentrations of 292 ng/g in carpet fibers and 427 ng/g in household textiles and upholstery (*Short-Chain Chlorinated Paraffins (SCCPs) and Other Chlorinated Paraffins Action Plan*, 2025). The US EPA also examined jacket products mainly manufactured in Europe and Asia and found PFAS concentrations ranging from 30 to 458×10^3 ng/m^2. Liu et al. (2014) reported a comprehensive review of the trends and concentrations of PFASs in clothing, carpets, and home textiles. Other than PFASs, pesticides were also identified in the textile products. The occurrence of the pyrethroid pesticide permethrin on indoor particles has been associated with the abrasion of carpet fibers (Berger-Preiß et al., 2002). In another report from the Danish EPA, bis(2-ethylhexyl) phthalate (DEHP) was identified in concentrations ranging from 2×10^3 and 8×10^3 ng/g in 20 samples obtained from textiles of wool, cotton, polyethylene terephthalate (PET), and flax (*Total Health Assessment of Chemicals in Indoor Climate from Various Consumer Products*, 2006). Moreover, the additives comprising siloxanes for textiles were also reviewed in other literature (Chen et al., 2011).

10.2.2 Furniture

Nowadays, the use of veneer on composite woods is increasingly common as an alternative to solid woods in the manufacturing of furniture items. Chairs, tables, dressers, desks, beds, and cabinets are generally manufactured using medium-density fiber-board or similar materials. Cushions utilized in furniture, beddings, and chairs are mainly made of synthetic foam commonly treated with flame retardants (Weschler, 2009). Chen et al. (2010) reported the presence of brominated flame retardants in bed mattresses, sofas, and pillows. Although polyurethane foams are used in carpet padding, which utilizes the PentaBDE mixture, the presence of polybrominated diphenyl ethers (PBDEs) was not identified in any of the samples. This shows the limited application of PBDEs in the production of such padding in order to meet furniture flammability standards in China. Nonetheless, the PBDEs were detected in the samples collected from vehicle parts such as plastic interiors, seat foam, and coating samples. The seat textile exhibited an extremely high concentration of BDE209 (32,611 ng/g), showing the excessive use of DecaBDE in textile manufacturing. In contrast, no PBDEs were found in any of the polyurethane foam samples collected from the furniture, which contradicts the assumption that PentaBDE formulations were typically utilized in any of these products. As analyzed by Stapleton et al. (2012), following the 2005 phase-out of PentaBDE, an increased amount of newer flame retardants started appearing in U.S. couches. In the reported work, the authors investigated approximately 102 samples collected from residential couches (purchased from 1985 to 2010 in the USA) made of polyurethane foam. Out of all collected samples, 41 samples bought before 2005 contained 39% BDE-99, BDE-47, and BDE-100, the primary components of the PentaBDE formulation, whereas 24% contained tris (1,3-dichloroisopropyl) phosphate (TDCPP). Lastly, the remaining 61 samples collected in 2005 or later demonstrated that TDCPP became the most common flame retardant (52%), along with a mixture of several non-halogenated organophosphate flame retardants (OPFRs), such as tris(phenyl) phosphate (TPhP) and tris(4-butylphenyl) phosphate (TBPP). Another research conducted by the group in 2009 showed the study on the existence of OPFRs in polyurethane foam, mainly utilized in various indoor and household items across the USA, such as couches, pillows, mattress pads, and chairs (Stapleton et al., 2009). The obtained results showed that TDCPP was one of the primary flame retardants present at levels between 1% and 5% by weight, followed by TCPP (from 0.5% to 22%). Another investigation examining flame retardants in toddler's furniture recognized another novel flame-retardant compound, namely V6 (2,2-bis(chloromethyl)-propane-1,3-diyltetrakis(2-chloroethyl) bisphosphate) in the majority of the collected foam samples, with the concentrations varying from 24.5×10^6 to 59.5×10^6 ng/g (Stapleton et al., 2011).

10.2.3 Electrical components

The ubiquitous presence of electronic devices in recent decades has led to a substantial increase in the release of toxic chemicals in the indoor environment (Weschler, 2009). Brominated and organophosphate-based flame retardants have been broadly studied for their presence in electronic components in recent years. In 2003, Kemmlein et al. recognized the emission of flame retardants from various electronic devices operated at 60°C, demonstrating that the release of flame retardants increases with temperature (Kemmlein et al., 2003). Under such operational temperatures, the unit-specific emission rates varied from 10 to 85 ng $unit^{-1}$ h^{-1} for organophosphate compounds, whereas 0.6–14.2 ng $unit^{-1}$ h^{-1} was observed for PBDEs.

The presence of PBDEs in raw electrical materials, as well as in electrical equipment such as computers and televisions (mainly produced in China), has been widely investigated by (Chen et al., 2010). PBDEs were obtained from Deca, Penta, and Octa mixtures, which were identified in 83.3%, 83.3%, and 58.3% of the television casings, respectively. The internal components of computers generally show higher concentrations of PBDEs, with nearly 279,965 ng/g, whereas the casing of computers and televisions shows significantly lower PBDE concentrations of <50 ng/g. The PBDE concentration was observed to be 3% of the total weight of the sample, while DecaBDE appeared to be dominant, comprising 80% of the total amount. In the same plastic samples, tetrabromobisphenol-A (TBBPA), polybrominated phenols (PBPs), and polybrominated biphenyls (PBBs) were also identified at 8.1×10^3 ng/g, 4.7×10^3 ng/g, and 0.25×10^3 ng/g, respectively. Furthermore, other components of the television, such as rear and front cabinet materials and associated circuit boards (from at least five TV sets) utilized in Japan, were investigated in 2008 (Takigami et al., 2008). In the reported study, the authors investigated rear cabinets and observed maximum concentrations of TBBPA and PBDEs of 48×10^6 ng/g and 19×10^6 ng/g, respectively.

Flame retardants, plasticizers, PBDEs, 1,2-Bis(2,4,6-tribromophenoxy) ethane (BTBPE), Tris (2-chloroisopropyl) phosphate (TCIPP), tris(phenyl)phosphate (TPhP), and tris(methylphenyl) phosphate (TMPP), were also identified in electronic waste with significantly higher concentrations (ranging between μg/g and mg/g) (Ballesteros-Gómez et al., 2013; Brandsma et al., 2014; Leslie et al., 2016). An early investigation revealed the presence of PFASs in electronic components obtained from Sweden, demonstrating that perfluorooctane sulfonate (PFOS) related substances are still used in numerous applications within semiconductor technology and photolithography (Herzke et al., 2012). PFOS-based chemicals are commonly used in the production of cell phones, digital cameras, printers, radar systems, and scanners (*Stockholm Convention on Persistent Organic Pollutants (POPs)*, n.d.).

Various investigations have reported an alarming situation regarding the release of toxic chemicals produced from the recycling of electronic waste. These studies not only highlight the major concerns associated with toxic metals but also report the levels of various organic contaminants. It has been observed that nearly 280 tons of polychlorinated biphenyls (PCBs) are released globally each year from electronic waste (assuming the global production of electronic waste is around 20 million tons annually). The recycling of transformers and condensers is identified as a key source of PCB release into the environments (Robinson, 2009).

10.2.4 Cosmetics and other products

Polyfluorinated compounds (PFCs) are identified in various household products, including cleaners, impregnating agents (such as waxes and floor polishes), conditioners, and lubricants. The reported study observed at least one PFC in 14 of the 26 product samples, along with 8:2 fluorinated alcohols (FTOH) detected as the dominating compounds (149 mg/mL). Additionally, perfluorooctanoic acid (PFOA) was also identified, with the concentrations 14.5 mg/mL, whereas PFOS was surprisingly not detected in any of the analyzed samples. FTOH, PFOS, and PFOA primarily appeared in lubricants and impregnating agents, but were absent from conditioners and cleaning agents. Impregnation products comprising FTOH exhibited identical proportions of 6:2, 8:2, and 10:2 FTOH. Similar FTOH ratios were also identified in lubricants containing PFCs. A 2006 survey conducted by the Danish EPA reported comparable findings (*Total Health Assessment of Chemicals in Indoor Climate from Various Consumer Products*, 2006). The spray impregnation of Leather and textiles was identified to contain perfluorooctane sulfonamide (PFOSA), with a concentration of 3.5×10^{-3} mg/mL, which aligned with the claim of the manufacturer (i.e., impregnation product contains fluorocarbon ingredients). PFOS was also identified in another spray product for camping items, including sleeping bags and tents, with a concentration of 2.12×10^{-1} mg/mL. In the case of liquid floor polishes (vinyl, linoleum, and cork), N-ethyl perfluorooctane sulfonamide (EtFOSA) was determined in one of the five collected samples, with a concentration of 0.1×10^{-1} mg/mL. Migration analysis confirmed the presence of PFOA and perfluoroheptanoic acid (PFHpA) in the impregnation agent, which is primarily used in liquid form in dry cleaning shops. In general, PFOA and PFHpA are released from similar impregnation agents that share identical features and applications. Eight types of PFAS were identified in shoe-related products, with low concentrations of PFHpA and PFOA, at 1.1×10^3 ng/g and 3.6×10^3 ng/g, respectively.

Capela et al. (2016) investigated nearly 123 healthcare and beauty products from Portugal and reported the presence of volatile methylsiloxanes in all examined product samples, with the body moisturizer containing the highest concentration of 7.54×10^5 ng/g. In another report by Lu et al. (2011), linear siloxanes were identified in 88% of the 158 healthcare products analyzed in China, with a maximum concentration of 52.6×10^6 ng/g in beauty products. Abelkop et al. (2015) identified these siloxanes as an additive compound in dry cleaning products. In addition to siloxanes, parabens are one of the most widely used SVOCs as preservatives in cosmetic products, with a mean value in the order of mg/L (Wang et al., 2017). Melo and Queiroz (2010) identified the presence of methyl paraben in body lotions with the maximum value of 1.65×106 ng/g.

10.2.5 Building materials

Organic pollutants are also released in the indoor environment from existing building materials. Various building materials, such as polyvinyl chloride (PVC) wires/cables, adhesive resin, and composite wood, were introduced after World War II. PVC pipes, which contain stabilizers and organic pollutants, are used to replace copper-made pipes in drain and water distribution systems(Weschler, 2009). The presence of flame retardant in insulating foam was first investigated by Kemmlein et al. (2003).

In the reported investigations, the content of HBCD and tris(2-chloro-1-methylethyl) phosphate (TCPP) was found to vary from 1% to 20%. Organophosphate and brominated compounds were identified in PVC wallpapers and insulating boards. An excessive concentration of TPhP (2.30×10^2 to 1.8×10^3 ng/g) and HBCDs (18×10^6 to 23×10^6 ng/g) were detected in various samples except for one PVC wallpaper, which contains HBCD as the most prevalent flame retardant. Due to its high chlorine content, PVC is inherently fire-resistant; therefore, the need for additional flame retardants was not necessary. Consequently, Kejriwal et al. reported that the PVC-based wallpapers play no role in the indoor flame retardant pollution (Kajiwara et al., 2011).

In 2016, Marie et al. reported the presence of bis(2-ethylhexyl) tetrabromophthalate (BEHTEBP), 2,3-dibromopropyl-2,4,6-tribromophenyl ether (DPTE), 1,2-bis(2,4,6-tribromophenoxy)ethane (BTBPE), 1,2-bis(pentabromodiphenyl)ethane (DBDPE), 2-ethylhexyl-2,3,4,5-tetrabromobenzoate (EHTBB), in various building materials including wallpaper, construction board, drainage pipes, polyurethane foam, vapor barrier, etc. (Marie et al., 2016). In 2025, Washburn et al. studied the presence of perfluorooctanoate (PFOA) in consumer products manufactured using fluorotelomer or fluoropolymers based products (Washburn, 2025). Out of the various tested products, PFOA was observed only in mill-treated carpeting, indicating the presence of PFOA at a very low concentration.

A Norwegian investigation reported the presence of PCBs in building façades, where concentrations of 2,4-dichlorobiphenyl (PCB7) varying from <1 to 2.9×10^5 ng/g in plaster and from <1 to 1.94×10^6 ng/g in paints. The concentration level varied based on the type and age of the building, being highest in buildings built during the 1950–60 period, while starting to drop in the 1970s (Andersson, Ottesen, & Volden, 2004). Limited application of dechlorane plus as a flame retardant has also been reported in epoxy and polyester systems, such as self-extinguishing phenolic resin laminated papers. Chlorinated paraffins are generally utilized across range of construction materials, such as varnished, adhesive, paints, and sealants (van Mourik, Gaus, Leonards, & Boer, 2016). Siloxanes are also utilized in paints and coatings to confer UV resistance, improving water repellency in concrete.

10.3 Difference between volatile organic compounds and semi-volatile organic compounds

Both VOCs and SVOCs are categorized as toxic and harmful chemicals, generally defined by their volatility. As discussed in previous chapters, VOCs easily evaporate at room temperature due to high vapor pressures

TABLE 10.1 Difference between VOCs and SVOCs.

VOCs (Pathak, 2023)	SVOCs (Lucattini et al., 2018)
High vapor pressure at room temperature (generally < 250°C)	Low vapor pressure and higher boiling points (generally 250°C–400°C)
Primary sources include paints, fuels, varnishes, adhesives, cleaning agents, building materials, etc.	Primary source furniture foam, electronics, cooking emissions, carpets, plastics, other household items
Quickly vaporize and disperse in the air	Gradually changes from a solid/liquid phase to air or dust, generally settling on the surfaces
Examples: Benzene, toluene, formaldehyde, xylene, acetone, and ethanol, etc.	Examples: Phthalates, hydrocarbons, flame retardants, organophosphate esters, PAHs, pesticides, phenol, phenolic compounds, etc.

and low boiling points, whereas SVOCs are comparatively less volatile, heavier, and remain in solid/liquid form. Although the SVOCs are less likely to vaporize, they can still pose significant health risks. Among various routes, the indoor dust generally works as a major sink for these compounds, which are widely present in consumer products. In order to control the exposure and set the safety parameters, WHO classified these compounds as indoor organic pollutants, typically varying from 240/260 to 380/L (Lucattini et al., 2018). The most common examples of SVOCs include benzyl alcohol, benzoic acid, organochlorine pesticides, phenolic compounds, and hydrocarbons. To draw a clear difference between VOCs and SVOCs, a comparative table is given below in Table 10.1.

10.4 Semi-volatile organic compounds in the indoor environment

10.4.1 Pesticides

Chlorinated pesticides and pyrethroids were identified in indoor dust samples collected from day-care, schools, and various homes in North Carolina during 2000–01. An increased level of pyrethroid was observed in the collected samples, with the concentration of 3.11×10^5 ng/g (Morgan, 2014). In 2003, Rudel et al. observed similar concentrations in the United States (1.72×10^5 ng/g) (Rudel, 2003), whereas another study carried out in 2013 by Trunnelle et al. reported the lower maximum values of 1.5×10^4 ng/g (Trunnelle, 2013). Chlorinated pesticide concentrations in North Carolina dust (4.8×10^3 ng/g) were found to be comparable to those reported in Danish homes, schools, and universities, with a maximum value of 19.03×10^3 ng/g (Bräuner, 2011). These concentration levels of pesticides exceeded those measured in house dust samples collected from Singapore (770 ng/g) and China (521 ng/g) (Tan, 2007).

10.4.2 Synthetic musks

Several investigations have been conducted to examine the presence of fragrance allergens and synthetic musks in indoor environments; all studies have reported the presence of these compounds in indoor air. In 2004, Fromme et al. examined the indoor air of kindergartens in Germany and identified mainly two musk compounds 1,3,4,6,7,8-hexahydro-4,6,6,7,8,8-hexamethylcyclopenta-γ-2-benzopyraan (HHCB) and acetylhexamethyltetralin (AHTN), showing minimum concentration levels of 0.10×10^3 and 44 ng/m^3, and maximum concentration levels reaching 0.3×10^3 and 0.11×10^3 ng/m^3, respectively (Fromme, 2004). Comparable concentrations (0.27×10^3 ng/m^3) of various synthetic musks were identified in Turkish schools and classrooms by Sofuoglu et al. in 2010 (Sofuoglu, 2010), and in 2016, Dallongeville et al. also found a similar concentration in the indoor environment of French households, with the concentration of 0.11×10^3 ng/m^3 (Dallongeville, 2016). Regueiro et al. reported the presence of HHCB and AHTN in samples collected from residential settings in northwestern Spain, with concentrations varying from 0.14×10^3 to 1.13×10^3 ng/m^3 for HHCB and 20–80 ng/m^3 for AHTN (Regueiro, 2009). These SVOCs were also detected in air sample collected from the resting zone within the laboratory, demonstrating concentrations of HHCB and AHTN of 60 and 20 ng/m^3, respectively.

10.4.3 Polychlorinated biphenyls

PCB concentration levels in indoor environments have been reported at various levels across various countries. In 2016, Melymuk et al. observed 0.14 ng/m^3 concentration level of PCB in bedroom air in the Czech Republic (Melymuk, 2016), while in French offices, apartments, and daycare centers, concentrations were reported to vary from 0.35 to 1.8 ng/m^3 (Moreau-Guigon & Chevreuil, 2014). A similar level of PCB concentrations was identified in Mexico City, where air samples collected from urban homes showed concentrations ranging from 0.210 to 0.840 ng/m^3, whereas in the semi-rural area, the concentrations were recorded as 0.1–0.32 ng/m^3 (Bohlin, 2008). Comparable PCB data were also obtained from private residential apartments and homes in Gothenburg (up to 1.6 ng/m^3) and in residences in Lancaster (0.15–2.1 ng/m^3) (Bohlin, 2008). The samples collected from the indoor environment of France also exhibited the PCB levels of 1.5 ng/m^3, which appeared to be consistent with values reported in previous studies, indicating broad indoor distribution of these compounds (Braouezec, 2016). However, a significantly higher PCB concentration level was recorded in urban areas of Bangkok, Thailand, with levels reaching 30 ng/m^3, while the rural and suburban areas showed much lower concentrations of 2.5 and 2.1 ng/m^3, respectively (Pentamwa & Oanh, 2008). Among these investigations, no direct indoor sources of PCBs were identified, suggesting that PCBs most likely entered from outdoor environments and accumulated indoors, specifically in spaces with limited ventilation, where degradation is slower due to reduced exposure to sunlight.

10.4.4 Polycyclic aromatic hydrocarbons

In 2014, Guigon and Chevreuil observed low concentration levels of Polycyclic aromatic hydrocarbons (PAHs) in French offices, daycare, and residential apartments, with values ranging from 0.5 to 1 ng/m^3 (Moreau-Guigon & Chevreuil, 2014). However, Kauneliene et al. reported a high concentration level of PAH in low-energy residential settings in Lithuania, ranging from 29.7 to 94 ng SPMD/day (Kaunelienė, 2016), while in the bedrooms of Czech Republic the concentration was comparatively dropped to 45 ng/m^3 (Melymuk, 2016), and in the living areas of Sweden and United Kingdom, it was reported to vary from 14 to 180 ng/m^3 and 8.5 to 60 ng/m^3, respectively (Bohlin, 2008). In Mexico City, PAH levels were reported to vary from 12 to 37 ng/m^3 and from 6.1 to 92 ng/m^3 in semi-rural and urban residences (Bohlin, 2008). Among all the reported literature, the highest PAH concentrations were observed in the indoor air of a Swedish alloy industry, where the concentration levels varied from 320 to 1900 ng/m^3 (Bohlin, 2008), and from an indoor turf facility in Connecticut, USA, the concentrations reached 341 ng/m^3 (Simcox, 2011).

10.4.5 Polyfluorinated alkyl substances

In 2007, Jahnke et al. investigated the air samples collected from an office in Hamburg and determined the presence of ionic PFAS, including PFOS and PFOA concentrations reaching 1.77 ng/m^3 (Jahnke, 2007). Later in 2012, Wu and Chang conducted an investigation on the air samples collected from an office in Singapore and detected a broad range of PFAS compounds, including various fluorooctane sulfonamides (FOSAs), fluorotelomer alcohols (FTOHs), and fluorooctane sulfonamidoethanols (FOSEs), with the concentration as high as 6.4 ng/m^3 (Wu & Chang, 2012). Recently, Crone et al. also reported the occurrence of per- and PFASs in source water (Crone, 2019).

10.4.6 Brominated flame retardants

In 2007, Saito et al. reported relatively higher concentrations of HBCD in the indoor environment of Tokyo, reaching nearly 29.5 ng/m^3 in office settings and 24 ng/m^3 in homes (Saito et al., 2007). A similar study also detected hexabromobenzene (HBB) reaching maximum values of 0.71 and 0.95 ng/m^3, 2,4,6-tribromophenol (TBPh) at concentration levels of 6.8 and 2.8 ng/m^3, and PBDEs at 5.9 and 36 ng/m^3 in residential and office samples, respectively. Compared to homes, the higher level of flame retardants in offices may be observed due to the greater use of flame-retardant sources (e.g., electronic equipment, fire-resistant pipes, ventilation units, and furnishings) (Saito et al., 2007). A low concentration level of PBDE was reported by Gevao et al. in Kuwait (Gevao, 2006). In 2016, Braouezec et al. conducted the investigation at the University Pierre et Marie Curie in Paris and reported an indoor air concentration of PBDE in indoor air, reporting a total concentration of 0.063 ng/m^3 (Braouezec, 2016). Ding et al. also reported similar concentrations in China (i.e., 0.54 ng/m^3 in households and 0.22 ng/m^3 in offices) (Ding, 2016), while in the Czech Republic, the concentrations in bedrooms reached 0.78 ng/m^3 (Melymuk, 2016), and at the Thai e-waste storage sites it was observed to be 0.35 ng/m^3 (Muenhor et al., 2010).

10.5 Photonic sensors for the detection of semi-volatile organic compounds

10.5.1 Detection of polyfluoroalkyl substances

A significant amount of research has been conducted to detect PFASs in the ambient environment (Rodriguez, 2020). These reported sensors offer several advantages, including low cost, in situ monitoring, high selectivity, rapid response time, and easy-to-use capabilities. In general, the lossy mode resonance (LMR) fiber sensor provides excellent sensitivity, a low detection limit, and remarkable reproducibility (Villar et al., 2025). Considering the advantages, Moro et al. utilized an LMR-based optical fiber sensor for label-free detection of PFOA (Moro, 2021). In reported work, human serum albumin (hSA) was utilized as a biological recognition layer for selective binding of PFOA in aqueous solution. Despite showing selective binding, the work highlights the need for further optimization of the LMR platform, demonstrates its potential to detect PFOA at low ng mL^{-1} levels, which aligns with limits set for PFAS in water. In the same year, Law et al. developed a self-assembled nanoporous interferometer for real-time monitoring of PFAS (Law, 2022). In the reported work, the authors fabricated an optical setup that combines perfluorosilane-functionalized nanoporous anodic alumina interferometers with reflectometric interference spectroscopy to enable real-time, label-free monitoring of PFOA self-assembled monolayers. By analyzing the variations in the effective optical thickness of nanoporous anodic alumina interferometers, the study investigated the real-time, in situ self-assembly of PFOA driven by fluorous interactions with perfluorosilane layers of various molecular lengths. Careful analysis of key parameters, including sensitivity, detection limit, and linear response, helps in identifying the optimal molecular length of perfluorosilanes for PFOA immobilization on functionalized surfaces. 1H,1H,2H,2H–perfluorooctyltriethoxysilane ($CF_3(CF_2)_5(CH_2)_2Si(OCH_2CH_3)_3$ (FOTS)-functionalized nanoporous anodic alumina interferometers provided a high sensitivity of 4.9 nm ($\mu g\ mL^{-1})^{-1}$, along with a low detection limit of 0.53 $\mu g\ mL^{-1}$, and excellent linearity of 0.97.

Faiz et al. reported the use of an optical fiber Fabry-Pérot interferometry (OFPI) coated with graphene oxide-doped alginate (Alg/GO) for the rapid identification of PFOA in water (Faiz, 2023). A uniform, thin layer of Alg/GO was obtained on the tip of the FPI sensor using the dip coating method, followed by cross-linking in calcium chloride. The absorption of PFOA on the deposited sensing material was validated using two separate analysis methods, i.e., analytical adsorption and Fourier-transform infrared (FTIR) spectroscopy. The obtained results suggested that Alg/GO adsorbs PFOA more efficiently than just alginate deposition. Similarly, the experimental results from the FPI sensor coated with Alg/GO exhibited significantly higher sensitivity towards PFOA, along with a low detection limit of 0.4 ppb and a rapid response time of 17 seconds. Later, enhanced selectivity was achieved by integrating surface plasmon resonance (SPR) with molecularly imprinted polymers (MIPs). In 2024, Pitruzzella et al. combined the properties of SPR and MIPs to develop a compact sensor for determining PFAS in river water (Pitruzzella, 2024). In the reported work, a 1 mm plastic optical fiber with a 980 µm PMMA core was integrated within a 3D-printed resin block (as shown in Fig. 10.2A) and polished to form a D-shape structure. Later, a 1.5 µm photoresist layer (Microposit S1813, $n \approx 1.6$) was spin-coated on top of the polished region to enhance plasmonic behavior and improve the adhesion of the gold film (60 nm). Fig. 10.2B illustrates the actual image of the fabricated sensor. The sensor utilized a wavelength interrogation method to analyze its response. The reported sensor exhibited excellent linearity in the presence of various PFOA concentrations, ranging from 5 to 1000 ppt, as shown in Fig. 10.2C and D.

Recently, Khitous et al. reported the deposition of MIP over a thin layer of gold nanofilm (60 nm) coated on top of D-shaped plastic optical fiber for the ultrasensitive detection of PFOA in aqueous water (Khitous, 2025). The approach enables access to target binding sites and enhances the light-matter interaction due to surface plasmons. By tuning the thickness of the MIP layer via controlled light exposure, the fabricated sensor provides an adjustable detection limit ranging from the attomolar to the nanomolar (53 a.m. to 6.3 p.m.), along with an average response time of 10 minutes. Due to the presence of target-specific cavities, the sensor coated with MIP also exhibited remarkable selectivity even at the concentrations 100 times higher

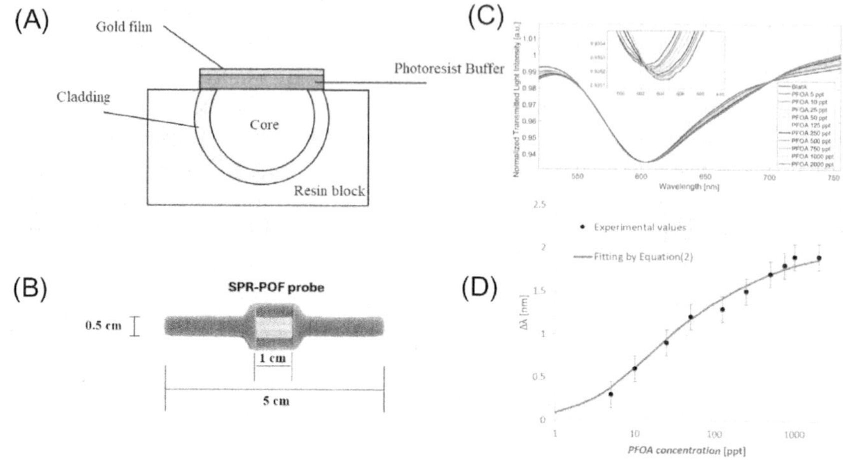

FIGURE 10.2 (A) Schematic diagram of sensor fabrication, (B) actual image of the fabricated sensor assembly, (C) shift in output spectra in the presence of various concentrations of PFOA, and (D) relationship between wavelength shift and PFOA concentration. *From Pitruzzella, R., Chiodi, A., Rovida, R., Arcadio, F., Porto, G., Moretti, S., ... & Cennamo, N. (2024). An MIP-based PFAS sensor exploiting nanolayers on plastic optical fibers for ultra-wide and ultra-low detection ranges—A case study of PFAS detection in river water. Nanomaterials, 14(21), 1764.*

than that of PFOA. Despite the average response time, the reported sensor appears as the first ever reported device to reach attomolar-level detection, making it a potential tool for environmental contaminant monitoring.

10.5.2 Detection of pesticides and insecticides

Pesticides are generally utilized to protect crops by destroying or repelling pests (Bernieri, 2019). A broad range of pesticides are used across various sectors, including herbicides, insecticides, fungicides, biocides, rodenticides, and similar chemical agents. These SVOCs are usually categorized based on their chemical composition and the types of pests they target (Kaur et al., 2019). Among them, the first major group of synthetic pesticides that has been extensively utilized is insecticides (Dalmolin, 2020). According to a report published by the United States Environmental Protection Agency, insecticides are mostly used in industrial, agricultural, commercial, and household applications, as well as in public health. A wide range of insecticides, e.g., carbamates, organophosphates, pyrethroids (or pyrethrins), neonicotinoids, and organochlorines, are categorized by their distinct chemical structures (Gupta, 2019).

Although these pesticides and insecticides play a crucial role across a range of applications, their excessive utilization can cause harmful effects on the environment and human health. Excessive utilization contributes

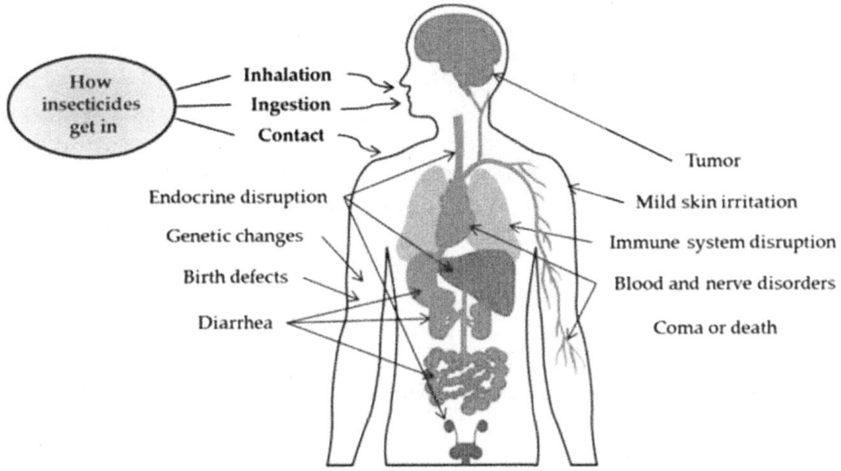

FIGURE 10.3 Schematic diagram showing the effect of insecticides on human health. *From Fauzi, N. I. M., Fen, Y. W., Omar, N. A. S., & Hashim, H. S. (2021). Recent advances on detection of insecticides using optical sensors. Sensors, 21(11), 3856.*

to soil, water, and air contamination, which ultimately reaches our plates in the form of food and water, posing a serious threat to human health (Bolzonella, 2019). This usually takes place when insecticides infiltrate soil and reach the groundwater, potentially contaminating drinking water resources. Furthermore, the spraying of insecticides over crops leads to the drift of their droplets and vapors in the air, which end up in the human body through inhalation. Oral exposure generally occurs by consuming contaminated food and water (Bolzonella, 2019). These chemicals can potentially accumulate within the human body over time. The effects of exposure to these insecticides on human health depend on the toxicity of the compound as well as the exposure time (Kim, 2017). Agricultural workers and their families are among the most vulnerable groups, experiencing direct and frequent exposure to these chemicals (Marrs, 1993). Exposure to these chemicals can cause various health effects such as skin irritation, gastrointestinal issues, endocrine disruption, genetic mutations, immune disruptions, tumors, neurological disorders, birth defects, and in some cases, death (Oaya, 2019). The schematic diagram shown in Fig. 10.3 illustrates the effect of insecticides on human health (Fauzi, 2011).

Considering the impact of insecticides on human health, various optical fiber sensors have been reported in the past decades for real-time detection and quantification. This method involves the utilization of colorimetric assays, fluorescence-based sensing, surface-enhanced Raman scattering (SERS), and chemiluminescent approaches. This section discusses various optical sensors reported in the past decades for identifying this contamination.

The first optical fiber-based pesticide sensor was reported by Bier et al. in 1992 (Bier, 1992). In the reported work, the authors utilized fluorescence-labeled aminohexylatrazine decorated over the end facet of a silica fiber to monitor triazines. It was observed that the fluorescence signal dropped significantly in the presence of triazines due to the binding of the antibody to functionalized fiber. The reported sensor exhibited a low detection limit of 0.1 ng/mL (for terbutryn). Later, Trettnak et al. reported a novel fiber optic sensor for the real-time detection of pesticides and related inhibitors of the enzyme acetylcholine esterase (AChE) (Trettnak, 1993). The sensors operated continuously and effectively detected AChE inhibitors from a distance of up to 100 m using a solid-state fiber-optic photometer. The fabricated sensor exhibited a low detection limit of 200 nM. In 1997, Andres et al. detected this pesticide using optical fibers immobilized with acetylcholinesterase (AChE) and thymol blue (Andres & Narayanaswamy, 1997). The reported sensor demonstrated a linear sensing response ($R^2 = 0.992$) to a wide range of concentrations of acetylcholine varying from 2.5 to 25 mM, along with the detection limits of 1.5×10^{-8} M (3.1 ppb) and 1.1×10^{-7} M (24.7 ppb) achieved for carbofuran and paraoxon, respectively. In the same year, Brummel et al. utilized an optical fiber probe for the selective detection of cyclodiene insecticides (Brummel et al., 1997). In the reported work, the authors immobilized anti-cyclodiene antibodies on a quartz optical fiber to selectively attach hexachlorocyclopentadiene-fluorescein cyclodiene. Kumar et al. utilized *Flavobacterium* sp. whole cells immobilized on glass fiber for selective detection of methyl parathion pesticide (Kumar et al., 2006). The fabricated sensor exhibited a low LOD of 0.3 μM along with a linear performance towards the methyl parathion concentration varying from 4 to 80 μM. The sensor was reported to be disposable, simple, direct, and single step, requiring a low amount of measurands. The performance of such sensors was further improved by Rajan et al. by applying a thin layer of silver film over silica fiber and later immobilizing the surface with AChE enzyme (Rajan et al., 2007). The sensor was based on a wavelength interrogation technique, where the SPR wavelength changes as the pesticide selectively binds to the enzyme. The sensing performance of the sensor was significantly enhanced by introducing SPR, which not only enhances the evanescent field but also improves the detection accuracy (Pathak et al., 2019).

In 2017, Arjmand et al. attempted to improve the sensing performance by reducing the diameter of the optical fiber (Arjmand, 2017). The fabricated tapered optical fiber sensor monitors changes in the ambient refractive index due to the selective binding of methyl parathion (as a model organophosphate pesticide) to the immobilized enzyme. The reported sensor exhibited a low detection limit of 2.4×10^{10} M for concentrations ranging from 10^{-10} M to 5×10^{-5} M. The reported sensor demonstrated excellent reproducibility for the pesticide and reactivating agent, with

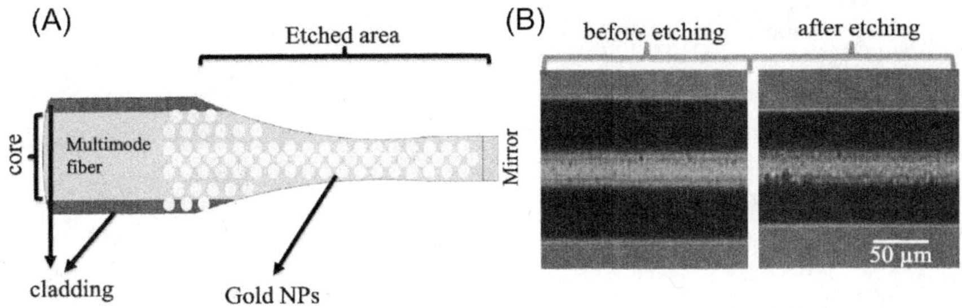

FIGURE 10.4 (A) Schematic diagram of the reported single-ended gold nanoparticle coated MMF. (B) Microscope image of MMF before and after tapering. *From Moslemi, A., Sansone, L., Esposito, F., Campopiano, S., Giordano, M., & Iadicicco, A. (2024). Optical fiber probe based on LSPR for the detection of pesticide Thiram. Optics & Laser Technology, 175, 110882.*

standard deviations of 2.36% and 2.44% (n = 6), respectively. The advancement in surface chemistry further helps in achieving improved performance by paving the way for uniform immobilization of the enzyme, leading to superior and selective binding of target pesticides. In 2020, Miliutina et al. reported the use of plasmon-assisted optical fibers immobilized with metal-organic frameworks (MOFs) for detecting organophosphorus pesticides in water (Miliutina, 2020). In the reported work, the SPR was achieved by depositing a thin layer of gold film over a multimode fiber (MMF). The MOF-5 was deposited on a gold surface to induce high affinity for target pesticides. The binding of organophosphorus pesticides leads to a change in the refractive index of MOF-5, which was monitored using Raman spectroscopy. The reported sensor was able to detect as low as 10^{-12} M concentration of pesticides, making it a promising tool for pesticide detection. Recently, Moslemi et al. reported a localized SPR (LSPR) based optical fiber sensor to detect Thiram pesticides (Moslemi, 2024). In the reported work, the authors fabricated a single-ended, tapered multi-mode fiber (MMF) coated with gold nanoparticles (~43 nm), with the end facet comprising a reflective mirror (as shown in Fig. 10.4). The sensor was operated in reflection mode, exhibiting two separate attenuation bands that facilitated the simultaneous monitoring of dual parameters. The author reported two sensitivities of 433 and 581 nm/RIU, corresponding to the first and second attenuation bands, respectively, due to changes in the refractive index of the ambient environment.

Integrated photonics have also garnered significant interest in developing compact, highly sensitive, and real-time systems for pesticide detection. Yadav et al. numerically optimized a silicon nitride-based hybrid dielectric–plasmonic sensor for real-time monitoring of pesticides in the environment and agricultural samples (Yadav, 2025). The authors used the finite difference time-domain (FDTD) method to optimize the structural parameters and analyze the performance of the designed sensor. Due to

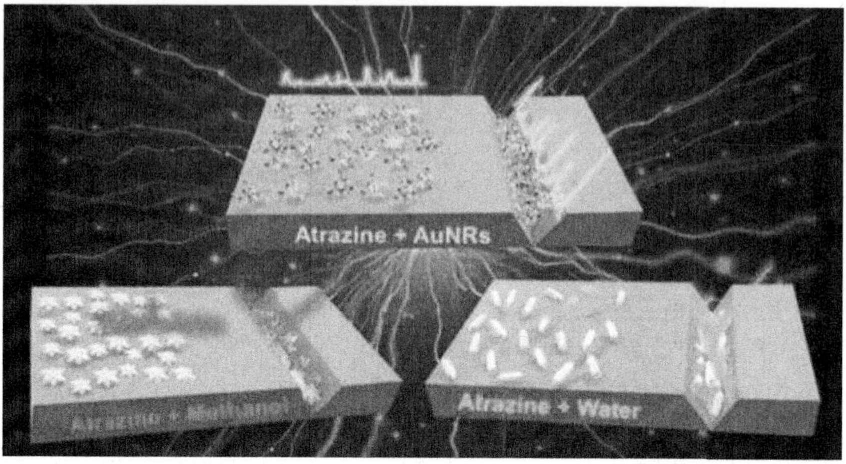

FIGURE 10.5 Schematic diagram illustrates the fabricated sensor and testing of Atrazin when mixed with methanol and water. *From Ferreira, M. G., de Barros, A., Shimizu, F. M., Sigoli, F. A., Bufon, C. C. B., & Mazali, I. O. (2025). Shape-Induced Enhanced Raman Scattering (SIERS) Platforms Modified with Gold Nanorods for Ultradiluted Atrazine Pesticide Detection. ACS Applied Nano Materials, 8(20), 10287-10296.*

the higher refractive index of silicon nitride, the proposed design enables exceptionally strong mode confinement and dual plasmonic resonances, exhibiting extremely high sensitivity of 2364 nm/RIU.

With the advancement of SERS, researchers found another way to further improve the performance of optical sensors for pesticide detection, offering label-free, rapid, and low detection limit compared to conventional optical detection methods. Tao et al. integrated silver nanocubes (SERS substrate) with the tapered optical fiber to develop SERS sensors to detect 4-Mercaptobenzoic acid (4-MBA) and thiram (Tao & Ge, 2023). The electromagnetic field intensity was significantly stronger for silver nanocubes as compared to the silver nanospheres due to the presence of more "hot spots." The SERS signal was enhanced by uniformly depositing silver nanocubes over the tapered region through electrostatic adsorption and the laser-induced method. The sensor achieved detection limits of 10^{-9} M for 4-mercaptobenzoic acid (4-MBA) and 10^{-8} M for thiram, with corresponding signal enhancement factors of approximately 1.64×10^5 and 6.38×10^4, respectively.

In another work, Ferreira et al. utilized the shape-induced enhanced Raman scattering (SIERS) effect to identify Atrazine pesticides (Ferreira, 2025). To analyze the performance, the enhancement in the SIERS signal was compared with the traditional Raman signal by performing measurements on a silicon (Si) substrate employing a V-shaped channel (Si-V) and the top of a flat Si substrate (flat-Si), as shown in Fig. 10.5. The obtained experimental results (free from Au nanorods) demonstrated a

7-times enhanced SIERS signal in the presence of the Atrazine molecule in comparison to the Raman signal. Whereas integrating SIERS with SERS signals utilizing gold nanorods deposited over Si-V and flat-Si significantly improved the limit of detection for atrazine over a wide concentration range from 1.6×10^{-8} to 1.6×10^{-20} mol L^{-1}.

10.5.3 Detection of dioxin and related compounds

Dioxins are another common, highly toxic, and persistent SVOC. Some common examples of dioxins include polychlorinated dibenzo-p-dioxins (PCDDs) and polychlorinated dibenzofurans (PCDFs), which have high molecular weight and low vapor pressures at higher temperatures. These dioxins are typically released into the atmosphere due to the formation of undesired compounds from various industrial and combustion processes (Breivik, 2004). The primary sources of dioxins include medical and municipal solid waste, waste incineration, chlorine bleaching, coal-fired power plants, forest fires, etc. Moreover, these SVOCs are also produced as pollutants from various herbicides, pesticides, and fungicides (Holt, 2010). Dioxins and related compounds are generally accumulated in plant and animal tissues when released accidentally into water or soil, and through this process, they reach human food and accumulate at higher concentrations, specifically in fatty tissues. These halogenated organic persistent organic pollutants are generally associated with the three families of molecules (Patrizi, 2019), i.e., PCDFs, PCDDs, and PCBs. The molecular structures of these molecules are shown below in Fig. 10.6.

Due to their long-lasting properties and high toxicity, researchers have invested significant effort in developing highly sensitive, rapid, responsive, and real-time monitoring of such pollutants at the emission sources (Gullett, 2012; Patrizi, 2014, 2019). However, continuous monitoring of all potential emission sources and the ability to identify each SVOC individually in a complex environment remain challenging. With advancements in photonics technology, researchers have successfully developed portable and highly sensitive detection systems. By integrating these photonic devices with the target-selective material, selectivity and rapid responsiveness can be precisely achieved.

Zhao et al. reported the fluorescence-based optical fiber immunosensors to detect PCB (Zhao, 1995). In the reported work, the authors utilized optical fiber sensors coated with partially purified polyclonal anti PCB antibodies. Detection of PCB was based on fluorescence from a TCPB-fluorescein conjugate that binds to the immobilized antibodies; PCBs competitively reduce this fluorescence. The sensor achieved a detection limit of 10 ppb for TCPB and detected 1 ppm Aroclors with good selectivity. Since the introduction of an optical dioxin detection system, researchers have put significant effort into improving the performance of such devices. In

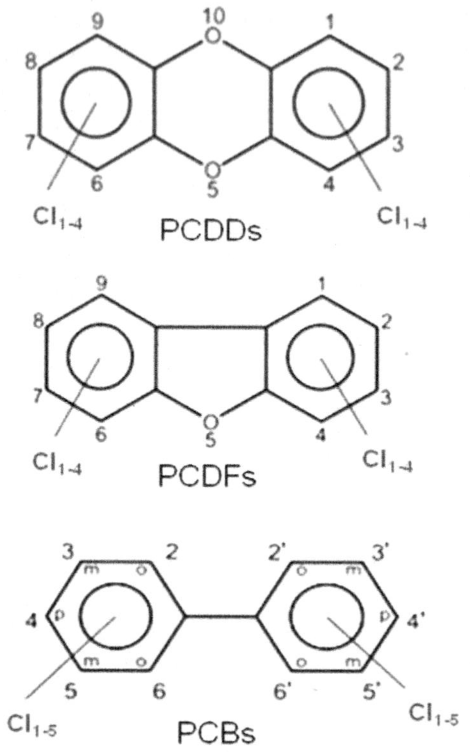

FIGURE 10.6 Molecular structures of PCDDs, PCDFs, and PCBs. *From Patrizi, B., Siciliani de Cumis, M., Viciani, S., & D'Amato, F. (2019). Dioxin and related compound detection: Perspectives for optical monitoring. International Journal of Molecular Sciences, 20(11), 2671.*

general, the optical sensor relies on the light-matter interaction, producing a significant change in the properties of propagated light, which is measured in the form of wavelength, intensity, phase, and polarization. A plasmonic sensor coated with a noble metal film of nanostructures enhances the evanescent field, facilitating a detectable resonance shift, high sensitivity, rapid response time, and improved detection accuracy. Utilizing the SPR phenomenon, Shimomura et al. reported a rapid-response optical sensor for atrazine, dioxins, and polychlorinated biphenyls (Shimomura, 2001). In the reported work, the authors utilized a competitive immunoassay to improve the sensing performance of the device. With this approach, the author reported selective detection of 2,3,7,8-TCDD and PCB within 15 minutes sampling time, along with low detection limits of 0.1 and 2.5 ng mL^{-1}. In another study, Soh et al. reported a comparative study between direct and competitive immunoassays to improve the sensing performance of the device for detecting dioxin precursors (Soh, 2003). To realize the SPR sensor, the author deposited a thin gold layer over the sensor chip which was later immobilized with anti-(2,4-dichlorophenol) antibody using a gold binding polypeptide (GBP) and protein G. The sensing performance of the sensor was monitored by analyzing the shift in the resonance peak arises from the interaction of the antigen-antibody in a flow setup which was filled with a 2,4-dichlorophenol solution. In the direct immunoassay using the modified sensor chip, the injection of a 25 ppm 2,4-dichlorophenol solution produced no notable SPR angle shift, remaining below 0.001 degree. To further enhance the sensitivity of the sensor, an indirect competitive immunoassay was employed. In this approach, 2,4-dichlorophenol was conjugated with bovine serum albumin (BSA). During the competitive assay, a solution containing 350 ppm of the 2,4-dichlorophenol-BSA conjugate, along with varying concentrations of 2,4-dichlorophenol varying from 10 to 250 ppb, was introduced into the sensing setup. The obtained sensitivity of the indirect immunoassay was observed to be significantly higher than that of the direct one, with a low detection limit of 20 ppb. High sensitivity, target selectivity, rapid response time, and low detection limits (sometimes in the ng/mL range) are the key advantages of these sensing systems (Shimomura, 2001). Notably, this type of plasmonic sensor generally responds only when molecules absorb on its surface; therefore, the selectivity of the device mainly depends on the material characteristics of the sensing material used on top of the metallic coating. Despite high sensitivity, the application of the SPR sensor in real-world settings is still limited as it experiences surface contamination due to continuous monitoring, which requires periodic maintenance, and in some cases replacement (silver and copper generally suffer oxidation) (Pathak, 2021).

SERS-based sensors follow similar mechanisms to SPR and are another widely utilized method to detect dioxins in the atmosphere. In 2009, Bantz and Haynes reported a silver film over nanospheres (AgFON) substrate for detecting and distinguishing PCBs (Bantz & Haynes, 2009). Among the five partition layers evaluated in the study, decanethiol (DT) provided the most efficient detection for PCB. Considering the response, an additional key sensing property was examined using the DT-modified SERS substrate. With the reported method, the author achieved a low detection limit of 50 pM without further optimization of the substrate or instrumentation. The sensing performance of SERS substrates was later improved by Zhu et al., who utilized a vertically arranged silver nanoplate film as a SERS substrate for the selective and sensitive detection of 3,3′,4,4′-tetrachlorobiphenyl (PCB-77) (Zhu, 2012). In the reported work, the authors achieved a thin film of vertically aligned silver nanoplates via spin-coating of silver seeds on top of the indium tin oxide (ITO), followed by electro-deposition in a mixture of citric acid and silver nitrate ($AgNO_3$). Due to the presence of various "hot spots" within the gaps between the adjacent nanoplates throughout the substrate, the silver nanoplate-assembled film demonstrated a strong SERS response along with superior reproducible signals. The author achieved a low detection limit of 10^{-6} M for PCB with a rapid response time. The performance of the device was further improved when the surface silver nanoplates were modified with DT to bind the target PCB-77 molecules effectively, hence improving the detection limit to 10^{-7} M.

Recently, Wang et al. utilized gold nano-cakes as an SERS substrate for real-time and onsite and ultra-high sensitivity of dioxins (Wang, 2023). Based on preliminary findings obtained from FDTD simulations, the enhancement factor of the gold nano-cakes SERS substrate was observed to be 10^{10}. By utilizing a portable Raman spectrometer, gold nano-cakes enable the detection of 1-chloro-dibenzo-p-dioxin (1-CDD), 2,8-dichloro-dibenzo-p-dioxin (2,8-DCDD), and 2,3,7-trichloro-dibenzo-p-dioxin (2,3,7-TrCDD) in clean water at detection limits of 5, 5, and 10 ng/mL, respectively.

10.5.4 Detection of hydrocarbons

The mid-infrared (MIR) region offers remarkable selectivity and sensitivity, allowing the identification of organic compounds from their characteristic fingerprint spectra. By utilizing the FTIR spectroscopy, the detection limit for such compounds can reach parts-per-billion (ppb) range (Kraft, 2000; Silva, 2011; Walsh, 1996). In 2013, Lu et al. introduced an attenuated total reflectance (ATR)-based optical sensor potentially suitable for in-situ and simultaneous monitoring of multiple chlorinated hydrocarbons present in water (Lu, 2013). In the reported study, the author investigated the presence of various hydrocarbons, including 1,3-dichlorobenzene, 1,2-dichlorobenzene, trichloroethylene, monochlorobenzene, perchloroethylene, and chloroform. The fabricated sensor comprises a polycrystalline silver halide fiber with a unique cylindrical design, coated with an ethylene/propylene copolymer, which works as a solid-phase extraction layer. This reported configuration was observed to significantly enhance the analytical signal, thereby improving device sensitivity. The sensor demonstrated excellent sensing performance in distinguishing various chlorinated hydrocarbons in water across a broad concentration range, varying from 5 to 700 ppb. For evanescent-field sensing, the mid-IR beam was launched into the sensing configuration through the cylindrical extension using a gold-coated off-axis parabolic mirror. In the MIR domain, different chemical bonds absorb specific infrared frequencies depending on the bonded atoms and their molecular environment, enabling precise chemical identification.

In the same year, Pejcic et al. further improved the performance by utilizing an optical sensor based on ATR Fourier transform infrared spectroscopy to determine the concentrations of various VOCs and SVOCs dissolved in water (Pejcic, 2013). The reported sensor comprises a zinc selenide (ZnSe) optical waveguide modified with poly(isobutylene) coating. The sensing performance of the sensor was optimized for various poly(isobutylene) thicknesses. The obtained result exhibited superior performance towards various concentrations of benzene, ethylbenzene, toluene, naphthalene, and xylenes in an oil-water mixture.

In 2016, Lu et al. reported another ATR sensor integrated with a polymer matrix for the selective identification of SVOCs in water samples (Lu, 2016). The author utilized a similar setup to characterize the performance of the sensor. In the reported sensor, the optical fiber was coated with a silver halide layer, followed by a hydrophobic ethylene-propylene copolymer in a 60:40 ratio. The polymer coating was able to selectively bind organic compounds while repelling water simultaneously. Based on the type of VOCs and monitoring conditions, IR spectra were obtained at appropriate frequencies, spectral resolutions, and scan times. These collected spectra were then utilized to identify the concentrations of each VOC along with other associated information. The reported approach demonstrated the ability to distinguish various VOCs simultaneously, at an extremely low concentration (10 ppb).

10.5.5 Detection of phenol and phenolic compounds

In 2018, Zhong et al. reported a rare-earth-doped optical sensor fabricated using a plastic optical fiber to selectively detect phenol concentration and its rate of degradation (Zhong, 2018). A 30 μm thick layer of Er^{3+}: $YAlO_3/SiO_2/TiO_2$ was deposited over the uncladed region of the optical fiber. The sensing layer incorporated the rare-earth phosphor Er^{3+}: $YAlO_3$ to produce the ultraviolet (UV) light under appropriate excitation. The generated UV emission was absorbed by the TiO_2 particles, producing electron-hole pairs, which allows the rapid monitoring of phenol compounds along with their degradation. By combining TiO_2 and Er^{3+}: $YAlO_3$, the author improved the refractive index of the sensing material, making it comparatively higher than that of the fiber core. This approach resulted in an increased evanescent wave penetration depth, improved surface luminescence, and enabled total internal reflection (TIR) efficiency within the sensing region, which collectively improve the performance of the sensor. In 2019, Zhong et al. developed a portable lab-on-chip device integrated with microalgal biofilm for real-time monitoring of the degradation of phenolic compounds (Zhong, 2019). The reported sensor comprises a compact chip-size device, combined with three different types of D-shaped optical fibers, i.e., (1) standard D-shaped optical fiber, (2) D-shaped fiber equipped with a glass fiber membrane to differentiate the target solution from Chlorella vulgaris, and (3) a normal D-shaped fiber and microchannel. The reported sensor was observed to selectively and accurately identify phenol compounds and monitor phenol degradation, as well as the progression of the deposited microalgal biofilm.

In 2019, the same research group, led by Zhong et al., reported an identical VOC sensor for identifying phenolic compounds in water samples

(Wang, 2019; Zhong, 2019). The reported work demonstrated the potential of plastic optical fibers for monitoring SVOCs (Wang, 2019). The sensing mechanism typically relies on the change in the properties of light reflected at the interface of the sensing material and the optical fiber. The sensing head was fabricated by selectively depositing a polymer membrane over a thin gold mirror and a SiO_2/nitrogen-doped TiO_2 (SNT) coating. Here, the gold layer acts as a reflective mirror for the launched light, allowing multiple TIRs and thereby enhancing the evanescent field at the interface. Here, the SNT coating improves light absorption, as nitrogen doping improves the relatively large bandgap of TiO_2, enabling stronger visible-light absorption. The addition of SiO_2 further enhances the quantum and photocatalytic performance of TiO_2. The sensing configuration also includes an inlet and outlet at the tip to facilitate the controlled flow of measurands. Once the launched light reaches to the fiber core, it interacts directly with the core-SNT interface, where it is diffracted and partially transmitted into the SNT layer because of the refractive-index difference between the fiber and SNT layer. The sensor exhibited a high selectivity towards phenolic compounds and notable sensitivity of 0.294×10^{-3} $(mg.L^{-1})^{-1}$, with a low detection limit of 30 μg/L. The sensing performance of the sensor was tested in the presence of various compounds, including pure phenol as well as various mixtures containing $FeCl_3$, $CuSO_4$, and glucose.

Mishra et al. utilized a long-period fiber Bragg grating (LPFBG) functionalized with an enzyme entrapped in polyacrylamide gel to identify a phenolic compounds in water (Mishra & Chiang, 2020). The author has used tyrosinase (polyphenol oxidase) as an enzyme with a thickness of 900 nm deposited around the sensing configuration. The basic principle of the sensor was based on the change in the refractive index of tyrosinase with the interaction of phenolic compounds leading to the shift in the grating wavelength. The results exhibit a good blue shift with an increasing concentration of the phenolic compounds in an aqueous solution. The fabricated device achieved a good sensitivity of 0.0088 nm/μM at a pH value of 7.0 in the concentration range varying from 0 to 800 μM of aqueous solution with a LOD of 6.8 μM. The selectivity of the sensor was also investigated using 4-chlorophenol, m-cresol, and phenol.

10.6 Future prospects and challenges

Although SVOCs are not as harmful as VOCs, their effect on human health makes it highly essential to identify and avoid them even in low traces. The advancement in materials science and photonics technology plays a key role in developing methods to detect these pollutants even in a complex environments. Despite several advancements, the detection of SVOCs is still challenging, and it's not only limited to the technological prospects but from the pollutant point of view too. As reported above, SVOCs possess low vapor pressure, so they generally accumulate on surfaces, aerosols, and dust particles, which makes the sampling and quantification process difficult. However, researchers have developed various tools that can identify some of these pollutants, such as PFASs and related compounds, dispersed in water sources. The optical sensor integrated with the molecular imprinting technology plays a vital role in selectively identifying these pollutants in water; further improvement in sensing material, such as combining plasmonics and specially engineered nanoparticles with the MIP, may further improve the performance and detection accuracy. The SVOCs include a broad range of pollutants such as phthalates, PBDEs, PFASs, PAHs, plasticizers, pesticides, etc. Therefore, it is challenging to develop a single sensor that can precisely identify and distinguish all compounds simultaneously, as they generally possess different physical and chemical characteristics, such as polarity, functional groups, and molecular weight. One possible solution to address this problem is to utilize an optical fiber Bragg grating (write multiple gratings in which each grating will reflect a specific wavelength) written on a single fiber, and the target-sensitive material (selected based on its affinity to a specific SVOC) deposited on each grating region. When a specific SVOC interacts with its corresponding sensing layer, it induces a change in the refractive index or physical properties of the deposited layer, which leads to a shift in the Bragg wavelength. Since each grating responds independently, the system can simultaneously monitor multiple SVOC classes with high selectivity and precision.

10.7 Conclusion

VOCs and SVOCs are recognized as the most harmful pollutants detected in the indoor environments of homes, offices, and schools. Although SVOCs are considered less hazardous compared to VOCs due to their lower vapor pressure, they exhibit another distinct feature of accumulating on surfaces, which can result in long-term health risks. This chapter has provided a detailed overview of SVOCs, including their emission sources, a brief comparison between VOCs and SVOCs, examples of SVOCs, and methods of detection.

References

Abelkop, A. D. K., Graham, J. D., & Royer, T. V. (2015). *Persistent, bioaccumulative, and toxic (PBT) chemicals*: Vol. 1. CRC Press https://www.taylorfrancis.com/books/mono/10.1201/9781351228893/persistent-bioaccumulative-toxic-pbt-chemicals-todd-royer-adam-abelkop-john-graham. doi:10.1201/9781351228893.

Andersson, M., Ottesen, R. T., & Volden, T. (2004). Building materials as a source of PCB pollution in Bergen, Norway. *Science of The Total Environment, 325*, 139–144. doi:10.1016/j.scitotenv.2003.11.014.

Andres, R. T., & Narayanaswamy, R. (1997). Fibre-optic pesticide biosensor based on covalently immobilized acetylcholinesterase and thymol blue. *Talanta*, 1335–1352. doi:10.1016/S0039-9140(96)02071-1.

Arjmand, M., et al. (2017). A sensitive tapered-fiber optic biosensor for the label-free detection of organophosphate pesticides. *Sensors and Actuators B: Chemical, 249*, 523–532. doi:10.1016/j.snb.2017.04.121.

Ballesteros-Gómez, A., Boer, J. D., & Leonards, P. E. G. (2013). Novel analytical methods for flame retardants and plasticizers based on gas chromatography, comprehensive two-dimensional gas chromatography, and direct probe coupled to atmospheric pressure chemical ionization-high resolution time-of-flight-mass spectrometry. *Analytical Chemistry, 85*(20), 9572–9580. doi:10.1021/ac4017314.

Bantz, K. C., & Haynes, C. L. (2009). Surface-enhanced Raman scattering detection and discrimination of polychlorinated biphenyls. *Vibrational Spectroscopy, 50*, 29–35. doi:10.1016/j.vibspec.2008.07.006.

Berger-Preiß, E., Levsen, K., Leng, G., Idel, H., Sugiri, D., & Ranft, U. (2002). Indoor pyrethroid exposure in homes with woollen textile floor coverings. *International Journal of Hygiene and Environmental Health, 205*(6), 459–472. doi:10.1078/1438-4639-00181.

Bernieri, T. (2019). Occupational exposure to pesticides and thyroid function in Brazilian soybean farmers. *Chemosphere, 218*, 425–429. doi:10.1016/j.chemosphere.2018.11.124.

Bier, F. F. (1992). Use of a fibre optic immunosensor for the detection of pesticides. *Sensors and Actuators B: Chemical, 7*, 509–512. doi:10.1016/0925-4005(92)80354-Z.

Bohlin, P., et al. (2008). Observations on persistent organic pollutants in indoor and outdoor air using passive polyurethane foam samplers. *Atmospheric Environment, 42*, 7234–7241. doi:10.1016/j.atmosenv.2008.07.012.

Bolzonella, C., et al. (2019). Is there a way to rate insecticides that is less detrimental to human and environmental health? *Global Ecology and Conservation, 20*, e00699. doi:10.1016/j.gecco.2019.e00699.

Brandsma, S. H., Boer, J., Velzen, M. J. M., & Leonards, P. E. G. (2014). Organophosphorus flame retardants (PFRs) and plasticizers in house and car dust and the influence of electronic equipment. *Chemosphere, 116*, 3–9. doi:10.1016/j.chemosphere.2014.02.036.

Braouezec, C. (2016). Cat serum contamination by phthalates, PCBs, and PBDEs versus food and indoor air. *Environmental Science and Pollution Research, 23*, 9574–9584. doi:10.1007/s11356-016-6063-0.

Bräuner, E. V., et al. (2011). Occurrence of organochlorine pesticides in indoor dust. *Journal of Environmental Monitoring, 13*, 522–526. doi:10.1039/C0EM00750A.

Breivik, K., et al. (2004). Primary sources of selected POPs: regional and global scale emission inventories. *Environmental Pollution, 128*, 3–16. doi:10.1016/j.envpol.2003.08.031.

Brummel, K. E., Wright, J., Eldefrawi, M. E., et al. (1997). Fiber Optic Biosensor for Cyclodiene Insecticides. *Journal of Agricultural and Food Chemistry, 45*, 3292–3298. doi:10.1021/jf9701190.

Butte, W., & Heinzow, B. (2002). Pollutants in house dust as indicators of indoor contamination. *Reviews of Environmental Contamination and Toxicology, 175*, 1–46. https://europepmc.org/article/med/12206053.

Capela, D., Homem, V., Alves, A., & Santos, L. Ú. (2016). Volatile methylsiloxanes in personal care products – using QuEChERS as a "green" analytical approach. *Talanta, 155*, 94–100. doi:10.1016/j.talanta.2016.04.029.

Chen, S., Chen, S., Jiang, S., Xiong, M., Luo, J., Tang, J., & Ge, Z. (2011). Environmentally friendly antibacterial cotton textiles finished with siloxane sulfopropylbetaine. *ACS Applied Materials and Interfaces*, 3(4), 1154–1162. doi:10.1021/am101275d.

Chen, S.-J., Ma, Y.-J., Wang, J., Tian, M. I., Luo, X.-J., Chen, D. A., & Mai, B. I.-X. (2010). Measurement and human exposure assessment of brominated flame retardants in household products from South China. *Journal of Hazardous Materials, 176*(1–3), 979–984. doi:10.1016/j.jhazmat.2009.11.138.

Crone, B. C., et al. (2019). Occurrence of per- and polyfluoroalkyl substances (PFAS) in source water and their treatment in drinking water. *Critical Reviews in Environmental Science and Technology, 49*(24), 2359–2396. doi:10.1080/10643389.2019.1614848.

Dallongeville, A., et al. (2016). Volatile and semi-volatile organic compounds of respiratory health relevance in French dwellings. *Indoor Air, 26*, 426–438. doi:10.1111/ina.12225.

Dalmolin, S. P., et al. (2020). Biomarkers of occupational exposure to pesticides: Systematic review of insecticides. *Environmental Toxicology and Pharmacology, 75*, 103304. doi:10.1016/j.etap.2019.103304.

Ding, N., et al. (2016). Brominated flame retardants (BFRs) in indoor and outdoor air in a community in Guangzhou, a megacity of southern China. *Environmental Pollution, 212*, 457–463. doi:10.1016/j.envpol.2016.02.038.

Faiz, F., et al. (2023). Graphene Oxide Doped Alginate Coated Optical Fiber Sensor for the Detection of Perfluorooctanoic Acid in Water. *IEEE Sensors Journal, 23*, 12861–12867. doi:10.1109/JSEN.2023.3271657.

Fauzi, N. I. M., et al. (2011). Recent Advances on Detection of Insecticides Using Optical Sensors. *Sensors, 21*, 3856. doi:10.3390/s21113856.

Ferreira, M. G., et al. (2025). Shape-Induced Enhanced Raman Scattering (SIERS) Platforms Modified with Gold Nanorods for Ultradiluted Atrazine Pesticide Detection. *ACS Applied Nano Materials, 8*, 10287–10296. doi:10.1021/acsanm.4c07289.

Fromme, H., et al. (2004). Occurrence of phthalates and musk fragrances in indoor air and dust from apartments and kindergartens in Berlin (Germany). *Indoor Air, 14*, 188–195. doi:10.1111/j.1600-0668.2004.00223.x.

Gevao, B., et al. (2006). Polybrominated diphenyl ethers in indoor air in Kuwait: Implications for human exposure. *Atmospheric Environment, 40*(8), 1419–1426. doi:10.1016/j.atmosenv.2005.10.053.

Gullett, B. K., et al. (2012). Near-Real-Time Combustion Monitoring for PCDD/PCDF Indicators by GC-REMPI-TOFMS. *Environmental Science & Technology, 46*, 923–928. doi:10.1021/es2027339.

Gupta, R. C., et al. (2019). Insecticides. *Biomarkers in Toxicology (Second Edition)* (pp. 455–475). Elsevier.

Herzke, D., Olsson, E., & Posner, S. (2012). Perfluoroalkyl and polyfluoroalkyl substances (PFASs) in consumer products in Norway – A pilot study. *Chemosphere, 88*(8), 980–987. doi:10.1016/j.chemosphere.2012.03.035.

Holt, E., et al. (2010). Polychlorinated Dibenzo- p -Dioxins and Dibenzofurans (PCDD/Fs) Impurities in Pesticides: A Neglected Source of Contemporary Relevance. *Environmental Science & Technology, 44*, 5409–5415. doi:10.1021/es903915k.

Hwang, H.-M., Park, E.-K., Young, T. M., & Hammock, B. D. (2008). Occurrence of endocrine-disrupting chemicals in indoor dust. *Science of the Total Environment, 404*(1), 26–35. doi:10.1016/j.scitotenv.2008.05.031.

Wu, Y., & Chang, V. W.-C. (2012). Development of analysis of volatile polyfluorinated alkyl substances in indoor air using thermal desorption-gas chromatography–mass spectrometry. *Journal of Chromatography A, 1238*, 114–120. doi:10.1016/j.chroma.2012.03.053.

Yadav, G. K., et al. (2025). On-chip plasmonic framework for scalable real time sensor networks. *Journal of the Optical Society of America B, 42*, 2535. doi:10.1364/JOSAB.576285.

Zhao, C. Q., et al. (1995). Fiber optic immunosensor for polychlorinated biphenyls. *Journal of Agricultural and Food Chemistry, 43*, 2308–2315. doi:10.1021/jf00056a064.

Zhong, N., et al. (2018). Optic fiber with Er3+:YAlO3/SiO2/TiO2 coating and polymer membrane for selective detection of phenol in water. *Sensors and Actuators B: Chemical, 273*, 1744–1753. doi:10.1016/j.snb.2018.07.092.

Zhong, N., et al. (2019). Monitoring Microalgal Biofilm Growth and Phenol Degradation with Fiber-Optic Sensors. *Analytical Chemistry*, 15155–15162. doi:10.1021/acs.analchem.9b03923.

Zhong, N., et al. (2019). A highly sensitive photocatalytic plastic optic-fiber sensor for selective detection of phenol in aqueous solutions. *Sensors and Actuators B: Chemical, 285*, 341–349. doi:10.1016/j.snb.2019.01.042.

Zhu, C., et al. (2012). Vertically aligned Ag nanoplate-assembled film as a sensitive and reproducible SERS substrate for the detection of PCB-77. *Journal of Hazardous Materials*, 389–395. doi:10.1016/j.jhazmat.2011.07.118.

Indoor Air Quality by Building Type, https://www.epa.gov/indoor-air-quality-iaq/indoor-air-quality-building-type

Kajiwara, N., Sueoka, M., Ohiwa, T., & Takigami, H. (2009). Determination of flame-retardant hexabromocyclododecane diastereomers in textiles. *Chemosphere, 74*(11), 1485–1489. doi:10.1016/j.chemosphere.2008.11.046.

Jahnke, A., et al. (2007). Development and application of a simplified sampling method for volatile polyfluorinated alkyl substances in indoor and environmental air. *Journal of Chromatography A, 1164*, 1–9. doi:10.1016/j.chroma.2007.06.068.

Kajiwara, N., Noma, Y., & Takigami, H. (2011). Brominated and organophosphate flame retardants in selected consumer products on the Japanese market in 2008. *Journal of Hazardous Materials, 192*(3), 1250–1259. doi:10.1016/j.jhazmat.2011.06.043.

Kaunelienė, V., et al. (2016). Indoor air quality in low energy residential buildings in Lithuania. *Building and Environment, 108*, 63–72. doi:10.1016/j.buildenv.2016.08.018.

Kaur, R., Mavi, G. K., Raghav, S., Khan, I., et al. (2019). Pesticides Classification and its Impact on Environment. *International Journal of Current Microbiology and Applied Sciences, 8*, 1889–1897. doi:10.20546/ijcmas.2019.803.224.

Kemmlein, S., Hahn, O., & Jann, O. (2003). Emissions of organophosphate and brominated flame retardants from selected consumer products and building materials. *Atmospheric Environment, 37*(39–40), 5485–5493. doi:10.1016/j.atmosenv.2003.09.025.

Khitous, A., et al. (2025). In situ synthesis of molecularly imprinted polymers by near-field photopolymerization for ultrasensitive PFOA plasmonic plastic fiber optic sensors. *Sensors and Actuators B: Chemical, 442*, 137992. doi:10.1016/j.snb.2025.137992.

Kim, K.-H. (2017). Exposure to pesticides and the associated human health effects. *Science of the Total Environment, 575*, 525–535. doi:10.1016/j.scitotenv.2016.09.009.

Kraft, M., et al. (2000). A Mid-Infrared Sensor for Monitoring of Chlorinated Hydrocarbons in the Marine Environment. *International Journal of Environmental Analytical Chemistry, 78*, 367–383. doi:10.1080/03067310008041354.

Kumar, J., Jha, S. K., & D'Souza, S. F. (2006). Optical microbial biosensor for detection of methyl parathion pesticide using Flavobacterium sp. whole cells adsorbed on glass fiber filters as disposable biocomponent. *Biosensors & Bioelectronics*, 2100–2105. doi:10.1016/j.bios.2005.10.012.

Law, C. S., et al. (2022). Real-time detection of per-fluoroalkyl substance (PFAS) selfassembled monolayers in nanoporous interferometers. *Sensors Actuators B: Chemical, 355*, 131340. doi:10.1016/j.snb.2021.131340.

Leslie, H. A., Leonards, P. E. G., Brandsma, S. H., de Boer, J., & Jonkers, N. (2016). Propelling plastics into the circular economy – weeding out the toxics first. *Environment International, 94*, 230–234. doi:10.1016/j.envint.2016.05.012.

Liu, X., Guo, Z., Krebs, K. A., Pope, R. H., & Roache, N. F. (2014). Concentrations and trends of perfluorinated chemicals in potential indoor sources from 2007 through 2011 in the US. *Chemosphere, 98*, 51–57. doi:10.1016/j.chemosphere.2013.10.001.

Lu, R., et al. (2013). Determination of Chlorinated Hydrocarbons in Water Using Highly Sensitive Mid-Infrared Sensor Technology. *Scientific Reports, 3*, 2525. doi:10.1038/srep02525.

Lu, R., et al. (2016). High-sensitivity infrared attenuated total reflectance sensors for in situ multicomponent detection of volatile organic compounds in water. *Nature Protocols, 11*, 377–386. doi:10.1038/nprot.2016.013.

Lu, Y., Yuan, T., Wang, W., & Kannan, K. (2011). Concentrations and assessment of exposure to siloxanes and synthetic musks in personal care products from China. *Environmental Pollution, 159*(12), 3522–3528. doi:10.1016/j.envpol.2011.08.015.

Lucattini, L., Poma, G., Covaci, A., de Boer, J., Lamoree, M. H., & Leonards, P. E. G. (2018). A review of semi-volatile organic compounds (SVOCs) in the indoor environment: Occurrence in consumer products, indoor air and dust. *Chemosphere, 201*, 466–482. doi:10.1016/j.chemosphere.2018.02.161.

Marie, F., Lars Schiøtt, S., Katrin, V., Lisbeth E., K., Jesper Bo, N., & Tom, W. (2016). Human exposure to novel flame retardants -from materials to humans (post.doc. project). NA. https://vbn.aau.dk/en/projects/human-exposure-to-novel-flame-retardants-from-materials-to-humans/.

Marrs, T. C. (1993). Organophosphate poisoning. *Pharmacology & Therapeutics*, 51–66. doi:10.1016/0163-7258(93)90066-M.

Melo, L. P., & Queiroz, M. E. C. (2010). Simultaneous analysis of parabens in cosmetic products by stir bar sorptive extraction and liquid chromatography. *Journal of Separation Science, 33*(12), 1849–1855. doi:10.1002/jssc.201000024.

Melymuk, L. (2016). Distribution of legacy and emerging semivolatile organic compounds in five indoor matrices in a residential environment. *Chemosphere, 153*, 179–186. doi:10.1016/j.chemosphere.2016.03.012.

Mercier, F., Glorennec, P., Thomas, O., & Bot, B. L. (2011). Organic contamination of settled house dust, a review for exposure assessment purposes. *Environmental Science and Technology, 45*(16), 6716–6727. doi:10.1021/es200925h.

Posner S., Perfluorinated compounds: Occurrence and uses in products. In Handbook of environmental chemistry (Vol. 17, pp. 25–39). Springer (2012), doi: 10.1007/978-3-642-21872-9_2.

Miliutina, E., et al. (2020). Plasmon-active optical fiber functionalized by metal organic framework for pesticide detection. *Talanta, 208*, 120480. doi:10.1016/j.talanta.2019.120480.

Mishra, S. K., & Chiang, K. S. (2020). Phenolic-compounds sensor based on immobilization of tyrosinase in polyacrylamide gel on long-period fiber grating. *Optics & Laser Technology, 131*, 106464. doi:10.1016/j.optlastec.2020.106464.

Moreau-Guigon, E., & Chevreuil, M. (2014). L'exposition humaine aux perturbateurs endocriniens via l'air ambiant : un risque sanitaire méconnuHuman exposure to endocrine disruptors via ambient air: An unknown health risk. *Archives des Maladies Professionnelles et de l'Environnement, 75*(1), 75–81. doi:10.1016/j.admp.2013.09.050.

Morgan, M. K., et al. (2014). Exposures of 129 Preschool Children to Organochlorines, Organophosphates, Pyrethroids, and Acid Herbicides at Their Homes and Daycares in North Carolina. *International Journal of Environmental Research and Public Health, 11*(4), 3743–3764. doi:10.3390/ijerph110403743.

Moro, G., et al. (2021). Nanocoated fiber label-free biosensing for perfluorooctanoic acid detection by lossy mode resonance. *Results in Optics, 5*, 100123. doi:10.1016/j.rio.2021.100123.

Moslemi, A., et al. (2024). Optical fiber probe based on LSPR for the detection of pesticide Thiram. *Optics & Laser Technology, 175*, 110882. doi:10.1016/j.optlastec.2024.110882.

Muenhor, D., Harrad, S., Ali, N., et al. (2010). Brominated flame retardants (BFRs) in air and dust from electronic waste storage facilities in Thailand. *Environment International, 36*, 690–698. doi:10.1016/j.envint.2010.05.002.

Oaya, C. S., et al. (2019). Impact of synthetic pesticides utilization on humans and the environment: an overview. *ACS Agricultural Science & Technology Journal, 11*, 279–286. doi:10.15547/ast.2019.04.047.

Pathak, A. K., Rahman, B. M. A, Kumari, S., & Singh, V. K. (2019). Sensitivity Enhancement of a Concave Shaped Optical Fiber Refractive Index Sensor Covered with Multiple Au Nanowires. *Sensors, 19*, 4210. doi:10.3390/s19194210.

Pathak, A. K., et al. (2021). A Highly Sensitive SPR Refractive Index Sensor Based on Microfluidic Channel Assisted With Graphene-Ag Composite Nanowire. *IEEE Photonics Journal, 13*, 1–8. doi:10.1109/JPHOT.2021.3069396.

Pathak, Akhilesh Kumar, et al. (2023). Recent Advances in Sensing Materials Targeting Clinical Volatile Organic Compound (VOC) Biomarkers: A Review. *Biosensors, 13*(1), 114. doi:10.3390/bios13010114.

Patrizi, B., et al. (2014). Characteristic vibrational frequencies of toxic polychlorinated dibenzo-dioxins and -furans. *Journal of Hazardous Materials, 274*, 98–105. doi:10.1016/j.jhazmat.2014.04.004.

Patrizi, B., et al. (2019). Dioxin and Related Compound Detection: Perspectives for Optical Monitoring. *International Journal of Molecular Sciences, 20*, 2671. doi:10.3390/ijms20112671.

Pejcic, B., et al. (2013). Direct quantification of aromatic hydrocarbons in geochemical fluids with a mid-infrared attenuated total reflection sensor. *Organic geochemistry, 55*, 63–71. doi:10.1016/j.orggeochem.2012.11.011.

Pentamwa, Prapat, & Oanh, Nguyen Thi Kim (2008). Levels of pesticides and polychlorinated biphenyls in selected homes in the Bangkok metropolitan region, Thailand. *Annals of the New York Academy of Sciences, 1140*, 91–112. doi:10.1196/annals.1454.005.

Pitruzzella, R., et al. (2024). An MIP-Based PFAS Sensor Exploiting Nanolayers on Plastic Optical Fibers for Ultra-Wide and Ultra-Low Detection Ranges—A Case Study of PFAS Detection in River Water. *Nanomaterials, 14*, 1764. doi:10.3390/nano14211764.

Rajan, Chand, S, & Gupta, B. D. (2007). Surface plasmon resonance based fiber-optic sensor for the detection of pesticide. *Sensors and Actuators B: Chemical, 123*, 661–666. doi:10.1016/j.snb.2006.10.001.

Rauert, C., & Harrad, S. (2015). Mass transfer of PBDEs from plastic TV casing to indoor dust via three migration pathways – A test chamber investigation. *Science of the Total Environment, 536*, 568–574. doi:10.1016/j.scitotenv.2015.07.050.

Regueiro, Jorge, et al. (2009). Development of a method based on sorbent trapping followed by solid-phase microextraction for the determination of synthetic musks in indoor air. *Journal of Chromatography A*, 2805–2815. doi:10.1016/j.chroma.2008.09.062.

Robinson, B. H. (2009). E-waste: An assessment of global production and environmental impacts. *Science of the Total Environment, 408*(2), 183–191. doi:10.1016/j.scitotenv.2009.09.044.

2025 Short-Chain Chlorinated Paraffins (SCCPs) and Other Chlorinated Paraffins Action Plan, https://www.epa.gov/assessing-and-managing-chemicals-under-tsca/short-chain-chlorinated-paraffins-sccps-and-other

Rodriguez, K. L., et al. (2020). Recent Developments of PFAS-Detecting Sensors and Future Direction: A Review. *Micromachines, 11*, 667. doi:10.3390/mi11070667.

Rudel, Ruthann A., et al. (2003). Phthalates, Alkylphenols, Pesticides, Polybrominated Diphenyl Ethers, and Other Endocrine-Disrupting Compounds in Indoor Air and Dust. *Environmental Science & Technology, 37*(20), 4543–4553. doi:10.1021/es0264596.

Saito, I., Onuki, A., Seto, H., et al. (2007). Indoor organophosphate and polybrominated flame retardants in Tokyo. *Indoor Air*, 28–36. doi:10.1111/j.1600-0668.2006.00442.x.

Shimomura, M., et al. (2001). Simple and rapid detection method using surface plasmon resonance for dioxins, polychlorinated biphenylx and atrazine. *Analytica Chimica Acta, 434*, 223–230. doi:10.1016/S0003-2670(01)00809-1.

Silva, A. M. S., et al. (2011). Evaluation of a PVC sensing phase for determination of chlorinated hydrocarbons in water by mid-infrared spectroscopy. *Journal of the Brazilian Chemical Society, 22*, 1470–1477. doi:10.1590/S0103-50532011000800010.

Simcox, N. J., et al. (2011). Synthetic Turf Field Investigation in Connecticut. *Journal of Toxicology and Environmental Health, Part A, 74*(17), 1133–1149. doi:10.1080/15287394.2011.586941.

Sofuoglu, A., et al. (2010). Polycyclic and nitro musks in indoor air: a primary school classroom and a women's sport center. *Indoor Air*, 515–522. doi:10.1111/j.1600-0668.2010.00674.x.

Soh, N., et al. (2003). A surface plasmon resonance immunosensor for detecting a dioxin precursor using a gold binding polypeptide. *Talanta, 60*, 733–745. doi:10.1016/S0039-9140(03)00139-5.

Stapleton, H. M., Klosterhaus, S., Eagle, S., Fuh, J., Meeker, J. D., Blum, A., & Webster, T. F. (2009). Detection of organophosphate flame retardants in furniture foam and U.S. house dust. *Environmental Science and Technology, 43*(19), 7490–7495. doi:10.1021/es9014019.

Stapleton, H. M., Klosterhaus, S., Keller, A., Ferguson, P. L., Bergen, S. V., Cooper, E., Webster, T. F., & Blum, A. (2011). Identification of flame retardants in polyurethane foam collected from baby products. *Environmental Science and Technology, 45*(12), 5323–5331. doi:10.1021/es2007462.

Stapleton, H. M., Sharma, S., Getzinger, G., Ferguson, P. L., Gabriel, M., Webster, T. F., & Blum, A. (2012). Novel and high volume use flame retardants in US couches reflective of the 2005 PentaBDE phase out. *Environmental Science and Technology, 46*(24), 13432–13439. doi:10.1021/es303471d.

Stockholm Convention on Persistent Organic Pollutants (POPs), https://chm.pops.int/Implementation/Alternatives/AlternativestoPOPs/ChemicalslistedinAnnexB/Perfluorooctanesulfonicacidandperfluorooctane/tabid/5869/Default.aspx

Takigami, H., Suzuki, G. O., Hirai, Y., & Sakai, S. (2008). Transfer of brominated flame retardants from components into dust inside television cabinets. *Chemosphere, 73*(2), 161–169. doi:10.1016/j.chemosphere.2008.06.032.

Technical Overview of Volatile Organic Compounds, https://www.epa.gov/indoor-air-quality-iaq/technical-overview-volatile-organic-compounds

2025 Text Version of the Indoor Air Quality House Tour, http://Text Version of the Indoor Air Quality House Tour

2006 Total Health Assessment of Chemicals in Indoor Climate from Various Consumer Products, http://chrome-extension://efaidnbmnnnibpcajpcglclefindmkaj/https://www2.mst.dk/Udgiv/publications/2006/87-7052-214-6/pdf/87-7052-215-4.pdf

Tan, J., et al. (2007). Selected organochlorine pesticide and polychlorinated biphenyl residues in house dust in Singapore. *Chemosphere, 68*(9), 1675–1682. doi:10.1016/j.chemosphere.2007.03.051.

Tao, P., Ge, K., et al. (2023). Fiber Optic SERS Sensor with Silver Nanocubes Attached Based on Evanescent Wave for Detecting Pesticide Residues. *ACS Applied Materials & Interfaces, 15*, 30998–31008. doi:10.1021/acsami.3c04059.

Trettnak, W., et al. (1993). Fiber-optic remote detection of pesticides and related inhibitors of the enzyme acetylcholine esterase. *Sensors and Actuators B: Chemical, 11*, 87–93. doi:10.1016/0925-4005(93)85242-3.

Trunnelle, K. J., et al. (2013). Pyrethroids in house dust from the homes of farm worker families in the MICASA study. *Environment International, 61*, 57–63. doi:10.1016/j.envint.2013.09.007.

van Mourik, L. M., Gaus, C., Leonards, P. E. G., Boer, J. de, et al. (2016). Chlorinated paraffins in the environment: A review on their production, fate, levels and trends between 2010 and 2015. *Chemosphere, 155*, 415–428. doi:10.1016/j.chemosphere.2016.04.037.

Villar, I. Del, Imas, J. J., Matias, I. R., et al. (2025). A review on lossy mode resonance-based sensors: Fundamentals and applications. *Sensors Actuators B: Chemical, 444*, 138421. doi:10.1016/j.snb.2025.138421.

Walsh, J. E., et al. (1996). Sensing of chlorinated hydrocarbons and pesticides in water using polymer coated mid-infrared optical fibres. *Analyst, 121*, 789. doi:10.1039/an9962100789.

Wang, F.-L., Zhao, Y.-G., Muhammad, N., Wu, S.-C., & Zhu, Y. (2017). Simultaneous determination of parabens and inorganic anions in cosmetics by a two-dimensional ultrahigh-performance liquid chromatography-ion chromatography valve-switching method. *RSC Advances, 7*(52), 32769–32776. doi:10.1039/C7RA00867H.

Wang, T., et al. (2023). Au nanocakes as a SERS sensor for on site and ultrafast detection of dioxins. *Vibrational Spectroscopy, 126*, 103518. doi:10.1016/j.vibspec.2023.103518.

Wang, Z., et al. (2019). Photochemical reflective optical fiber sensor for selective detection of phenol in aqueous solutions. *Applied Optics*, 2091. doi:10.1364/AO.58.002091.

Washburn, S. T., et al. (2025). Exposure Assessment and Risk Characterization for Perfluorooctanoate in Selected Consumer Articles. *Environmental Science & Technology, 39*(11), 3904–3910. doi:10.1021/es048353b.

Weil, E. D., & Levchik, S. V. (2008). Flame retardants in commercial use or development for textiles. *Journal of Fire Sciences, 26*(3), 243–281. doi:10.1177/0734904108089485.

Weschler, C. J. (2009). Changes in indoor pollutants since the 1950s. *Atmospheric Environment, 43*(1), 153–169. doi:10.1016/j.atmosenv.2008.09.044.

CHAPTER

11

Next-gen VOC sensing: AI-driven non-invasive detection

11.1 Introduction

Real-time volatile organic compound (VOC) monitoring plays a crucial role in a broad range of applications, including industrial processes, environmental monitoring, building health, chemical processing, educational and office environments, and noninvasive human health monitoring. For instance, VOC sensors help protect workers in chemical factories and deep mining and assist in diagnosing medical conditions based on variations in the concentration levels of VOCs present in exhaled breath (Padvi et al., 2021; Pathak & Viphavakit, 2022). Artificial intelligence (AI) involves the development of computational tools that can perform tasks typically associated with human intelligence, such as analyzing images (Krizhevsky et al., 2017), recognizing speech (Nassif et al., 2019), making decisions (Duan et al., 2019), and more, by learning from the obtained data. AI has revolutionized the industrial sector by automating various challenging tasks and has become a powerful tool across numerous important fields, such as uncovering patterns within large datasets, driving innovation, and performing various complex tasks that typically require human intelligence. Machine learning (ML) is a well-known branch of AI that allows systems to learn from data, identify patterns, and improve performance over time without explicit programming. A more advanced subset, deep learning (DL), utilizes artificial neural networks (ANNs), defined as a structure composed of interconnected processing nodes, to automatically extract complex features from large datasets. These foundational principles form the backbone of a broad range of AI-based applications,

Photonic Sensors for Detection of VOC Biomarkers: Technological Principles and Medical Applications
DOI: https://doi.org/10.1016/B978-0-443-29166-1.00009-3

Copyright © 2026 Elsevier Inc. All rights are reserved, including those for text and data mining, AI training, and similar technologies.

including fraud detection (West & Bhattacharya, 2016), virtual assistants (Pereira et al., 2023), autonomous vehicles (Atakishiyev et al., 2024), and healthcare (Hamet & Tremblay, 2017). For instance, AI-powered image recognition tools can detect diseases from medical scans with high accuracy, whereas natural language processing systems are utilized in customer service and market research to analyze customer sentiment (Sharma et al., 2025). These advancements in AI highlight its role in improving accuracy, enhancing efficiency, and enabling unique solutions to real-world challenges across various fields. In the context of VOC detection, integrating AI with sensor technologies significantly boosts the capability of sensing devices to detect and distinguish between multiple VOCs through intelligent signal processing and pattern recognition. This intelligent sensing approach enhances sensitivity, selectivity, and response time while also facilitating the development of innovative applications in industry, environmental monitoring, and healthcare.

In clinical laboratories, a broad range of disease-related biomarkers can be tested, including urine routine, hemagglutination index, blood routine, infectious index, cytokine indicators, and blood gas analysis indicators. When reverse transcription polymerase chain reaction (RT-PCR) tests fail to confirm a diagnosis in patients with positive symptoms, identifying other biomarkers becomes an essential task. These additional tests help to reach a decision and effectively prevent delays in both treatment and isolation. The advancement of AI technology is considered appropriate for handling large sample studies, minimizing human-related bias, and therefore improving the accuracy of diagnosis and prediction assessments. Several articles have been reported in the past decades highlighting the role of ML in advancing sensing technologies (Cui & Zhou, 2020; Singh et al., 2024; Taha et al., 2020). Recently, Low et al. (2025) also reported an AI-integrated microfluidic paper-based multiplexed surface-enhanced Raman scattering (SERS) biosensor with electrophoretic removal and electrical modulation for accurate acute myocardial infarction (AMI) diagnosis and prognosis. In the reported work, the authors utilized a wax hydrophobic barrier stamped over a paper using a PDMS stamp along with a paper-based microfluidics integrated with electrophoresis, in order to remove unbound molecules, therefore, improving the selectivity of the on-chip SERS signal collection. ML, along with sensors, allows accurate quantification of biomarkers including high-density lipoprotein, cardiac troponin I, and low-density lipoprotein with low root mean squared error (RMSE). Yaqoob and Younis (2021) reported a comprehensive review over chemiresistive and field-effect transistor (FET)-based gas sensors to fulfill the demands of healthcare professionals, emphasizing the impact of material advancements and ML-based feature extraction in effectively improving the absorption efficiency, along with the sensitivity, specificity, response time, and overall smartness of the sensor system. Ye et al. (2021) published a extensive reviewe to describe the potential of ML in improving the performance of electronic nose (E-Nose) systems, specifically for odor identification, by enabling both quantitative and qualitative and analyses, and by minimizing sensor drift, which enhances the detection accuracy, adaptability, and reliability across diverse fields, such as food engineering, robotics, environmental monitoring, and healthcare applications, while elucidating the role of manual versus learned feature extraction. Similarly, Yang's study focused on the application of ML in improving the performance of single-gas sensors and sensor arrays, mainly targeting metal- or semiconductor-oxide-based sensor and their applications in pollution detection and VOC concentration monitoring, while also providing in-depth information regarding ML-based feature extraction and classical algorithms that improve the device stability and selectivity, and reduce sensor drift (Yang et al., 2023). Readers are also advised to study these published literatures, which provide crucial information on the target facets of ML's role in gas sensing applications.

This chapter is divided into seven sections, in which Section 11.1 describes the introduction and importance of AI in health monitoring. Section 11.2 describes the fundamentals of ML algorithms. Section 11.3 discusses the AI methods generally used in gas sensing, whereas Section 11.4 discusses the AI-integrated photonic devices reported so far for VOC sensing. Section 11.5 provides a brief case study describing how AI has become a game changer in medical applications and how AI is implemented to improve the quality of human life. Finally, Sections 11.6 and 11.7 discuss challenges and opportunities, and the conclusion, respectively.

11.2 Fundamentals of machine learning algorithms

With the increasing volume of complex process data acquired by both the distributed and discrete sensors, there is a growing demand for more sophisticated data preprocessing techniques to extract meaningful information. ML, a key domain within AI, is capable of handling data with complex characteristics such as multidimensionality and nonstationary behaviors. It also offers a powerful tool for interpreting complex data generated by sensors (Argyris, 2022). Various types of ML algorithms have been successfully implemented across a broad range of applications, providing outcomes that are more accurate and efficient compared to those obtained from traditional methods (Jing et al., 2021; Ma et al., 2022). This section focuses on several types of ML algorithms that have been utilized in optical sensing, such as logistic regression (LR) (Nusinovici et al., 2020), linear regression (Tung Ho & Ané, 2014), decision tree algorithms (Kotsiantis, 2013), Gaussian process regression (GPR) (Jing et al., 2021),

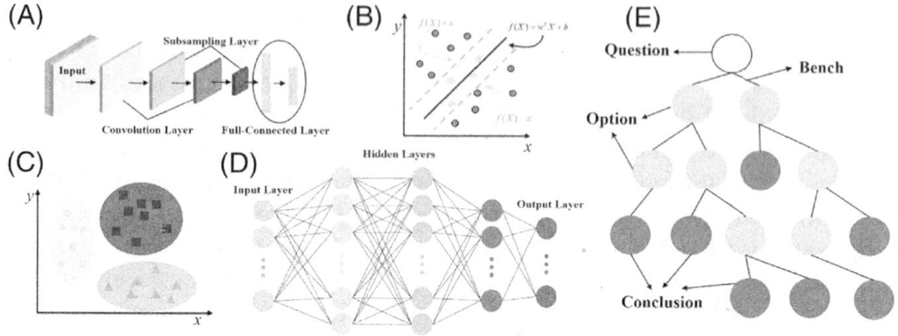

FIGURE 11.1 ML algorithms. (A) Convolutional neural network (CNN), (B) SVM, (C) K-means, (D) Artificial neural network (ANN), and (E) DT. *From Zhou, Y., Zhang, Y.-N., Yu, Q., Ren, L., Liu, Q., & Zhao, Y. (2024). Application of machine learning in optical fiber sensors.* Measurement, 228. *https://doi.org/10.1016/j.measurement.2024.114391.*

support vector machines (SVM) (Suzuki et al., 2024), random forest (RF) (Hajjem et al., 2014), K-means clustering (Ikotun et al., 2023), and ANNs (Zhang et al., 2014). Fig. 11.1 exhibits schematic representations of various types of ML approaches (Zhou et al., 2024). This section has been categorized into three major classes of ML, namely classification/clustering, regression, and ANN. By highlighting these methods, this section aims to enhance understanding of both the underlying concepts of ML and their specific applications in the field of gas sensing.

The above methods have been well defined in several published studies from various perspectives (Schmarje et al., 2021; Tian et al., 2023; Wang et al., 2023b). For instance, in (Schmarje et al., 2021), the authors divided the ML algorithms based on their training process, i.e., unsupervised learning, semi-supervised learning, and self-supervised learning. In contrast, Wang et al. (2023b) divided the ML methods according to their implementation purposes, i.e., the reported work divided the methods into regression, classification, clustering, and related categories. Meanwhile, Tian et al. (2023) divided these methods in a more precise way by categorizing them from the perspective of technology implementation, emphasizing how different ML algorithms address specific challenges in triboelectric-based intelligent sensing systems. Their framework not only highlights the features of the general ML algorithms but also explains how these approaches improve the data analysis and broaden the applications of triboelectric nanogenerators. Understanding the potential of each method is extremely crucial for researchers, specifically those who lack a background in ML algorithms. Previous studies have extensively detailed ML methods and their key functions in the development of AI-driven systems (Chang et al., 2018; Ikotun et al., 2023; Zhang et al., 2014; Zhou et al., 2024). Therefore, this chapter only summarizes the AI techniques employed in gas sensing and reviews the different types of AI-integrated VOC/gas sensors reported to date.

11.3 Popular artificial intelligence methods in gas sensing

Typically, the gas sensors produce measurable signals that require rigorous processing and interpretation to quantify and detect the target gases. AI methods are particularly considered an effective way to perform such tasks and offer robust solutions for several signal processing and interpretation functions (Chowdhury & Oehlschlaeger, 2025). AI technologies have significantly enhanced the adaptability of sensing systems by enabling them to automatically process data and optimize sensor performance in real time. This generally includes adjustments to sensor operations based on signal characteristics, environmental conditions, or other variations. Before the advancement of DL, gas sensing generally relied on chemometric technologies, e.g., Multiple Component Analysis (MCA), Principal Component Analysis (PCA), Independent Component Analysis (ICA), Partial Least Squares (PLS), and Self-Modeling Curve Resolution (SMCR) (Chowdhury & Oehlschlaeger, 2025). These methods, along with empirical models based on regression analysis or calibration curves, formed the backbone of early sensor data interpretation. Researchers started distinguishing these traditional approaches from ML in the late 1980s due to their effectiveness, simplicity, and interpretability. Despite several advantages, they often struggled with complex, nonlinear interactions and required manual preprocessing. In contrast, modern DL models excel in capturing complex nonlinear patterns and perform automated feature extraction, offering improved predictive accuracy. Nevertheless, ML and DL models also have several limitations and require careful validation. Traditional methods, meanwhile, still hold value in scenarios involving limited data or when model transparency is essential.

The output signal from the gas sensors is produced in various forms and can be categorized into two types, i.e., (1) time-series signals, typically associated with metal-oxide semiconductor (MOS) gas sensors and (2) spectral signals, commonly produced by sensors that employ direct absorption spectroscopy. Although these spectral data and time series are fundamentally 1D, various sensing and signal processing systems generate higher-dimensional data. For example, spectral information can be transformed into 2D representations using Gramian-based techniques or structured as 2D time-resolved spectra (Tan et al., 2023). In more complex systems, such as time-resolved tomographic measurements, the data may extend into 2D, 3D, and even 4D. The volume and type of data produced by the model are highly dependent on the target objective and associated experimental setup. For example, in gas monitoring (using the MOS sensor as a representative case), time-series or spectral absorption profiles are generally obtained by varying the gas concentration. Researchers generally obtain huge sets of data by varying the environmental conditions, such as pressure, temperature, or interference from other gases, in order to confirm the robustness of the ML models. Separating input data typically involves dividing it into different subsets for training, validation, and testing, with care taken to ensure that each subset reflects the full range of experimental conditions. A wide range of AI methods can be utilized in gas sensing applications for classification, feature extraction, and data interpretation. Prior to the application of AI-based models, sensor signals typically undergo preprocessing using non-AI approaches. For instance, the sensor signal processing may include drift correction, baseline correction, amplitude normalization, time alignment, cropping, smoothing, segmentation, and filtering. This preprocessing is considered crucial to improve the signal quality, ensure consistency, and prepare the data effectively for analysis utilizing AI, ML, or DL methods.

For the detailed information and the historical development of AI, ML, and DL over the past decades, readers are directed to comprehensive reviews and text books authored by other researchers such as Hopgood (2021), Bishop (n.d.), and Wang et al. (2023a) for AI, as well as by Alpaydin (2021), Murphy (2022), Roberts et al. (2022), Goodfellow et al. (2016), and Zhang et al. (2023a) for ML and DL. In reference to gas sensing, supervised learning techniques have been more extensively applied compared to the unsupervised and reinforcement learning approaches. However, unsupervised approaches such as hierarchical clustering, K-means clustering, and self-organizing maps have been utilized for detecting patterns in sensor data without the need for labeled input (Yaqoob & Younis, 2021). Additionally, unsupervised algorithms, such as PCA, have been utilized to identify outliers or faults in sensor outputs (Praveenchandar et al., 2022). PCA and t-distributed stochastic neighbor embedding (t-SNE) have been utilized to reduce the complexity of high-dimensional gas sensor data (Yaqoob & Younis, 2021).

Supervised learning approaches have been extensively employed in gas detection, with several examples presented in Section 11.4 (AI-integrated VOC sensors). These techniques can be categorized into classification, ensemble methods, regression, anomaly detection, dimensionality reduction, and time series forecasting techniques. Various types of standard ML classification algorithms have been extensively applied to identify patterns in gas sensor data for the purpose of gas detection. These generally include SVM, LR, decision trees (DT), k-nearest neighbors (kNN), RFs, NNs, and DL approaches. To estimate gas concentrations, conventional regression methods have also been employed, such as polynomial and linear regression, support vector regression (SVR), ridge and lasso regression, and regression models based on DT and NNs. Additionally, ensemble learning techniques have also been applied for

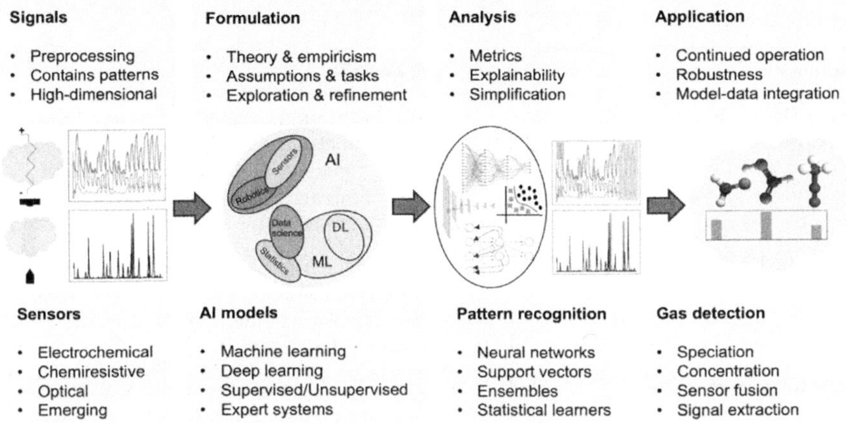

FIGURE 11.2 AI integration in gas sensing: Schematic diagram illustrates the simplified workflow for AI integration in gas sensing. *From Chowdhury, M. A. Z., & Oehlschlaeger, M. A. (2025). Artificial intelligence in gas sensing: A review.* ACS Sensors, *10(3), 1538–1563. https://doi.org/10.1021/acssensors.4c02272.*

classification and regression, including boosting (e.g., gradient boosting, AdaBoost, and XGBoost), bagging (e.g., RF), and stacking. Table 11.1 summarizes some popular AI methods along with their application.

Fig. 11.2 outlines a workflow for AI model development and implementation in gas sensing. Measured signals from gas sensors or sensor arrays contain identifiable latent patterns that allow various sensing tasks or applications, such as speciation, concentration prediction, signal extraction (denoising or low-dimensional representation), or sensor data fusion. These applications can be carried out based on sensor signal patterns identified using AI and through the analysis of AI formulations and the signals themselves. Thus, developing an AI model for a particular sensor is initiated by formulating these assumptions and hypotheses, which are consolidated. Specific protocols vary across studies; a general workflow involves systematic data collection, cleaning, and annotation (for classification), followed by exploratory analysis to understand the patterns and outliers before feeding the data into ML pipelines. Reliable AI models are the product of the crucial interplay between experimental rigor and computational methodologies.

11.4 Artificial intelligence-integrated volatile organic compounds sensors

Pambudi et al. (2018) reported the detection of VOCs using an ANN algorithm. The reported sensor incorporated a target-sensitive material to

TABLE 11.1 Overview of common ML, DL, and AI methods for gas detection and quantification.

Method	Description
Linear Discriminant Analysis	A classification method that finds a linear combination of features to separate classes (e.g., gases, combinations of gases) by maximizing the ratio of between-class variance to within-class variance
Quadratic Discriminant Analysis	An extension of LDA that allows each class to have its own covariance matrix, enabling the decision boundaries to be quadratic rather than linear
Ridge Regression	Applies L2 regularization to linear regression to control the complexity of the model by penalizing large coefficients in order to obtain a simpler model
Lasso Regression	Applies L1 regularization to linear regression, allowing some coefficients to shrink to zero for the purpose of feature selection
Logistic Regression	Classifies inputs by applying the logistic function to a weighted sum of the features, modeling the probability of different categories of gases
Support Vector Machine	Seeks an optimal boundary that separates classes of gases by maximizing the margin between them
Support Vector Regression	Uses the margin concept from SVM to minimize prediction errors for continuous outputs
Decision Trees	Repeatedly splits data based on feature thresholds, forming a hierarchical set of if–then rules for gas detection or concentration quantification
Ensemble Methods	Includes bagging (e.g., Random Forests), boosting (e.g., Gradient Boosting, AdaBoost, and XGBoost), and stacking, which combine outputs of base models through a meta-learner
Random Forest	Creates an ensemble of decision trees to aggregate predictions and reduce variance

(*continued on next page*)

TABLE 11.1 Overview of common ML, DL, and AI methods for gas detection and quantification—cont'd

Method	Description
K-Nearest Neighbors	Classifies or regresses a point based on the majority vote or average among its nearest neighbors
Neural Networks	Learns complex patterns by passing inputs through interconnected layers of weighted nodes
Deep Learning	Extends neural networks with multiple hidden layers to capture more intricate feature representations

From Chowdhury, M. A. Z., & Oehlschlaeger, M. A. (2025). Artificial intelligence in gas sensing: A review. ACS Sensors, 10*(3), 1538–1563. https://doi.org/10.1021/acssensors.4c02272.*

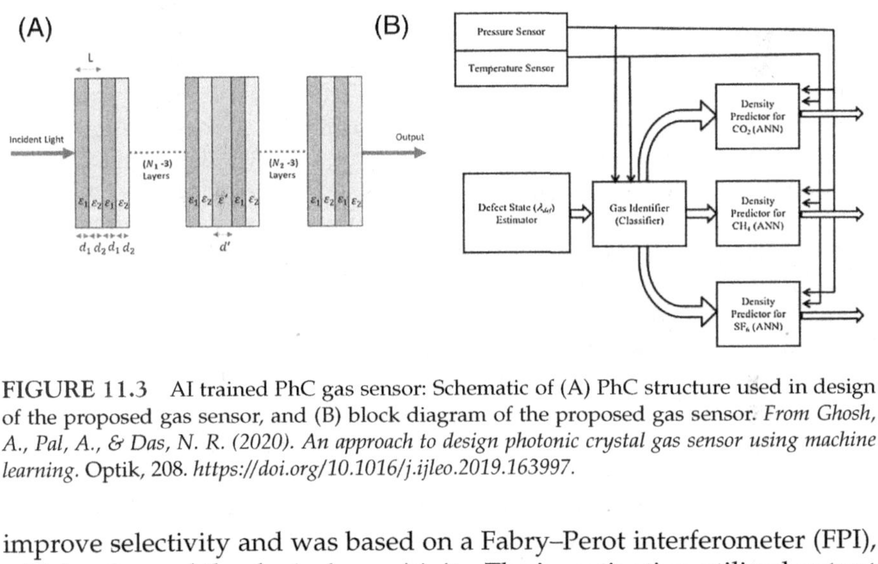

FIGURE 11.3 AI trained PhC gas sensor: Schematic of (A) PhC structure used in design of the proposed gas sensor, and (B) block diagram of the proposed gas sensor. *From Ghosh, A., Pal, A., & Das, N. R. (2020). An approach to design photonic crystal gas sensor using machine learning.* Optik, 208. *https://doi.org/10.1016/j.ijleo.2019.163997.*

improve selectivity and was based on a Fabry–Perot interferometer (FPI), which enhanced the device's sensitivity. The investigation utilized output data obtained from the sensor, which was converted into digital form via an analog-to-digital converter; data processing and detection were subsequently performed by a computer. The results showed that the sensor array could generate distinct patterns corresponding to each VOC. The neural network-based pattern recognition system accurately detected the type of VOC with a 100% success rate. Ghosh et al. (2020) utilized an ML-based method to design, optimize, and analyze a photonic crystal (PhC) based VOC sensor to precisely estimate three greenhouse gases, i.e., sulfur hexafluoride, methane, and carbon dioxide (CO_2). Fig. 11.3A illustrates the schematic diagram of the proposed PhC design comprising defect layers with N_1 and N_2 periods on both sides. Each period of the PhC consists of two dielectric layers of thicknesses d_1 and d_2 with dielectric constants of ε_1 and ε_2, respectively. Fig. 11.3B represents a simple block diagram of the proposed gas sensor developed using the optimized PhC structure. Initially, the temperature and pressure information of the gases, along with the data bank, were fed to the gas identifier, which includes a pre-trained classifier with optimized hyperparameters. Later, the density of each gas was estimated using the ANN trained separately for each gas. SVM and RF classifiers were used for the performance evaluation, with hyperparameter optimization. The RF classifier yielded the highest accuracy of ~100% in the gas identification stage. The results show that the present approach helps design a precision gas sensor even with a simple PhC structure.

In 2021, Chalasani et al. reported a convolutional neural network (CNN) based highly sensitive optical fiber gas sensor to selectively detect various gases, including nitrogen (N_2), helium (He), and CO_2 (Rao et al., 2021). Square-shaped dielectric rods arranged in a cubic lattice configuration were utilized to confirm the uniform spacing and area among the individual rods. In order to improve the performance of the CNN, the flower pollination algorithm (FPA) was utilized to tune its hyperparameters precisely. The designed CNN-based optical sensing configuration included three resonant nanocavities: one nanocavity of refractive index 2.6, along with two additional nanocavities with a refractive index of 2.1, aligned along the input and output waveguide paths. The performance of the designed sensor was investigated for three certified gas sample cylinders and a high-precision mass flow controller. Moreover, a clean air generator was also connected to the same pipeline in order to dilute the hydrocarbon flow as required. The proposed sensor was evaluated under gas concentrations ranging from 60 to 900 ppm. In the reported work, 80% of the data was used for training and 20% for testing during model training and evaluation. The obtained results showed a 7.5% increment in training accuracy with the optimized CNN. When benchmarked against three alternative optimization methods, the proposed CNN-based optical sensor approach exhibited excellent sensing performance. Ganesh Babu et al. (2022) reported an optimized convolutional neural network (OCNN)-based optical sensor to detect multiple gases and VOCs (i.e., N_2, methane, and carbon monoxide) simultaneously using PhCs. The sensor design comprised a dielectric rod with a square cross-section and a cubic lattice arrangement. The reported OCNN multi-gas optical sensing system included three nanocavities, among which one cavity with the refractive index of 2.6 was located at the center of the PhC, whereas the other two, comparatively smaller cavities, were positioned at the input and output path of the PhC waveguide. In order to improve the performance of the model and enable precise tuning of hyperparameters, a gray wolf optimization (GWO) algorithm was utilized in this work. To experimentally validate the response of the OCNN optical sensor, the device was integrated with a clean air generator (to dilute the concentrations of the hydrocarbons), a flow controller, and three calibrated gas cylinders (for different gases). The reported device was tested under varying concentrations of the gases, ranging from 60 to 200 ppm. With the optimized CNN model, the sensor exhibited a 7.06% improvement in performance.

Naku et al. (2023) detected 11 different types of VOCs by combining an array of optical fiber sensors based on FPIs along with ML. The work employed three optical fiber tip sensors with different surfaces to form an array to improve the selectivity and accuracy in detecting VOC samples. Among the three sensors, one fiber had a flat cleaved face and was used without modification, whereas the endfaces of the other two fibers were immobilized with hydrophobic and hydrophilic materials. The dynamic evaporation of droplets from the tip of sensor arrays was recorded during controlled immersion tests involving 11 different organic liquids. The recorded time-dependent signals were then processed by utilizing the continuous wavelet transform to convert them into image representations. The obtained images were then utilized as input to train the CNNs to classify and identify volatile liquids. The sensor array, comprising three different optical fiber tips, provided a diverse dataset, which helped in improving the performance of the ML models. In the reported work, the authors applied various data fusion strategies to integrate output data obtained from the sensor array within the ML framework, analyzing their effect on the reliability of the model. The findings demonstrated that the ML algorithms were highly effective in achieving high accuracy in differentiating various volatile liquids based on the unique characteristics of their evaporation rates. In the same year, Liu et al. reported an AI-based "Photonic Nose" as an augmented sensing platform for VOC detection in the mid-infrared (MIR) range (Liu et al., 2024). The reported work employed a suspended waveguide structure based on a sub-wavelength grating cladding, as shown in Fig. 11.4, aiming to improve the sensing performance of the device. To precisely analyze the absorption spectra of VOC mixtures, including acetone and isopropyl alcohol, the sensor was integrated with two ML models. A CNN was first utilized to classify 19 different binary mixture ratios, achieving an accuracy of 93.57%. Following this, a multilayer perceptron (MLP) regressor was applied to perform spectral decomposition and reconstruct the original spectra of the pure components. An average root-mean-square error (RMSE) of 2.44 vol% was achieved in predicting VOC concentrations using this approach. The obtained results highlight the potential of the reported "photonic nose" for accurate and quantitative analysis of complex VOC/gas mixtures in integrated spectral sensing systems. Xie et al. (2024) also reported an AI-integrated "Photonic Nose" for MIR spectroscopic analysis of low-concentration VOC mixtures. To improve the sensing performance in terms

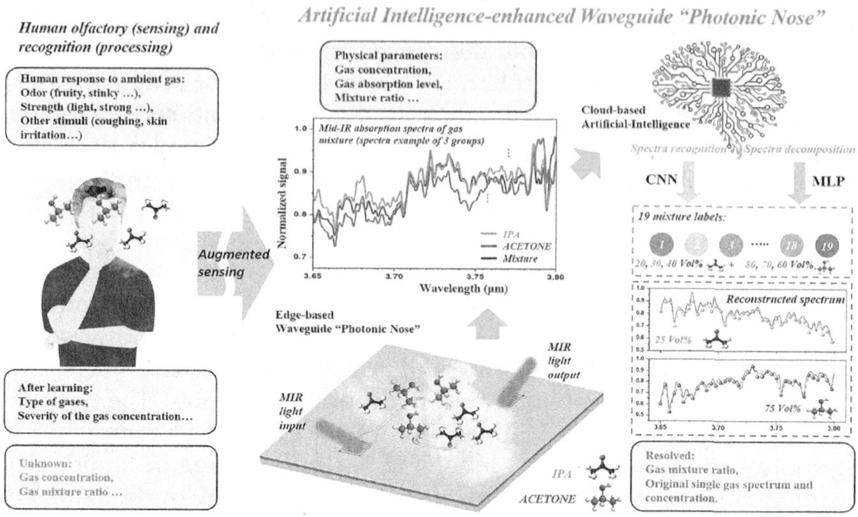

FIGURE 11.4 MIR photonics waveguide sensor: Schematic of artificial intelligence-enhanced MIR photonic waveguide sensor for gas mixture sensing and analysis. *From Liu, X., Zhang, Z., Zhou, J., Liu, W., Zhou, G., & Lee, C. (2024). Artificial intelligence-enhanced waveguide "photonic nose"– augmented sensing platform for VOC gases in mid-infrared.* Small, 20(34). *https://doi.org/10.1002/smll.202400035.*

of sensitivity and bandwidth, the "photonic nose" was coupled with multi-resonant plasmonic nanoantennas to cover the MIR spectral region of molecular fingerprints. In the reported work, the authors deposited metal-organic frameworks (MOFs) on the sensing configuration, which enables selective adsorption of target gases. The reported sensor achieved a low detection limit of 1.99, 3.43, and 9.82 ppm for isopropanol, ethanol, and acetone, respectively. Through the application of ML algorithms, the device demonstrated exceptional classification performance, reaching 100% accuracy across 125 unique mixing ratios (isopropanol and ethanol each evaluated at five concentration ranging from 0 to 130 ppm, and acetone crossed across five concentrations ranging from 0 to 201 ppm). Additionally, component concentrations for isopropanol and ethanol were predicted with high precision, showing RMSEs below 10 ppm even in the presence of 50 ppm acetone interference. This approach offers a promising pathway toward smart, adaptive sensing systems for environmental and industrial applications.

Zhang et al. (2023b) also developed a novel, ML-integrated, fiber-optic portable Raman sensor for real-time monitoring of a panel of pathogen-specific molecular fingerprint VOCs. In the reported work, the authors employed ML algorithms to extract molecular information from the raw Raman spectra, enabling highly accurate predictions of complex VOC mixtures even at high dilution levels (up to a 100-fold dilution). The reported ML-enhanced Raman sensing platform demonstrated a significant improvement in sensor performance, along with rapid response capabilities and on-site detection of target VOCs in a mixed-gas environment.

Hashemitaheri also employed an ML approach to precisely and simultaneously identify benzene, toluene, ethylbenzene, and xylenes (BTEX) concentrations based on an absorption spectroscopy gas sensing system (Hashemitaheri et al., 2024). In the reported work, the authors utilized CNN to detect the absorbance spectra for each VOC and estimate their concentrations within a mixed gaseous environment. A synthetic dataset was generated via a series of physics-based simulations to develop and train the predictive model. The dataset comprised the overall absorbance of various random BTEX mixtures over time, each modeled using varying proportions based on their permissible exposure limits (PELs). Notably, benzene possesses a significant detection challenge because of its low absorbance and strict PEL range (typically from 1 to 5 ppm). To resolve the issue, a three-stage approach was proposed and implemented to effectively differentiate BTEX components present at any concentration level. The model demonstrates strong predictive accuracy, achieving R^2 values exceeding 0.99 for toluene, ethylbenzene, and o-xylene, and over 0.96 for benzene.

Wekalao performed a numerical investigation on a graphene-integrated plasmonic sensor to achieve high sensitivity towards VOCs present in wastewater (Wekalao, 2025). The proposed sensing configuration comprises a circular resonator with a diameter of 7 μm, coated with silver to improve its resonant properties and detection accuracy. The area surrounding the resonator was considered to be a square in structure with dimensions of 13 by 12 μm, which serves as a secondary resonator element, contributing to the overall sensor response. Both the circular resonator and the square ring are surrounded by a rectangular structure measuring 13 by 1 μm, coated with a thin gold film. The rectangular and square ring resonators help enhance the electromagnetic response and improve the sensitivity of the device. The entire structure was positioned on top of a graphene-coated larger square base layer, measuring 17 by 17 μm. The resonator array was then placed on top of a SiO_2 substrate. The sensing performance of the proposed design was investigated by varying key design parameters, including the structural dimensions of the resonator, the chemical potential of graphene, and the incident angles. The reported sensor exhibited a maximum sensitivity of 227 GHz/RIU along with a figure of merit of 3.852 RIU^{-1}, and a quality factor of 7.746, tested over a wide frequency range from 0.1 to 1 THz. Additionally, the detection capabilities of the sensor were observed to be significantly enhanced by utilizing stacking ensemble ML methods, which exhibited an impressive prediction accuracy with regression values ranging from 91% to 100%.

The sensor also demonstrated compatibility with a broad refractive index range of 1.3256 to 1.542 and a consistent full width at half maximum (FWHM) of 0.059 THz. Recently, Boonruang et al. (2025) reported an array of Ta_2O_5:WO_3 composite integrated guided mode resonance (GMR) sensors to selectively identify and discriminate VOCs in a complex and mixed gaseous environment at room temperature. A 3 × 3 GMR sensor array was fabricated using a co-sputtering method with a shadow mask to alter the composition of the Ta_2O_5:WO_3 waveguide film on top of an imprinted one-dimensional grating. The composite film mainly possesses two major roles, i.e., (1) guiding the incident light along the waveguide and (2) acting as an absorptive sensing layer for selective interaction with VOC molecules. An in-house nine-channel CMOS-based spectrometer was utilized to produce distinct VOC fingerprints via the cross-reactivity of the sensors. The sensing performance of the device was experimentally investigated under various exposures to VOC vapors and a ternary alcohol mixture comprising ethanol, isopropanol, and methanol. The results demonstrated variations in the signal matrix corresponding to each VOC. Notably, the WO_3-functionalized sensor exhibited an enhanced selectivity toward methanol, while the integration of Ta_2O_5 with the WO_3 matrix at a 60:40 ratio resulted in a signal enhancement of up to 2.5-fold. The resonance data were measured and further analyzed using ML techniques to show the potential of the sensor to accurately identify and discriminate VOCs. Utilizing the Classification Learning tool in MATLAB with a KNN algorithm, the model achieved an excellent accuracy of 91.3%, where 15-fold cross-validation of 160 tested data (10 replicates of each of the 16 samples) was used to train the model.

Wang et al. (2025) reported an artificial "photonic nose" integrated with MOFs to identify trace hazardous gases and VOCs via ML integration. The sensing system comprises a 3D PhC integrated with MOFs, as shown in Fig. 11.5. The sensing principle of the nose was based on gas-induced modulation of the refractive index within the MOFs' 3D PhC array. Each 3D PhC structure, functionalized with a specific MOF, exhibits different adsorption behavior, resulting in distinct optical responses that serve as an optical fingerprint for 13 different toxic gases at a low concentration of 3 ppm under ambient conditions (25°C and 45% relative humidity). The reported photonic nose demonstrated a rapid response of less than 1 second, along with excellent reproducibility. The optical fingerprints corresponding to the 13 gases were analyzed using the KNN algorithm, achieving an average classification accuracy of 97.8%. Furthermore, the performance of the sensor was evaluated under elevated levels of humidity conditions (80% RH) at room temperature. The system also enables quantitative estimation of unknown gas concentrations, offering a promising approach for detection in toxic and hazardous environments.

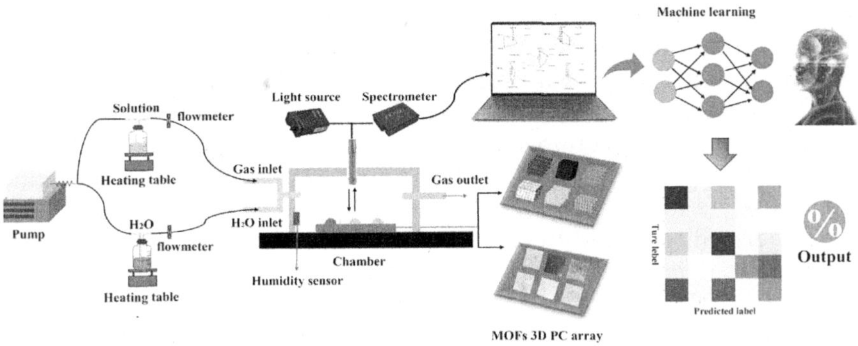

FIGURE 11.5 Photonic nose: Schematic diagram of the photonic nose and the analysis process of gases. *From Wang, Y., Li, R., Xin, S., Wang, Z., Yan, J., Yong, X., Wang, Q., & Chen, Y. (2025). An artificial optical nose integrated by metal-organic frameworks three-dimensional photonic crystal array for identification of trace hazardous gases through machine learning integration.* Journal of Environmental Chemical Engineering, 13*(3). https://doi.org/10.1016/j.jece.2025.117043.*

Zhou et al. (2025) reported an AI-enhanced nanophotonics waveguide-based MIR lab-on-a-chip (LoC) platform for real-time detection of VOC mixtures. In the reported work, the authors integrated MIR-compatible nanophotonics structure with a zero-bias graphene photodetector and performed real-time monitoring of acetone, ethanol, and isopropanol across various volume percentage (vol%) mixtures in water. Additionally, the authors combined the potential of ML with the fabricated LoC, achieving successful classification of collected spectra from the 27 mixed-VOCs samples in aqueous solutions, along with a remarkable accuracy of 95.77%. Furthermore, to predict the concentration of the individual VOCs in the mixture, a convolutional regression network-based regression model was utilized for spectrum decomposition. The reported method achieved 83.33% accuracy for single-component concentration predictions within a 1 vol% error range, with an average root-mean-squared error of 1 vol% for mixture concentration predictions.

In addition to the above-discussed photonic devices, several other optical VOC sensors have been reported over the past decades, utilizing ML to precisely identify VOC concentrations. Among several, laser-based VOC detection has gained significant interest in a wide range of applications, including environmental monitoring, the industrial sector, scientific research, and medical diagnostics. This method relies on the analysis of the unique spectral signatures of gaseous molecules, typically referred to as tunable laser absorption spectroscopy (TLAS) or tunable diode laser absorption spectroscopy (TDLAS). Laser absorption spectroscopy (LAS) offers several advantages over traditional VOC detection methods, including rapid response, high sensitivity, non-contact operation, target selectivity, and a long operation lifespan, making it highly suitable for real-time, in-situ, and online monitoring (Pathak, 2025). Detailed information about TDLAS and its working principle can be found in other articles published in recent years (Lin, 2022; Sun, 2024). Despite several advantages, the TDLAS has traditionally been confined to laboratory environments, and challenges remain with its broader application due to issues such as background noise, baseline fitting, and signal drift. A TDLAS gas sensing configuration integrated with end-to-end 1D CNNs has been widely adopted for accurate gas detection and precise quantification of VOC concentrations. Tian et al. (2021) reported TDLAS systems integrated with CNN-based signal processing for the selective detection of acetylene and methane, demonstrating their potential and reliability in detecting various other gaseous molecules.

Separating overlapping absorption features is crucial for the simultaneous classification and quantification of various gaseous components from the absorption spectra. Despite the advantages, the application of DL models for identifying multicomponent gas mixtures from infrared absorption spectra has not been extensively studied. In a recent work, Chowdhury and Oehlschlaeger reported a 1D CNN-based model, derived from HITRAN data (Chowdhury et al., 2022). Their approach enabled the detection of various small molecules across flexible spectral windows. The model exhibited high classification performance for several crucial gases, including H_2O, CO_2, N_2O, O_3, CO, CH_4, NO, SO_2, NO_2, and NH_3. Sun et al. (2022) reported a gas sensing system to identify methane and carbon monoxide by utilizing TDLAS integrated with 1D CNNs trained and tested on simulated datasets. The ML approach in the reported systems effectively addresses limitations related to the spectral cross-interference arising from overlapping spectral lines. Guan et al., (2024) developed a cavity-enhanced direct frequency comb spectroscopy system combined with various ML models for the detection of carbon monoxide and CO_2 with high precision. The reported work demonstrated that the particle swarm optimization (PSO) achieved strong predictive accuracy while maintaining the compact size of the sensor and its suitability for outdoor environments.

11.5 Case studies

Recently, a study demonstrated that AI detected breast cancer years before diagnosis from screening mammograms (Thomas, n.d.). In the reported finding, all mammogram results were passed through AI analysis by utilizing INSIGHT MMG, a commercially available AI algorithm. Notably, the INSIGHT MMG was not originally designed to evaluate future cancer risk and had not been optimized for such a task. The algorithm provided a continuous variable, referred to as the cancer detection score, ranging from 0 to 100. Higher detection scores reflect a greater risk of a positive mammogram. The maximum AI scores and the score differences were compared between the breasts of women who screened positive for cancer, those who screened negative, and those later diagnosed with cancer during the routine screening interval. Notably, the AI demonstrated significant potential in executing such advanced tasks with high precision.

AI has greatly advanced the medical and health sector, specifically disease diagnosis via breath analysis to identify metabolic disorders, respiratory conditions, and the early detection of cancers, establishing a new benchmark in modern diagnostics. These technologies are integrated with gas sensors to selectively detect target VOC biomarkers present in exhaled breath, therefore, offering a non-invasive approach to traditional diagnostic methods. For example, breath analysis has been successfully utilized over the past decades to diagnose various health conditions such as asthma and chronic obstructive pulmonary disease by detecting associated VOCs with lung inflammation. In cancer diagnostics, AI-enhanced VOC detection systems have shown significant potential in distinguishing between healthy and malignant tissues by analyzing exhaled breath samples. Research has demonstrated that these gas sensing systems can recognize various VOC signatures emitted by tumors, aiding in the early detection of lung and colorectal cancers.

By exploiting ML algorithms, breath analyzer tools can significantly improve the diagnostic accuracy by learning and interpreting extensive breathomics datasets (Gallos et al., 2023). These AI models are capable of identifying subtle VOC biomarkers associated with specific diseases. For example, a diabetic patient may exhibit an elevated level of acetone in exhaled breath, which serves as a key biomarker for diabetes. ML-based sensing systems can precisely monitor such variations in real time, thus allowing continuous health monitoring and enabling rapid medical interventions (Kaloumenou et al., 2022). This approach not only supports early disease detection but also opens the path for more personalized healthcare by adapting treatment plans based on the VOC profiles of patients. Overall, AI-driven breath analysis offers a promising approach in medical diagnostics, merging advanced computational technology with practical clinical use to improve patient outcomes. Various clinical studies and research initiatives have exploited the AI-integrated VOC sensing technology to continuously monitor and predict early-stage disease diagnosis.

AI-based smart sensors integrated with the Internet of Medical Things (IoMT) have been widely adopted for continuous monitoring of patients oxygen levels and respiratory gases. By analyzing the obtained data in real time, these smart sensors can quickly identify early signs of hypercapnia or hypoxia (Manickam et al., 2022). These AI-driven systems also alert healthcare professionals in case of any abnormalities, therefore improving

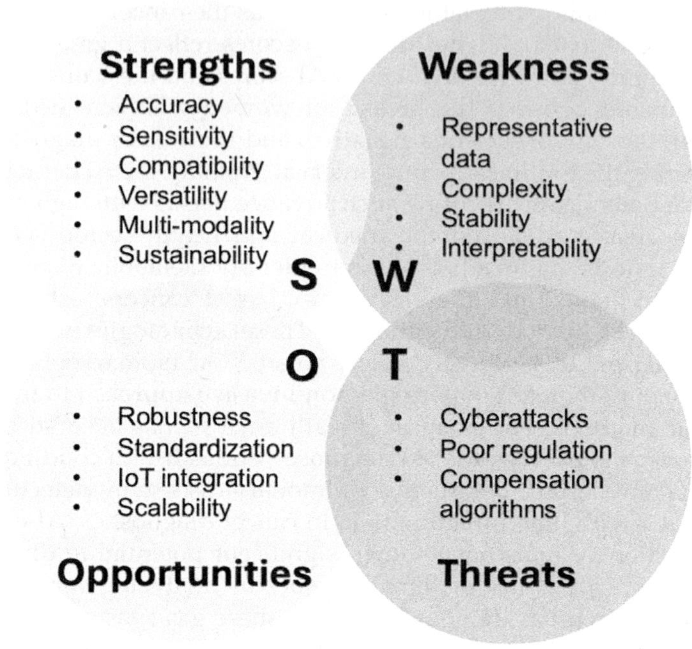

FIGURE 11.6 Overview of AI in gas sensing: SWOT analysis of the AI-integrated sensors. *From Chowdhury, M. A. Z., & Oehlschlaeger, M. A. (2025). Artificial intelligence in gas sensing: A review.* ACS Sensors, 10*(3), 1538–1563. https://doi.org/10.1021/acssensors.4c02272.*

the quality of life by reducing hospitalization costs. Initial investigations have utilized AI to study patterns in CO_2 and oxygen levels in patients suffering from sleep apnea or chronic obstructive pulmonary disease. These AI-integrated tools help anticipate exacerbations, enabling earlier and effective medical responses (Tsvetanov, 2024).

11.6 Challenges and opportunities for artificial intelligence in gas sensing

Integration of AI along with the gas sensing technology can significantly improve the performance and accuracy of the sensor, making it more autonomous and potentially capable of real-time monitoring. Despite several advantages, this approach presents several challenges that need to be addressed in order to realize its full potential. Fig. 11.6 exhibits a summary diagram of strengths, weaknesses, opportunities, and threats of AI-based gas sensing technology, and also Table 11.2 summarizes the major challenges and potential solutions associated with AI-integrated gas sensors.

TABLE 11.2 Challenges and potential solutions for AI-based gas sensors.

Area	Challenges	Potential solutions
Gas sensor data	• Acquisition difficulties: Variabilities due to environmental conditions, sensor drift, and other factors. Sparse features, noisy signal, and irrelevant information	• Develop data-independent models and leverage prior learning from previously trained models
	• Data quality and volume: Collecting a large and diverse data set representing all possible scenarios. Poor sampling, insufficient data, and missing critical outliers	• Implement advanced statistical sampling methods to improve data quality
	• Multi-modality and bias: Poorly sampled or biased datasets and the integration of multimodal data	• Employ data augmentation strategies, including synthetic data generation, transformation, and feature space augmentation
		• Employ few-shot learning models and long-tail learning approaches to enhance generalization from limited or imbalanced datasets
		• Leverage hybrid sensor arrays to improve accuracy and robustness
Learning algorithms	• Feature learning: Learning relevant local and nonlocal features effectively	• Utilize advanced AI methods such as generative models, attention mechanisms, transformer architectures, and physics-driven models
	• Outlier detection: Identify outliers and ensure that models are not overly dependent on initial conditions	• Explore semi-supervised, unsupervised, and self-supervised learning techniques
	• Spatiotemporal integration: Integrating spatiotemporal patterns into predictive modeling frameworks	• Implement foundation models, multimodal learning architectures, and reinforcement learning strategies

(*continued on next page*)

TABLE 11.2 Challenges and potential solutions for AI-based gas sensors—cont'd

Area	Challenges	Potential solutions
		• Use causal inference-based feature engineering and transfer learning approaches
AI–sensor integration	• Scalability: Scaling AI models to different dataset sizes while maintaining performance	• Implement multimodal learning and adaptive learning techniques, such as reinforcement learning
	• Sensor drift and dynamic offsets: Sensor drift and dynamic offsets can compromise long-term stability, leading to performance inconsistencies	• Develop real-time drift compensation algorithms and periodic recalibration strategies
	• Environmental adaptation: Adapting to rapidly changing or harsh environmental conditions leading to unpredictable sensor behaviors	• Use sensor fusion techniques, such as Kalman filters or decision-level fusion
		• Incorporate environmental compensation models to adjust sensor readings based on environmental conditions
Interpretability and trustworthiness	• Explainability: Ensuring that AI models and their decisions are interpretable	• Design modular models that allow each component to be inspected individually
	• Model complexity: Balancing the complexity of AI models with the need for interpretability	• Use interpretability techniques, such as SHAP values, LIME, Grad-CAM, and saliency maps
	• Transparency: Providing clear explanations for model predictions and identifying potential biases	• Implement bias mitigation strategies, such as reweighting or adversarial training
		• Develop user-friendly visualization tools to understand the inner workings of AI models

(*continued on next page*)

TABLE 11.2 Challenges and potential solutions for AI-based gas sensors—cont'd

Area	Challenges	Potential solutions
Scalability and environmental factors	• Resource constraints: Limited computational resources on sensor devices can restrict the complexity of AI models	• Implement edge computing strategies
	• Integration complexity: Integrating AI models with existing infrastructure and systems can be challenging	• Develop lightweight and energy-efficient AI models optimized for deployment in resource-constrained environments
	• Environmental Influences: Temperature, humidity, and pressure can affect sensor performance	• Employ sensor fusion and adaptive filtering techniques
		• Establish standardization protocols and middleware frameworks across the industry
Regulatory and safety	• Compliance and certification: Adhering to safety and regulatory standards essential for deployment in critical applications	• Develop dedicated AI-specific regulatory standards and certification procedures
	• Lack of standardization: The absence of standardized guidelines for AI integration can hinder reproducibility and validation	• Implement systematic model documentation, traceability, and auditing mechanisms
		• Integrate explainable AI techniques

From Chowdhury, M. A. Z., & Oehlschlaeger, M. A. (2025). Artificial intelligence in gas sensing: A review. ACS Sensors, *10(3), 1538–1563. https://doi.org/10.1021/acssensors.4c02272.*

Rapid advancements in the field of AI suggest that AI can play a crucial role in identifying optimal sensing materials and accelerating the design and optimization of these materials to enhance sensitivity, selectivity, and stability (Yuan et al., 2024). A key aspect of advancing the AI-based gas sensing technologies lies in the integration of AI techniques with emerging developments in materials science. By exploiting ML for experimental design, high-throughput analysis, and advanced data interpretation, researchers can efficiently explore a broad range of chemical spaces as well as identify critical structures that might otherwise remain hidden (Coto et al., 2023). The productive potential of AI not only accelerates the discovery process but also contributes to the refinement and optimization of experimental procedures and computational models by closing the feedback loop between theoretical predictions and practical applications (Schmidt et al., 2019).

With continued advancements, generative models hold significant potential for designing entirely new type of sensing materials from scratch, precisely tailored in order to achieve specific performance requirements (Coto et al., 2023; Fuhr & Sumpter, 2022). Simultaneously, explainable AI techniques are becoming increasingly valuable tools for revealing the rationale behind model predictions, therefore improving both transparency and scientific understanding (Barredo Arrieta et al., 2020; Zhong et al., 2022). This integrative approach could potentially open new pathways for the development of next-generation gas/VOC sensors that exhibit not only high accuracy but also adaptability to dynamic and complex real-world applications.

11.7 Conclusion

In conclusion, the integration of AI tools into gas and VOC sensing systems has emerged as a rapidly advancing field of research, as evidenced by the numerous publications in recent years. AI holds the potential to significantly enhance sensor capabilities, including improvements in the sensing accuracy, device sensitivity, target selectivity, and adaptability to environmental conditions. By enabling complex multispecies detection, AI-based methods are establishing the foundation for the next generation of smart sensor systems that are multifunctional, precise, and adaptable to a broad range of applications, including industrial process control, environmental monitoring, and medical diagnostics. This chapter has emphasized the key role of AI in advancing gas sensor technologies, highlighting innovative integration strategies, key challenges, and emerging opportunities for future development.

References

Alpaydın, E. (2021). *Machine learning*. The MIT Press. doi:10.7551/mitpress/13811.001.0001.

Argyris, A. (2022). Photonic neuromorphic technologies in optical communications. *Nanophotonics, 11*(5), 897–916. http://www.degruyter.com/view/j/nanoph?rskeyCC=PZEKdQ&result=1&q=Nanophotonics. doi:10.1515/nanoph-2021-0578.

Atakishiyev, S., Salameh, M., Yao, H., & Goebel, R. (2024). Explainable artificial intelligence for autonomous driving: A comprehensive overview and field guide for future research directions. *IEEE Access, 12*, 101603–101625. doi:10.1109/ACCESS.2024.3431437.

Barredo Arrieta, A., Díaz-Rodríguez, N., Del Ser, J., Bennetot, A., Tabik, S., Barbado, A., Garcia, S., Gil-Lopez, S., Molina, D., Benjamins, R., Chatila, R., & Herrera, F. (2020). Explainable artificial intelligence (XAI): Concepts, taxonomies, opportunities and challenges toward responsible AI. *Information Fusion, 58*, 82–115. http://www.elsevier.com/inca/publications/store/6/2/0/8/6/2/index.htt. doi:10.1016/j.inffus.2019.12.012.

Bishop, C. M. (2006). *Pattern recognition and machine learning*. Springer US.

Boonruang, S., Srisuai, N., Chananonnawathorn, C., Noikorn, P., Tantiwanichapan, K., Horprathum, M., & Mohammed, W. S. (2025). Spectral interrogation Ta_2O_5: WO_3 composite guided mode resonance sensor array: Detection and discrimination of VOC mixture. *Sensors and Actuators B: Chemical, 426*, 137086. doi:10.1016/j.snb.2024.137086.

Chang, C.-W., Lee, H.-W., & Liu, C.-H. (2018). A review of artificial intelligence algorithms used for smart machine tools. *Inventions, 3*(3). doi:10.3390/inventions3030041.

Chowdhury, M. A. Z., & Oehlschlaeger, M. A. (2025). Artificial intelligence in gas sensing: A review. *ACS Sensors, 10*(3), 1538–1563. doi:10.1021/acssensors.4c02272.

Chowdhury, M. A. Z., Rice, T. E., & Oehlschlaeger, M. A. (2022). VOC-net: A deep learning model for the automated classification of rotational THz spectra of volatile organic compounds. *Applied Sciences (Switzerland), 12*(17). www.mdpi.com/journal/applsci/. doi:10.3390/app12178447.

Coto, A. G., Precker, C. E., Andersson, T., Laukkanen, A., Suhonen, T., Rodriguez, P. R., & Muños-Landín, S. (2023). The use of generative models to speed up the discovery of materials. *Computer Methods in Material Science, 23*(1). doi:10.7494/cmms.2023.1.0802.

Cui, F., & Zhou, H. S. (2020). Diagnostic methods and potential portable biosensors for coronavirus disease 2019. *Biosensors and Bioelectronics, 165*. doi:10.1016/j.bios.2020.112349.

Duan, Y., Edwards, J. S., & Dwivedi, Y. K. (2019). Artificial intelligence for decision making in the era of big data – evolution, challenges and research agenda. *International Journal of Information Management, 48*, 63–71. doi:10.1016/j.ijinfomgt.2019.01.021.

Fuhr, A. S., & Sumpter, B. G. (2022). Deep generative models for materials discovery and machine learning-accelerated innovation. *Frontiers in Materials, 9*. journal.frontiersin.org/journal/materials. doi:10.3389/fmats.2022.865270.

Gallos, I., Tryfonopoulos, D., Shani, G., Amditis, A., Haick, H., & Dionysiou, D. (2023). Advancing colorectal cancer diagnosis with AI-powered breathomics: Navigating challenges and future directions. *Diagnostics, 13*(24), 3673. doi:10.3390/diagnostics13243673.

Ganesh Babu, R., Chellaswamy, C., Geetha, T. S., & Ramesh, R. (2022). Fibre optic sensor based multi-gas detection using optimized convolutional neural network. *Journal of Modern Optics, 69*(7), 403–417. doi:10.1080/09500340.2022.2041753.

Ghosh, A., Pal, A., & Das, N. R. (2020). An approach to design photonic crystal gas sensor using machine learning. *Optik, 208*, 163997. doi:10.1016/j.ijleo.2019.163997.

Goodfellow, I., Bengio, Y., & Courville, A. (2016). *Deep learning*. MIT Press http://www.deeplearningbook.org.

Guan, G., Liu, A., Wu, X., Zheng, C., Liu, Z., Zheng, K., Pi, M., Yan, G., Zheng, J., Wang, Y., & Tittel, F. K. (2024). Near-infrared off-axis cavity-enhanced optical frequency comb spectroscopy for CO_2/CO dual-gas detection assisted by machine learning. *ACS Sensors, 9*(2), 820–829. http://pubs.acs.org/journal/ascefj. doi:10.1021/acssensors.3c02146.

Hajjem, A., Bellavance, F., & Larocque, D. (2014). Mixed-effects random forest for clustered data. *Journal of Statistical Computation and Simulation, 84*(6), 1313–1328. doi:10.1080/00949655.2012.741599.

Hamet, P., & Tremblay, J. (2017). Artificial intelligence in medicine. *Metabolism: Clinical and Experimental, 69*. doi:10.1016/j.metabol.2017.01.011.

Hashemitaheri, M., Ebrahimi, E., Silva, G., & Attariani, H. (2024). Optical sensor for BTEX detection: Integrating machine learning for enhanced sensing. *Advanced Sensor and Energy Materials, 3*(3), 100114. doi:10.1016/j.asems.2024.100114.

Hopgood, A. A. (2021). *Intelligent systems for engineers and scientists*. CRC Press. doi:10.1201/9781003226277.

Ikotun, A. M., Ezugwu, A. E., Abualigah, L., Abuhaija, B., & Heming, J. (2023). K-means clustering algorithms: A comprehensive review, variants analysis, and advances in the era of big data. *Information Sciences, 622*, 178–210. http://www.journals.elsevier.com/information-sciences/. doi:10.1016/j.ins.2022.11.139.

Jing, H., Zhao, C., & Gao, F. (2021). Non-stationary data reorganization for weighted wind turbine icing monitoring with Gaussian mixture model. *Computers & Chemical Engineering, 147*, 107241. doi:10.1016/j.compchemeng.2021.107241.

Kaloumenou, M., Skotadis, E., Lagopati, N., Efstathopoulos, E., & Tsoukalas, D. (2022). Breath analysis: A promising tool for disease diagnosis—the role of sensors. *Sensors, 22*(3), 1238. doi:10.3390/s22031238.

Kotsiantis, S. B. (2013). Decision trees: A recent overview. *Artificial Intelligence Review, 39*(4), 261–283. doi:10.1007/s10462-011-9272-4.

Krizhevsky, A., Sutskever, I., & Hinton, G. E. (2017). ImageNet classification with deep convolutional neural networks. *Communications of the ACM, 60*(6), 84–90. doi:10.1145/3065386.

Lin, Shan, et al. (2022). Improvement of the Detection Sensitivity for Tunable Diode Laser Absorption Spectroscopy: A Review. *Frontiers in Physics, 10*. doi:10.3389/fphy.2022.853966.

Liu, X., Zhang, Z., Zhou, J., Liu, W., Zhou, G., & Lee, C. (2024). Artificial intelligence-enhanced waveguide "photonic nose"–augmented sensing platform for VOC gases in mid-infrared. *Small, 20*(34). http://onlinelibrary.wiley.com/journal/10.1002/. doi:10.1002/smll.202400035. (ISSN)1613-6829.

Low, J. S. Y., Teh, H. F., Thevarajah, T. M., Chang, S. W., & Khor, S. M. (2025). An AI-assisted microfluidic paper-based multiplexed surface-enhanced raman scattering (SERS) biosensor with electrophoretic removal and electrical modulation for accurate acute myocardial infarction (AMI) diagnosis and prognosis. *Biosensors and Bioelectronics, 270*. doi:10.1016/j.bios.2024.116949.

Ma, X., Mo, J., Zhang, J., & Huang, J. (2022). Optical fiber vibration signal recognition based on the fusion of multi–scale features. *Sensors, 22*(16), 6012. doi:10.3390/s22166012.

Manickam, P., Mariappan, S. A., Murugesan, S. M., Hansda, S., Kaushik, A., Shinde, R., & Thipperudraswamy, S. P. (2022). Artificial intelligence (AI) and internet of medical things (IoMT) assisted biomedical systems for intelligent healthcare. *Biosensors, 12*(8). http://www.mdpi.com/journal/biosensors/. doi:10.3390/bios12080562.

Murphy, K. P. (2022). *Probabilistic machine learning*. The MIT Press.

Naku, W., Nambisan, A. K., Roman, M., Zhu, C., Gerald, R. E., & Huang, J. (2023). Identification of volatile organic liquids by combining an array of fiber-optic sensors and machine learning. *ACS Omega, 8*(5), 4597–4607. pubs.acs.org/journal/acsodf. doi:10.1021/acsomega.2c05451.

Nassif, A. B., Shahin, I., Attili, I., Azzeh, M., & Shaalan, K. (2019). Speech recognition using deep neural networks: A systematic review. *IEEE Access, 7*, 19143–19165. doi:10.1109/ACCESS.2019.2896880.

Nusinovici, S., Tham, Y. C., Chak Yan, M. Y., Wei Ting, D. S., Li, J., Sabanayagam, C., Wong, T. Y., & Cheng, C. Y. (2020). Logistic regression was as good as machine learning for predicting major chronic diseases. *Journal of Clinical Epidemiology, 122*, 56–69. www.elsevier.com/locate/jclinepi. doi:10.1016/j.jclinepi.2020.03.002.

Padvi, M. N., Moholkar, A. V., Prasad, S. R., & Prasad, N. R. (2021). A critical review on design and development of gas sensing materials. *Engineered Science, 15*, 20–37. doi:10.30919/es8d431.

Pambudi, D. S. A., Rivai, M., & Arifin, A. (2018). Indonesia Detection of organic solvent compounds using optical fiber interferometer array and neural network pattern recognition. *International Conference on Information and Communications Technology, ICOIACT 2018* (pp. 477–482). Institute of Electrical and Electronics Engineers Inc. http://ieeexplore.ieee.org/xpl/mostRecentIssue.jsp?punumber=8343062. doi:10.1109/ICOIACT.2018.8350681.

Pathak, A. K. (2025). Advances in clinical chemistry volatile organic compound sensing: Conventional to advanced approaches. *Advances in Clinical Chemistry, 129*.

Pathak, V., & Viphavakit, C. (2022). A review on all-optical fiber-based VOC sensors: Heading towards the development of promising technology. *Sensors and Actuators A: Physical, 338,* 113455. doi:10.1016/j.sna.2022.113455.

Pereira, R., Lima, C., Pinto, T., & Reis, A. (2023). Virtual assistants in Industry 4.0: A systematic literature review. *Electronics (Switzerland), 12*(19). doi:10.3390/electronics12194096.

Praveenchandar, J., Vetrithangam, D., Kaliappan, S., Karthick, M., Pegada, N. K., Patil, P. P., Rao, S. G., & Umar, S. (2022). IoT-based harmful toxic gases monitoring and fault detection on the sensor dataset using deep learning techniques. *Scientific Programming, 2022.* https://www.hindawi.com/journals/sp/. doi:10.1155/2022/7516328.

Roberts, D. A., Yaida, S., & Hanin, B. (2022). *The principles of deep learning theory: An effective theory approach to understanding neural networks.* Cambridge University Press. doi:10.1017/9781009023405.

Schmarje, L., Santarossa, M., Schroder, S.-M., & Koch, R. (2021). A survey on semi-, self- and unsupervised learning for image classification. *IEEE Access, 9,* 82146–82168. doi:10.1109/access.2021.3084358.

Schmidt, J., Marques, M. R. G., Botti, S., & Marques, M. A. L. (2019). Recent advances and applications of machine learning in solid-state materials science. *NPJ Computational Materials, 5*(1). https://www.nature.com/npjcompumats/. doi:10.1038/s41524-019-0221-0.

Sharma, N. A., Ali, A. B. M. S., & Kabir, M. A. (2025). A review of sentiment analysis: Tasks, applications, and deep learning techniques. *International Journal of Data Science and Analytics, 19*(3), 351–388. doi:10.1007/s41060-024-00594-x.

Singh, I., Gupta, A., Gupta, C., Mani, A., & Basu, T. (2024). AI-driven improvements in electrochemical biosensors for effective pathogen detection at point-of-care. *Engineering Proceedings, 73*(1). doi:10.3390/engproc2024073005.

Subba Rao, C., Geetha, T. S., Chellaswamy, C., & Arul, S. (2021). Optimized convolutional neural network-based multigas detection using fiber optic sensor. *Optical Engineering, 60*(12). http://www.spie.org/x867.xml. doi:10.1117/1.OE.60.12.127108.

Sun, J., Chang, J., Wei, Y., Zhang, Z., Lin, S., Wang, F., & Zhang, Q. (2022). Dual gas sensor with innovative signal analysis based on neural network. *Sensors and Actuators B: Chemical, 373,* 132697. doi:10.1016/j.snb.2022.132697.

Sun, Jiachen, et al. (2024). Tunable diode laser absorption spectroscopy for detection of multi-component gas: a review. *Applied Spectroscopy Reviews, 59.* doi:10.1080/05704928.2024.2302608.

Suzuki, T., Hasebe, T., & Miyazaki, T. (2024). Quantum support vector machines for classification and regression on a trapped-ion quantum computer. *Quantum Machine Intelligence, 6*(1). https://www.springer.com/journal/42484. doi:10.1007/s42484-024-00165-0.

Taha, B. A., Mashhadany, Y. A., Mokhtar, M. H. H., Zan, M. S. D. B., & Arsad, N. (2020). An analysis review of detection coronavirus disease 2019 (Covid-19) based on biosensor application. *Sensors (Switzerland), 20*(23), 1–29. doi:10.3390/s20236764.

Tan, A., Wang, B., Zhao, Y., Wang, Y., Zhao, J., & Wang, A. X. (2023). Near-infrared spectroscopy analysis of compound fertilizer based on GAF and quaternion convolution neural network. *Chemometrics and Intelligent Laboratory Systems, 240.* doi:10.1016/j.chemolab.2023.104900.

Thomas L., AI detects breast cancer years before diagnosis from mammograms. https://www.news-medical.net/news/20241008/AI-detects-breast-cancer-years-before-diagnosis-from-mammograms.aspx.

Tian, L., Sun, J., Chang, J., Xia, J., Zhang, Z., Kolomenskii, A. A., Schuessler, H. A., & Zhang, S. (2021). Retrieval of gas concentrations in optical spectroscopy with deep learning. *Measurement, 182,* 109739. doi:10.1016/j.measurement.2021.109739.

Tian, Z., Li, J., Liu, L., Wu, H., Hu, X., Xie, M., Zhu, Y., Chen, X., & Ou-Yang, W. (2023). Machine learning-assisted self-powered intelligent sensing systems based on triboelectricity. *Nano Energy, 113,* 108559. doi:10.1016/j.nanoen.2023.108559.

Tsvetanov, F. (2024). Integrating AI technologies into remote monitoring patient systems. *Engineering Proceedings, 70*(1). www.mdpi.com/journal/engproc. doi:10.3390/engproc2024070054.

Tung Ho, L., & Ané, C. (2014). A linear-time algorithm for gaussian and non-gaussian trait evolution models. *Systematic Biology, 63*(3), 397–408. doi:10.1093/sysbio/syu005.

Wang, H., Fu, T., Du, Y., Gao, W., Huang, K., Liu, Z., Chandak, P., Liu, S., Van Katwyk, P., Deac, A., Anandkumar, A., Bergen, K., Gomes, C. P., Ho, S., Kohli, P., Lasenby, J., Leskovec, J., Liu, T. Y., Manrai, A., & Zitnik, M. (2023a). Scientific discovery in the age of artificial intelligence. *Nature, 620*(7972), 47–60. https://www.nature.com/nature/. doi:10.1038/s41586-023-06221-2.

Wang, Y., Adam, M. L., Zhao, Y., Zheng, W., Gao, L., Yin, Z., & Zhao, H. (2023b). Machine learning-enhanced flexible mechanical sensing. *Nano-Micro Letters, 15*(1). https://www.springer.com/journal/40820. doi:10.1007/s40820-023-01013-9.

Wang, Y., Li, R., Xin, S., Wang, Z., Yan, J., Yong, X., Wang, Q., & Chen, Y. (2025). An artificial optical nose integrated by metal-organic frameworks three-dimensional photonic crystal array for identification of trace hazardous gases through machine learning integration. *Journal of Environmental Chemical Engineering, 13*(3), 117043. doi:10.1016/j.jece.2025.117043.

Wekalao, J. (2025). Design and optimization of a graphene-enhanced plasmonic sensor with machine learning integration for organic compound detection in wastewater. *Plasmonics*. https://www.springer.com/journal/11468. doi:10.1007/s11468-025-02858-z.

West, J., & Bhattacharya, M. (2016). Intelligent financial fraud detection: A comprehensive review. *Computers and Security, 57*, 47–66. doi:10.1016/j.cose.2015.09.005.

Xie, J., Ren, Z., Zhou, H., Zhou, J., Liu, W., & Lee, C. (2024). Artificial intelligence-enhanced "photonic nose" for mid-infrared spectroscopic analysis of trace volatile organic compound mixtures. *Advanced Optical Materials, 12*(33). http://onlinelibrary.wiley.com/journal/10.1002/. doi:10.1002/adom.202401582. (ISSN)2195-1071.

Yang, S., Lei, G., Xu, H., Lan, Z., Wang, Z., & Gu, H. (2023). A review of the high-performance gas sensors using machine learning. *Machine Learning for Advanced Functional Materials*, 163–198. doi:10.1007/978-981-99-0393-1_8.

Yaqoob, U., & Younis, M. I. (2021). Chemical gas sensors: Recent developments, challenges, and the potential of machine learning—A review. *Sensors, 21*(8). https://www.mdpi.com/1424-8220/21/8/2877/pdf. doi:10.3390/s21082877.

Ye, Z., Liu, Y., & Li, Q. (2021). Recent progress in smart electronic nose technologies enabled with machine learning methods. *Sensors, 21*(22). doi:10.3390/s21227620.

Yuan, Z., Luo, X., & Meng, F. (2024). Machine learning-assisted research and development of chemiresistive gas sensors. *Advanced Engineering Materials, 26*(20). http://onlinelibrary.wiley.com/journal/10.1002/(ISSN)1527-2648. doi:10.1002/adem.202400782.

Zhang, A. J. S. A., Lipton, Z. C., & Li, M. (2023a). *Dive into deep learning*. Cambridge University Press.

Zhang, B., Asad Rahman, M. D., Liu, J., Huang, J., & Yang, Q. (2023b). Real-time detection and analysis of foodborne pathogens via machine learning based fiber-optic Raman sensor. *Measurement, 217*, 113121. doi:10.1016/j.measurement.2023.113121.

Zhang, H., Wang, Z., & Liu, D. (2014). A comprehensive review of stability analysis of continuous-time recurrent neural networks. *IEEE Transactions on Neural Networks and Learning Systems, 25*(7), 1229–1262. http://ieeexplore.ieee.org/xpl/RecentIssue.jsp?punumber=5962385. doi:10.1109/TNNLS.2014.2317880.

Zhong, X., Gallagher, B., Liu, S., Kailkhura, B., Hiszpanski, A., & Han, T. Y. J. (2022). Explainable machine learning in materials science. *NPJ Computational Materials, 8*(1). https://www.nature.com/npjcompumats/. doi:10.1038/s41524-022-00884-7.

Zhou, J., Liu, X., Zhou, H., Xu, S., Xie, J., Xu, C., Liu, W., Zhang, Z., & Lee, C. (2025). Artificial-intelligence-enhanced mid-infrared lab-on-a-chip for mixture spectroscopy analysis. *Laser and Photonics Reviews, 19*(1). http://onlinelibrary.wiley.com/journal/10.1002/. doi:10.1002/lpor.202400754. (ISSN)1863-8899.

Zhou, Y., Zhang, Y., Yu, Q., Ren, L., Liu, Q., & Zhao, Y. (2024). Application of machine learning in optical fiber sensors. *Measurement, 228*, 114391. doi:10.1016/j.measurement.2024.114391.

Index

Page numbers followed by "*f*" and "*t*" indicate, figures and tables respectively.

9780443291661